# ROUTLEDGE HANDBOOK OF ENERGY IN ASIA

*The Routledge Handbook of Energy in Asia* presents a comprehensive review of the unprecedented growth of Asian energy over the past quarter of a century. It provides insightful analysis into variation across the continent, whilst highlighting areas of cross-learning and regional cooperation between the developed and developing countries of Asia. Prepared by a team of leading international experts, this book not only captures East Asian domination, particularly that of China, but also highlights the growing influence of South Asia and the ASEAN.

Organised into four parts, the sections include:

- the demand for energy in the region and its main drivers at the sector level;
- developments in energy supply, including fossil fuels and renewable energy sources;
- energy policies and issues such as sector reform and climate change;
- the transition to a low carbon pathway.

This handbook offers a complete picture of Asian energy, covering supply and demand, as well as contemporary challenges in the sector. As such, it is a valuable resource for students and scholars of energy policy, Environmental Studies, and Asian Studies.

**Subhes C. Bhattacharyya** is Professor of Energy Economics and Policy in the School of Engineering and Sustainable Development at De Montfort University, UK.

# ROUTLEDGE HANDBOOK OF ENERGY IN ASIA

Edited by Subhes C. Bhattacharyya

Routledge
Taylor & Francis Group

LONDON AND NEW YORK

First published 2018
by Routledge

2 Park Square, Milton Park, Abingdon, Oxfordshire OX14 4RN
52 Vanderbilt Avenue, New York, NY 10017

*Routledge is an imprint of the Taylor & Francis Group, an informa business*

First issued in paperback 2020

*British Library Cataloguing-in-Publication Data*
A catalogue record for this book is available from the British Library

*Library of Congress Cataloging-in-Publication Data*
Names: Bhattacharyya, Subhes C., editor.
Title: Routledge handbook of energy in Asia / edited by
Subhes C. Bhattacharyya.
Description: Abingdon, Oxon; New York, NY: Routledge, 2018. |
Includes bibliographical references and index.
Identifiers: LCCN 2017027526| ISBN 9781138999824 (hardback) |
ISBN 9781315656977 (ebook)
Subjects: LCSH: Energy consumption–Asia. | Energy development–Asia. |
Energy industries–Asia. | Energy policy–Asia.
Classification: LCC HD9502.A782 R68 2018 | DDC 333.79095–dc23
LC record available at https://lccn.loc.gov/2017027526

ISBN: 978-1-138-99982-4 (hbk)
ISBN: 978-0-367-66001-7 (pbk)

Typeset in Bembo
by Sunrise Setting Ltd, Brixham, UK

# CONTENTS

# FIGURES

# TABLES

# CONTRIBUTORS

**Venkatachalam Anbumozhi** is a Senior Energy Economist at the Economic Research Institute for Association of South East Asian Nations (ASEAN) and East Asia (ERIA), Indonesia Economic Research Institute for ASEAN and East Asia (ERIA), Indonesia. His previous positions include Senior Capacity Building Specialist at Asian Development Bank Institute, Assistant Professor at the University of Tokyo, Senior Policy Researcher at the Institute for Global Environmental Strategies and Assistant Manager in Pacific Consultants International, Tokyo. He has published several books, authored numerous research articles and produced many project reports on energy policies, energy infrastructure design, and private sector participation in Green Growth.

**Subhes C. Bhattacharyya** is Professor of Energy Economics and Policy at the School of Engineering and Sustainable Development of the Faculty of Technology at De Montfort University, Leicester, UK. He has more than 30 years of experience in the energy sector in various capacities. He has extensively worked on energy issues in Asia and has widely published on the subject. He is the author of *Energy Economics: Concepts, Issues, Markets and Governance* (2011) and editor of two other books on rural electrification.

**Rob Byrne** is Lecturer in the SPRU (Science Policy Research Unit) at the University of Sussex and co-convenes the Energy and Climate Research Domain of the ESRC STEPS Centre and its Africa Sustainability Hub. He is particularly experienced in energy and development in an African context. Rob's recent book *Sustainable Energy for All: Technology, Innovation and Pro-Poor, Green Transformations* (Routledge, with David Ockwell) draws policy lessons from the most in-depth historical analysis to date of the success of the off-grid solar PV market in Kenya.

**Christopher M. Dent** is Professor of East Asia's International Political Economy, Department of East Asian Studies, University of Leeds, UK. His research interests centre on the international political economy of East Asia and the Asia–Pacific, particularly issues relating to energy, trade, development, climate change strategy and regional economic co-operation and integration. He has acted as a consultant advisor to the British, Australian, Chilean, German, Lao PDR and United States governments, as well as the Asian Development Bank, European Commission, ASEAN Secretariat, APEC Secretariat, Secretariat for Central American Economic Integration and Nike Inc.

**Tilak K. Doshi** is Managing Consultant with Muse, Stancil & Co. (Asia). An industry expert with over 25 years of international experience in leading oil and gas companies, his previous appointments include Senior Fellow and Program Director, King Abdullah Petroleum Studies and Research Centre (Riyadh, Saudi Arabia); Chief Economist, Energy Studies Institution, National University of Singapore; Executive Director for Energy, Dubai Multi Commodities Centre; Specialist, Saudi Aramco (Dhahran, Saudi Arabia); Chief Asia Economist, Unocal Corporation (Singapore). Doshi is the author of many articles and books, the most recent of which is *Singapore Chronicles: Energy* (2016).

**Xinling Feng** is a research assistant at the Institute for Global Environmental Studies in Japan, has a PhD of Environmental Policy Science from Waseda University Graduate School of Environment and Energy Engineering in Japan, and has worked for Waseda University Faculty of Science and Engineering for three years. She is engaged in research projects on environment and energy policies.

**Xiang Gao** is an associate professor at the Energy Research Institute, Beijing, China. He has eight years experience working as a researcher in the field of energy, environmental protection and climate change issues, and is familiar with the relevant policies of China and of major economies. He has also participated in the United Nations Framework Convention on Climate Change (UNFCCC) negotiations since 2009 as a member of the China delegation, focusing on the mitigation and transparency issues. Gao has also served as member of the Compliance Committee of the Kyoto Protocol, and as expert reviewer for the national reports submitted by Parties under the UNFCCC and the Kyoto Protocol.

**Shabbir H. Gheewala** is a professor at the Joint Graduate School of Energy and Environment, Thailand where he heads the Life Cycle Sustainability Assessment Lab. He also holds an adjunct professorship at the University of North Carolina Chapel Hill, USA, and a Distinguished Adjunct Professor position at the Asian Institute of Technology, Thailand. His research focuses on sustainability assessment of energy systems; sustainability indicators; and certification issues in biofuels and agro-industry. He is on the editorial boards of the *International Journal of Life Cycle Assessment*, the *Journal of Cleaner Production, Energy for Sustainable Development*, and the *Journal of Sustainable Energy and Environment*.

**Xiumei Guo** is Senior Research Fellow, Curtin University Sustainability Policy Institute, Curtin University, Australia. She conducts research and publishes in the fields of sustainability studies, energy and environmental sciences, innovation and technology policy.

**Tooraj Jamasb** is Chair in Energy Economics at Durham University Business School, Durham University. He previously held a post as SIRE Chair in Energy Economics, Heriot-Watt University and was Senior Research Associate, University of Cambridge. He is a Research Associate at the Energy Policy Research Group (University of Cambridge); the Centre for Energy and Environmental Policy Research (Massachusetts Institute of Technology); and the Oviedo Efficiency Group (University of Oviedo). He is co-editor of inter-disciplinary books including *The Future of Electricity Demand: Customers, Citizens and Loads* (2011). In addition, he is an Associate Editor of the *Energy Strategy Reviews* journal.

**Nandakumar Janardhanan** is an academician and policy specialist. He specialises on energy and climate policy. Dr Janardhanan has over 13 years of experience in working for policy think tanks and universities of national and international repute in India, Japan and the UK. His areas of interest include energy security and geopolitics, low carbon development, climate change and nuclear power.

**Noah Kittner** is a PhD candidate in the Energy and Resources Group at UC Berkeley and researcher in the Renewable and Appropriate Energy Laboratory. After graduating with a BS in Environmental Science from UNC-Chapel Hill (highest honours), Noah was a Fulbright Fellow at the Joint Graduate School of Energy and Environment in Bangkok, Thailand researching technical and policy aspects of solar electricity and sustainability assessment. Recently, he co-authored a Thai Solar PV Roadmap with colleagues at Chulalongkorn University, Thailand.

**Sivanappan Kumar** is Professor in Energy Studies at the Department of Energy, Environment and Climate Change, School of Environment, Resources and Development, Asian Institute of Technology. He has extensive experience in carrying out research in South Asia and in South East Asia on renewable energy technologies, energy efficiency, Technology Needs Assessment for greenhouse gas mitigation, energy access, low carbon cities, and low carbon energy systems and green growth. He has published more than 65 monographs and book chapters and more than 100 peer-reviewed journal articles. He is a Guest Professor at Tongji University, Shanghai.

**Akihisa Kuriyama** has worked for the Institute for Global Environmental Strategies (IGES) to implement capacity building programmes for mitigation activities in ASEAN countries such as Cambodia, Lao PDR, the Philippines and Indonesia. He also conducts quantitative research on the Kyoto Mechanisms and $CO_2$ emissions in electricity sectors.

**Dora Marinova** is Professor and Director of Curtin University Sustainability Policy Institute, Curtin University, Australia. She is a specialist in sustainability and technology policy, and has published extensively in these fields.

**Bijon Kumer Mitra** has worked at the Natural Resource and Ecosystem Services Area of the Institute for Global Environmental Strategies (IGES), Japan since 2009. He has ten years of experience in the field of natural resource management. His current research interest is the water-energy-food-climate nexus. He uses a quantitative assessment framework to assess resource allocation trade-offs, aiming to provide guidance for optimal decision-making. He holds a PhD in the Science of Biotic Environment from Iwate University, Japan.

**Ryoko Nakano** is a policy researcher at the Institute for Global Environmental Strategies in Japan. Ms. Nakano obtained her MA degree from the Faculty of Energy, Resources and Environment at the School of Advanced International Studies, Johns Hopkins University and currently specializes in energy efficiency and behaviour change.

**Jatin Nathwani** is the founding Executive Director, Waterloo Institute for Sustainable Energy and holds the prestigious Ontario Research Chair in Public Policy for Sustainable Energy at the University of Waterloo. His current focus is on implementing a global change initiative: he is the Co-Director, with Professor Joachim Knebel (Karlsruhe Institute of Technology, Germany), of the consortium 'Affordable Energy for Humanity (AE4H): A Global Change Initiative' that comprises 90+ leading energy access researchers and practitioners from 23 institutions and ten countries. Prior to his appointment at the University in 2007, Professor Nathwani worked in a leadership capacity in the Canadian energy sector.

**Rabindra Nepal** is Lecturer in Economics at CDU Business School, Charles Darwin University in Australia. He is an internationally reputed and recognised economist specialising in the field of natural resource (energy) and environment.

**Artie W. Ng** is currently Principal Lecturer and Deputy Director with the School of Professional Education & Executive Development at The Hong Kong Polytechnic University. He co-edited the book *Paths to Sustainable Energy*, in collaboration with Professor Jatin Nathwani. He is an International Associate with the Centre for Social and Environmental Accounting Research at the University of St Andrews and a member with Waterloo Institute for Sustainable Energy at the University of Waterloo. He has been invited by the United Nations Economic and Social Commission for Asia and the Pacific on the topic of Technology Facilitation for Sustainable Development.

**David Ockwell** is Reader in Geography at the University of Sussex and Deputy Director of Research in the ESRC STEPS Centre. He is also a Senior Research Fellow in the Sussex Energy· Group, a Fellow of the Tyndall Centre for Climate Change Research and sits on the board of the Low Carbon Energy for Development Network. His most recent book *Sustainable Energy for All: Technology, Innovation and Pro-Poor, Green Transformations* (Routledge, with Rob Byrne) draws policy lessons from the most in-depth historical analysis to date of the success of the off-grid solar PV market in Kenya.

**Debajit Palit** has around 20 years of experience working in the field of clean energy access, technology adaptation, resource assessment and energy planning; project implementation; rural electrification policy and regulation; gender and energy; impact assessment of energy sector projects and capacity building. He has vast national and international experience working in projects for UN organisations, the World Bank, Asian Development Bank and National Governments across countries in South and South East Asia and Sub-Saharan Africa. Mr Palit has written widely on energy access and rural electrification issues, particularly on South Asia, and has published around 40 research papers in peer-reviewed journals, conference proceedings and books.

**Binu Parthan** is the Principal Consultant at Sustainable Energy Associates (SEA), a global consulting and advisory company, and has over 20 years of professional experience in financing, policy and technology aspects of clean energy and climate change. Prior to SEA, he was the Deputy Director General for the Renewable Energy and Energy Efficiency Partnership (REEEP) and also the executive director of IT Power India where he led the energy and environment practice. Parthan holds a Doctorate in Electrical Engineering (Low-carbon Energy) from the Technical University of Graz and has also authored or co-authored five books and 18 publications and papers apart from over 50 professional reports.

**Tsani Fauziah Rakhmah** is a Research Associate at the Economic Research Institute for ASEAN and East Asia (ERIA), Indonesia. During her work at ERIA, Tsani has been engaged in research on low carbon technologies, economic impacts on fossil fuel subsidy removal and the integrated electricity market in ASEAN. Her areas of research interest include low carbon development, climate change, and renewable energy policy. Tsani graduated in Environmental Management and Development, Australian National University in 2015. Previously, she worked at the Coordinating Ministry of Economic Affairs (Indonesia) to facilitate Special Economic Zone establishment in Indonesia.

**Ronald D. Ripple** is the Mervin Bovaird Professor of Energy Business and Finance in the School of Energy Economics, Policy, and Commerce in the Collins College of Business at The University of Tulsa. Dr Ripple has studied oil and natural gas markets for over 36 years, getting his start in the Office of the Governor of Alaska. He recently authored a chapter on the Geopolitics of Australia Natural Gas Development for the joint Harvard-Rice Geopolitics of Natural Gas Study and has published numerous peer-reviewed journal articles, trade press articles, and reports, typically focusing on oil and natural gas markets.

**Anupama Sen** is Senior Research Fellow at the Oxford Institute for Energy Studies (OIES). In addition to OIES Papers, her research has appeared in academic journals such as the *Energy Journal* and *Oxford Review of Economic Policy*, professional publications such as *Gas Matters*, as well as several book chapters and Op-Eds. Dr Sen has been a Fellow of the Cambridge Commonwealth Society since 2009 and was previously a Junior Research Fellow and then Visiting Fellow at Wolfson College, Cambridge. She is also a Region Head on the Asia Pacific Desk at Oxford Analytica.

**Xunpeng Shi** is Principal Research Fellow at the Australia–China Relations Institute, University of Technology Sydney and an Adjunct Senior Research Fellow at the Energy Studies Institute (ESI), National University of Singapore. Xunpeng is also serving as President of the Chinese Economics Society Australia (CESA) and an Associate Editor for the *Journal of Modelling for Management*. He has worked with leading energy institutes in China, Indonesia, Brunei and Singapore. His areas of expertise include: the Chinese economy; natural gas pricing; coal industrial policy; economics and policy of energy market integration and connectivity; renewable energy; energy efficiency with a regional focus on ASEAN and East Asia.

**Ashish Shrestha** is a Consultant with the World Bank's Energy & Extractives Global Practice, where he works on the energy transition to low carbon pathways. He also serves as a technical expert for the Energy Sector Management Assistance Program (ESMAP) Global Facility for Mini-grids, providing advisory and operational support for mini-grid development in Asia and Africa. Ashish was previously a researcher with the World Bank's Development Economics Research Group, where his research focus was on the nexus of clean energy and climate change, including bio-energy and forest carbon, as well as sustainable transportation.

**Ming Su** is Associate Professor at the Energy Research Institute, Beijing, China. He is engaged in the research of energy and environment economics, energy development strategy and planning, and focuses on the issues about China's green development, energy transition and environment improvement. His paper 'An economic analysis of final consumption and carbon emissions responsibilities' was selected in "China Economics 2010" as one of the Annual Excellent Economic Theses. He also was awarded 2014's Annual Excellent Youth of Academy Macroeconomic Research, National Development and Reform Commission, China.

**Beni Suryadi** is Acting Manager of the, Policy and Research Analytics (PRA) Programme of the ASEAN Centre for Energy. His recent works are, among others, the RE Outlook for ASEAN – a REmap Analysis in cooperation with IRENA, ASEAN ESCO Report, and now he is leading the team to develop the 5th ASEAN Energy Outlook which will be launched at the 35th ASEAN Ministers on Energy Meetings (AMEM), September 2017 in Manila and will provide policy makers with an understanding of the energy trends and challenges being faced by the region up to the year 2040.

**Kentaro Tamura** received a PhD in International Relations from the London School of Economics and Political Science. He had been a lecturer at the Eco-Technology Laboratory at Yokohama National University and joined the Institute for Global Environmental Strategies (IGES) in 2003. His research interests include the implementation of the Paris Agreement and policy-making processes in major economies.

**Govinda R. Timilsina** is a Senior Research Economist in the Development Research Group of the World Bank, Washington, DC. His key expertise includes energy economics and planning, macroeconomic (general equilibrium) and sectoral modelling, the economics of biofuels and other clean and renewable energy resources, carbon pricing and climate change mitigation and urban transportation policies. Prior to joining the Bank, Dr Timilsina was a Senior Research Director at the Canadian Energy Research Institute. At present he is leading a number of World Bank studies including on sustainable urban planning in the Middle East and North Africa.

**Tania Urmee** is a Senior Lecturer at the School of Engineering and Information Technology, Murdoch University and a leading expert in the Renewable Energy field with a focus on rural electrification. She has undertaken research and project developments in a wider socio-cultural and geographical context including Bangladesh, Uganda, South Africa, Kenya, and Thailand. She is experienced in the use of stakeholder research techniques on community engagement in climate change and sustainable energy policy for developing countries. Tania holds a PhD in Renewable Energy from Murdoch University, Australia and a Masters in Energy Technology from the Asian Institute of Technology, Thailand.

**Sanjayan Velautham** was appointed as the Executive Director of the ASEAN Centre for Energy (ACE) in January 2015. A registered professional engineer (PEng), with a doctoral degree in Engineering, about 15 years of experience in the energy industry and ten years of research/teaching experience, he worked in Singapore initially with the Agency of Science, Technology and Research (A*STAR) as a Deputy Director and then with General Electric as Country Manager for the Power Generation Services business. He started his career with Tenaga Nasional Bhd. in Malaysia within the Power Generations Division. He also served as the National Project Manager for the BioGen Project for the United Nations Development Prgramme (UNDP – Malaysia).

**Vlado Vivoda** is Research Fellow at the Centre for Social Responsibility in Mining, the Sustainable Minerals Institute, the University of Queensland. He was previously based at the Griffith Asia Institute at Griffith University. Vlado has published extensively on a wide range of topics related to energy and minerals. His particular focus is on the international political economy of investment and trade in strategic energy commodities. He has authored numerous articles that were published with high profile journals, including *New Political Economy* and the *Journal of East Asian Studies*. His book (published with Routledge in 2014) examined Japan's energy security challenges after Fukushima.

**Takako Wakiyama** is a research fellow at the Institute for Global Environmental Strategies (IGES), and a PhD student in the Integrated Sustainability Analysis (ISA) team in the University of Sydney. Currently she is working on Japanese energy system analysis and energy economics, and the development of Japan's multi-regional input output (MRIO) model. She has developed a career as a researcher by conducting research on climate and energy policies and economics in Japan and Asian countries.

**Arthur A. Williams** completed a PhD at Nottingham Trent University in 1992. He has been involved in the development and dissemination of low-cost technologies for rural electrification, and has visited various Asian countries in connection with this work. He has been a lecturer in Electrical and in Mechanical Engineering at Nottingham Trent University before transferring to the University of Nottingham in 2007. He lectures in the field of sustainable energy and is course director for the interdisciplinary MSc in Sustainable Energy Engineering. His research is in Renewable Energy systems.

**Yanrui Wu** is Professor in Economics, Business School of the University of Western Australia, Australia. His research interests include energy and environmental economics, the economics of innovation, economic growth and productivity analysis. He has published extensively in these fields.

**Songli Zhu** joined the Energy Research Institute (ERI), National Development and Reform Commission in 1999 and has been working on climate mitigation policy and low carbon development strategy analysis to inform policy making and political dialogue for more than 15 years. As the team leader, she led the Greenhouse Gas inventory development in energy activities for China's Second National Communication during 2008–2011, and is now working on inventory development for the Third National Communication and First Biennial Update Report. She is now also leading one of the key projects in 13th Five-Year Plan, focusing on air pollution abatement in China and its key areas.

**Eric Zusman** is a senior policy researcher/area leader at the Institute for Global Environmental Studies in Hayama, Japan. Dr Zusman holds a Bachelor's degree in Mandarin Chinese from Rutgers University, a dual Master's degree in public policy and Asian studies from the University of Texas at Austin and a PhD in political science from the University of California, Los Angeles. For much of the past decade he has worked on environmental issues in Asia.

# PREFACE

The story of Asia's economic development over the past three decades or so is a fascinating one. The steady economic growth, led by China since the 1990s and India since the new millennium, has transformed the region pulling millions of people out of poverty. The export-led economic growth model has spread from the newly industrialised countries to a wider range of countries and the region as a whole benefited from the relocation of industrial activities from the developed world. But the vast internal market of the region and the growing affluence of the population have also ensured that the economy of the region is resilient to withstand economic shocks. The region is thus shifting from the factory of the world to a major consumer market that the world cannot ignore.

Reliable and affordable supply of energy has underpinned the economic performance and the growing demand for energy in the region has resulted in a greater influence on the global energy market over the past two decades. The region boasts of unprecedented developments in this sector that the world cannot ignore. The exceptional expansion of the coal industry to feed the electric power generation, the rapid growth in electric power generation capacity over the past two decades, and the phenomenal developments in renewable energy industries such as solar and wind power have transformed the region into the world leader in these areas. The aggressive participation of the region's National Oil Companies in foreign oil and gas exploration as well as in the acquisition of assets from around the world did not go unnoticed either. As the petroleum market dynamics have changed in the past few years with the Shale Explosion, producers and exporters started to look eastwards for rescue and the region has provided the necessary support.

Yet, there are still areas for further development. The region still has a large number of people without access to electricity or clean cooking energy. Despite improvements in recent times, millions of people lack access to electricity and billions still rely on solid fuels including traditional energies for cooking. The lack of access to energy hinders economic development and affects the poor disproportionately. Similarly, high reliance on fossil fuels to support economic development has caused severe environmental damage, particularly in urban areas, that imposes significant economic and social costs. The region has to grow economically to eradicate poverty and to ensure better living conditions for its growing population. But the region cannot continue with its legacy energy system in view of environmental and climate change concerns. The system has to change to a smart, low carbon path, which in turn requires careful governance and investment in infrastructure development.

Collectively, the region has vast experience of transforming economies and improving conditions. Countries of the region will aspire to move to the next level of economic development by becoming high income economies. The developed countries of the region provide living examples of such transformation. On that journey, the countries will face changing economic structures as well as social and demographic changes. This will have a tremendous impact on the energy needs of the economies and the region will have to manage the process effectively to emerge victorious. Learning from one another through cooperation and better regional integration of markets and infrastructure will be essential in this new phase of development, which will certainly become another fascinating story for the future.

This handbook offers reviews and reflections by academic leaders in Asian energy from all over the world covering a range of issues and developments. It recounts the progress made so far and explores the way forward, particularly the path towards a low carbon energy future. This book highlights the tremendous growth and improvement over the past two to three decades and provides insightful analysis of the drivers, contexts and issues. The book captures the collective wisdom of these experts in their respective areas and I hope it will serves as a valuable reference for anyone interested in Asian energy studies.

Clearly, it was a tremendous challenge to put together such a collection of works authored by these extremely busy contributors. The process of commissioning external reviewers and completing the internal and external review was also time-consuming. I am extremely grateful to all the contributors for their generous voluntary time contribution and support for this book project. I know they all had to reorganise their schedules to make room for this work and respond to my requests for information, clarification and details at short notice. Without their continued support despite their busy schedules this work could not have been completed.

I would also express my gratitude to all the reviewers who graciously accepted my request to review the chapters and provided constructive criticisms to improve the quality of the book. I acknowledge their voluntary time contribution for the review work and their professionalism in delivering the reviews in a timely manner. I appreciate their efforts.

I must also thank my current employer, De Montfort University, for allowing me to pursue this project and for granting time to complete the work as part of my research activities. Without this generous support, I would not have been in a position to complete this task on time. Last but not the least, I thank my spouse Debjani and daughter Saloni for bearing with me during the preparation of the manuscript over the past few months. With their support and sacrifice, the manuscript preparation continued into the evenings and nights and I owe them my gratitude.

# ACKNOWLEDGEMENTS

The contribution of the following reviewers who have reviewed one or more chapters of the handbook is greatly acknowledged. Despite their busy schedules, these experts have provided valuable comments to improve the chapters and to ensure the high quality of the handbook.

Dr Jiwan Acharya, Senior Energy Specialist, India Resident Mission, Asian Development Bank, New Delhi, India

Dr Amela Ajanovic, Vienna University of Technology, Energy Economics Group (EEG), Vienna, Austria

Dr Jian Chen, Associate Professor, Beijing Normal University, Zhuhai, China

Dr Jon Cloke, Research Associate, Geography Department, Loughborough University, Loughborough, UK

Dr Shyamasree Dasgupta, Assistant Professor, School of Humanities and Social Sciences, Indian Institute of Technology, Mandi, Himachal Pradesh, India

Dr Amar Doshi, School of Economics and Finance, QUT Business School, Queensland University of Technology, Brisbane, Australia

Dr Terry Van Gevelt, Lecturer, University of Cambridge, Cambridge, UK

Dr Obindah Gherson, Department of Economics and Development Studies, Covenant University, Nigeria

Dr VVN Kishore, Former Professor and Head, Department of Energy and Environment, TERI University, New Delhi, India

Dr Xuanli Liao, Senior Lecturer, University of Dundee, Dundee, UK

Dr Bundit Limmeechokchai, Sirindhorn International Institute of Technology, Thammasat University, Thailand

Dr Ken'ichi Matsumoto, Associate Professor, Graduate School of Fisheries and Environmental Sciences, Nagasaki University, Nagasaki, Japan

Dr Arabinda Mishra, Senior Social Scientist, International Centre for Integrated Mountain Development, Kathmandu, Nepal

Dr Xiaoyi Mu, Reader in Energy Economics, University of Dundee, Dundee, UK

Dr Pallav Purohit, Research Scholar, International Institute for Applied Systems Analysis, Austria

Dr Gopal K Sarangi, Assistant Professor, TERI University, New Delhi, India

Prof. Benjamin Sovacool, Professor of Energy Policy, University of Sussex, UK

Dr Yris Fondja Wandji, Paris Dauphine University, Paris, France

# ABBREVIATIONS

| | |
|---|---|
| AEC | ASEAN Economic Community |
| AEPC | Alternative Energy Promotion Centre |
| AIIB | Asian Infrastructure Investment Bank |
| AJEEP | ASEAN-Japan Energy Efficiency Partnership |
| APAECASEAN | Plan of Action for Energy Cooperation |
| APG | ASEAN Power Grid |
| ASEAN | Association of South East Asian Nations |
| BEE | Bureau of Energy Efficiency |
| BOM | Build-Own-Maintain |
| BOOM | Build-Own-Operate-Maintain |
| BOOT | Build-Own-Operate-Transfer |
| BP | British Petroleum |
| BPDB | Bangladesh Power Development Board |
| BRT | Bus Rapid Transit |
| CCS | Carbon Capture and Storage |
| CFSP | Cambodian Fuelwood Saving Project |
| CIF | Climate Investment Funds |
| CIL | Coal Indian Limited |
| CLP | China Light and Power |
| CMA | Central Mining Administrations |
| CNOOC | China National Offshore Oil Corporation |
| CNPC | China National Petroleum Corporation |
| CREDA | Chhattisgarh Renewable Energy Development Agency |
| CREP | Community Rural Electrification Programme |
| CRIB | Climate Relevant Innovation-system Builders |
| CSP | Concentrated Solar Power |
| CSPO | Certified Sustainable Palm Oil |
| CSR | Corporate Social Responsibility |
| DBT | Direct Benefit Transfer |
| DDG | Decentralized Distributed Generation |
| DISCOMs | Distribution Companies |

| | |
|---|---|
| DME | Dubai Mercantile Exchange |
| DNA | Designated National Authority |
| ECA | Energy Conservation Act |
| EEO | Energy Efficiency Obligation |
| ERIA | Economic Research Institute for ASEAN and East Asia |
| EROI | Energy Return on Investment |
| ESCO | Energy Service Companies |
| ESPO | East Siberia – Pacific Ocean |
| ESSPA | Energy Supply Security Planning for ASEAN |
| FiT | Feed-in-tariff |
| FLNG | Floating LNG |
| FYP | Five Year Plan |
| GBEP | Global Bioenergy Partnership |
| GCF | Green Climate Fund |
| GEF | Global Environmental Facility |
| GOI | Government of India |
| GSCI | Goldman Sachs Commodity Index |
| GWEC | Global Wind Energy Council |
| HELE | High Efficiency Low Emission |
| HPS | Husk Power Systems |
| IBF | Input-Based Franchises |
| ICE | Inter Continental Exchange |
| IDA | International Development Association |
| IDCOL | Infrastructure Development Company Limited |
| IGCC | Integrated Gasification Combined Cycle |
| IHA | International Hydropower Association |
| INDC | Intended Nationally Determined Contributions |
| INE | Shanghai International Energy Exchange |
| IRENA | International Renewable Energy Agency |
| ITMO | Internationally Transferred Mitigation Outcomes |
| JBIC | Japan Bank for International Cooperation |
| JCC | Japanese Customs Cleared (Japan Crude Cocktail) |
| JNOC | Japanese National Oil Company |
| KDB | Korea Development Bank |
| KEPCO | Korea Electric Power Corporation |
| KfW | Kreditanstalt für Wiederaufbau |
| KNOC | Korea National Oil Corporation |
| MHP | Micro Hydro Power |
| MLP | Multi-Level Perspective |
| MNRE | Ministry of New and Renewable Energy |
| MOEF | Ministry of Environment and Forestry, India |
| MOEJ | Ministry of the Environment, Japan |
| MOF | Ministry of Finance |
| MoP | Ministry of Power, India |
| NAMA | Nationally Appropriate Mitigation Actions |
| NAPCC | National Action Plan of Climate Change |
| NBCI | National Biomass Cookstoves Initiative |

| | |
|---|---|
| NDRC | National Development and Reform Commission |
| NEA | Nepal Electricity Authority |
| NEA | National Energy Administration |
| NEC | National Energy Commission |
| NOC | National Oil Company |
| NORAD | Norwegian Agency for Development Cooperation |
| NPBD | National Program of Biogas Development |
| NPC | National People's Congress |
| NPIC | National Program on Improved Cookstoves |
| NSI | National Systems of Innovation |
| OECF | Overseas Economic Cooperation Fund |
| ONGC | Oil and Natural Gas Corporation |
| PALECO | Palawan Electric Cooperative |
| PAYG | Pay-As-You-Go |
| PCC | Political Consultative Conference |
| Petronas | Petroliam Nasional Berhad |
| PLN | Perusahaan Listrik Negara |
| PO | Partner Organisations |
| PRA | Price Reporting Agencies |
| PSF | Private Sector Facility |
| PTA | Power Trade Agreement |
| PTC | Power Trading Corporation |
| QTP | Qualified Third Party |
| REA | Renewable Energy Act |
| REC | Rural Electrification Corporation, India |
| REDD | Reduced Emissions from Deforestation and Forest Degradation |
| REEE | Renewable Energy and Energy Efficiency |
| REEEP | Renewable Energy and Energy Efficiency Partnership |
| REP | Rural Electricity Policy |
| REST | Rural Electricity Supply Technology |
| RISE | Readiness for Investment in Sustainable Energy |
| RPS | Renewable Portfolio Standard |
| SARI | South Asia Regional Energy Initiative |
| SCGC | South China Grid Corporation |
| SDG | Sustainable Development Goals |
| SE4ALL | Sustainable Energy for All |
| SEB | State Electricity Board |
| SEC | Singapore Environment Council |
| SERC | State Electricity Regulatory Commission |
| SGC | State Grid Corporation |
| SHFE | Shanghai Futures Exchange |
| SHS | Solar Home Systems |
| SOE | State Owned Enterprises |
| SPC | State Power Corporation |
| SREP | Scaling up Renewable Energy Programme |
| SRI | Socially Responsible Investment |
| TAGP | Trans-ASEAN Gas Pipeline |

| TIS | Technological Innovation System |
| TPES | Total Primary Energy Supply |
| TVE | Town and Village Enterprises |
| UNESCAP | United Nations Economic and Social Commission for Asia and the Pacific |
| UNFCCC | United Nations Framework Convention on Climate Change |
| UNIDO | United Nations International Development Organisation |
| USD | United States Dollar |
| VGF | Viability Gap Funding |
| WBREDA | West Bengal Renewable Energy Development Agency |
| WTI | West Texas Intermediate |
| WTO | World Trade Organisation |

# PART 1

# Energy use in Asia

# 1

# INTRODUCTION

*Subhes C. Bhattacharyya*

## Asian economic development

Asia has been at the centre of global attention over the past three decades or so. According to the Asian Development Bank (ADB, 2017), developing Asia has transformed itself from a predominantly low-income region in the 1990s to a middle-income one in terms of per capita income measured in purchasing power parity (PPA). In 1991, more than 90% of the region's population lived in low-income countries but by 2015, the majority of the population of the region live in middle-income countries. This dramatic transformation has happened fairly quickly by sustaining high levels of economic growth across the region.

Before the 1980s, the strong economic development was limited to Four Tigers, namely Hong Kong, Korea, Singapore and Taiwan. Malaysia and Thailand joined the high growth path in the 1980s. Led by industrial development and economic growth in Japan, the 'flying geese' pattern, a catching-up process, initiated the Asian Miracle (see Kojima, 2000 for details on the flying geese model of growth). However, it is the double digit growth in China since its adoption of the open door policy and economic reform in late 1970s that has transformed the Asian economy beyond recognition. Between 1990 and 2015, China's gross domestic product (GDP) grew 11-fold in constant 2005 dollar terms. Other Asian countries have also been influenced by this miraculous growth pattern and many joined the bandwagon. The region as a result grew faster than the world economy, with an average growth rate difference of 1.5% sustained over the period between 1990 and 2015, except for a short period during the 1997 Asian Financial Crisis (see Figure 1.1). The Asian economy continued to follow a relatively high growth path even after the global financial crisis of 2008. The share of Asian GDP in the global scene is thus showing a growing trend: from 19% in 1990, it has reached 27% in 2015 in constant 2005 US dollar terms. In purchasing power parity, the share becomes much higher.

Faster economic growth has resulted in growing per capita income across the board. A plot of per capita income in 1990 versus that in 2015 (in logarithmic scale) shows the trend clearly (see Figure 1.2). Only a few countries are below US$1000 per capita income level, most of the countries are between US$1000 and US$10,000 range, while a handful are above the US$10,000 range. It can be seen that most of the countries are above the 45 degree line, showing that their 2015 per capita GDP has improved compared to the 1990 level.

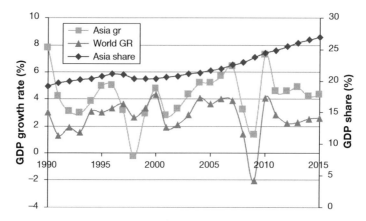

*Figure 1.1*   GDP growth rate and share of the Asian economy

*Data source*: International Energy Agency (2015).

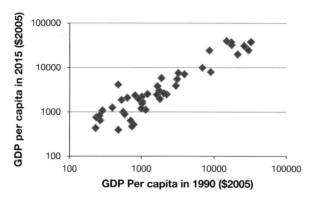

*Figure 1.2*   Scatter plot of GDP per capita in Asian countries

*Data source*: International Energy Agency (2015).

Asian countries achieved the above high growth path following two main approaches: by being the factory of the world and by driving domestic demand through the expansion of the middle-income class (Nakaso, 2015). The first driver was supported by the gradual relocation of industrial activities from developed countries to Asia to take advantage of the availability of a low-cost and skilled workforce, as well as a weak environmental and regulatory environment. This was further supported by trade liberalisation which increased trade volumes and led to growth in foreign direct investments. The expansion of export-oriented industrial activities created opportunities for better income that attracted rural agricultural labour force to factory sites in urban areas, resulting in large-scale migration and rapid urbanisation. The rise in income also expanded the size of the middle-income class, which became the second driver of economic expansion.

## Unprecedented transformation of Asia

The economic growth has transformed the Asian way of living and has brought unprecedented changes. Four main transformations are worth mentioning, namely, a remarkable reduction in poverty incidence, the rise of the middle-income class, rapid urbanisation and demographic transition. Asia was infamous for its high incidence of poverty which is a common feature of low-income economies. However, the economic transformation has been successful in pulling millions of people out of poverty. Between 2002 and 2013, 707 million people in Asia and the Pacific have moved out of extreme poverty based on US$1.90 a day poverty line (2011 Purchasing Power Parity, PPP). East Asia has recorded the highest reduction in poverty level (from 31.9% in 2002 to 1.8% in 2013) but most of the sub-regions have managed to reduce poverty by 20% over this period. By 2013, only 9% of the region's population (or 330 million) were living in extreme poverty condition, most of them concentrated in South Asia (ADB, 2016).

Simultaneously, the emergence and expansion of the middle class has changed the societal complexion beyond recognition. According to Kharas (2017), the first billion people of the middle class, reached around 1985, took more than 150 years from the Industrial Revolution but the next billion was added only in 21 years and the third billion was added in only nine years. Almost one half of the three billion middle class people lived in Asia by 2015 and almost 90% of the next billion entering this class will be in Asia. The better-off section of the population started to consume goods and services adopting the international trends, thereby offering a large potential consumer base which was hard to ignore. It is estimated that the middle-income class consumed US$35 trillion (2011 PPP value) in 2015 (Kharas, 2017). This represents a large domestic market for goods and services which thus offered the second impetus to growth. The size of the middle class is expected to grow rapidly over the next 15 years or so and by 2030, Asia is expected to account for 65% of the global middle-income class of 5.4 billion people.

Rapid urbanisation has accompanied high economic growth in Asia. As Table 1.1 indicates, between 1990 and 2014, Asia has added more than 900 million urban people, 50% of whom came from East Asia. China alone added almost 450 million urban people. India and China account for 30% of the global urban population and more than a billion urban people. Asia had 17 megacities by 2016 (i.e. cities with more than ten million habitants), of which six megacities were in China and another five in India (UN, 2016). Between 2014 and 2050, India will add 404 million more while China will add another 292 million urban people. Overall, Asia will add 1.1 billion urban people during that period and will reach an urbanisation rate of 62% (UN-DESA, 2014).

Moreover, Asian demography is in transition. The population is expected to grow slowly compared to the previous decades. For example, East Asia will see a much slower growth in population and China will see a marginal fall in its population size by 2050 (from 1.376 billion in 2015 to 1.348 billion in 2050). On the other hand, Central Asia will maintain a high population

*Table 1.1* Urbanisation in Asia

| Country/region | Urban population | | | Urban share of population (%) | | |
|---|---|---|---|---|---|---|
| | *1950* | *2014* | *2050* | *1950* | *2014* | *2050* |
| East Asia | 467 | 960.2 | 1250.2 | 34 | 59 | 78 |
| China | 308.2 | 758.4 | 1049.9 | 26 | 54 | 76 |
| South Asia | 284.3 | 551.9 | 1129.2 | 25 | 33 | 51 |
| India | 222 | 410.2 | 814.4 | 26 | 32 | 50 |
| South East Asia | 140.2 | 294.4 | 507.7 | 32 | 47 | 64 |
| Indonesia | 54.6 | 134 | 227.8 | 31 | 53 | 71 |
| West (Central) Asia | 31.6 | 35.8 | 57.7 | 48 | 44 | 56 |
| Asia | 923 | 1842 | 2945 | 30 | 45 | 62 |

*Data source*: UN-DESA (2014).

growth rate but because of its low population base, the overall impact will be less noticeable. India will displace China to become the world's most populous country by 2022 and by 2050 India is likely to have 1.7 billion people. Indonesia, Pakistan and Bangladesh are other major countries where population will grow. As life expectancy improves, the share of the population living over 65 years will gradually increase. By 2050, China will have an old-age dependency ratio (i.e. ratio of the population aged 65 and over to the population between 15 and 64) of 46.7% whereas the ratio will be higher than 70% in Japan. But Asia will continue to benefit from its youthful population structure and the ratio of working population to dependent population will remain favourable over the next decades. For example, India and Indonesia will have only 21% of their population aged 65 and over per 100 persons of working age (UN-DESA, 2015). Thus, a growing population and changes in the population structure will influence economic development and future development of Asia.

## Energy in Asian development

Energy has played a major role in Asia's economic and social transformation over the past three decades. Faster economic growth in the region resulted in rapid growth in energy demand and supply of energy in the region. Between 1990 and 2012, Asia's share in global primary energy demand increased from 26% to over 40%. As Asia became the factory of the world, the industrial sector received greater attention, which in turn required reliable, cheap and adequate supply of energy. The energy policy in the region thus aimed at competitive supply and energy security by focusing on a range of resources. Realising the importance of energy for economic development, most of the countries of the region favoured state control of and participation in the energy sector development, at least in the initial phase of development.

Initially the focus was on local resource development, particularly development of fossil fuels and hydropower resources. Coal is very accessible in Asia and being a major resource of the region and a cheaper source of energy, it has played an important role in supporting economic development of the region. The coal industry has grown rapidly to support the economic activities, particularly of industry and of the electricity sector. Since the first crisis of oil in 1973, the region has moved away from oil for electricity generation and coal became the main fuel for electricity generation in the region in due course. However, high dependence on coal has brought environmental degradation as a major developmental challenge but low cost and widespread availability of coal in Asian countries implies that the region is likely to rely on coal in the future. Thus striking a balance in coal use and managing the environmental challenges will remain a major issue.

Industrialisation, urbanisation and the rise of the middle-income class have also influenced the energy use in Asian economies. Urban areas account for almost three-quarters of the gross domestic product of countries as most of the economic activities take place in and around urban areas. As a result, they also consume 75% of the primary energy demand of a country. Better access to electricity and other forms of energy in urban areas and the ability to afford a better life-style supports much higher consumption of energy on a per capita basis in urban areas than in rural areas. Buildings (both residential and commercial) and transport systems account for a significant share of urban energy use. The rapid growth in energy use in Asia is partly attributable to the high growth in urban energy use. But the fossil fuel dependency of the region brings environmental challenges and most of the urban areas suffer from poor air quality due to high levels of pollutants such as particulate matters, sulphur oxides and nitrogen oxides, carbon monoxide and lead.

The region has already taken some initiatives to address the above challenge. The policy of diversification of the fuel mix for electricity generation has been pursued for quite some time.

For example, countries with limited resource endowments such as Japan and Korea have relied on imported natural gas and nuclear power. Other countries such as China and India also have plans for expansion of nuclear power. However, the nuclear accident in Japan in 2011 has forced countries to rethink their nuclear strategy. Similarly, hydro-resources of the region have also been developed since 1950s, although there still remains significant potential for further development. But large-scale development of renewable energies has been the main instrument for managing the environmental and climate change challenge. The region has shown strong commitment to renewable energies. In addition, energy efficiency improvement has also been taken up on a priority basis.

Despite rapid economic growth and socio-economic transformations, Asia still faces challenges in terms of universal energy access. Millions of people still do not have access to clean energy. Similarly, the effective energy sector governance remains an issue in the region. Although most countries have undertaken some form of sector reform, many have not completed the process and the state influence on the energy sector remains considerable. Asian countries need strong, forward-looking energy sector governance to support their economic development as well as a transition to a low carbon energy pathway.

## Purpose and content of the book

The purpose of this handbook is to provide a review of the status of the energy sector in Asia capturing the recent developments, issues, challenges and the future outlook. The book is divided into four parts: Part 1 is devoted to energy use in Asia and considers the regional picture as well as sector level energy use patterns, it contains six other chapters in addition to this introduction. Part 2 deals with main energy supply in Asia, including oil, natural gas, electricity, and renewable energies. This part contains seven chapters. Part 3 focuses on various energy policy issues, including energy sector deregulation, security of supply, funding for infrastructure development and climate change response. This part contains six chapters. Part 4 focuses on the efforts towards low carbon transition and captures various dimensions in six chapters. Finally, the concluding remarks are presented in the last chapter. A brief introduction to each chapter is provided below.

### Part 1: Energy use in Asia

Chapter 2 by Bhattacharyya provides a review of the energy situation in Asia. The chapter considers the period between 1990 and 2012 and describes the energy use trends in four regions of Asia, namely East, West-Central, South and South East. The regional differences in primary energy use at present and the future outlook is captured in the chapter. Su and Zhu provide a review of energy situation in China in Chapter 3. They consider the developments in energy use pattern in China since the new millennium and capture the recent slow-down in energy use in China. They also highlight the need for the transition to a low carbon future. In Chapter 4, Bhattacharyya and Palit introduce the energy access challenge in Asia. The chapter offers an overview of the problem and discusses the initiatives in different countries to enhance electricity and clean cooking energy access. The chapter emphasises the possibility of learning from one another to improve energy access in the region.

Chapters 5, 6 and 7 focus on energy use at the sector level. Bhattacharyya presents a review of industrial energy use in Asia in Chapter 5. Timilsina and Shrestha present the trends in transport energy use in the region in Chapter 6. The residential energy use is covered in Chapter 7 by Bhattacharyya. Chapters 5 and 6 use decomposition analysis to analyse the drivers of change in energy use in the industrial and transport sectors. Similarly, Chapter 7 provides a review of residential energy use in four regions of Asia and presents an outlook for the future.

## Part 2: Energy supply in Asia

In Chapter 8, Doshi presents a review of the oil supply in Asia focusing on demand, supply, trade and pricing. The chapter also discusses the role of national oil companies in the region and presents the refining activities in the continent. Ripple presents the natural gas trade and supply in Asia in Chapter 9. He considers the demand and supply gap in the region and discusses the changing trade scenario in the region due to global gas market adjustments arising from new supply options from shale gas in the US. He also presents the gas supply outlook for the region.

The next two chapters (Chapters 10 and 11) focus on coal and electricity supply in the region. Bhattacharyya presents a review of the role of coal in Asia in Chapter 10 and its importance for electricity generation in the region. A brief history of coal sector development in three main coal suppliers of Asia, namely China, India and Indonesia is also presented. Chapter 11 on the other hand provides electricity profiles of Asian countries and discussions on the regional power markets. The outlook of electricity supply in the region completes the chapter.

Chapters 12 to 14 present the status of renewable energy supply in Asia. Urmee and Kumar present the developments in grid-connected solar power supply in Asia in Chapter 12. They discuss technologies (namely solar photovoltaic and concentrated solar power technologies), policy initiatives in the region to promote solar energy, and the future potential of solar power in Asia particularly as a climate mitigation measure. Dent reviews the wind power development in Chapter 13. He presents an overview of wind energy development and analyses various challenges for wind energy in Asia considering technologies, policies, interventions, infrastructure challenges and institutional and socio-technical aspects. Williams reviews the hydropower development in Asia in Chapter 14. The chapter provides a brief history of hydropower in the region and discusses the impacts of hydropower. He also considers pumped storage as well as small-scale hydropower technologies and discusses their roles for universal energy supply.

## Part 3: Energy policy issues in Asia

Sen and colleagues provide an analysis of electricity sector reform in selected Asian countries in Chapter 15. They present the standard electricity sector reform model and introduce the challenge of balancing economic growth and environmental impacts. The conflicts between these objectives in the case of three south Asian countries, namely India, Bhutan and Nepal are assessed in the chapter. Wu and colleagues review the electricity sector deregulation and market integration in China in Chapter 16. They consider the power sector development in China, various reform initiatives in the sector and policy implications of deregulated electricity industry in the country. In Chapter 17, Zhu and colleagues present the reform initiatives in the Chinese energy sector, with a focus on initiatives for low carbon pathways to steer the country from its coal dependence. They consider the institutional reform, pricing reform and policies to promote renewable energies and energy efficiency in the country. Vivoda discusses the energy security issues in Asia in Chapter 18. He focuses on major energy security challenges in the region arising from competitive security approaches, market fragmentation, fossil fuel subsidies and environmental sustainability. He suggests increased cooperation among countries for better energy security outcomes.

In Chapter 19, Ng and Nathwani argue the need for smart and sustainable energy infrastructure development in Asia in the next phase of its development. This is required to move away from the legacy technologies to low carbon sustainable options. They suggest that such a shift will require a coordinated effort and they propose a framework for sustainable energy infrastructure investment in Asia. Janardhanan and Mitra focus on Asia's climate change response and find that although Asian countries have taken significant initiatives to mitigate climate change, their efforts are fragmented and inadequate to meet the international targets of climate mitigation. Power sector integration and up-scaling the renewable energy adoption are suggested as key options for a better response.

## Part 4: Energy in a carbon-constrained world

In Chapter 21 Wakiyama and colleagues present a survey of climate policies in four Asian countries (China, India, Indonesia and Japan) and undertake a decomposition analysis to identify the main drivers influencing carbon emissions. They find that the policies vary across countries and the country case studies suggest the reasons for such variations. They argue that the international climate policy regime has to accommodate such national variations but offer guidance for achieving a common goal.

Anbumozhi and colleagues review the energy situation in the ASEAN and analyse the issue of low carbon transition in the region in Chapter 22. They assess the current policy initiatives and consider a range of options to allow a feasible transition by 2030. They suggest a list of actions at the regional level to implement an ASEAN-wide low carbon energy transformation.

In Chapter 23 Kuriyama and Tamura consider the issue of decarbonisation of the electricity system of Greater Mekong sub-region and suggest that individual country policies are inadequate to support a region-level objective. They argue that implementation of the regional grid offers greater benefits for the electricity system decarbonisation and can be a better alternative to national actions. Gheewala and colleagues analyse the costs and benefits of biofuels in the Asian context in Chapter 24. They review the status of biofuels in the region, consider the policy environment and assess the costs and benefits of biofuel development. They also highlight issues surrounding biofuel development in Asia.

In Chapter 25 Parthan reviews the financing options for enhancing energy access in Asia. He reviews several country experiences and considers a range of commonly used and new options that could be used to support energy access in the region. Ockwell and Byrne offer a socio-technical innovation systems perspective for enhancing energy access in developing countries in Chapter 26. They review the recent research on energy access and identify the insights from socio-technical innovations perspective that can shape the policy debate in the area. They also offer a theoretical framework to support energy access interventions.

The concluding chapter summarises the main points of the book.

## Regional definition

It is important to define the regional coverage of Asia used in the handbook. Although there is some variation in the regional coverage according to sources of information, the definition used by the Asian Development Bank has been followed in general. The following groups or regional classification are used:

ASEAN (Association of South East Asian Nations): This group consists of the following countries Brunei Darussalam, Cambodia, Indonesia, the Lao People's Democratic Republic, Malaysia, Myanmar, the Philippines, Singapore, Thailand and Viet Nam.

Central Asia (or West Asia) comprises of Armenia, Azerbaijan, Georgia, Kazakhstan, the Kyrgyz Republic, Tajikistan, Turkmenistan and Uzbekistan.

East Asia comprises of Japan, the People's Republic of China, Hong Kong, China; the Republic of Korea; Democratic Republic of Korea, Mongolia and Taipei, China.

South Asia consists of Afghanistan, Bangladesh, Bhutan, India, the Maldives, Nepal, Pakistan and Sri Lanka. However, due to data limitations, Afghanistan, Bhutan and the Maldives are included in Other Asia.

South East Asia comprises of Brunei Darussalam, Cambodia, Indonesia, the Lao People's Democratic Republic, Malaysia, Myanmar, the Philippines, Singapore, Thailand and Viet Nam. Due to data limitations, Lao has been included in Other Asia in most cases.

Other Asia consists of the following: Afghanistan, Bhutan, Cook Island, East Timor, Fiji, French Polynesia, Kiribati, Laos, Macau (China), Maldives, New Caledonia, Palau, Papua New Guinea, Samoa, Solomon Islands, Tonga and Vanatu.

Japan, the Republic of Korea, Hong Kong, Taipei, China and Singapore are the developed countries of the region and the rest are considered as developing countries. Note that the above definition excludes the Middle East, although they are physically located in Asia.

Although the above definition has been followed generally, contributing authors of different chapters have used data from different sources. In such cases, the authors have clarified the specific country coverage in their chapters.

## References

ADB, 2016, Key Indicators for Asia and the Pacific 2016, Asian Development Bank, Manila.

ADB, 2017, Asian Development Outlook 2017: Transcending the middle-income challenge, Asian Development Bank, Mandaluyong City, Philippines.

Kharas, H., 2017, The Unprecedented Expansion of the Global Middle Class: An Update, Working paper 100, Global Economy and Development at Brookings, Brookings, Washington, DC.

Kojima, K., 2000, The "flying geese" model of Asian economic development: origin, theoretical extensions, and regional policy implications, Journal of Asian Economics, 11:375–401.

Nakaso, H., 2015, Asian Economy: Past, Present and Future, Keynote address at the Securities Analysts Association of Japan International Seminar, April 24, Bank of Japan (www.boj.or.jp/en/announcements/press/koen_2015/data/ko150424a1.pdf).

UN, 2016, The World's Cities in 2016, Data Booklet, United Nations, New York.

UN-DESA, 2014, World Urbanization Prospects: The 2014 Revision, Highlights, ST/ESA/SER-A/352, United Nations – Department of Economic and Social Affairs, Population Division, New York.

UN-DESA, 2015, World Population Prospects: The 2015 Revision, Volume II: Demographic Profiles, United Nations, Department of Economic and Social Affairs, Population Division, New York.

# 2

# A REVIEW OF THE ENERGY SITUATION IN ASIA

*Subhes C. Bhattacharyya*

## Introduction

Asia, the most populated continent in the world, is a major energy user. Between 1990 and 2012, the total primary energy use has increased from 2312 Mtoe to 5358 Mtoe, recording an annual average growth of 3.9%. As the centre of gravity of global economic activity is moving eastwards to Asia, so is the energy use. However, due to its large and growing population, Asia's energy requirement per person remains low compared with the global average and it is the home of a large number of people without access to electricity and clean cooking energies (see Table 2.1).

This chapter provides an overview of the energy situation in Asia. As the geographical coverage tends to vary depending on the source of information, it is important to specify the geographical coverage used in this chapter. Four regions, namely East Asia, West Asia, South Asia and South East Asia, have been considered in this chapter to represent Asia. As time series data is not readily available for smaller countries and island nations of the region, they are grouped under a catch all "Other Asia" category. The regional country focus of the chapter is as follows:

East Asia: Peoples Republic of China including Hong Kong, Republic of Korea (or South Korea), Democratic Republic of Korea, Mongolia, Chinese Taipei and Japan.

Central (or West) Asia: Armenia, Azerbaijan, Georgia, Kazakhstan, Kyrgyzstan, Tajikistan, Turkmenistan, and Uzbekistan.

South Asia: Bangladesh, India, Nepal, Sri Lanka and Pakistan.

South East Asia: Brunei Darussalam, Cambodia, Indonesia, Malaysia, Myanmar, Philippines, Singapore, Thailand, and Vietnam.

Other Asia: Afghanistan, Bhutan, Cook Island, East Timor, Fiji, French Polynesia, Kiribati, Laos, Macau (China), Maldives, New Caledonia, Palau, Papua New Guinea, Samoa, Solomon Islands, Tonga and Vanatu.

The coverage is in line with International Energy Agency data publications, from where the energy data has been used in the chapter.

The second section presents the overview of primary energy supply in the region, which is followed by a review of the energy situation at the sub-region level in the third section. The fourth section presents a review of energy conversion to electricity while the fifth section reviews

*Table 2.1* Key energy information

|  | 1990 | 2012 |
|---|---|---|
| Primary energy requirement (Mtoe) |  |  |
| Asia | 2312.2 | 5358.0 |
| East | 1495.9 | 3747.3 |
| West | 192.9 | 175.5 |
| South | 383.3 | 842.7 |
| South East | 233.2 | 578.5 |
| Other Asia | 6.9 | 14.9 |
| World | 8780.2 | 13371.0 |
| Asian share (%) | 26.3 | 40.1 |
| People having energy access (%) |  |  |
| Electricity – East Asia | 93 | 99 |
| Clean cooking fuel – East Asia | 36 | 56 |
| Electricity – West Asia | 89 | 93 |
| Clean cooking energy – West Asia | 82 | 95 |
| Electricity – South Asia | 52 | 79 |
| Clean cooking fuel – South Asia | 22 | 36 |
| Electricity – South East Asia | 70 | 90 |
| Clean cooking energy – South East Asia | 25 | 54 |

*Data source*: International Energy Agency (2015b) and International Energy Agency and the World Bank (2015).

the final energy use. Environmental concerns related to energy use and forecasts of energy demand are considered in the last two sections. It needs to be mentioned that at the time of writing this chapter, consolidated data for 2012 was most commonly available due to delays in data reporting by international agencies. However, the economic slowdown in China since 2014 has resulted in a slower growth in energy demand. This has been indicated in the chapter as appropriate. In addition, following the energy accounting framework convention, it is assumed that primary energy supply is equal to primary energy demand in any given year for any country. Accordingly, supply and demand terms are often used interchangeably.

## Overview of primary energy supply

Over the past two decades, Asia emerged as a major player in the global energy scene. In 1990, the Asian share in the global primary energy supply was 26% but this had increased to 40% by 2012. Asian primary energy demand grew twice as fast as the global energy demand growth during this period. As Figure 2.1 indicates, China clearly dominated the Asian energy supply scene during this period. China's share in the regional energy supply was about 38% until 2002 but since then, the share climbed rapidly and by 2012, it accounted for 54% of Asian energy supply.

Expectedly, the sub-regional distribution of primary energy supply is heavily skewed towards East Asia. East Asia accounted for 65% of the regional primary energy supply in 1990 and between 1990 and 2012, its supply has grown two and a half times. West Asia on the other hand used the least amount of energy in the region and its primary energy supply remained almost stagnant over the past two decades. Consequently, the share of East Asia has gradually increased from 65% in 1990 to 70% in 2012, whereas West Asia's share has fallen from 8% in 1990 to 3% in 2012. South East Asia and South Asia have managed to maintain a steady share during this period (see Figure 2.2).

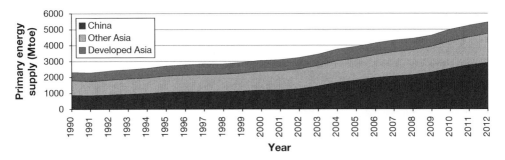

*Figure 2.1*　Evolution of Asian primary energy supply

*Data source*: International Energy Agency (2015).

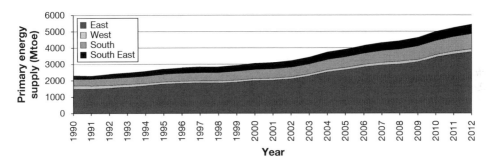

*Figure 2.2*　Regional distribution of primary energy supply in Asia

*Data source*: International Energy Agency (2015b) and UNStats (2017).

A wide variation in energy consumption per person is clearly visible in the region. South Asia has the lowest average primary energy use per person (just above 500 toe/person in 2012) whereas East Asia recorded the highest average at about 2300 toe/person in 2012. South Asian energy consumption per person was almost one half of that of South East Asia while East and West Asia enjoyed almost twice the energy consumption level per person of South East Asia in 2012 (see Figure 2.3). Note that West Asia suffered a loss of per capita energy use between 1990 and 1995. The fall of the Soviet Union and consequent economic turmoil in the Former Soviet Union countries was responsible for this decline. However, at the country level, the difference is even more glaring: Bangladesh, Myanmar and Tajikistan all consumed less than 300 toe/person in 2012 whereas the Republic of Korea and Singapore had high consumption at above 4700 toe/person. However, the growing trend of energy use in the region is clearly visible.

The energy–economy relationship among Asian countries (see Figure 2.4) clearly shows that as countries become wealthier (i.e. their gross domestic product (GDP) per person increases) their energy use also increases proportionally up to a per capita income of 5000 USD (constant 2005 values). There is some variation in energy demand within this income range but the spread is small. This suggests that economic growth of developing countries in the region depends on intensification of energy use initially as they try to industrialise. However, as income goes beyond 5000 USD level, the energy demand appears to stagnate. The data fits a typical S curve quite well and suggests that countries like Japan (JP on the graph) and South Korea (SK on the graph) have managed, to some extent, to delink GDP growth and energy use. Brunei Darussalam (BN on the graph) has recorded the highest energy use per person in the sample, although its GDP per capita is

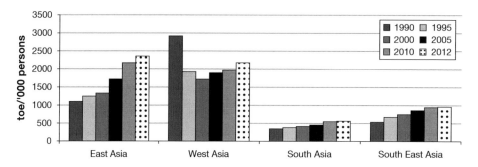

*Figure 2.3*   Growing trend and disparity in per capita energy use in Asia

*Data source*: International Energy Agency (2015b) and UNStats (2017).

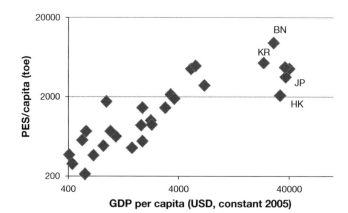

*Figure 2.4*   Energy–economy correlation in Asia

*Data source*: International Energy Agency (2015b) and UNStats (2017).

lower than that of some other countries. The diagram also suggests that economic prosperity of countries may be possible through a less energy-intensive development pathway.

The region has made significant progress in terms of energy efficiency improvements. Figure 2.5 indicates that energy intensity has been declining rapidly in West Asia as it is catching up with other sub-regions. In fact, a regional convergence in energy intensity is clearly visible. East Asia however remains the most energy efficient area in Asia whereas South Asia and South East Asia have recorded practically similar energy intensity in recent times. However, as discussed below, there is still significant scope for improvement. The West Asian case remains interesting here: as indicated in previous figures, West Asia consumed the least amount of energy in Asia but because the size of the economy and the population is small, the energy intensity was high early in the 1990s. But due to somewhat declining energy demand and doubling of economy size, the energy intensity has fallen rapidly. Note also the contrast between per capita energy use trend and the energy intensity trend. While per capita energy use is showing a rising trend, the energy intensity is falling. This suggests that energy demand is growing faster than the population growth rate but significantly slower than the economic growth of the region.

However, the regional energy intensity masks the large variation in efficiency level in the region. The country level trend indicates that Japan has maintained the lowest level of energy intensity in

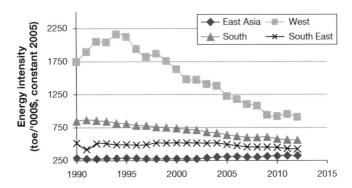

*Figure 2.5*   Energy intensity trend at the regional level in Asia

*Data source*: International Energy Agency (2015b) and UNStats (2017).

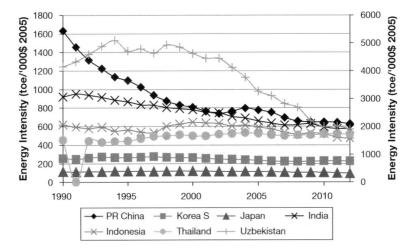

*Figure 2.6*   Energy intensity of selected Asian countries

*Data source*: International Energy Agency (2015b) and UNStats (2017).

the region, followed by South Korea (see Figure 2.6). While China, India, Indonesia and Thailand have reached almost similar energy intensity levels in recent times, they consume almost six times higher the amount of energy used by Japan for a dollar worth of output. This clearly indicates the serious energy efficiency gap existing in the region and the potential for significant improvement.

## Proven reserves of fossil fuels

Asia is not well-endowed with petroleum resources. According to British Petroleum (2016), only about 2% of the global oil reserves (or 5.2 billion tonnes) and about 6% of the gas reserves (or 11.9 trillion cubic metres) were located in Asia at the end of 2014. China, India, Indonesia, Malaysia and Vietnam account for 93% of the total oil reserves and 82% of gas reserves in the region. The size of the gas reserve has improved slightly over time due to intensive exploration activities but oil reserves remained unchanged. However, it must be noted that the estimates for oil and gas reserves vary from one agency to another due to differences in the definitions, geographical coverage, and level of

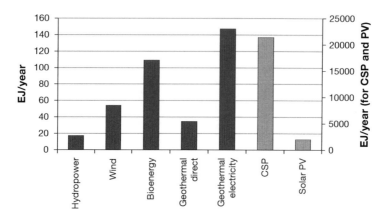

*Figure 2.7*  Renewable energy potential in Asia

*Data source*: Hoogwijk and Graus (2008).

uncertainty considered in the estimates. For example, the estimates of conventional oil reserves for Asia vary between 4.6 billion tonnes and 7.2 billion tonnes (Rogner et al., 2012).

On the other hand, about a quarter of the global coal reserves (or 212 billion tonnes) are located in Asia. Almost 96% of the coal reserves is shared between China, India and Indonesia, with China alone accounting for more than 50% of the regional coal reserves. In terms of reserves to production ratio, China could maintain its current level of coal production for another 31 years, while India and Indonesia can sustain for 89 and 71 years respectively. The ratio falls close to 12 and 18 years for oil and between 27 and 50 years for natural gas (British Petroleum, 2016).

Asia has significant renewable energy potential as well. The region is endowed with solar energy potential, as well as geothermal, biomass, wind and hydropower potential. Due to high solar irradiance received by the region, the solar energy potential is relatively high (see Figure 2.7). It is estimated in Rogner et al. (2012) that the region has the potential of generating 2391 EJ/year using concentrated solar power and 2026 EJ/year using solar PV (photo voltaic). The geothermal energy potential comes next, which can support 147 EJ/year through electricity generation and 35 EJ/year through direct heating. The region can also produce 109 EJ/year from biomass, more than 50 EJ/year from wind and 17 EJ/year from hydropower (Rogner et al., 2012).

The sub-regional distribution is shown in Table 2.2. China because of its size dominates in renewable energy potential. Almost 50% of the CSP (Concentrated Solar Power) potential comes from China. In fact, 74% of the CSP potential belongs to East Asia. South Asia comes next with another 23% potential. The rest of Asia contributes only 4% to the overall CSP potential of the region. In contrast, South Asia and East Asia have an almost equal share (48% each) in solar PV potential. Wind energy potential is shared between East Asia and South East Asia, with East Asia contributing more than 68% of the potential. All the sub-regions have some hydropower and bioenergy potential. China (and hence East Asia) dominates in these resources as well but both South Asia and South East Asia make some contribution.

### Indigenous production, import/export

Asia produces a significant share of its primary energy supply. Indigenous production of energy as a share of primary energy supply remained around 80% since 1990, which is a significant achievement considering the rapid growth in primary energy supply in the region. As Figure 2.8

*Table 2.2* Annual distribution of renewable energy resources in Asia

| Sub-region | Hydropower (EJ/yr) | Solar PV (EJ/yr) | Wind (EJ/yr) | CSP (EJ/yr) | Bioenergy (EJ/yr) | Geothermal direct heating (EJ/yr) | Geothermal Electricity (EJ/yr) |
|---|---|---|---|---|---|---|---|
| China | 8.91 | 938.2 | 33.8–36.2 | 11,271 | 37 | 11.8 | 52 |
| Other East Asia | 0.82 | 30 | 2.8–3.0 | 4526 | 12 | 0.8 | 3.4 |
| India | 2.38 | 379.3 | < 0.1 | 1756 | 12 | 3.5 | 15 |
| Other South Asia | 2 | 596 | < 0.1 | 3171 | 5 | 9.3 | 38 |
| Japan | 0.49 | 3.7 | 3.8–4.0 | 10 | 3 | 0.5 | 2.9 |
| South East Asia | 2.89 | 79.5 | 12.6–13.6 | 645 | 40 | 8.8 | 36 |

*Data source*: Rogner et al. (2012).

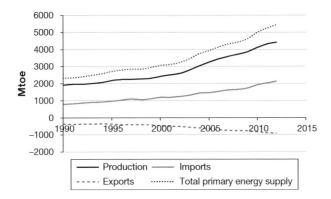

*Figure 2.8*   Indigenous production, imports and exports of energy in Asia

*Data source*: International Energy Agency (2015b).

indicates, there was a sharp rise in indigenous production since 2000 but this was also accompanied by a noticeable growth in import and export demands. Between 1990 and 2012, the indigenous production grew about 4% while exports and imports grew at 4.3% and 4.6% respectively. Export is shown as a negative flow as this quantity of energy goes out of the national boundary and the volume of export has increased over time. Notice that the total primary energy supply bears a close similarity with the production trend.

The sub-regional picture (see Table 2.3) indicates that East Asia saw the highest increase in local production, followed by supply expansion in South East Asia and South Asia. On the other hand, imports grew more rapidly in South Asia (at an average rate of 9.3% between 1990 and 2012) and South East Asia than East Asia. In terms of shares in regional production and import, East Asia accounted for about two-thirds while South Asia and South East Asia shared almost equally the rest, with a minor contribution from West Asia and other Asian countries. On the other hand, South East Asia was dominant exporter from the region, with West Asia playing a significant role.

## Fuel mix of primary energy supply

At the regional level, coal has consolidated its leadership in the energy mix since 1990 by increasing its share from 36% in 1990 to above 50% in 2012. The share of natural gas has

*Table 2.3* Sub-regional distribution of indigenous production, imports and exports of energy in Asia

| Regions | 1990 | | | 2012 | | |
|---|---|---|---|---|---|---|
| | Production (Mtoe) | Imports (Mtoe) | Exports (Mtoe) | Production (Mtoe) | Imports (Mtoe) | Exports (Mtoe) |
| East Asia | 1020.9 | 526.0 | −54.7 | 2651.8 | 1391.5 | −158.8 |
| South Asia | 346.4 | 48.6 | −3.5 | 652.1 | 345.6 | −69.1 |
| South East Asia | 305.8 | 115.7 | −169.9 | 744.1 | 363.7 | −476.1 |
| West Asia | 230.0 | 93.7 | −128.3 | 353.5 | 22.9 | −197.7 |
| Other Asia | 8.5 | 3.5 | 4.8 | 16.2 | 8.9 | −10.5 |
| Total | 1911.7 | 787.5 | 361.3 | 4471.8 | 2132.7 | −912.2 |
| | Share in production % | Share in imports % | Share in exports % | Share in production % | Share in imports % | Share in exports % |
| East Asia | 53 | 66 | 15 | 60 | 65 | 17 |
| South Asia | 18 | 6 | 1 | 15 | 16 | 8 |
| South East Asia | 16 | 15 | 47 | 17 | 17 | 52 |
| West Asia | 12 | 12 | 36 | 8 | 1 | 22 |
| Other Asia | 0 | 1 | 1 | 0 | 0 | 1 |

*Data source*: Based on International Energy Agency (2015b).

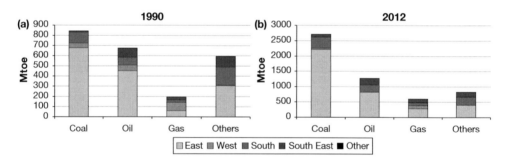

*Figure 2.9*  Regional primary energy supply mix

*Data source*: International Energy Agency (2015b).

marginally improved from 8% in 1990 to about 11% in 2012 while oil share has marginally fallen from 29% in 1990 to 23% in 2012. The share of other energies (including biomass and modern renewable energies) has fallen to about 15% in 2012 from above 26% in 1990 (see Figure 2.9a & b). As the figures show, primary coal supply (and use) has increased from 844 Mtoe in 1990 to 2710 Mtoe in 2012 recording a three-fold growth. Natural gas supply has also more than trebled during the same period but from a much lower base of 195 Mtoe in 1990. Accordingly, despite the supply expansion, the share of natural gas remains low in the overall energy mix. Other energies have recorded the slowest growth during this period – mainly due to near stagnation of biomass energy supply in the region. Although modern renewable energies are gaining momentum, the contribution during the period of study was limited.

Within the region, both East Asia and South Asia maintained a similar rate of growth in coal use during the period and more than trebled the primary coal supply. South East Asia and the rest

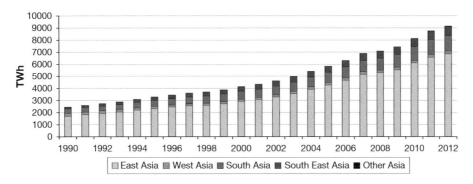

*Figure 2.10*  Electricity output in Asia

*Data source*: International Energy Agency (2015b).

of Asia recorded a much higher growth in coal use during this period but from a very low base. However, since 2014 coal demand in China has slowed down and in 2015 coal demand in China was 1.5% lower than that in 2014 (British Petroleum, 2016). This is attributed to slower than expected economic activity in the country. As a result, the regional energy demand growth has decelerated but it still remains much higher than the global average.

The primary oil demand has increased, almost doubled, between 1990 and 2012. East Asia accounted for about 64% of the demand in 2012, while South Asia and South East Asia accounted for 16.5% each. In terms of natural gas use, the shares are more evenly distributed. East Asia accounts for 47.5% of the gas demand, while South East Asia contributed 20% of the demand. The remaining was shared between South Asia and West Asia.

In terms of renewable energy use, the region has made significant progress. Almost 40% of global hydroelectricity was produced in the region in 2015. Almost 72% of the regional hydropower came from China, while India, Japan and Vietnam contributed another 18% of the hydropower generation. Thus only four countries accounted for 90% of the regional hydropower generation (British Petroleum, 2016). In respect of other commercial renewable energy use, the region has a 28% share in the global market. China leads here as well, followed by India and Japan. By the end of 2015, China had installed 43 GW in solar PV capacity, the highest capacity in the world ahead of Germany and the US. Japan has also installed more than 35 GW in PV capacity by 2015. Similarly, China has installed 145 GW of wind power capacity by 2015 while India has installed another 25 GW. Almost 40% of the global wind power capacity was installed in Asia.

## Electricity generation

Electricity generation was one of the major energy conversion activities undertaken in the region. Between 1990 and 2012, electricity output has increased from 2453 TWh to 9151 TWh (see Figure 2.10), recording about four-fold increase during the period (or an annual growth rate of 6.2%). The average regional growth masks an annual growth rate of 6.5% in China as well as negative growth rate in West Asia. The regional distribution of electricity output shows that electricity output in East Asia grew from 1710 TWh in 1990 to 6852 TWh in 2012, most of which is attributable to China. Accelerated growth in output since 2000 can also be noticed. Output grew from 342 TWh in South Asia in 1990 to 1288 TWh in 2012. Electricity output in South East Asia is catching up with South Asia and by 2012, the region produced close to 752 TWh.

Due to rapid growth in electricity output in East Asia, its share in the regional electricity output increased from 70% in 1990 to 75% in 2012. West Asia has seen a drastic reduction in its share – just 3% in 2012 compared to 10% in 1990. South Asia maintained its market share while South East Asia has improved its share marginally.

Almost 80% of electricity in the region comes from fossil fuels. 78% of electricity in China in 2012 came from fossil fuel sources, while the rest came from hydropower, wind and solar energies. Similarly, 81% of electricity in India in 2012 came from fossil fuel sources. Coal remains the main fuel for electricity in East and South Asia, mainly due to dominant contribution of China and India. Natural gas is the second most important fuel while hydropower is the third source of electricity.

## Sub-regional trends in primary energy supply

This section provides further details at the sub-regional level.

### *East Asia*

East Asia is composed of PR China, Hongkong, Mongolia, Republic of Korea, DR Korea, Japan and Taipei. The region had a population of 1.36 billion in 1990 but this had increased to 1.59 billion in 2012. This is the second most populous sub-region of Asia, after South Asia. The region's economic output was $5.2 trillion (2005 constant $) in 1990 and this had increased to $11.65 trillion in 2012.

Primary energy use in East Asia increased from about 1496 Mtoe in 1990 to 3722 Mtoe in 2012, recording an annual growth of 4.2% (see Figure 2.11). Coal is the dominant primary energy in this region and accounted for 45% of the primary energy demand in 1990, increasing to 60% by 2012. A rapid growth in coal supply post-2002 is clearly visible. Petroleum products were the second important source of primary energy, accounting for about 22% of the primary demand in 2012.

About 70% of the primary energy demand was met indigenously in the sub-region and despite the growing demand the share of domestic production remained steady. This implies that the energy extraction activities in this sub-region, particularly in China, have expanded to meet the needs. However, countries such as Japan and South Korea are more import dependent than others

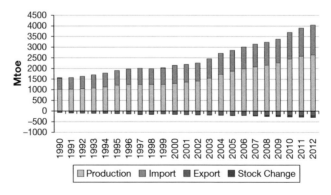

*Figure 2.11*   Trend of primary energy demand in East Asia

*Data source*: International Energy Agency (2015b).

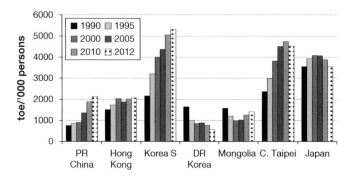

*Figure 2.12*  Variation in primary energy use per person in East Asia

*Data source*: International Energy Agency (2015a) and International Monetary Fund (2015).

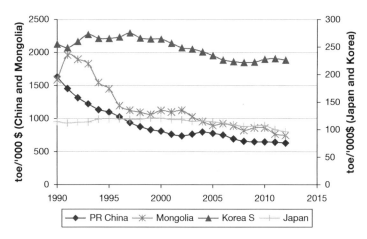

*Figure 2.13*  Trend of primary energy intensity in East Asia

*Data source*: International Energy Agency (2015b) and International Monetary Fund (2015).

due to limited endowment of energy resources. Similarly, there is significant variation in terms of energy use per person. Despite rapid growth in primary energy demand, China on average consumes almost one half of that consumed by Japan or South Korea (see Figure 2.12). South Korea has the highest energy use per person in the sub-region. While China and South Korea have recorded an increasing trend in per capita energy use, DR Korea shows a continuous declining trend and Japan is showing signs of demand reduction since 2005. The political isolation of DR Korea and limited economic growth are responsible for rapid deterioration of energy use in the country.

On the other hand, in terms of primary energy intensity (i.e. primary energy use per unit of GDP), Japan is most efficient in the sub-region. South Korea is also highly energy efficient but has not managed to reach the Japanese level. China and Mongolia on the other hand have shown significant improvements in recent times but still they consume five times more energy than Japan to produce the same level of economic output (see Figure 2.13). This shows the potential for energy efficiency improvements in the region.

## West Asia

West Asia is composed of Armenia, Azerbaijan, Georgia, Kazakhstan, Kyrgyzstan, Tajikistan, Turkmenistan and Uzbekistan. The region had a small population of 66 million in 1990 which increased to almost 81 million in 2012. The economic output was $0.1 trillion in 1990 and this increased to almost $0.2 trillion in 2012. The economic downturn after the collapse of the former Soviet Union has affected these countries significantly, resulting in a limited growth in energy demand. In 1990, the primary energy demand of the sub-region was 192 ktoe whereas in 2012, this fell to 175 ktoe. Consequently, the energy use per person either stagnated or fell in most of the countries in the area (see Figure 2.14). The dramatic decline shows the effect of economic contraction on the energy use.

The composition of primary energy supply shows that this sub-region produces more than its domestic needs (see Figure 2.15). During the period of economic crisis, both indigenous production and exports fell dramatically. Local production and exports have started to gain momentum in the new millennium and by 2012, the region exported more than what it consumed locally.

West Asia has the highest primary energy intensity in Asia. Although most countries have recorded significant improvement in recent times (see Figure 2.16), there is scope for further improvement. For example, Georgia has recorded the lowest energy intensity in this sub-region

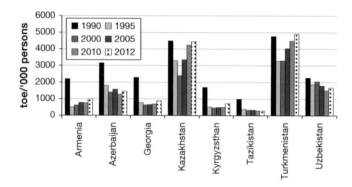

*Figure 2.14*   Per capita primary energy use in West Asia

*Data source*: International Energy Agency (2015b) and International Monetary Fund (2015).

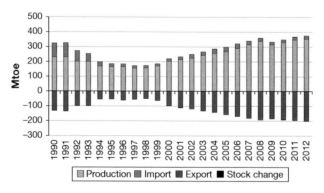

*Figure 2.15*   Trend of primary energy supply in West Asia

*Data source*: International Energy Agency (2015b).

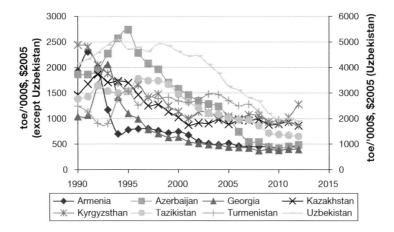

*Figure 2.16*  Energy intensity trend in West Asia

*Data source*: International Energy Agency (2015b) and International Monetary Fund (2015).

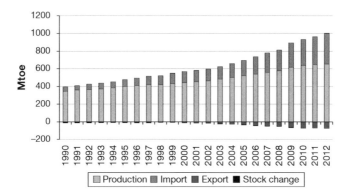

*Figure 2.17*  Trend of primary energy supply in South Asia

*Data source*: International Energy Agency (2015b).

but compared to Japan, it remains four times more intensive. Uzbekistan on the other hand is about five times more energy intensive compared with Georgia. Therefore, there is potential for learning from one another in the sub-region and surely from other better performing countries.

## South Asia

For the South Asia region, Bangladesh, India, Nepal, Sri Lanka and Pakistan were included as data for other countries were not readily available. The population of this sub-region has increased from 1.09 billion in 1990 to 1.6 billion in 2012, making it the most populous sub-region of Asia. The sub-regional economic output grew from $0.45 trillion in 1990 to $1.66 trillion (2005 constant $) in 2012. The primary energy supply of this sub-region has seen significant growth: in 1990, 383 Mtoe was consumed but in 2012, this rose to 928 Mtoe. Although indigenous production remains important, the growing energy need has resulted in higher import dependence (see Figure 2.17). Imports have grown from 48 Mtoe in 1990 to 346 Mtoe in 2012, recording an

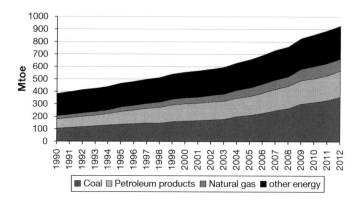

*Figure 2.18*  Trend of primary energy mix in South Asia

*Data source*: International Energy Agency (2015b) and International Monetary Fund (2015).

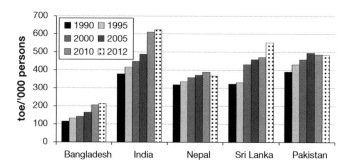

*Figure 2.19*  Comparison of primary energy use per person in South Asia

*Data source*: International Energy Agency (2015b) and International Monetary Fund (2015).

average growth rate of above 9% per year for the above period. India is the dominant energy user in this area and accounted for about 85% of the South Asian primary energy demand in 2012.

Coal remains the main energy source of the sub-region but other energy, particularly biomass remains an important energy source. Despite some increase in biomass use over the years, its share in the primary energy has fallen from 46% in 1990 to 28% in 2012. Coal consumption has more than trebled during the period but natural gas has recorded a higher growth rate compared to coal and petroleum products (see Figure 2.18).

The region has one of the lowest primary energy consumption per person (see Figure 2.19). High population density, slower economic growth compared to East Asia and lower penetration of modern energies are responsible for lower energy use. For example, India's per capita primary energy use is almost one-fourth of that of China. While high primary energy use per person is not a desirable attribute of sustainable development, access to a minimum level of energy supply is essential. Lack of energy access in South Asia (discussed in detail in Chapter 4) is a major challenge.

In terms of energy intensity, the sub-region has recorded an improving trend in recent times. Most of the countries of the region have similar levels of energy efficiency (see Figure 2.20), with Sri Lanka and Bangladesh leading the trend, followed by India and Pakistan. While the sub-region has performed relatively well in this respect, further improvements are possible taking inspiration from the Japanese example.

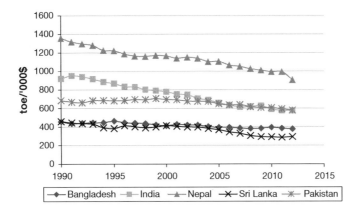

*Figure 2.20*   Trend of energy intensity in South Asia

*Data source*: International Energy Agency (2015b) and International Monetary Fund (2015).

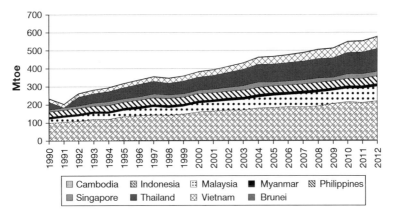

*Figure 2.21*   Primary energy trend in South East Asia

*Data source*: International Energy Agency (2015b).

## South East Asia

Within South East Asia, nine countries are covered (Brunei, Cambodia, Indonesia, Malaysia, Myanmar, Philippines, Singapore, Thailand and Vietnam). The population of this sub-region increased from 434 million in 1990 to 603 million in 2012. The economic output in 1990 was $0.45 trillion (2005 constant dollar terms) but this increased to $1.4 trillion in 2012. The sub-region consumed 223 Mtoe in 1990, and 578 Mtoe in 2012.

Indonesia is the largest primary energy user of this sub-region, followed by Thailand and Malaysia. Three of them accounted for 73% of the primary energy demand of the sub-region in 2012 (see Figure 2.21).

The primary energy mix in South East Asia is quite different from other regions. Coal plays a minor role but petroleum products account for a large share of the supply. Natural gas has also made significant progress in this sub-region but other energies including biomass are the most dominant fuel (see Figure 2.22).

The sub-region shows significant variation in energy use per person (see Figure 2.23). On one hand, a person in Cambodia on average consumed 0.370 toe in 2012, while a consumer in Brunei

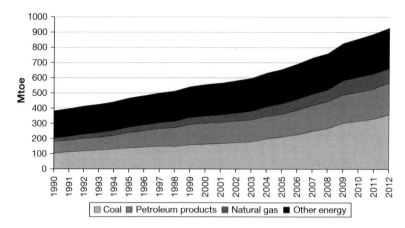

*Figure 2.22*  Primary energy fuel mix in South East Asia

*Data source*: International Energy Agency (2015b).

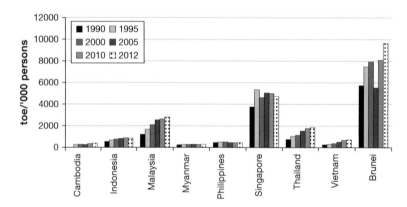

*Figure 2.23*  Comparison of energy use per person in South East Asia

*Data source*: UN International Energy Agency (2015b).

consumed more than 9.650 toe. Singapore comes second to Brunei in terms of per capita energy use. Thailand and Malaysia fall in the middle range (between 1800 and 2800 toe/'000 persons) while Indonesia has per capita energy use similar to India.

However, the performance shows less variation in terms of energy intensity. Vietnam has the highest energy intensity in the sub-region (excluding Myanmar) whereas Singapore is the most efficient economy in the area. The improvement in energy efficiency appears to be less vigorous than other sub-regions and in some cases (such as Vietnam and Brunei), the energy intensity has increased in recent times (see Figure 2.24).

## Total final energy demand

The total final energy consumption in the region has grown from 1729 Mtoe in 1990 to 3433 Mtoe in 2012, recording an annual average growth of 3.2%. In 1990, the residential sector was the

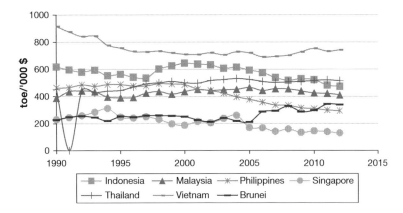

*Figure 2.24*   Energy intensity trend in South East Asia

*Data source*: International Energy Agency (2015b) and International Monetary Fund (2015).

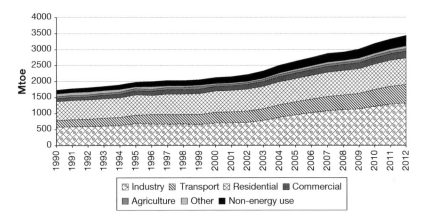

*Figure 2.25*   Trend of final energy use in Asia

*Data source*: International Energy Agency (2015b).

most dominant energy user in the region but by 2012, industry has emerged as the most important energy using sector at the regional level (see Figure 2.25). Industrial final energy demand has grown at 4.5% per year during this period, whereas the residential sector has seen a slower growth at 1.6% per year. The transport sector was however the fastest growing sector in final energy demand, recording a growth rate above 6% per year. As a result, the sectoral final energy mix has significantly changed during this period. In 1990, the residential sector accounted for 34% of final energy demand, followed by industry with 33% and transport at 12%. In 2012, the industrial sector had a 38% share in final energy demand, followed by the residential sector (24%) and transport sector (17%). The share of agriculture sector has fallen from 4% in 1990 to 2% in 2012, while non-energy use and commercial sectors have seen some growth in their shares (see Figure 2.26)

The contribution of each sub-region to the total final energy demand follows the primary energy demand pattern and is dominated by East Asia due to overwhelming Chinese demand (see Figure 2.27). South Asia comes next, with South East Asia taking the third position.

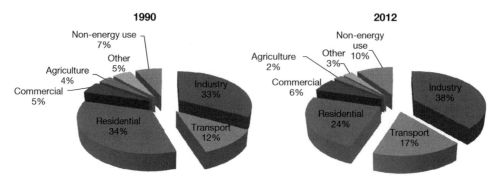

*Figure 2.26*   Sectoral share of final energy demand in Asia in 1990 and 2012

*Data source*: International Energy Agency (2015b).

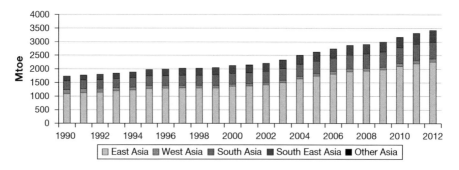

*Figure 2.27*   Regional distribution of final energy demand in Asia

*Data source*: International Energy Agency (2015b).

West Asia and the rest of Asia make only a minor contribution. However, it should be noted that due to faster demand growth in East Asia, its share in the regional final energy demand increased from 63% in 1990 to 66% in 2012. South Asia's share has remained unchanged at 18%. South East Asia has seen a minor rise in its market share (from 10% in 1990 to 12% in 2012) while West Asia has recorded the largest fall in share (9% in 1990 to 3% in 2012).

## Carbon emission trends from fuel combustion

Growing energy use in Asia has been responsible for growing emissions in the region. According to data provided by the International Energy Agency (2015a), Asia's $CO_2$ emission almost trebled between 1990 and 2013: from 5.25 billion tonnes of $CO_2$ emission in 1990, it increased to 14.89 billion tonnes in 2013. As a result, Asia's share of $CO_2$ emission has also risen dramatically in the world: from 25% in 1990 to 46% in 2013. The sub-regional trend closely follows the energy use pattern (see Figure 2.28). East Asia, with major contributions from China, is the most dominant emitter in the region: almost three-quarters of the emission come from this sub-region. China alone accounts for 60% of the carbon emissions of the region. In contrast, India accounts for only 12% of the regional carbon emissions. Clearly acceleration in emissions can be noticed from the year 2000 onwards.

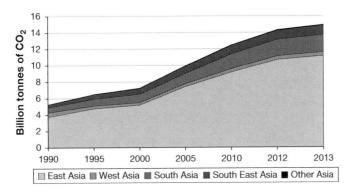

*Figure 2.28* Trend of $CO_2$ emission in Asia

*Data source*: International Energy Agency (2015a).

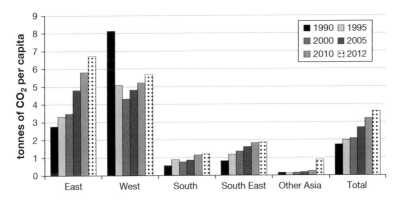

*Figure 2.29* Trend of $CO_2$ emissions per person in Asia

*Data source*: International Energy Agency (2015a).

A large share of the emissions comes from coal: for example, coal use in China contributed 7.5 billion tonnes of carbon emissions in 2013 (or ~84% of its carbon emissions). 72% of India's emissions came from coal. Because electricity production uses a significant amount of coal in the region, it is a major contributor to carbon emissions in the region. Industry, transport and residential users are other major contributors to carbon emissions.

In terms of emissions per person, a clearly increasing emission trend is visible all over the region (see Figure 2.29). The regional average of 3.61 tonnes of $CO_2$ per person is below the global average of 4.47 tonnes in 2012 but the spectacular growth in emission load in East Asia is noticeable as well. The emission load in this sub-region was above the world average in 2012 but was 30% lower than the average for the Organisation for Economic Co-operation and Development (OECD) countries (International Energy Agency, 2015a). West Asia also has a relatively high level of emissions per person but the economic downturn in the region has lowered the emission load since 1990. Lower per capita emissions in South Asia and South East have still maintained the regional average low but continued reliance on fossil fuels in the future will be a cause of concern.

However, in terms of carbon emission intensity (i.e. carbon emissions per dollar of output), West Asia remains the worst offender. East Asia performs comparatively better within the region

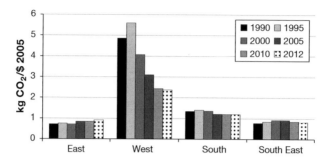

*Figure 2.30* Sub-regional trend in carbon emission intensity

*Data source*: International Energy Agency (2015a) and International Monetary Fund (2015).

whereas South Asia remains relatively carbon intensive with respect to economic output (see Figure 2.30). The high value of the economic output of Japan (contributing about 40% of the East Asian output in 2012) has a mitigating effect on the East Asian outcome. China had an emission intensity of 1.85 kg $CO_2$/\$ (constant 2005) in 2012, which is much higher than the sub-regional average. The emission intensity of OECD countries in 2012 was 0.3 kg $CO_2$/\$ (constant 2005) – which is much lower than the Asian average. This suggests that Asian countries could aim for significant carbon intensity reduction in the future.

## Future energy needs

The centre of gravity of global economic activity has shifted to Asia and consequently, countries in the region are making progress economically. The fast rate of economic growth is reducing poverty in the region and allowing better lifestyles for the population. As Asia industrialises, the economic activities are shifting to urban areas, prompting migration of rural population in search of better living conditions and income opportunities. Rapid urbanisation is transforming the landscape and is influencing energy demand significantly. The combined effect of urbanisation and income growth is likely to create a section of affluent population who will have significant paying capacity to lead modern lifestyles. In addition, the population is still growing in most of these countries due to a larger share of younger population. While better income and improved living conditions will affect population growth, it will take some time before the population stabilises in the region.

The Regional Outlook prepared by Asian Development Bank (Asian Development Bank, 2013) suggests that in the business as usual scenario the energy demand will grow at 2.1% annually for the forecast period up to 2035. Using this growth rate figure and applying this to the regional energy demand in 2012 indicated earlier produces a demand of 8642 Mtoe. The above growth rate is slower than the historical growth rate encountered in the region. Maturity of the economies and saturation of demand can justify such a slow-down in energy demand. Recent economic slow-down in China and weaker economic growth globally would also influence economic performance of Asian countries. Accordingly, slower energy demand growth is quite possible.

While industrial energy demand drove Asia's past energy demand growth, in the future it may change. A shift from energy intensive activities in China and improvements in energy efficiency due to policy influence is likely to reduce the share of industrial sector. As indicated in Chapter 3, China has taken a number of measures to curb its industrial energy requirements. The residential sector on the other hand is likely to dominate due to changing lifestyles of the population, particularly in urban areas where the trend of following western lifestyles is clearly visible. Increased use of energy-intensive appliances such as air-conditioners, dishwashers, washing machines, microwave ovens, etc. will put increasing pressure on electricity and therefore on primary energy. The share of the transport sector will also improve as affluence will make motorised travel more affordable. Asia is likely to rely on coal as the major source of energy but its share in the primary energy may reduce due to slower demand growth in China. As countries in the region try to cope with environmental challenges and keep their international pledges for emission reduction, natural gas and renewable energies will be favoured and are likely to grow faster than coal or oil.

## Concluding remarks

The review of the energy situation in Asia presented in this chapter shows that the region has emerged as a leading energy user in the world, consuming more than 40% of world primary energy in 2012. With a demand growth rate of 3.9% per year between 1990 and 2012, the region has recorded the fastest energy demand growth. However, as the region houses millions of people without access to electricity and clean cooking energies and as the population of the region is growing, the average per capita consumption remains low even compared with the global average.

There is significant sub-regional variation in energy demand: East Asia, with a fast growing Chinese economy, has consolidated its leading position accounting for about 70% of the region's energy demand. South Asia and South East Asia make similar contributions but the West Asian region has lost its share drastically during the post-Soviet Union era. But in terms of per capita energy use and energy intensity, West Asia ranks highest in the region, showing the existence of inefficient energy services and practices. Energy use per person is growing rapidly in East Asia but due to the influence of Japan, South Korea and more recently China, the energy intensity in this sub-region has been declining. However, there is still significant scope for energy efficiency improvements in the region and significant learning potential from the regional best practices.

The region has the distinction of depending heavily on coal for its energy supply. Large coal reserves have helped the region to maintain a high level of indigenous primary energy supply. But the rising climate change concerns and local pollution challenges are likely to influence future reliance on coal for powering Asia. Rapid industrialisation of East Asia, particularly China, has catapulted the industry sector as the main final energy user in the region. However, the residential energy demand remains strong and the transport energy demand is growing rapidly.

The growing population, sustained economic growth, and the emergence of a sizeable middle income group are influencing the demand for energy in the future. Asia is likely to demand more energy in the future to maintain its economic growth but to ensure the region's commitment to sustainable development, the transition to a low carbon pathway is essential. This remains a major challenge for the future but the varying stage of development of Asian countries allows them to take advantage of the vast experience available within the region to decouple energy and economic growth and reduce their carbon footprints.

## References

Asian Development Bank, 2013. *Energy Outlook for Asia and the Pacific 2013*, Manila: Asian Development Bank.

British Petroleum, 2016. *BP Statistical Review of World Energy, 2016*, London: British Petroleum.

Hoogwijk, M. & Graus, W., 2008. *Global Potential of Renewable Energy Sources: A Literature Assessment, Background Paper*, Utrecht, The Netherlands: ECOFYS.

International Energy Agency, 2015a. *CO₂ emissions from Fossil Fuel Combustions: Highlights*, Paris: International Energy Agency.

International Energy Agency, 2015b. *World Energy Balance (via UK Data Service)*, Paris: International Energy Agency.

International Energy Agency and the World Bank, 2015. *Sustainable Energy for All 2015 – Progress Toward Sustainable Energy*, Washington, DC: The World Bank.

International Monetary Fund, 2015. *World Economic and Financial Surveys: Regional Economic Outlook: Sub-Saharan Africa Navigating Highwinds*, Washington, DC: International Monetary Fund.

Rogner, H. *et al.*, 2012. Chapter 7. Energy Resources and Potentials. In *Global Energy Assessment – Toward a Sustainable Future*, Cambridge and New York: Cambridge University Press.

UNStats, 2017. *United Nations Statistics Division*. [Online] Available at: https://unstats.un.org/unsd/snaama/dnlList.asp [Accessed 25 April 2017].

# 3

# REVIEW OF THE OVERALL ENERGY SITUATION IN CHINA

*Ming Su and Songli Zhu*

## Introduction

China, as one of the biggest countries in the world, has been increasingly influential in global energy issues. In terms of population, China, being the most populous nation in the world, had 1371.22 million persons by mid- 2015, accounting for 18.66% of the world's total. The country's Gross Domestic Product (GDP) on the basis of exchange rate stood at 10866.44 billion USDs at the end of 2015, accounting for 14.8% of the global GDP and ranking the second in the world since 2010. On the basis of purchasing power parity (PPP), the share of China's GDP increased to 15.8% in the same year, still ranking the second, but with a narrowed gap compared to the US.[1] The country is urbanizing rapidly, leading to a large rural-to-urban migration and half of the population currently lives in cities. Because more than 75% of the Chinese population is expected to reside in cities within the next 20 years, the establishment of new cities and extension of the existing ones is expected to continue (He et al., 2016). The "opening and reforming" polices, initialized at the beginning of the 1980s and deepened more in recent years, have prompted China to actively participate in the globalization process, taking an important role in the field of economy, trade and energy development.

This chapter presents an overview of the energy situation in China, particularly focusing on the recent period starting from 2000 onwards. First, energy production is briefly presented, followed by discussions on energy consumption, energy mix, energy conversion, end-use profile, the energy–economy relationship and finally environmental issues.

## Overall picture of energy production

China is generally rich in energy sources, particularly coal, but has relatively limited oil and gas reserves. The basic energy production profile is decided by the resource endowment to a great extent.

Since 2000, the production of energy has expanded rapidly in response to China's fast-growing economy. Between 2000 and 2015, primary energy production grew from 1390 million tons coal equivalent (Mtce) to 3620 Mtce,[2] increasing by 6.6% annually (see Figure 3.1). The annual growth rate between 2002 and 2005 was over 10%, and it was nearly 10% between 2009 and 2011. Nevertheless, a constrained energy supply was still observed during these two periods. In other

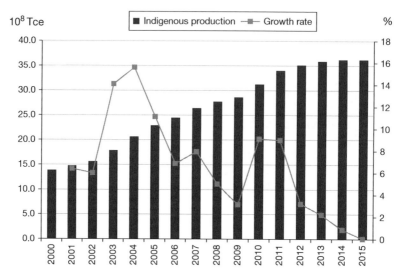

*Figure 3.1*  Energy production in China in 2000–2015

*Data source*: Authors.

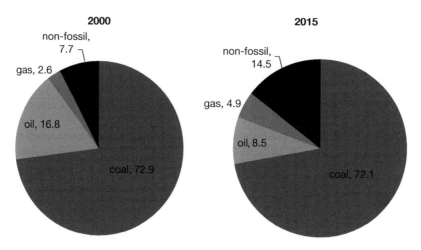

*Figure 3.2*  China's energy mix in 2000 and 2015

*Data source*: Authors.

years, the annual growth rate of energy supply was lower than 8%, with the rate being as low as 1% during the period 2013–2015, growing only marginally.

Coal is the most important primary energy product, sharing over 70% of the overall production. In physical terms, coal production increased from 1380 million tons (Mt) in 2000 to almost 4000 Mt in 2013 (half of the world production), slightly decreasing in 2014 and 2015. Crude oil production has been maintained at 200 Mt since 2000, accounting for 8.5% of overall production in 2015. Fast growth has been observed in natural gas production, from 27.2 billion cubic meters ($m^3$) in 2000 to 124.4 billion $m^3$ in 2015, increasing at an average rate of 10.7% annually. Consequently, the share of gas in the energy supply grew from 2.6% to 4.9% (see Figure 3.2). The production scale of non-fossil fuel (hydro, wind, solar and nuclear) rose dramatically, from 107 Mtce and 7.7% in 2000 to 525 Mtce and 14.5% in 2015.

## Overall picture of energy consumption

The total energy consumption of China grew fast and steadily during the period of 2000–2015, increasing from 1469 Mtce to 4300 Mtce by an average growth rate of 7.4% per year. Up to 2015, as the biggest energy consumer in the world, China shared 20% of the world total; energy consumption per head amounted to 3.1tce, catching up and surpassing the world average which is around 2.6 tce.

China had two notably different phases from 2000 in its energy consumption trend: 2003–2011 and 2012–2015. In the former phase, the annual average of growth rate (AAGR) amounted to 8.6% with the higher-end at nearly 17% in 2003–2004, as shown in Figure 3.3. This meteoric growth meant, on average, over 240 Mtce was stocked onto the level in the previous year during this period (even 300 Mtce in particular years), which is almost the total energy demand in UK. In contrast, the latter phase saw the rate dropping to an average of 2.7% per year and further decline is expected. The preliminary data for 2015 shows the energy consumption almost stayed flat, growing by 0.9% compared with 2014 (NBS, 2016).

Unbalanced regional development and consequent energy consumption is observed across China. The eastern area, which produces the majority of the GDP, has consumed around 50% of the total energy since 2000. Nevertheless, the growth rate of energy consumption in this area decreased significantly in recent years, and its share fell to 46.8% in 2014 (see Figure 3.4). In contrast, energy consumption in the western area moved upwards, owing to the "Go West" Program initiated in 2000, from 22.6% in 2000 to 27.1% in 2014. The middle region of China kept stable with a slight down trend. Regarding energy density (i.e. energy consumption per square kilometer (km$^2$)), the eastern area consumes the highest amount, 1942tce/km$^2$,

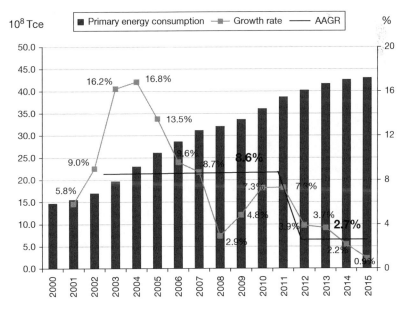

*Figure 3.3* China's energy consumption and growth rate trend in 2000–2015

*Data source*: Authors.

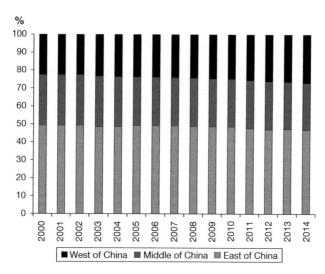

*Figure 3.4*  Share of energy consumption by regions

*Data source*: Authors.

*Note*: Eastern area includes: Beijing (BJ), Tianjin (TJ), Hebei (HB), Liaoning (LN), Shanghai (SH), Jiangxu (JX), Zhejiang (ZJ), Fujian (FJ), Shandong (SD), Guangdong (GD) and Hainan (HN); Middle area includes: Shanxi (SX), Jilin (JL), HeilongJiang (HLJ), Anhui (AH), Jiangxi (JX), Henan (HN), Hubei (HB) and Hunan (HN); Western area includes: Sichuan (SH), Chongqing (CQ), GuiZhou (GZ), Yunan (YN), Tibet (XZ), Shaanxi (SAX), Ganxu (GS), Qinghai (QH), Nixiang (NX), XiJiang (XJ), Guangxi (GX) and Inner Mongolia (NMG).

followed by the middle region (688tce/km$^2$); whereas the western area, because of its vast land space, has the lowest value, 173tce/km$^2$, less than one tenth of the eastern area.

An enlarged demand and supply gap has resulted as China's energy demand has grown faster than energy production. China has had to increase its energy imports continuously. In 2000, the total energy import was 50 Mtce and the rate of external dependence was only 3.4%. By 2015, the total net import had exceeded 700 Mtce and the external dependence rate surpassed 16.0%. Of that total, the net import of oil rose from 76 Mt in 2000 to 340 Mt in 2015, and the external dependence rate soared to 63.5%. China began to import natural gas from 2006, and in 2015, the net import had increased to 58 billion m$^3$, with an external dependence rate of 32.0%. Besides vast domestic production, China's coal imports also increased dramatically. It started importing coal from 2009 and in 2015 increased the amount to over 200 Mt.

## Trend of energy mix

Because of high domestic availability, China's energy mix is dominated by coal. In general, coal met over 70% of the energy demand until recently. As early as the 1990s, policies for optimizing energy mix were formulated to reduce the over-dependence on coal and to promote the contribution of oil and gas. In the new century, a specific target on non-fossil fuel (15% up to 2020) was officially announced in 2009, followed by the long-term target (20% up to 2030) set in 2014. Nevertheless, the leading position of coal remains unchanged so far. During the time period of 2000–2013, coal consumption grew from 1360 Mt to 4250 Mt, climbing to an unprecedented average of 9.2% per year. The consequence is that the share of coal in the energy mix continued to grow in the first ten years of the new century.

Fortunately, driven by technological progress in non-fossil fuels and an aggressive policy push, the terrible trend was turned around gradually after 2010 (Wang and Yang, 2015; Wang and Liu, 2014; Price et al., 2011). The installed capacity of hydro power was increased to 320GW and its electricity production amounted to 1.1 TWh by the end of 2015. China's wind energy capacity continuously ranks first in the world in terms of both newly installed and accumulated capacity which reached 130GW by 2015. The highest growth was observed in solar photovoltaic (PV) systems whose capacity was 43GW in 2015, compared with 19MW in 2000. Figure 3.5 shows synchronous growth of wind and solar power in China and the world.

Nuclear is also an important alternative option and its capacity was over 26GW in 2015. By the end of 2015, the overall share of non-fossil fuels in the energy mix increased to 12.0% (see Figure 3.6), whereas it was just 7.3% in 2000. Regarding power generation, electricity generated from renewable energy (RE) accounted for 23.9% of the 2015 total compared to 16.6% in 2000 (NSB, 2016). Meanwhile, the scale of natural gas is expanding. As mentioned before, the domestic production of natural gas amounted to 124.4 billion m$^3$, increasing by an average of 10 billion m$^3$

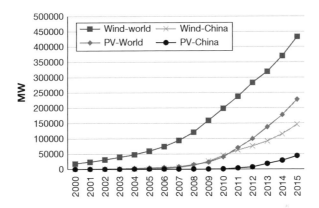

*Figure 3.5*   Development of wind and solar power in China and the world

*Data source*: Authors.

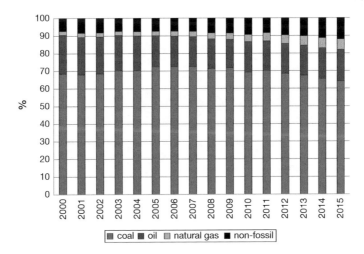

*Figure 3.6*   Trend of energy mix in China 2000–2015

*Data source*: Authors.

per year from 2000. Along with the completion of the Central Asia–China gas pipeline, the China–Myanmar pipeline and a series of LNG stations in the south-east coastal area, 58 billion m³ was imported in 2014 and more in 2015. All these contributed to the noticeable promotion of natural gas in primary energy consumption, measuring 5.9% in 2015 against 2.2% in 2000. In the year of 2013–2014 coal consumption in physical units dropped 2.9%, and dropped a further 3.7% in 2014–2015, hailed as an historic event after China's meteoric growth in the 2000s (Green and Stern, 2015).

## Fossil energy conversion and processing activities

Since this century, China's energy conversion and processing, including thermal power, centralized heating, coking and oil refining, witnessed a rapid growth. The input of thermal power energy rose from 422 Mtce in 2000 to 1,320 Mtce in 2014, up 8.5% year on year. Along with the installation of a large number of efficient coal-fired units and closure of small ones, the power generating efficiency rose constantly from 32.4% in 2000 to 39.8% in 2014. The input in centralized heating grew from 72 Mtce in 2000 to 178 Mtce in 2014, up 6.7% year on year, with heating efficiency around 70%; coking input rose from 139 Mtce in 2000 to 554 Mtce in 2014, mainly driven by the dramatic development of the iron and steel industry, up 10.4% year-on-year, with coking efficiency around 96.5%; oil refining output rose from 290 Mtce in 2000 to 722 Mtce in 2014, up 6.7% year on year, with refining efficiency about 97%.

## Energy end-use profile

In this century, with the continuous and swift development of the economy and the upgrading of living standards, China's final energy use grew quickly, from 1,062 Mtce in 2000 to 3,139 Mtce in 2014, with an average annual growth rate of 8.05% (see Figure 3.7).[3] Manufacturing's share of final energy use was still close to 70% in 2014, with both transportation and residential sharing almost 11%, respectively. The fastest growth rate was observed in the commercial and transportation sector during 2000–2014, up to 8.85% and 8.93% per year, respectively.

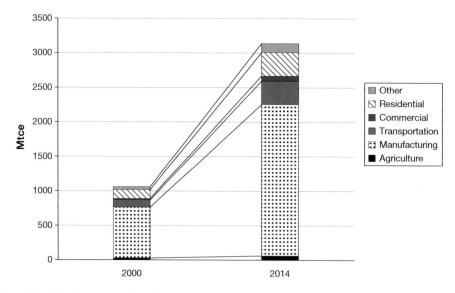

*Figure 3.7*   China's energy end-use by sector in 2000 and 2014

*Data source*: Authors.

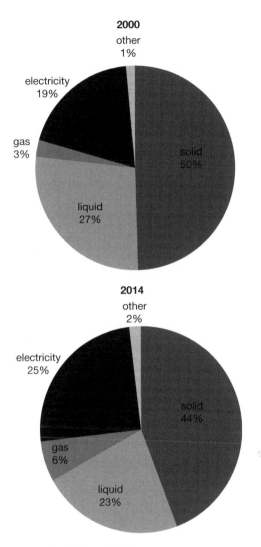

*Figure 3.8* Mix of final energy use in 2000 and 2014

*Data source*: Authors.

Regarding the fuel mix of final energy use, from 2000 to 2014, coal-related products were always leading, but its proportion declined constantly from 49.6% to 44.3%; petroleum products also experienced a substantial decline from 27.2% to 22.7% (see Figure 3.8). At the same time, the consumption of electricity increased continuously, from 19.1% to 24.9%, about 5% up; consumption of natural gas grew rapidly from 2.6% to 6.3%, rising about 4%. The share of other energy sources (commercialized biomass, waste incineration for energy purpose) was relatively stable at about 1.5%.

## Relationship between energy development and economic growth

Energy intensity (energy use per GDP) and energy elasticity (the ratio between annual growth rate of energy consumption and that of GDP) are two commonly used parameters to show the relationship between energy and economic growth.

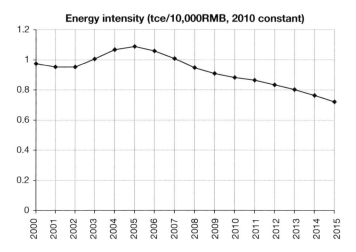

*Figure 3.9*  Trend of energy intensity since 2000 in China

*Data source*: Authors.

Between 1970 and 2001, China was able to limit energy demand growth to less than half of its GDP. Energy use per GDP has declined by approximately 5% per year during the period (Zhou et al., 2010). However, the period of 2001–2005 saw a dramatic reversal of the historic relationship between energy use and GDP growth, for the sake of the accelerated evolution of industrialization and urbanization. Energy use per unit of GDP increased by an average of 3.8% per year between 2002 and 2005 (see Figure 3.9).

The global impact of this dramatic increase is significant. By 2006, nearly 50% of global energy demand growth was due to growth in China (Zhou et al., 2010). In recognition of the unsustainable pace of energy demand growth and its associated adverse consequences, stringent energy conservation and emission reduction policies and measures (PAMs) were initiated in November 2005 and implemented progressively in the Eleventh five-year plan (FYP, 2006–2010) and the Twelfth FYP (2011–2015). The effectiveness of these policies was significant, as energy use per GDP dropped 19.1% after the Eleventh FYP and further dropped 18.2% after the Twelfth FYP. Specifically, it dropped from 1.089 tce/10000RMB in 2005 to 0.722 tce/10000RMB in 2015 (in 2010 constant price).

Nevertheless, based on the exchange rate in 2015, China's energy intensity was almost 55% higher than the world average, nearly 2.2 times that of the USA, over 2.5 times that of Japan, and even higher than other emerging economies such as Brazil. Particularly, the energy efficiency of major energy-intensive industrial products was 15–40% lower than the advanced level in the world (Jiang et al., 2014). Expressed in an alternative way, China's GDP is about 60.5% of USA in 2015 (in terms of exchange rate), whereas energy consumption is 30% higher. Compared also with Japan, China's GDP is 2.6 times bigger, but its energy consumption is 6.5 times higher than that of Japan. In the global context, China shares 14.5% of GDP output, but consumes 23% of global energy demand.[4]

A similar trend was also observed in terms of energy elasticity which climbed from 2000 until 2005, and then declined (Figure 3.10). In 2003 and 2004, the parameter was as high as an unprecedented 1.6. A difference was observed during the international financial crisis in 2008, when the parameter decreased to about 0.3, yet soon rebounded after strong investment from the central government to stimulate the economy. Along with the emergence of the "new norm"

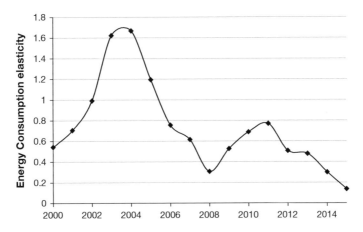

*Figure 3.10*   Trend of energy elasticity since 2000

*Data source*: Authors.

status of economic development since 2011(Green and Stern, 2015), energy elasticity declined once again, to around 0.3 in 2014, and even lower than 0.2 in 2015.

Similar to the general trend of the country, regional energy intensity from 2000 to 2005 increased by a different scale, and began to decrease, pushed by the rigorous energy-saving policies since 2006. Compared to the level in 2000, energy intensity in the eastern, middle and western areas of China dropped in 2014 by 40.4%, 41.2% and 30.4%, respectively. In terms of physical units, 2014 energy intensity in these areas was 0.578tce/10000RMB, 0.737tce/10000RMB and 0.945tce/10000RMB, respectively. The major reason for the rebounding of energy intensity and large energy elasticity observed at the beginning of the new century was the fast growth of heavy industries during the industrialization process, and the export of industrial products on a larger scale after China joined the WTO in 2002 (Yang et al., 2014). Meanwhile, the inferior energy efficiency in China's manufacturing industries was very relevant to the rapid expansion of energy demand. Along with the basic completion of industrialization and urbanization, the energy elasticity dropped back to around 0.5 and even lower, indicating a more healthy relationship between energy and economic growth. The overall trend was consistent with that of other major developed countries at a similar evolutionary stage.

## The environmental issues related to energy development

The large and fast growing energy industry in China has caused significant environmental degradation, including land subsidence, water intrusion and, particularly, air pollution. By the end of 2014, the area of land subsided, destroyed or occupied by coal-mining and related activities amounted to over one million hectares, and the area suffering from severe water and soil erosion caused by coal-mining has amounted to over 200 square kilometers in the North-west region (CAE, 2011). Oil and gas production in North China is highly related to the formation of a cone of depression of groundwater in that area. Intensive hydropower development changed significantly the ecological conditions of related rivers, lakes and wetlands in specific regions. Last but not least, the issues that have arisen from uranium resource exploitation, such as solid residual disposal and waste water discharge, should not be neglected either.

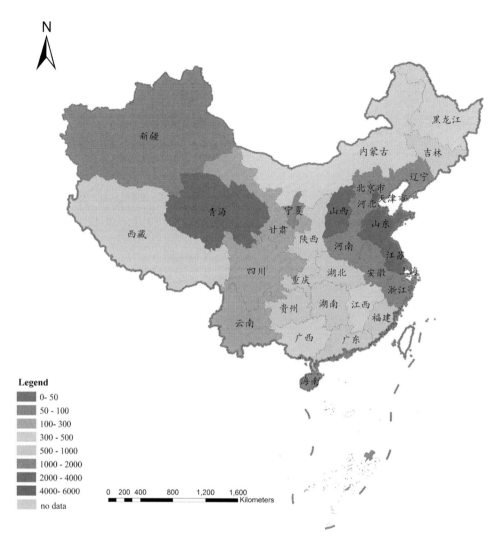

*Figure 3.11* Coal consumption density in China

*Data source*: Authors.

Most of the air-borne pollutants come from fossil fuel combustion. China has been the largest emitter of $SO_2$, NOx, soot/dust, anthropogenic mercury and $CO_2$ since 2006. The super-high coal consumption density (in terms of coal consumption per square kilometer) in middle and east China, (see Figure 3.11), is the major reason for the haze that has troubled China since the winter of 2012. It is estimated that coal burning contributes 50–60% to the concentration of $PM_{2.5}$, the number one primary pollutant in many cities (CCCPT, 2014). Taking the Beijing-Tianjin-Hebei Metropolitan Region, the Yangtze River Delta and the Pearl River Delta as examples, the coal density is as high as 1794, 2267 and 981 tons/$km^2$ (see Figure 3.11), respectively, in these three regions where the worst haze was monitored as well. Regarding $CO_2$ emissions from fuel combustion and cement production, China shared 28.7% of total emissions in the world in 2015, and accounted for 68% of the incremental emissions for the period of 2000–2015 (GCB, 2016).

Particularly, road transport is one of the major sources with fast increasing trend. According to the Initial and Second National Communication of China submitted to the United Nation Framework of Climate Change Convention (NDRC, 2004; 2012), in 1994, $CO_2$ emission from road transportation was estimated as 105Mt, accounting for 3.8% of $CO_2$ emission from fuel combustion, and 2.6% of overall emission (not including land use, land use change and forestry, LULUCF); in 2005, $CO_2$ emission from road transportation was estimated as 373 Mt, accounting for 6.4% of $CO_2$ emission from fuel combustion, and 4.5% of overall emission (not including LULUCF). Though the share of road transport emission overall is still quite low, compared with the level in developed countries, the annual growth rate between 1994 and 2005 was as high as 12.2%, and the growth continued after 2005 – it was estimated by the authors that the $CO_2$ emission from road transportation in 2015 was over 700 Mt, more that double of 2005. Although the growth rate has slowed with the progress of exhaust gas emission standards, road transport is also the major source of NOx. In 2014, 6.28 Mt NOx, 30.2% of the total, came from this sector (Ministry of Environment Protection (MEP), 2015), compared to 5.83 Mt in 2009 (Ministry of Environment Protection (MEP), 2010). Generally, the eastern area which is more economically active, populous and mobilized than the middle and western regions tends to generate more emissions (Cai et al., 2011; Hao et al., 2014).

## Summary

This chapter provides a general picture of China's energy development since 2000. The indigenous energy supply is expanding rapidly, especially coal and natural gas; however, this has failed to meet the fast growing energy demand. Up to 2015, about 16% of total energy was imported from overseas, mainly oil and gas. The energy mix is still dominated by coal, although with a general downwards trend, owing to a strong policy to save energy and the expansion of low-carbon energy. From end-use perspectives, the manufacturing industry still accounted for nearly 70% of final energy use in 2014, transportation accounts for 10.9%. There is however a strong regional difference in energy use in China, with the eastern region consuming a major share.

Two different phases, 2003–2011 and 2012–2015, were particularly pronounced. In the former phase, a much higher annual energy demand, growth rate (>10%) and elasticity (>1.0), and rebounded energy intensity were observed, warning the country of the concerns on energy security, the over-consumption of coal, GHG emissions and air pollutant emissions. With moderated economic expansion and strong policy intervention, the latter phase saw a much relieved situation. It seems energy demand in China has come to a period of low growth, along with the "new norm" status of economic development.

China announced its Nationally Determined Contribution (NDC) under the United Nations Framework of Climate Change Convention (UNFCCC) in June 2015 and ratified the Paris Agreement in October 2016, promising to peak its GHG emission around 2030 and, possibly, to pursue an even earlier peak. Furthermore, to achieve the 2°C target, a study from the Inter-governmental Panel of Climate Change (IPCC) concludes that world emissions must return at least to the level of 2010 by 2030, and be cut by 40% or more by 2050, and reach near-zero emission by the end of this century (IPCC, 2014). This challenges the whole world energy system especially in China, which still heavily depends on coal. Although coal consumption would appear to have already peaked in 2013 (Asuka, 2016), contributing significantly to the surprising declining of world $CO_2$ in 2015 and 2016 (Robert et al., 2016; Nature editor, 2016), the uncertainty is still large. China needs to transfer more speedily to a real low-carbon system.

# Notes

1 All data above is cited from World Bank database: http://datacatalog.worldbank.org/
2 Data regarding gross energy production/consumption is generated by the method of coal equivalent conversion, including non-fossil fuel energy.
3 Non-fossil fuel energy is converted by a calorific value calculation method before it is included in final energy consumption.
4 Most of the data in this paragraph comes from WDI2015; BP Statistical Review of World Energy (2016).

# References

Asuka J. 2016. Assessment of China's Greenhouse gas emission reduction target for 2030: possibility of earlier peaking. *Journal of Contemporary China Studies*. Vol. 5(1): 57–68.

BP Statistical Review of World Energy. 2016. Available at: www.bp.com/en/global/corporate/energy-economics/statistical-review-of-world-energy.html (June 17, 2016).

CAE (Chinese Academy of Engineering). 2011. Medium and long-term energy development strategy of China (2030 and 2050) (中国能源中长期发展战略研究). Beijing: Science Press (in Chinese).

Cai B., Cao D., Liu L., Zhou Y. and Zhang Z. 2011. China Transport $CO_2$ Emission Study (中国交通二氧化碳排放研究). *Advances in Climate Change Research*. Vol. 7(3): 197–203 (in Chinese).

China Coal Cap Program Team (CCCPT). 2014. Contribution of coal to air pollution in China (煤炭使用对中国大气污染的贡献). Available at: www.nrdc.cn/coalcap/console/Public (May 20, 2016).

Global Carbon Budget (GCB). 2016. An annual update of global carbon budget and trends. Available at: www.globalcarbonproject.org/carbonbudget/16/files/GCP_budget_2016_v1.0_FinalRelease.pdf (Nov. 4, 2016).

Green F. and Stern N. 2015. China's 'New Normal': Better Growth, Better Climate. Available at: www.lse.ac.uk/GranthamInstitute/wp-content/uploads/2015/03/Green-and-Stern-policy-paper-March-2015b.pdf (Jul. 8, 2016).

Hao H., Geng Y., Wang H. and Ouyang M. 2014. Regional disparity of urban passenger transport associated GHG (greenhouse gas) emissions in China: a review. *Energy*. Vol. 68(8): 783–793.

He G., Mol APJ. and Lu Y. 2016. Wasted cities in urbanizing China. *Environmental Development*. Vol. 18: 2–13.

IPCC. 2014. Assessing Transformation Pathways, In: Climate Change 2014: Mitigation of Climate Change. Contribution of Working Group III to the Fifth Assessment Report of the Intergovernmental Panel on Climate Change. Clarke L., Jiang K., Akimoto K., Babiker M., Blanford G., Fisher-Vanden K., Hourcade J., Krey V., Kriegler E., Löschel A., McCollum D., Paltsev S., Rose S., Shukla P., Tavoni M., van der Zwaan B. and van Vuuren D. (eds.). IPCC, Geneva, Switzerland.

Jiang L., Folmer H. and Ji M. 2014. The drivers of energy intensity in China: A spatial panel data approach. *China Economic Review*. Vol. 31: 351–360.

Ministry of Environment Protection (MEP). 2010. China vehicle emission control annual report (2010年中国机动车污染防治年报). 2011. Available at: www.gov.cn/gzdt/2010-11/04/content_1738150.htm (Nov. 4, 2011) (in Chinese).

Ministry of Environment Protection (MEP). 2015. China vehicle emission control annual report (2015年中国机动车污染防治年报). 2016. Available at: http://wfs.mep.gov.cn/dq/jdc/zh/201601/P020160115523794855203.pdf (Jan. 9, 2016) (in Chinese).

National Bureau of Statistics of China (NBS). 2016. China Statistical Abstract (2016 中国统计摘要). Beijing: China Statistics Press.

National Development and Reform Commission (NDRC). 2004. Initial National Communication on Climate Change of the People's Republic of China. Available at: http://unfccc.int/resource/docs/natc/chnnc1e.pdf (Dec. 10, 2004).

National Development and Reform Commission (NDRC). 2012. Second National Communication on Climate Change of the People's Republic of China. Available at: http://unfccc.int/resource/docs/natc/chnnc2e.pdf (Nov. 10, 2012).

Nature, Seven Days (News in Brief). 2016. Carbon dioxide emissions stay stable. *Nature*. Vol. 539: 334.

Price L., Levine MD., Zhou N., et al. 2011. Assessment of China's energy-saving and emission-reduction accomplishments and opportunities during the 11th Five Year Plan. *Energy Policy*. Vol. 39(4): 2165–2178.

Robert BJ., Josep GC., Corinne LQ., Robie MA., Jan IK., Glen PP. and Nebojsa N. 2016. Reaching peak emissions. *Nature Climate Change*. Vol. 16(1): 7–10.

Wang K. and Liu Y. 2014. Prospect of China's energy conservation and emission reduction during the remaining years of the 12th Five-Year Plan period. *International Journal of Global Energy Issues*. Vol. 39(1/2):18–34.

Wang Z. and Yang L. 2015. Delinking indicators on regional industry development and carbon emissions: Beijing–Tianjin–Hebei economic band case. *Ecological Indicators*. Vol. 48: 41–48.

Yang L., Wang JM. and Pan H. 2014. Relationship between energy consumption, economic development and carbon emissions in China. *Environmental Engineering & Management Journal*. Vol. 13(5): 1173–1180.

Zhou N., Levine MD. and Price L. 2010. Overview of current energy efficiency policies in China. *Energy Policy*. Vol. 38(11): 1–37.

# 4

# ENERGY POVERTY IN ASIA

*Subhes C. Bhattacharyya and Debajit Palit*

## Introduction

Energy plays a crucial role in any economy and countries with better energy access are generally found to achieve better human development indicators. Likewise, countries with higher income generally report better energy access (Bhattacharyya, 2012). Lack of access to reliable and affordable energy supply inhibits development by restraining economic activities and denying the population the opportunity to develop their human capital (Bhattacharyya, 2012). Furthermore, sustainable development as defined in *Our Common Future* (WCED, 1987) (or the Brundtland Report[1]), cannot be achieved without access to affordable, reliable and clean energy services (Bhattacharyya, 2013). With 1.2 billion people globally lacking access to electricity in 2014 and 2.7 billion still using traditional solid fuels for cooking (International Energy Agency, 2015), access to clean and reliable energy is an important challenge for sustainability and economic growth of developing countries.

Recognising the challenge, the Sustainable Energy for All initiative was launched by the UN Secretary General in 2012 to crystallize global attention on addressing this global challenge and to ensure access to modern energy services for the entire global population by 2030. The Sustainable Development Goals, adopted in 2015 and launched in 2016, also specifically included targets for universal energy access by 2030 under Goal 7. Given that Asia,[2] particularly South Asia, hosts a large number of people without energy access and there are many more affected by unreliable energy supply, ensuring universal energy access remains a challenge. Yet, countries in the region have made significant progress in promoting access to electricity and clean cooking energies. There is significant learning potential from one another for collectively addressing the challenge in an effective way.

This chapter aims to provide a brief overview of the energy access status in Asia and present successful and not-so-successful cases to draw lessons for future interventions in energy access area. The chapter is organised as follows: the second section presents a brief overview of energy access status in the region; the third section presents case studies from different countries, the fourth section highlights the lessons others can learn from the case studies and finally, the concluding section offers some suggestions for further actions.

# Overview of energy access in Asia

## *Brief review of energy access status*

According to the International Energy Agency (2015), 526 million people (or 14% of the region's population) lacked access to electricity in Asia in 2013 while about 1.9 billion people (or 51% of the population) relied on traditional energies for cooking purposes. Developing Asia accounted for almost 50% of the global population without electricity access and 70% of those lacking access to clean cooking energy globally in 2013. The challenge faced by Asian countries therefore cannot be underestimated.

There is however significant variation in energy access within the region. As shown in Figure 4.1, while all areas within Asia have recorded significant improvements in energy access between 1990 and 2012, East Asia stands out. In respect of electricity access, East Asia has almost reached universal electrification, while in cooking energy access, about 56% of the population relies on non-solid fuels. South Asia on the other hand has 355 million people without electricity access and 1.1 billion people without clean cooking energy. In fact, 67% of people lacking electricity access in Asia live in South Asia, while for cooking energy it is about 60%. South East Asia comes in the middle – in terms of electricity access, about 90% of the population has access to electricity while 54% of have access to modern cooking energies. However, on a positive side in East Asia and South Asia, the increase in access to clean cooking was more than the population increase by 12 million and 1 million respectively during the 2010–12 (GTF, 2015).

Table 4.1 presents the list of countries with major concentrations of population without energy access in Asia. It can be noticed that the list is shorter for electricity and many countries in the region have successfully managed to provide electricity access. Many countries such as Vietnam, Indonesia, Sri Lanka and Pakistan have made significant progress between 1990 and 2012 (see Figure 4.2) (ADB, 2015). However, cooking energy access remains an issue in a wider set of countries. Five countries (India, China, Bangladesh, Pakistan and Indonesia) account for 88% of the unserved population in cooking energy and would require more focused attention.

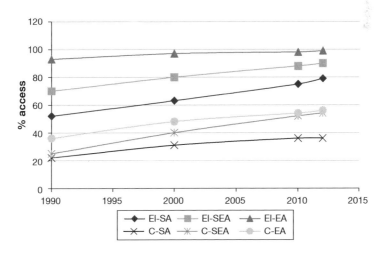

*Figure 4.1* Improvements in energy access in Asia

*Data source*: International Energy Agency and the World Bank, "Sustainable Energy for All 2015—Progress Toward Sustainable Energy," The World Bank, Washington, DC, 2015.

*Note*: El – electricity, C – cooking, SA – South Asia, SEA – South East Asia, EA – East Asia.

*Table 4.1* Major concentrations of population without energy access in Asia in 2013

| Country | Population without electricity (million) | Country | Population without clean cooking energy (million) |
|---|---|---|---|
| India | 237 | India | 841 |
| Bangladesh | 60 | China | 450 |
| Pakistan | 50 | Bangladesh | 140 |
| Indonesia | 49 | Pakistan | 105 |
| Myanmar | 36 | Indonesia | 98 |
| Philippines | 21 | Philippines | 53 |
| DPR Korea | 18 | Myanmar | 49 |
| Cambodia | 10 | Vietnam | 42 |
| Nepal | 7 | Nepal | 22 |
| Vietnam | 3 | Sri Lanka | 15 |
| Laos | 1 | Thailand | 15 |
| | | Cambodia | 13 |
| | | DPR Korea | 12 |
| | | Laos | 4 |
| | | Mongolia | 2 |
| Total | 492 | Total | 1861 |

*Data source*: International Energy Agency (2015).

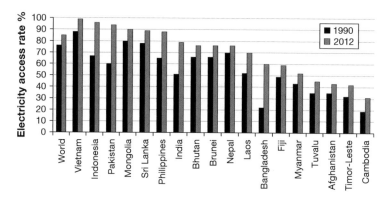

*Figure 4.2*   Progress in electricity access in Asian developing countries between 1990 and 2012

*Data source*: ADB, "Sustainable Energy for All – Tracking progress in Asia and the Pacific: A Summary Report," Asian Development Bank, Manila, 2015.

However, the data on energy access remains weak. The information provided by International Energy Agency varies significantly from the Global Tracking Framework data due to differences in the methodology, definition and source of information. Accordingly, the reality can be different from what is indicated by the available statistics.

Further, the energy access problem is a predominantly rural issue, although it exists in urban areas in some countries. As shown in Table 4.2, most urban areas in the region have access to electricity and to a large extent to modern cooking energies. However, the situation changes dramatically for rural areas, particularly for cooking energy. Only 2 to 4% of rural households have access to clean cooking energy in Afghanistan, Bangladesh, Laos, Myanmar or

*Table 4.2* Rural-urban gap in energy access

| | Electricity (EL) | | Cooking (C) | |
|---|---|---|---|---|
| | EL-Rural | EL-Urban | C-Rural | C-Urban |
| Afghanistan | 32 | 83 | 4 | 72 |
| Bangladesh | 49 | 90 | 2 | 44 |
| Bhutan | 53 | 100 | 49 | 98 |
| Cambodia | 19 | 91 | 4 | 49 |
| China | 100 | 100 | 17 | 84 |
| DPR Korea | 14 | 41 | 3 | 11 |
| India | 70 | 98 | 12 | 72 |
| Indonesia | 93 | 99 | 29 | 82 |
| Laos | 55 | 98 | 2 | 11 |
| Mongolia | 70 | 99 | 10 | 54 |
| Myanmar | 31 | 95 | 2 | 19 |
| Nepal | 72 | 97 | 14 | 70 |
| Pakistan | 91 | 100 | 14 | 84 |
| Philippines | 82 | 94 | 22 | 64 |
| Sri Lanka | 86 | 100 | 15 | 66 |
| Thailand | 100 | 100 | 62 | 86 |
| Vietnam | 98 | 100 | 36 | 82 |

*Data source*: International Energy Agency and the World Bank (2015).

*Note*: EL-Rural: Electrification rate in rural areas (%); EL-Urban: Urban electrification rate (%); C-Rural: Cooking energy access in rural areas; C-Urban: Cooking energy access in urban areas.

DPR Korea. The situation is not very encouraging in China, India, Pakistan, Nepal or Sri Lanka where 10 to 17% of the rural population have access to clean cooking energies. Given the large population base of some of these countries, the magnitude of the issue can be imagined.

## Main regional energy access challenges

Due to varying levels of social and economic development, geographical conditions and population factors, countries in the region face different challenges to ensure universal energy access (see United Nations Development Programme, 2013). In terms of electricity access, while most countries have primarily relied on grid extension, it is now realised that off-grid solutions can be effective as well, particularly for remote locations and sparsely populated areas where low initial demand may not justify grid extension. Archipelagos in the region (e.g. Indonesia and the Philippines) face the added challenge of cross-island grid extension and instead local grid systems and local resource-based distributed generation systems deserve more attention. Unfavourable political conditions have affected economic development and energy access in countries like Myanmar, DPR Korea or Afghanistan and rapid catch-up in energy access through strong planning, international collaboration and investment programmes would be essential. On the other hand, countries like India where grid is expanding rapidly and the size of population without electricity access is falling rapidly, the issues are quite different: integration of off-grid systems to the grid, connections to poor households in areas with grid coverage, standardisation of the systems and improving the reliability of supply and operational performance of the electricity distribution companies are major challenges.

On the cooking energy front, the entire region faces a more significant challenge as a large section of the population is still dependent on solid fuels. The transition to modern cooking fuels has not progressed well in many countries and there is limited analysis of strategies or options for accelerating the transition to a cleaner cooking energy pathway. Given that solid fuel dependence is likely to remain significant in the future, there is a realisation that attention to improved cookstoves is a necessary element of any improvement strategy but there remain significant barriers to accelerated adoption of improved cookstoves (Palit & Bhattacharyya, 2014). While many interventions have taken place in the past to promote improved cook-stoves, it appears that there has been an overemphasis on new technology development without necessarily considering the user perspectives. The behavioural aspects of using a new technology and its ability to meet the local needs have received limited attention, which has affected successful promotion of improved cookstoves. Affordability is another factor hindering their promotion and suitable financing mechanisms to support adoption has not been widely explored (Palit & Bhattacharyya, 2014).

Considering the magnitude of the problem, the energy access will need a large amount of investment over the next 14 years. It is often-quoted that an investment of US$50 billion per year is required until 2030 to achieve universal energy access globally but such a figure for the regional investment requirement is not easily available. It is indicated in (SE4ALL, 2015) that India needs between US$8 and US$10 billion per year, Bangladesh needs about US$2 billion while Myanmar, Afghanistan and the Philippines need about US$1 billion per year. A reasonable regional estimate is between US$15 and US$20 billion per year. This is not a small investment target, particularly when countries need to sustain this level of investment over a long period of time.

The funding challenge also manifests differences in the perspectives of funding agencies and project developers in the field. A stakeholder survey reported in IRENA (2015) indicates that the perspective of the financial institutions and funding agencies differs from that of the project developers. While the funding agencies indicate there is no shortage of funding, the developers find it difficult to access the available funding. The funders need to follow certain procedures and checks before approving funds whereas the developers find these procedures and compliance with them difficult and time consuming. Accordingly, there is a problem of funds reaching the projects at the time when they need them. While the finance sector wants to engage, they do not see the critical mass for this market and do not find reliable data and bankable models. On the other hand, developers, being small and new to this area, lack an understanding and appreciation of the instruments and find access to funds difficult.

Other challenges that hinder energy access and economic development include: poor quality of electricity supply that restricts productive use of energy; limited purchasing power of the poorer section of the population that hinders purchase of efficient appliances and equipment; and inhibiting urban infrastructure particularly in informal urban settlements (i.e. slum areas) that prevents access to the network (Larsen et al., 2016). Energy access policies need to be more broadly embedded in the development agenda and it is essential to create an enabling environment for promoting energy access investment.

A review of scores from the RISE (Readiness for Investment in Sustainable Energy) framework, developed by the World Bank Group, for Asian countries shows that in the case of energy access, the region is well placed compared with Sub-Saharan African counterparts: India scored 94 while Nepal received 80 and Mongolia 68. Planning is widely adopted in these countries but there are weaknesses related to policies and regulations and their implementation as well as in terms of pricing and subsidies (World Bank, 2014). While investment readiness makes private investment attractive, favourable governance and policy environment is essential to ensure the running of viable businesses. Serving the bottom of the pyramid consumers requires ensuring affordability of energy access services, which in turn depends on the availability of low cost finance for investment and operation of the business activities. Lack of adequate and appropriate support in this area remains a challenge for the scaling-up of energy access initiatives. While unlocking the climate fund has been suggested as a solution (Rai et al., 2016), so far the climate fund has not reached the target countries. It remains uncertain to what extent such funds would be able to meet the financing gaps.

## Selected case studies of energy access interventions in Asia

In this section, a selected set of country examples are presented to showcase how different countries in the region have attempted to improve energy access in the past. Due to time and space constraints, an attempt has been made to capture only selected notable cases.

### *Chinese electrification example*

China's achievement in electrification is an inspiring example for any developing country. Reaching 100% electrification in a country of more than 1.3 billion people spread over 9.6 million square kilometres of area is no mean achievement. China managed to provide electricity access to more than 900 million people over a period of 50 years (Peng & Pan, 2006) and only one million people are estimated to lack access in some remote areas (International Energy Agency, 2015). The government plans to provide electricity to them by 2020 through off-grid means.

In 1949, when the People's Republic was founded, more than 90% of rural households did not have access to electricity (Bie & Lin, 2015). During the period of central planned economy (1949–1977), a decentralised approach to rural electrification through small hydropower development at the local level received attention. These projects were funded and managed by rural residents and collectives and the power plants operated on a stand-alone basis. Electricity access made steady progress and by the end of this period, 63% of the rural population had access to electricity (Bie & Lin, 2015). A step change in rural electrification occurred in the second phase of development when the country abandoned Maoist policies and adopted economic reforms in 1978. Industrial activities through town and village enterprises were promoted during this period. Decentralised operation of local grids was continued but in addition to hydropower, thermal power based on coal started to gain importance. There was a shift towards an integrated energy strategy during this period which focused on the promotion of four fuels. Within a decade, 78% of the rural population had access to electricity. Between 1988 and 1997, electricity access was made almost universal and by 1997, 97% of the rural population had access (Bhattacharyya, 2013). In the final phase, China focused on consolidation of rural electricity system as well as universal electrification. The local grid system was upgraded and a national network was developed. In addition, off-grid approaches were also used in remote areas.

In contrast to most developing countries China adopted a pragmatic approach. Rural electrification in China used multiple technologies (small hydro, coal, renewable energies) and multi-mode delivery options (central grids, local grids and hybrid systems), placed a strong emphasis on rural development and allowed a strong role for local level government (Bhattacharyya & Ohiare, 2012). China also followed a phased development approach where the local grid was initially developed in the rural areas to cater to low demand but as demand grew, the system was then renovated to integrate to the central grid system. The phased development strategy along with early recognition of an electrification–rural development link ensured appropriate management of financial resources, initial demand creation and a self-reliant system (Bhattacharyya, 2012).

Three phases of electrification of rural China were characterised by different features:

1    severe shortage of supply during the first phase restricted household use of electricity mainly to lighting;
2    a rapid expansion of electricity system during the second phase allowed widespread use for irrigation as well as for lighting and other household uses (e.g. TV, fans);
3    the third phase has further intensified electricity use in highly energy intensive appliances in rural areas (such as air-conditioning, washing machines, refrigerators, etc.), promoting urban-style living in rural areas (Bhattacharyya & Ohiare, 2012).

China's success with rural electrification can be attributed to the decentralised approach to development where local resource development and management was closely linked with agricultural development, development of town and village enterprises and rural poverty alleviation programmes. Unlike many other developing countries where electrification depended on state subsidies, China adopted the cost recovery principle for rural infrastructure development. Local engagement and resource mobilisation played an important role. In addition, the phased development approach was pragmatic and allowed demand creation before extending the national grid. Although issues like technology compatibility due to poor system standardisation, unreliable supply and high power losses emerged, the benefits of phased development appear to have outweighed the costs.

## Indian rural electrification example

The government of India has been making conscious efforts to improve the infrastructure for electricity provision since the beginning of planned economic development in the country in 1951. However, more than 400 million people were without access to electricity in 2005. Bhattacharyya observes that though, a number of programmes (such as the Kutir Jyoti Programme, Minimum Needs Programme, and Accelerated Rural Electrification Programme) were launched at different times to enhance electricity access either as part of overall rural development or by specifically targeting rural electrification, the multiplicity of programmes made funding for each of them inadequate and the programme implementation was not properly coordinated or managed (Bhattacharyya, 2006). With rural electrification becoming a political priority during the early 2000s, the Indian government started to create the necessary enabling environment, through the REST (Rural Electricity Supply Technology) Mission in 2001, the Electricity Act 2003, National Electrification Policy 2005, and Rural Electrification Policy 2006, to address the rural electrification issue (Palit et al., 2014). For example, in 2001, the government declared the objective of "power for all" by 2012 under the REST mission and continued it with the launch of a large-scale electrification effort, the Rajiv Gandhi Grameen Vidyutikaran Yojana (RGGVY), in April 2005 (the scheme has now been renamed as Deen Dayal Upadhyay Gram Jyoti Yojana with effect from December 2014). Thereafter, the country has made significant progress in the provisioning of electricity in rural areas. In 2001, around 43% of the rural population in India had access to electricity for lighting, which increased to 66% by end of 2015.

   The RGGVY was launched by merging all other existing schemes of rural electrification, with the goal of electrifying all un-electrified villages/hamlets, providing access to electricity to all households, and providing 23.4 million free connections to households below the national poverty line. The scheme attempted to address some of the chronic ailments of rural electrification in India, such as poor distribution networks, lack of maintenance, low load density with high transmission losses, rising costs of delivery, and poor quality of power supply (Palit et al., 2014).

Under the scheme, the government of India provided 90% capital subsidy for construction of Rural Electricity Distribution Backbone (REDB) and creation of Village Electrification Infrastructure (VEI) and Decentralized Distributed Generation (DDG). The balance of 10% of the project cost was soft financed by the Rural Electrification Corporation (REC), the nodal agency for implementation of the programme. The state electricity utilities were made responsible for implementing the work of rural electrification in their respective states.

A key feature of the programme was to ensure revenue sustainability of rural electricity provisioning for which, electricity distribution franchise system was introduced. The franchisees, selected and operating under the distribution utilities, were made responsible for metering, billing, and revenue collection (MBC) for designated territories. In some cases, input-based franchises (IBF) were also introduced where these franchisees procure electricity in bulk from the distribution utility and distribute it in their operational areas- typically areas covered by one or more feeders or distribution transformers. At the same time, DDG was brought under the mainstream power sector to electrify villages where grid extension was techno-economically daunting. For the first time, the DDG scheme also attempted to address the issue of the operational sustainability of off-grid projects through an annuitized subsidy delivery mechanism and by providing an operational subsidy—to address the gap between the high cost of generation in remote areas and the revenue inflow in such projects.

According to an evaluation done by Planning Commission, Government of India, 93.3% households were electrified in 15 sample states during the period 2006–2012 (Anon., 2014). Five states (out of the 15 sample states), namely, Andhra Pradesh, Haryana, Himachal Pradesh, Karnataka, and Uttar Pradesh, achieved 100% of the target. The scheme has been quite successful in meeting its objective of providing electricity to villages and hamlets in rural areas and bringing electricity connection to households. The same evaluation also pointed out some bottlenecks in implementation of the scheme. For example, in case of intensification of village electrification, the performance was not satisfactory at 53% achievement. Quality of power supply, sustainability of infrastructure, revenue generation from the connected households, and the contribution of this initiative to overall rural development were also highlighted as concerns.

With renaming of the RGGVY to DDUGJY, the scope has been further enhanced. Specifically, the DDUGJY has two major additional components i.e. feeder separation and Power for All by 2019, that were not there in the RGGVY. The Indian government is working towards providing power to the last mile standing in the country and has commenced taking various measures for the same. The target is to complete all village electrification by March 2018 and all household electrification by 2019. The DDUGJY scheme also aims to improve the power supply in rural households as well as reduction of the peak loads.

In addition to the grid extension model, India also implemented a number of off-grid projects, both in the private sector as well as by government agencies. The two most successful models of mini-grids implemented by government agencies in India are those implemented by WBREDA and CREDA (GNESD, 2014). WBREDA has set up more than 20 mini-grids based on solar power plants with an aggregated capacity of around 1MWp supplying stable and reliable electricity to around 10,000 households in West Bengal. CREDA, on the other hand, has electrified around 35,000 households across more than 1400 villages and hamlets with low capacity solar mini-grids in Chhattisgarh. While mini-grids in Chhattisgarh are mostly based on micro solar PV plants (typically <10kWp capacity), the solar mini-grids in the Sunderbans are in the range of 25–150kWp (Palit, 2013).

Though most mini-grids in India are implemented and operated following the Village Energy Committee model, CREDA and WBREDA have developed their own service delivery mechanisms, by directly taking care of the operation and maintenance (O&M) through a multi-tier system of maintenance. The village energy committees are assigned a limited role, such as local oversight and grievance redress forum. The projects being publicly funded, mostly flat tariff is charged, set based on the O&M cost, including the salary of the operators and other related aspects such as the consumers' ability to pay. The tariff is not necessarily determined based on economic cost of generation and distribution. For instance, Ulsrud et al. (2011) observe that WBREDA's designing of tariffs for mini-grids in the Sunderbans has considered several factors such as government funding, the customer base, income distribution, availability of anchor customers, acceptability of measures to put a ceiling on consumption, and equity issues in finding the right balance between affordability and the economic cost of electricity generation. A disadvantage of this flat tariff system, however, is overloading by some house-holds (connecting additional electrical points to those authorised), which puts extra pressure on the entire system.

## Example from Nepal

Nepal, characterized by scattered population and located in a geographically difficult terrain, has lately achieved significant progress in electrifying its rural areas with an average 76% national electrification rate and 72% in the rural areas. The Nepalese Government recognizing the fact that that providing electricity to a rural population is a major challenge due to the scattered nature of the population in remote mountainous areas, thus envisaged a combination of grid extension, isolated generation and reliance on alternative approaches.

The Community Rural Electrification Programme (CREP), a grid-based rural electrification programme, contributed significantly to the progress in rural electrification. The CREP was launched in 2003–2004 to enhance the grid-based rural electrification in the country, primarily using energy from hydro projects. The community involvement in rural electrification was done to bring in operational efficiency in the distribution sector, which was witnessing high system losses and poor revenue collection over the years. Consumer associations, typically in the form of cooperatives, take the responsibility of managing, maintaining, and expanding the rural distri-bution of electricity (Palit & Chaurey, 2011). In addition, several off-grid initiatives have been undertaken to electrify remote sparsely populated and geographically difficult terrain. The off-grid electrification efforts in the country could be traced back to the early 1970s, but it was more intensively promoted after 1990s using micro-hydro as well as solar PV (Sarangi et al., 2014). An interesting feature in Nepal is that smaller capacity SHS (locally called *solar tuki*) with capacity between 2.5Wp and 10Wp have also been widely disseminated (Palit & Chaurey, 2011).

Both grid and off-grid electrification models are centred on community participation. Although the grid-based CREP programme is managed by the Nepal Electricity Authority (NEA) and substantially funded by government, communities take a lead role in providing part funding of the project and addressing non-technical losses within their distribution zones. For instance, communities provide 20% of the investment costs in terms of cash and kind, and the rest of the funding comes through government sources. Similarly, in case of off-grid systems, com-munities play a crucial role in the management of the systems with private manufacturer and installer companies carrying out the survey, design and installation of the systems. The off-grid projects are funded through government subsidies, community equity, contribution from local government, and contribution from other organizations. The government subsidies are mobi-lized from the funding support extended by several donor agencies. Organizationally, while CREP is managed by NEA, the off-grid energy projects are controlled by the Alternative Energy Promotion Centre (AEPC), the nodal agency responsible for all renewable energy activities in Nepal.

## Example from Bangladesh

The urban electrification rate in Bangladesh is around 90%. However, less than half of the population in rural areas have access to electricity. Though the rural household electrification level is low, the country has the distinction of having the most successful solar home system (SHS) programme globally for electrifying its off-grid habitations. The SHS programme implementation is coordinated by Infrastructure Development Company Limited (IDCOL), a state-owned financial institution. The programme, since its inception in 2003, has installed SHSs of different capacities in more than four million households with more than two-thirds of the installation done during the last three years. The success of the programme can be ascertained from the fact that IDCOL has over-achieved its targets on several occasions. In 2003 it decided to finance 50,000 SHSs by June 2008, but it achieved the target almost three years in advance in September 2005; it again achieved the target of 200,000 SHS seven months ahead of schedule in May 2009. The millionth figure was achieved by 2010 against the set target of 2012 (Palit, 2013). While the programme initially relied on subsidies, capital as well as institutional grant, they were gradually phased out as system prices declined owing to technological advances and economies of scale (Sadeque et al., 2014).

IDCOL works with development partners, local grassroots organisations, which are called partner organizations and the suppliers of SHS. While IDCOL sets the technical specifications, certifies products and components, and selects partner organizations based on clear eligibility criteria, the major role of the partner organizations is to select potential customers, offer micro-lending, install systems, and provide the after-sales support (Palit, 2013). The partner organization acts as the financial intermediary under the IDCOL programme. IDCOL provides two different types of grant support—institutional development grant and system buy-down grant— to its partner organizations. The programme made the SHS affordable through a combination of affordable credit mechanism and (phase wise declining) grants.

The institutional development grant has been instrumental in creating the necessary rural infrastructure for service delivery in terms of both dissemination and post-installation maintenance of the systems by partner organizations. Palit (2013) observes that this grant also enabled the partner organization to build its capacity by hiring staff and training employees and credit monitoring. Both these grants are also intended to enable the partners to purchase the technology below prevailing market rates and provide micro-loans to customers, in essence lowering the price of SHS and increasing the institutional strength for last-mile distribution and maintenance. However, to promote competition, the capital grants have also been reduced over time as more SHS capacity was installed (Sovacool & Drupady, 2011). The institutional development grant has also been phased out except for new and smaller partner organizations that are provided a nominal grant of US$3 per system. The partner organization reportedly also extends a buy-back guarantee that gives customers an option to sell their system back at a depreciated price if the household obtains a grid connection within the year of purchase (Sadeque et al., 2014).

The financial model works such that all stakeholders in the value chain benefit in the process. IDCOL offers refinancing through soft loans to smaller partner organizations and channel grants to reduce SHS costs and support the institutional development of partner organizations. For the larger partners, the interest charged for refinance is little higher, with other above-mentioned terms and conditions remaining the same. The partner organizations provide microfinance loans to households, who are also required to make a down payment equivalent to 10%–20% of the total system cost. The reminder is repaid in monthly instalments over a period of two–three years at the prevailing market interest rates. The cost of the system covers both the interest and the maintenance cost.

While the IDCOL SHS programme has been very successful and has the potential to replicate in other electricity access deficit countries, some aspects of the programme may be unique to Bangladesh and difficult to replicate in other countries (Sadeque et al., 2014). For example, the programme has benefitted from a strong pre-existing network of microfinance institutions with a wide reach in rural areas. Other factors that have contributed to the programme's success are:

1   the high density of Bangladesh's rural population, which fostered competition and economies of scale;
2   rising rural incomes and remittances from abroad, which stimulated demand for the off-grid solar systems;
3   the existence of entities interested in doing business with rural customers and the country's entrepreneurial culture.

The most important aspect that seems to contribute immensely to the success is the tangible benefit to all stakeholders across the programme. While Bangladesh government could improve the electricity access in the country, IDCOL and the partner organizations benefitted commercially through lending, the local solar industry developed due to the huge scale of the programme, and the users of SHS benefitted by use of clean energy systems and paying for them in small instalments (Sarangi et al., 2016).

## *Example from Indonesia*

Electrification in Indonesia, an archipelago of more than 17000 islands, is very challenging due to the remoteness and low population density of smaller islands. However, despite the geographical challenge, the country has made significant progress in recent times: IEA reports 81% of the population had access to electricity in 2013 (International Energy Agency, 2015) whereas ADB reports 84% households having electricity access by end of 2014 (ADB, 2016). About 12.5 million households (or 48.8 million people) still lack access to electricity (ADB, 2016). There is a significant variation in the electrification level in the country: PLN, which is the national electric utility, mostly focused on densely populated urban areas of Java-Bali-Sumatra-Kalimantan-Sulawesi areas, whereas the rural areas where the poor live remained poorly connected.

The national electrification rate increased from 7% in 1980 to 43% in 1995 but the financial crisis in 1997 affected the power sector badly and PLN concentrated on maintaining its urban-centric electricity supply in Java-Bali area and in economically viable areas of Sumatra. This was adopted to minimise commercial loss and socially oriented electricity access programmes suffered for almost a decade (Bhattacharyya, 2013). The electricity sector development plan of 2003 recognised the problem and set a target of extending access to 90% of the population by 2020 (The World Bank, 2005). The decentralised solutions were suggested as an alternative as they offer distinct advantages in Indonesia due to availability of natural resources and spatial issues related to grid extension (The World Bank, 2005). However, despite the experimentation in the 1980s and pilot projects in later years, the country did not record much success in off-grid electrification. PLN accounts for 97% of all household connections and only 3% were carried out through other means (ADB, 2016).

Since the middle of the new millennium, PLN has been pursuing the electrification agenda more vigorously. Over the past decade it managed to connect 20 million new customers, including 3.7 million in 2013 alone. This has improved the prospects of electrification in the country and now the government aims near-universal electrification by 2020 (ADB, 2016). However, Indonesia faces several challenges in reaching near-universal electrification by 2020.

First, the present level of investment in electrification is significantly unlikely to achieve the target of universal electrification by 2020 and an accelerated investment programme is required. Second, the present electrification programme is complex and is leading to assets that are not recognised in the books of the government or PLN. This is because the programme is publicly funded and procured under government rules and implemented by PLN. The assets are transferred to PLN as in-kind equity but because of complex auditing and transfer procedures have led to indeterminate status of certain assets. Third, the off-grid electrification programme is run on an ad-hoc basis, requires project-by-project approval and suffers from a clear subsidy mechanism. Accordingly, this programme is not replicable (ADB, 2016). Reaching the remaining population without electricity will require a significant coordinated effort and Indonesia needs a strong government support and commitment to achieve this on time.

## Experience with clean cooking energy

As indicated before, Asia has the largest population in the world lacking access to clean cooking energy. The common strategies used in enhancing clean cooking energy in the region include promotion of clean conventional (e.g. petroleum fuels) and renewable cooking energies (e.g. biogas) and technological intervention to improve the efficiency of solid fuel use.

Many countries have supported liquid or gaseous hydrocarbons to displace solid cooking fuels. Kerosene and liquefied petroleum gas (LPG) have been widely promoted to reduce the environmental damage at the point of use. When these alternative fuels were promoted, they were often subsidised by the government to ensure that consumers switch to modern fuels but because energy is always used in conjunction with appliances, which tend to require upfront capital investment, the subsidised fuel benefited the relatively richer section of the population. In addition, the prolonged use of price subsidy imposed a significant financial burden on the government as it proved to be politically difficult to eliminate such subsidies. Rural consumers who depend heavily on solid fuels did not receive adequate attention: the supply chain was not adequately developed which resulted in unreliable supplies; limited market development hindered the scale up of improved and efficient stoves based on local fuels; the standard sized LPG bottles often imposed significant financial burden on the users and poor financial viability due to high transaction costs in rural business attracted limited private sector participation (Bhattacharyya, 2012).

Alongside petroleum fuel, Asian countries have also promoted biogas as an alternative cooking energy. The economic rationale is compelling: biogas reduces import dependency and saves foreign exchange, enhances local resource use and provides clean energy (Bhattacharyya, 2012). China is the world leader in biogas production. While the above options aim at fuel switching, it is realised that solid fuels will continue as an important cooking fuel in the future and one way to contain the adverse effects is to try to improve its use through better cookstoves.

In India, the Ministry of New and Renewable Energy has been running the programmes such as the National Program of Biogas Development (NPBD) and National Program on Improved Cookstoves (NPIC) since the early 1990s for providing clean cooking energy solutions to the rural population. The NPBD was broadened and rechristened as the National Biogas and Manure Management Program (NBMMP) in 2010. NBMMP mainly caters to setting up family-size biogas plants to provide fuel for cooking purposes and organic manure to rural households. A cumulative total of 4.75 million family type biogas plants had been installed in the country as of 31 March 2014 against an estimated potential of 12 million plants (MNRE, 2016). On the other hand, MNRE discontinued NPIC in 2002 as the programme did not achieve much success and put the responsibility of promoting improved cookstoves on the local government institutions. At the time of discontinuation, around 33.8 million smokeless stoves (which were not necessarily improved as far as the thermal efficiency is concerned), which was around 27% of the target, were reportedly distributed in rural India. However, researchers argue that national biogas and cookstoves programmes were mostly technology focused, with dissemination being the objective and numbers deployed as the target for measuring success (Balachandra, 2010). A proper market development approach was never conceived. As a result, despite the large market potential, there are a limited number of players in the market. MNRE re-launched the National Biomass Cookstoves Initiative (NBCI) with a target to disseminate 2.75 million improved cookstoves during the period 2013–2017 through different agencies, including the private sector.

Another country that has achieved considerable success is Cambodia in Southeast Asia, where the NGO GERES introduced the New Lao Stove (NLS) in 2003 in response to the growing demand for charcoal for domestic cooking in urban areas of the country. The NLS project was initiated through the Cambodian Fuelwood Saving Project (CFSP) in collaboration with the Ministry of Industry, Mines and Energy. It was developed as an attractive alternative to the traditional stoves by creating an improved grate design and insulation while retaining similarity with respect to traditional Southeast Asian cooking stoves. Quality control, standardization system including a quality seal system and traceability procedures were built in within the dissemination supply chain. More than 3.6 million improved stoves – New Lao Stove and Neang Kongrey Stove – have reportedly been disseminated between the start of the project and the end of 2014, primarily following a commercialized supply chain approach (Anon., 2016).

GERES worked with stove producers, distributors and end consumers to develop a highly demanded product as well as an optimal price structure in order for all involved parties to gain from the new improved cookstove. A micro-credit fund was also set up to help establish the production and distribution network. They supported the creation of the association of producers and distributors of improved cookstoves in Cambodia, which have now 253 independent business owners (84 stove producers and 171 distributors) from 11 provinces as members. While initially the project received funding support from the European Union, the Union withdrew its start-up support in 2006. Thereafter, the maintenance of the project activity was taken care of using additional funding from selling Verified Emission Reductions (VERs). The carbon finance was also used to solidify the market position, maintain the quality assurance system and scale up dissemination up to market saturation (GNESD, 2016).

The Nepal Biogas Support Programme (BSP) is another successful initiative, which focussed on technological innovation, financial engineering and market development, to extend the biogas as a clean cooking technology to the rural areas of Nepal. The programme is managed by the AEPC with the support of Biogas Sector Partnership Nepal (BSP-Nepal). The principal objective of the BSP is to promote the wide-scale use of biogas as a substitute for wood, agricultural residues, animal dung and kerosene that are presently used for the cooking and lighting needs of most rural households in Nepal. Since its inception in 1992, around 0.3 million biogas plants out of the total technical possibility of 1.3 million have been installed in all 75 districts and more than 2800 village development committees of Nepal (as of the end of 2011/12) (AEPC, 2016). Along with household biogas plants, around 20 community and 300 institutional biogas systems have been already set up in different locations. At the initiation of the Programme, there was essentially only one state-owned company, the Gobar Gas Company (GGC), producing biogas systems. As a direct result of the market development program, currently more than 39 rural micro enterprises are involved for construction of biogas plants and provision of after-sale services (Bajgain & Shakya, 2005). One of the important features of the BSP has been its innovative financial engineering and judicious application of consumer subsidies to help develop the market for biogas systems. The government provides a subsidy that covers about 30% of the overall cost of the plants. In addition, there is also a biogas credit fund which is a revolving fund to support credit facilities channelled through over 260 rural micro-finance institutions for biogas users. With increased availability of micro credit and provision of additional subsidy for the disadvantaged population, BSP-Nepal has been following an inclusive but market based model for the dissemination. Some of the other key success differentials are: strengthening the capacity of the biogas construction companies for meeting strict production quality and service standards for their biogas systems, training for successful operation and maintenance and guarantee of after sales service to ensure the success of technology.

## Lessons from case studies

These case examples indicate that grid extension is the main mode of electrification adopted in Asian countries. Countries like China, India, Indonesia, Thailand, and Vietnam have successfully extended grid to reach a majority of their populations. Decentralised, off-grid solutions have mostly been used as a temporary solution in cases where grid is unlikely to reach. This has caused an uneven playing field for off-grid electrification. Similarly, the cooking energy options have paid greater attention to fuel-switching with a subsidised programme but locally relevant sustainable solutions received lesser attention except in China and to a lesser extent in India.

In all successful cases, the state has played an important role. The state has funded most of the capital costs of electrification projects and in some cases, the state has provided additional support to allow consumers use a minimum amount of electricity. However, in all cases it was harder to electrify the final quarter than the rest. Countries trying to ensure universal electrification in Asia will find it more challenging to achieve this unless they take adequate preparation. Given the predominance of grid extension for electricity access, the quality and reliability of supply will assume greater importance going forward. The integration of off-grid systems to the grid will also emerge as an important issue and appropriate technical standards and procedures will have to be agreed on to avoid stranded asset issue in the future.

The case studies suggest that it is critical to develop an ecosystem of innovation beyond 'physical access.' The rate of success is directly dependent on the government's commitment to create an enabling environment that includes among others, having a clear-cut policy framework and milestones, systems for defining and enforcing appropriate benchmarks and standards, financial support mechanisms, and support towards developing human capital. Further, in order to fully achieve universal access and make sure that energy needs are being met, the studies point to the need for energy policies to specifically target the under-privileged population. The case studies also point to the need for a well-defined institutional structure with clear roles and responsibilities between various agencies for successful outcomes.

The renewable energy electricity generation projects are successful where an organized/structured delivery model has been followed. While many off-grid projects are implemented using a community driven model, this can sometimes be a limitation, as implementation metrics and operational practices differ from organization to organization and agencies are not able to benefit from a standardized set of implementation guidelines. It is, however, also important to note that buy-in and acceptance by the community and the ability to see the benefits have helped in the success of projects.

Case studies also indicate that it is important to avoid distorted incentives to ensure sustainable projects. For example, Chinese government supported infrastructure development in electricity but the pricing system ensured cost recovery to sustain the projects. On the other hand, subsidised supply in India crippled the electric utilities and ultimately hindered supply of energy to the electrified villages in many areas. Similarly, the explicit recognition of energy access and rural development link at the project/programme level is essential to ensure long-term success.

## Concluding remarks and the way forward

Asia remains one of the two most important regions in the world with a serious energy access challenge. South Asia is the worst affected area in Asia but all countries in the region have made significant progress in enhancing electricity access. However, there are still some challenges to overcome and ensure that people are using the electricity, which in turn is contributing to the economic development. The status in the case of cooking energies is more alarming as more than 1.8 billion people still lack access to clean cooking energies. Significant efforts will thus be required to achieve the target of universal energy access by 2030 and sustainable use of energy.

Asian countries can learn from each other in ensuring sustainable energy access for all. The countries lagging can learn lessons from countries that have made excellent progress in electrification and cross-learning can help eradicate the electricity access problem. Co-operation and experience sharing along with developing consumer friendly techno-institutional-financial models in clean cooking energy provision is also required to address the challenge.

Lack of access to energy holds the countries back in terms of economic and social progress and Asian countries, especially countries in South Asia, need to overcome the problem to achieve a more equitable and durable development.

## Notes

1 The definition given here is as follows: "Sustainable development is development that meets the needs of the present without compromising the ability of future generations to meet their own needs" (WCED, 1987).
2 In this chapter, Asia does not include the Pacific. See Chapter 1 for details about the regional coverage.

## References

ADB, 2015. *Sustainable Energy for All – Tracking Progress in Asia and the Pacific: A Summary Report*, Manila: Asian Development Bank.
ADB, 2016. *Achieving Universal Electricity Access in Indonesia*, Mandaluyong City, Philippines: Asian Development Bank.
AEPC, 2016. *National Rural and Renewable Energy Programme (NRREP) and Biogas*. [Online] Available at: www.aepc.gov.np/ [Accessed 18 July 2016].
Anon., 2014. *Evaluation Report on Rajiv Gandhi Grameen Vidyutikaran Yojana (RGGVY)*, New Delhi: Planning Commission, Government of India.
Anon., 2016. *Facilitating and Promoting Sustainable Cooking Practices*. [Online] Available at: www.geres.eu/en/dissemination-of-improved-domestic-cookstoves-in-cambodia#action-description [Accessed 15 June 2016].
Bajgain, S. & Shakya, I., 2005. *The Nepal Biogas Support Program: A Successful Model of Public Private Partnership for Rural Household Energy Supply*, Netherlands: SNV – Netherlands Development Organisation, Biogas Sector Partnership – Nepal.
Balachandra, P., 2010. *Climate Change Mitigation as a Stimulus for Expanding Rural Energy Access in India*. Report Submitted to the Belfer Center for Science and International Affairs, Cambridge, MA: Harvard Kennedy School, Harvard University.
Bhattacharyya, S., 2006. Energy access problem of the poor in India: Is rural electrification a remedy? *Energy Policy*, 34, pp. 3387–3397.
Bhattacharyya, S., 2012. Energy access programmes and sustainable development: A critical review and analysis. *Energy for Sustainable Development*, 16(3), pp. 260–271.

Bhattacharyya, S., 2013. *Rural Electrification through Decentralised Off-Grid Systems in Developing Countries*, London: Springer-Verlag.

Bhattacharyya, S. C. & Ohiare, S., 2012. The Chinese Electricity Access Model for Rural Electrification: Approach, Experience and Lessons for Others. *Energy Policy*, 49, pp. 676–687.

Bie, Z. & Lin, Y., 2015. An overview of rural electrification in China: History, technology and emerging trends. *IEEE Electrification Magazine*, 2015(March), pp. 1–12.

GNESD, 2014. *Renewable Energy Based Rural Electrification: The Mini-Grid Experience from India*, New Delhi: The Energy and Resources Institute (TERI) for the Global Network on Energy for Sustainable Development (GNESD).

GNESD, 2016. *The New Lao Cook Stove Project*. [Online] Available at: http://energy-access.gnesd.org/cases/36-the-new-lao-cook-stove-project.html [Accessed 16 June 2016].

GTF, 2015. *Progress Towards Sustainable Energy: Global Tracking Framework Summary Report*, Vienna: Sustainable Energy for All.

International Energy Agency and the World Bank, 2015. *Sustainable Energy for All 2015—Progress Toward Sustainable Energy*, Washington, DC: The World Bank.

International Energy Agency, 2015. *World Energy Outlook 2015*, Paris: International Energy Agency.

IRENA, 2015. *Accelerating Off-Grid Renewable Energy: IOREC 2014, Key Findings and Recommendations*, Abu Dhabi: International Renewable Energy Agency.

Larsen, T., Ackom, E. & Mackenzie, G., 2016. *Sustaining Energy Access: Lessons from Energy Plus Approach and Productive Use in Developing Countries; Summary for Policymakers*, Copenhagen: Global Network on Energy for Sustainable Development (GNESD).

MNRE, 2016. *National Biogas and Manure Management Programme*. [Online] Available at: http://mnre.gov.in/schemes/decentralized-systems/schems-2/ [Accessed 16 June 2016].

Palit, D., 2013. Solar energy programs for rural electrification: Experiences and lessons from South Asia. *Energy for Sustainable Development*, 17(3), pp. 270–279.

Palit, D. & Bhattacharyya, S., 2014. Adoption of cleaner cookstoves: Barriers and way forward. *Boiling Point*, 64(64), pp. 6–9.

Palit, D., Bhattacharya, S. & Chaurey, A., 2014. Indian Approaches to Energy Access. In Bhattacharyya, S. C. and Palit, D. (eds.) *Energy PovertyGlobal Challenges and Local Solutions*, New York: Oxford University Press.

Palit, D. & Chaurey, A., 2011. Off-grid rural electrification experiences from South Asia: Status and best practices. *Energy for Sustainable Development*, 15, pp. 266–276.

Peng, W. & Pan, J., 2006. Rural electrification in China: History and institutions. *China and World Economy*, 14(1), pp. 71–84.

Rai, N., Best, S. & Soanes, M., 2016. *Unlocking Climate Finance for Decentralised Energy Access*, London: International Institute for Environment and Development.

Sadeque, Z., Rysankova, D., Elahi, R. & Soni, R., 2014. Scaling up access to electricity: The case of Bangladesh. *Live Wire*. http://documents.worldbank.org/curated/en/699721468003918010/pdf/88702-REPF-BRI-PUBLIC-Box385194B-ADD-SERIES-Live-wire-knowledge-note-series-LW21-New-a-OKR.pdf (accessed on 17 August 2017).

Sarangi, G. K., Palit, D., Jain, R. & Goswami, A., 2016. Energy Access for All, chapter 12 in Dhar Chakrabarty, P. G. (ed.) *People, Planet and Progress Beyond 2015*, New Delhi: TERI Press.

Sarangi, G. K. *et al.*, 2014. Poverty Amidst Plenty: Renewable Energy Based Mini-Grid Electrification in Nepal. In *Mini-Grids for Rural Electrification of Developing Countries*, New Delhi: Springer.

SE4ALL, 2015. *SE4ALL Advisory Board's Finance Committee Report on Scaling Up Finance for Sustainable Energy Investments*, Vienna: Sustainable Energy for All.

Sovacool, B. & Drupady, I., 2011. *The Radiance of Soura Shakti. Energy Governance Case Study 8*, Singapore: Lee Kyuan Yew School of Public Policy, National University of Singapore.

The World Bank, 2005. *Electricity for All: Options for Increasing Access in Indonesia*, Washington, DC: The World Bank.

Ulsrud, K., Winther, T., Palit, D., Rohracher, H. & Sandgren, H., 2011. The solar transitions research on solar mini-grids in India: Learning from local cases of innovative socio-technical systems. *Energy for Sustainable Development*, 15, pp. 293–303.

United Nations Development Programme, 2013. *Achieving Sustainable Energy for All in the Asia-Pacific*, Bangkok: United Nations Development Programme, Asia Pacific Research Centre.

WCED, 1987. *Our Common Future*, Oxford: Oxford University Press.

World Bank, 2014. *RISE: Readiness for Investment in Sustainable Energy – A Tool for Policymakers*, Washington, DC: The World Bank.

# 5

# INDUSTRIAL ENERGY USE IN ASIA

*Subhes C. Bhattacharyya*

## Introduction

Industry, particularly manufacturing industry, forms an important economic activity in Asia.[1] In the developing countries of Asia, the value addition in the manufacturing industries has increased from $315 billion in 1990 to $2362 billion in 2014 (in constant 2005 dollars) (UNIDO, 2016), recording an average annual growth rate of about 8.8% over this period. This industrial activity was the main driving force of economic growth in the region and in most cases the sector grew more rapidly than the overall economic growth of the region. China's manufacturing industry accounted for more than 33% of its gross domestic product (GDP) in 2013 and generated more than $1100 per person. Manufacturers also made a high contribution to the GDP of Thailand (34%) and South Korea (29%) in 2013. Manufacturing exports constituted the most important exportable goods of these countries and in many cases accounted for more than 90% of their respective export income (see Table 5.1). Because of its increasing importance, employment in the manufacturing sector is growing in developing Asia: between 1990 and 1994, this sector accounted for 13.4% of total employment but between 2010 and 2013, this increased to 15.7% (UNIDO, 2016).

In comparison, 21% of Germany's GDP comes from industry and 88% of its exports came from this sector in 2013. Only 12% of US GDP comes from industry and 75% of its exports originate from the industrial sector. Consequently, industry in Asia is a major final energy user: between 1990 and 2012, the energy demand in the sector almost trebled, recording an annual average growth of 4.5% per year.[2] In 1990, the region consumed 467.5 Mtoe while in 2012, the demand increased to 1238 Mtoe. However, as shown in Figure 5.1, the demand grew more rapidly in the new millennium: between 2002 and 2012, the demand grew at above 7% per year, as opposed to 2% per year between 1990 and 2000. There is a clear sub-regional variation in industrial energy demand: East Asia dominates and has consolidated its domination as the leader in industrial energy use accounting for more than 70% of regional demand in 2012, while South Asia and South-East Asia together account for about a quarter of the demand. The acceleration of industrial energy demand in East Asia after 2002 was mainly responsible for the near doubling of energy demand within a decade.

Like the regional level, the sub-regional composition of industrial energy demand is dominated by a few countries. For example, China is the most important player in East Asia with

*Table 5.1* Importance of industry in selected Asian economies

| Description | Share of manufacturing value added in GDP (2013) % | Per capita manufacturing value added (2005$) | Share of manufacturing export in total export (%) |
|---|---|---|---|
| China | 33 | 1142.6 | 96.6 |
| India | 14 | 161.7 | 83.1 |
| Republic of Korea | 29 | 7180.7 | 97.2 |
| Indonesia | 25 | 451.3 | 60.1 |
| Thailand | 34 | 1168.4 | 88.0 |
| Kazakhstan | 11 | 605.9 | 20.8 |
| Japan | 21 | 7820.7 | 91.8 |

*Data source*: UNIDO (2016).

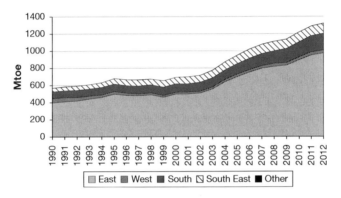

*Figure 5.1* Trend of industrial energy demand in Asia

*Data source*: International Energy Agency (2015).

an 83% share in 2012 while Japan and South Korea contributed another 13% to industrial demand in this sub-region. India accounted for 87% of South Asian industrial demand in 2012. Kazakhstan accounted for 63% of West Asian industrial demand in 2012. Only South-East Asia offered a more diversified picture: Indonesia, Malaysia, Thailand and Vietnam contributed 88% of the industrial demand in the sub-region but Indonesia remains the leader, closely followed by Thailand. Nine countries mentioned above accounted for 85 to 91% of industrial energy demand in Asia between 1990 and 2012 (see Figure 5.2). As the figure shows, China alone contributed 68% of the group's demand (or 61% of the region's industrial demand), and has registered a rapid growth since 2002. This chapter focuses on these countries in some detail below.

In terms of fuel mix, coal is the dominant fuel in the Asian industrial sector. In 1990, coal had a 51% share but this has reduced over time and by 2012, only 46% of energy used in the industrial sector came from coal. In physical terms, coal consumption in the sector has increased from 291 Mtoe in 1990 to 610 Mtoe in 2012, recording an overall annual growth of 3.4%. Natural gas has emerged as the fastest growing energy in the industrial sector, recording an annual average growth of above 7% during this period; but because natural gas started from a low base, despite a fourfold growth in demand, its share in the overall industrial energy mix remains low. Electricity, on the other hand, has recorded a significant improvement in terms of growth rate and

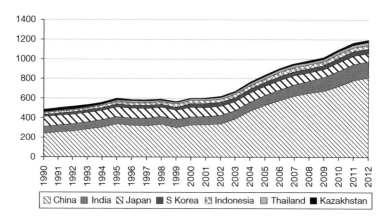

*Figure 5.2*   Major contributors to industrial energy demand in Asia

*Data source*: International Energy Agency (2015).

share: in 1990, electricity had a share of 17%, but by 2012 its share had increased to 28% (see Figure 5.3). The volume of electricity consumption increased more than fourfold during this period, recording an annual demand growth of above 6%. Oil comes third in terms of importance but other energies (biomass and other renewable energies) continue to play some role in this region.

There is some variation in industrial sector fuel mix at the sub-regional level. For example, 50% of industrial sector demand in East Asia in 2012 was met by coal, with electricity satisfying another 31%. The fuel mix is more diversified in South Asia where coal supplied 42% of industrial final demand, electricity 19%, other energies 18%, oil 11% and natural gas 10% (see Figure 5.4). The resource availability and the nature of industrial activities seem to influence the fuel choice to a large extent.

## Composition of industrial energy demand in selected Asian countries

Given the relative importance of seven countries in Asian industrial demand, this section analyses the industrial energy demand of each of these countries considering different industry categories. This also allows development of a comparative position across these countries.

### *China*

As indicated earlier, the industrial sector has been the engine of growth in China and the leading economic activity, accounting for 30% of China's total employment (Ouyang & Lin, 2015). Industrial energy demand in China can be grouped into four major areas, namely the iron and steel industry, the chemical industry, the non-metallic mineral industry and the final category which encompasses all other industries. Industrial demand has undergone a profound change between 1990 and 2012. The non-metallic minerals industry was the most important energy consuming activity in 1990, but the iron and steel industry recorded the highest growth in energy demand between 1990 and 2012 with an average growth rate of 9.2%. As a result, from third position in 1990, the iron and steel industry has emerged as the most energy demanding activity in 2012. Similarly, China is the world leader in cement production and accounted for 51% of global production (4.6 billion tonnes) in 2015. China produced almost nine times more

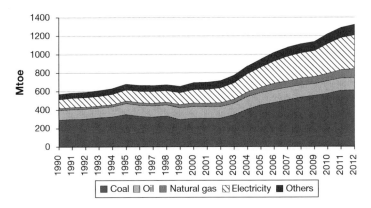

*Figure 5.3* Fuel mix in the industrial sector

*Data source*: International Energy Agency (2015).

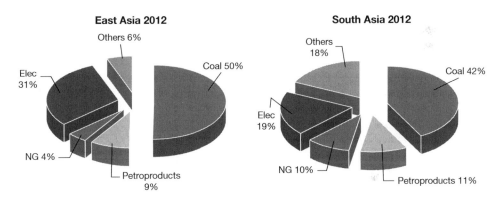

*Figure 5.4* Diversified fuel mix in the industrial sector of Asian sub-regions

*Data source*: Author based on International Energy Agency (2015).

cement than India in 2015. The non–metallic minerals and chemical industries recorded an almost fourfold increase in energy demand during the same period but accelerated demand growth in the iron and steel sector displaced these industries into second and third place respectively (see Figure 5.5).

It is important to mention that China revised its energy balance data in 2010 retrospectively for the period between 1996 and 2008. This revision was based on the Second National Economic Census of 2008, which indicated a significant underestimation of energy consumption at the business and enterprise level. Consequently, coal consumption was revised upward and the annual primary energy consumption was revised upward by 5% on average (Ke et al., 2012).

During this period, the value addition in the manufacturing industry grew 12.5% per year on average, recording a faster growth than the overall economic output. The GDP elasticity of manufacturing energy demand for this period was 0.54, reflecting much slower energy demand growth compared to value addition. Consequently, industrial energy intensity fell rapidly in China while energy demand was growing (see Figure 5.6). A study (Luken & Piras, 2011) noted

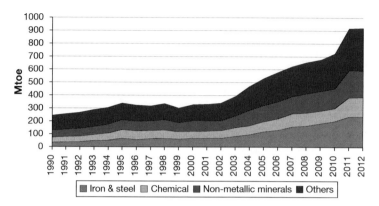

*Figure 5.5* Chinese industrial energy demand by major industry groups

*Data source*: International Energy Agency (2015).

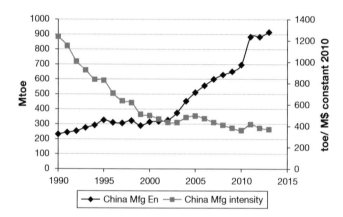

*Figure 5.6* China's industrial energy use and intensity

*Data source*: International Energy Agency (2015).

that various initiatives taken by the Chinese government (namely the Eleventh Plan target of 20% reduction in energy by 2010 compared to 2005 baseline), including energy intensity reduction in ten key projects and a focus on top-1000 energy consuming enterprises, were responsible for the rapid improvement in energy intensity in the country. Price et al. (2010) observed that if the top-1000 enterprise programme continues to realise the energy saving achieved in 2006, it will save 100 tce by 2010.

The index decomposition method has been used to analyse the drivers of change in energy demand. In the additive formulation, the change in energy demand between two periods is broken down into a set of additive components:

$$E = Q.\text{EI} = Q.(E/\text{VA}).(\text{VA}/Q),$$

where $E$ = energy, $Q$ = economic output, VA = sector output, $E/\text{VA}$ = energy intensity, $\text{VA}/Q$ = sectoral share ($S$) in economic output (or structural share).

The Log-Mean Divisia Index (LMDI) approach suggested by (Ang, 2005) uses the following relationships:

$$\Delta E_Q = \frac{(E_T - E_0)}{\ln(E_T) - \ln(E_0)} \times \ln\left(\frac{Q_T}{Q_0}\right)$$

$$\Delta E_{EI} = \frac{(E_T - E_0)}{\ln(E_T) - \ln(E_0)} \times \ln\left(\frac{EI_T}{EI_0}\right)$$

$$\Delta E_S = \frac{(E_T - E_0)}{\ln(E_T) - \ln(E_0)} \times \ln\left(\frac{S_T}{S_0}\right)$$

$$\Delta E = \Delta E_Q + \Delta E_{EI} + \Delta E_S$$

Applying the additive LMDI method to the aggregate industrial demand in China shows that the output effect has significantly contributed to an increase in energy demand but the intensity effect acted in the opposite direction and mitigated the demand growth. The structural effect was not significant in its contribution in most cases (see Figure 5.7). It is also interesting to note that as industrialisation intensified, the improvement in energy intensity slowed down compared to the gains achieved in the last decade of the last century. This therefore highlights the need for further attention on energy efficiency improvements in the sector.

## India

Manufacturing sector output constitutes about 12 to 14% of overall economic output in India and its share has remained at this level over the past two decades. Compared to China, India has a much lower level of industrial activity.

The industrial energy consumption in India can be grouped under five activities: non-specified activities appear as the most important energy user (which may be due to data reporting issues); the iron and steel industry is otherwise the most important energy user. This is followed by the non-metallic minerals and chemicals industries respectively. The steel industry recorded the highest growth in energy demand, averaging above 7% per year between 1990 and 2012.

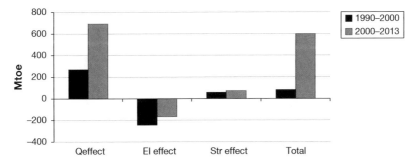

*Figure 5.7*   Decomposition of aggregate industrial energy demand in China

*Data source*: Author based on International Energy Agency (2015) and UNIDO (2016).

India is the fourth largest steel producer in the world and produced 81 million tonnes in 2013, recording a 5% growth over its 2012 production. Increased steel output was responsible for the energy demand growth in this sector. India is also the second largest producer of cement in the world with a 5.9% share in the global production in 2015. But compared to China, India's steel and cement outputs are much smaller. The non-metallic industry comes second in terms of energy demand growth with above 6% growth during the period. Other activities recorded a modest demand growth – between 2.5% and 4% per year (see Figure 5.8).

While both India and China have seen a rapid growth in energy demand for their respective iron and steel industries, China's energy demand in this sector was five times higher than that of India in 2012. According to the World Steel Association, China produced 731 Mt of crude steel in 2012, and was ranked first in the world. India produced 77 Mt in the same year (World Steel Association, 2014). It is natural that with a smaller steel industry, India's energy demand will be less. However, the energy used per ton of crude steel production appears to be higher in China than in India: this results from the fact that 90% of China's crude steel was produced using an oxygen converter process and about 9% using an electric arc furnace. On the other hand, 68% of India's steel was produced using an electric arc furnace (World Steel Association, 2014). Energy use per unit of steel-making in an electric arc furnace is less, making the process energy efficient.

India also shows much slower energy efficiency improvement compared with China. In fact, India's industrial energy intensity has stagnated in recent times even showing signs of an increasing trend (see Figure 5.9). As a result, India's manufacturing energy intensity was almost twice that of China in 2013, which supports its increasing energy demand trend as well. This is also reflected in the GDP elasticity of industry energy demand: India in recent times had recorded elasticity greater than one, thus requiring more energy per unit of economic growth.

Using the LMDI approach, the decomposition analysis of changes in India's industrial energy demand suggests that all three elements played some role in the overall energy demand. But the activity effect has dominated the industrial energy demand over the past decade. The intensity effect has slowed down considerably compared to the last decade of the twentieth century (see Figure 5.10). The rising intensity trend shown in Figure 5.9 supports this observation. The decomposition on a yearly basis shows considerable fluctuations in terms of the contribution of each effect but the output or activity effect was always positive while the intensity effect was negatively influencing the demand in most years until 2007. Reddy and Ray (2010) found a

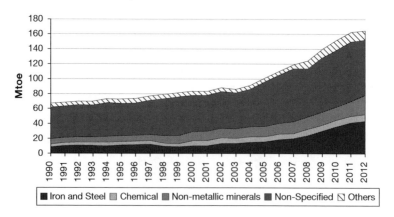

*Figure 5.8*   Indian industrial energy demand by activities

*Data source*: International Energy Agency (2015).

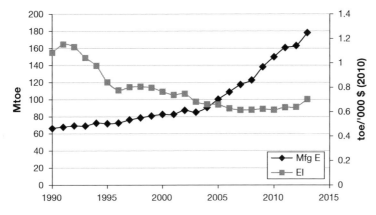

*Figure 5.9*  India's industrial energy demand and intensity

*Data source*: Author based on International Energy Agency (2015) and UNIDO (2016).

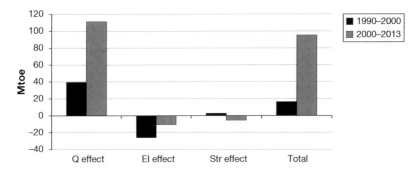

*Figure 5.10*  Decomposition of India's industrial energy demand

*Data source*: Author based on International Energy Agency (2015) and UNIDO (2016).

similar pattern of effects in their study. India has initiated significant energy efficiency improvement policies including Perform, Achieve and Trade (PAT) and it is expected to reduce energy intensity in due course.

## Japan

About 18% of Japan's GDP is generated by manufacturing activities. Japan is seen as the role model of Asian industrialisation and many countries in the region have been influenced by its strategies, policies and overseas investments.

Japanese energy use in industry can be effectively grouped into five sub-activities: the iron and steel industry, the chemical and petrochemical industries form two major components, non-metallic minerals and machinery are two substantial energy users and the rest is dispersed across a large number of smaller activities including some non-specified. However, miscellaneous industries constitute the most significant energy demand in the country, although this demand is showing a downward trend (see Figure 5.11). Japan is the world's second largest producer of steel after China. In 2012, Japan produced 107 Mt crude steel which is almost one-seventh of China's production. But China consumed 12 times more energy than Japan in producing its steel.

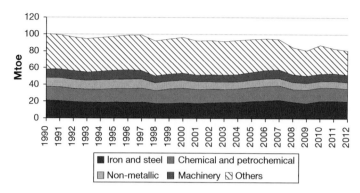

*Figure 5.11*   Industrial energy demand by sub-sector in Japan

*Data source*: International Energy Agency (2015).

This can be attributed to better technical efficiency and differences in the steelmaking processes between the two countries. Japan produced 77% of its steel using the oxygen process and the remaining using electric arc furnace process. Accordingly, China follows in Japanese footsteps in steelmaking but its processes are not as efficient as Japan's.

Unlike India and China, where industrial energy demand is growing, Japan shows a declining energy demand trend despite the growing manufacturing value addition. Japan has managed to sustain energy efficiency improvements over a long time and between 1990 and 2013 the country achieved an annual reduction of 2.1% of energy intensity on average (see Figure 5.12). Rapid improvements in energy efficiency have contributed to a reduction in industrial demand. This has also helped reduce GDP elasticity of demand to 0.36.

Figure 5.12 also indicates that change in industrial energy demand in Japan was less dramatic than in China or India. The decomposition analysis confirms that the intensity effect was the prominent factor behind the change in industrial energy demand (see Figure 5.13). In fact, Japan has intensified its energy efficiency improvements in recent times which contributed a reduction of 30 Mtoe between 2000 and 2013. The structural effect on the other hand has contributed positively to energy demand, reflecting the improvement in the industrial sector's contribution to overall economic output.

## South Korea

Industrial value added in South Korea increased at 7% per year on average between 1990 and 2013 and manufacturing activities accounted for 26.5% of the country's economic output in 2013.

Like Japan, South Korea's industrial energy demand can be grouped into five activities: iron and steel production, the chemical industry, non-metallic mineral production, machinery and a catch-all group called "others". It is the latter sector, "others", which consumes most energy – showing the diversity of industrial activities in the country (see Figure 5.14). Four other sectors namely iron and steel, chemicals, non-metallic minerals and machinery accounted for about two-thirds of industrial energy demand in 2012. The iron and steel industry consumed about 23% of industrial energy demand, the chemical industry 18%, while the other two activities contributed 25%.

Similar to China and India, South Korea also shows an increasing energy demand trend in the manufacturing sector but the impacts of the financial crises of 1997 and 2008 are clearly visible. The rapid growth in industrial energy demand in the 1990s came to a halt after the financial crisis

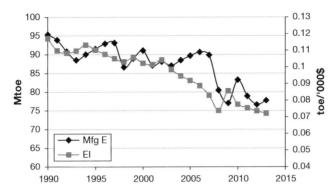

Figure 5.12 Industrial energy use and energy intensity trend in Japan

Data source: Author based on International Energy Agency (2015) and UNIDO (2016).

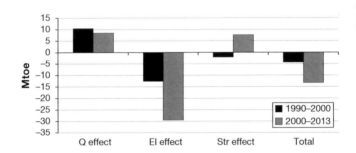

Figure 5.13 Decomposition of industrial energy demand in Japan

Data source: Author based on International Energy Agency (2015) and UNIDO (2016).

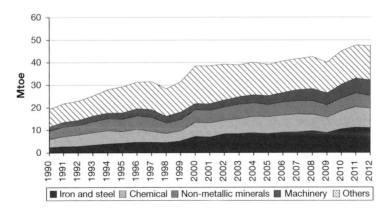

Figure 5.14 Korea's industrial energy demand by activities

Data source: International Energy Agency (2015).

of 1997 and even after the recovery, the demand grew less rapidly. This was due to a rapid fall in energy intensity during this period and the efforts towards energy efficiency improvements did not slow down due to the economic crisis of 1997. However, the same cannot be said about the economic crisis of 2008. Energy efficiency improvements in the industrial sector appear to have stagnated and, consequently, energy demand has started to grow rapidly from 2010 onwards (see Figure 5.15).

Whereas Japan has achieved an industrial energy intensity of 0.07 toe/000$ in the industrial sector, South Korea still needs almost double the amount of energy per unit of value addition. This clearly shows that despite the improvements in energy efficiency, South Korea may still have further potential to reduce its industrial energy needs following the Japanese model. In terms of GDP elasticity of energy demand in this sector, South Korea has recorded an elasticity of 0.5, indicating the decoupling of energy demand growth from GDP growth; but compared to Japan, South Korea still has a higher elasticity.

The decomposition of energy intensity in the manufacturing industry clearly shows the influence of various factors during different periods (see Figure 5.16). In the 1990s, when energy demand was growing fast, the activity effect was clearly the main driver. The role of energy efficiency and structural change in the industry was less visible during this period. However, the

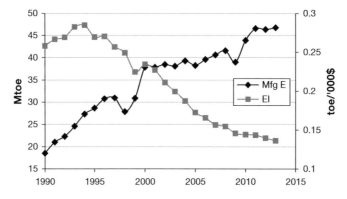

*Figure 5.15*   Industrial energy demand and energy intensity trends of South Korea

*Data source*: Author based on International Energy Agency (2015) and UNIDO (2016).

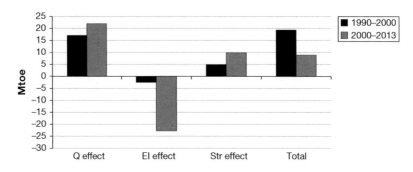

*Figure 5.16*   Decomposition of industrial energy demand change in Korea

*Data source*: Author based on International Energy Agency (2015) and UNIDO (2016).

importance of factors changed after the financial crisis of 1997. The structural and activity effects contributed to an increase in energy demand and both factors played relatively important roles in this respect. The high contribution of structural effect is clearly visible during the second period, which reflects the growing importance of manufacturing industry in the economy. On the other hand, energy intensity effect has mitigated the demand growth by producing a strong counter-balancing effect.

A study by Oh et al. (2010) indicated that the replacement of blast furnaces by electric furnaces and the displacement of coal in the cement industry has improved the energy intensity of energy intensive industries in the late 1990s. Moreover, the structural change brought by the faster growth of less energy intensive industries (such as car production, machinery production and the electronics industry) compared with the energy-intensive ones (such as cement or steel) has significantly improved the overall intensity of the manufacturing sector.

### Indonesia

Indonesia's industrial value addition grew by 5.7% per year on average (constant 2010 dollar terms) between 1990 and 2013. This has resulted in a moderate improvement in the contribution of industry to the overall economic output (from about 21% in 1990 to about 25% in 2013).

Indonesia's industrial sector consumes the most energy in South-East Asia but compared to other major Asian players, Indonesia has a relatively small industrial sector. Most of the industrial energy (about 66%) is consumed in non-specified activities, indicating weak data availability. Non-metallic mineral production accounts for about 12% of energy demand and another 10% is consumed by other industries. The rest is used by the chemical industry and textile manufacturing, with textiles taking a minor share. There was a relatively modest growth in demand after the financial crisis in 1997 and it took almost a decade to regain the momentum of industrial demand growth (Figure 5.17).

The GDP elasticity of industrial energy demand in Indonesia for this above period is 0.75 – this suggests that for every 1% growth in industrial value addition, energy demand in industry grows by 0.75%. This is almost double the rate of Japan and 50% higher than South Korea. Industrial energy demand has been affected by the economic and political crises faced by Indonesia during

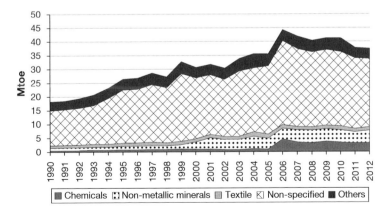

*Figure 5.17*  Indonesia's industrial demand trend

*Data source*: International Energy Agency (2015).

the period of study. Prior to 1997, the demand growth was about 7% per year but the financial crisis resulted in a fall in energy demand in 1998. The political instability in the country led to a volatile energy demand in the early part of the new millennium. Although the financial crisis of 2008 did not affect economic output significantly, energy demand has started to show a declining trend, with some stability emerging in recent times. This fall in demand appears to be driven by a significant reduction in energy intensity (see Figure 5.18), making Indonesia's manufacturing energy intensity quite comparable with that of South Korea.

The results of the decomposition analysis show that the activity effect has positively influenced energy demand but that the effect was stronger during the first decade of the new millennium. Energy intensity on the other hand has positively influenced energy demand, creating a significant improvement in energy efficiency in the industrial sector which is clearly visible in the second period. The influence of structural change was less strong compared to the other two effects but a clear pattern before and after the 1997 crisis is clearly visible. Prior to the 1997 crisis, the structure was supporting energy demand growth, implying an increased emphasis on energy-demanding activities. Post crisis, the structure contributed to a reduction in energy demand, thus promoting less energy-intensive activities (see Figure 5.19). The reduction in energy intensity and a shift towards less intensive use of energy in industry has resulted in a much slower growth in energy demand in the sector in recent times.

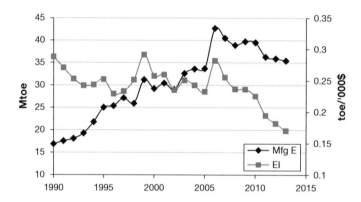

*Figure 5.18*   Indonesia's industrial energy demand and energy intensity

*Data source*: Author based on International Energy Agency (2015) and UNIDO (2016).

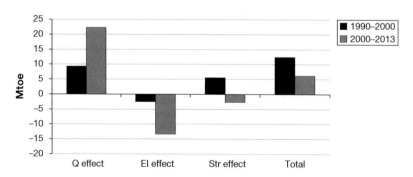

*Figure 5.19*   Decomposition of changes in industrial energy demand in Indonesia

*Data source*: Author based on International Energy Agency (2015) and UNIDO (2016).

## *Thailand*

The industrial value addition grew almost 5% per year on average in Thailand between 1990 and 2013, but the manufacturing sector's share did not change much in constant dollar terms. The manufacturing sector's share of Thailand's GDP was 25.6% in 1990 and increased to 31% in 2010 but reduced to about 29% in 2013.

Thailand's industrial energy demand grew at 5.6% per year on average, recording a threefold increase between 1990 and 2013. While the energy demand grew steadily, the effects of the economic crises of 1997 and 2008 were clearly visible. The composition of industrial energy demand in Thailand is very different from other Asian countries. While non-metallic mineral production has emerged as a major player in 2012, it was the food industry in the 1990s that dominated industrial demand. The food industry continues to be a major energy user in the country. The balance of energy demand is shared by various industries but no single industry plays any significant role (see Figure 5.20).

The economic crises have shifted the demand trend to a lower trajectory but Thailand has managed to retain relatively high industrial energy demand growth (see Figure 5.21). In contrast to most of the other countries considered in this chapter, Thailand did not really benefit from any significant energy efficiency gains. In fact, in the 1990s, the industrial energy intensity grew rapidly until the economic crisis of 1997. After the crisis, a reduction in intensity was visible but this did not continue for long and energy intensity stagnated until the next crisis of 2007/08. Since 2010, energy intensity has started to increase rapidly again due to the changing structure of industrial activities in the country.

A study by Chontanawat et al. (2014) notes that the economic and trade liberalisations of 1980 led to the rapid industrial growth of Thailand, but the financial sector reforms of the 1990s did not greatly benefit the manufacturing sector. However, Thailand's labour-intensive and export-oriented manufacturing activities helped the country's economic recovery after the 1997 crisis. In terms of industrial structure, the share of low-intensity activities (such as food and textile) in the industrial output declined over time while that of energy-intensive activities (such as the chemical industry, basic metal production and the manufacture of fabricated materials) increased. The structural change has resulted in an increase in industrial sector energy intensity.

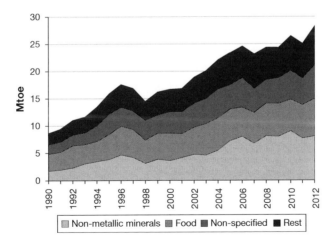

*Figure 5.20*   Trend of industrial energy demand in Thailand

*Data source*: International Energy Agency (2015).

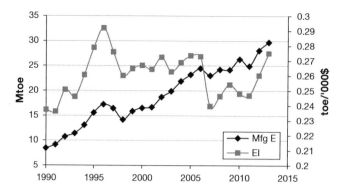

*Figure 5.21* Industrial energy demand and energy intensity trend of Thailand

*Data source*: Author based on International Energy Agency (2015) and UNIDO (2016).

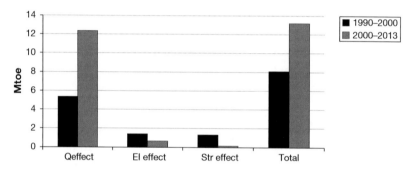

*Figure 5.22* Decomposition of changes in industrial energy demand in Thailand

*Data source*: Author based on International Energy Agency (2015) and UNIDO (2016).

The decomposition of changes in industrial energy demand confirms these observations (see Figure 5.22). The activity effect is the most dominant driver of industrial energy demand in Thailand and this has become a more influential factor in the past decade. The intensity effect and structural effect have also contributed to energy demand growth in this sector but their contribution remains limited compared to the activity effect. The drivers for Thailand were very different compared to the other countries considered in this chapter and there exist significant opportunities to reduce industrial energy demand through energy efficiency improvements. The shift towards energy-intensive industrial activities has, quite clearly, affected the energy demand of this sector.

## Kazakhstan

Kazakhstan, has, on the other hand, registered a very different industrial demand pattern compared to the other countries examined in this chapter. The rapid decline in industrial energy demand, arising from the economic crisis due to the collapse of the Soviet Union, only started to reverse in the new millennium (Figure 5.23). Non-specified energy activities remained the most important demand user in the country. However, the demand demonstrated peaks around 2005 and 2007 and the clear effect of the global economic recession of 2008. The fluctuating demand pattern perhaps reflects the fragile economic condition of the country.

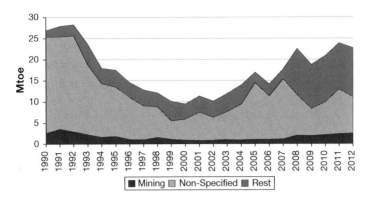

*Figure 5.23*  Industrial energy demand trend in Kazakhstan

*Data source*: International Energy Agency (2015).

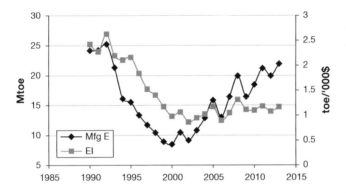

*Figure 5.24*  Industrial energy demand and energy intensity in Kazakhstan

*Data source*: Author based on International Energy Agency (2015) and UNIDO (2016).

The industrial energy intensity in Kazakhstan appears to follow a similar pattern as sectoral demand. When industrial demand fell rapidly, energy intensity followed suit but when industrial demand started to rise, energy intensity started to stagnate and remained practically unchanged (see Figure 5.24). While the current level of energy intensity is almost one-half of that of the 1990s, there is still further scope for improvement as Kazakhstan consumes almost 15 times more energy in industry than Japan for every dollar of output.

A study by Gomez et al. (2014) indicates that Kazakhstan has one of the highest energy intensities in the world. It suggests that the iron and steel industry and non-metallic minerals industry are comparatively inefficient (40% higher compared to international benchmarks) and offer significant energy saving potential. The study estimates that the industrial sector can reduce energy demand by more than 30% compared to its current demand.

The decomposition analysis of changes in industrial energy demand shows that both activity effect and intensity effect contributed to demand reduction in the 1990s when the country was passing through an economic recession. In the new millennium, the situation reversed and both these factors contributed to energy demand growth, although the activity effect was the dominant influence (see Figure 5.25). The structural effect contributed positively to energy demand in the

1990s but there appears to have been a structural shift in recent times which has contributed towards demand reduction.

The energy use in manufacturing industries throughout all the countries discussed in this chapter accounted for more than 90% of the industrial energy demand, with construction and mining activities playing a minor role. The energy intensity in the manufacturing sector, however, shows a significant variation. On one hand, Japan has successfully reduced its intensity from its already low level of 117 toe/Million $ (constant 2010) in 1990 to 73 toe/Million $ in 2012. Japan's manufacturing industry is the lowest energy-intensive sector in the region and sets the benchmark for energy-efficiency aspiration for all countries in the region. On the other hand, Kazakhstan has the most energy-intensive manufacturing sector in the region. While energy intensity has fallen significantly over the period, from over 2400 toe/Million $ in 1990 to close to 1100 toe/Million $ in 2012, its manufacturing energy intensity remains almost 15 times higher than that of Japan. South Korea comes closer to Japan in terms of energy intensity in the manufacturing sector but with an energy intensity of 140 toe/Million $ in 2012, manufacturing industry in South Korea consumes twice as much energy as that of Japan. China and India, two major industrial energy users of the region, also have significant potential for improvements in energy efficiency. While both the countries had similar levels of energy efficiency in the 1990s,

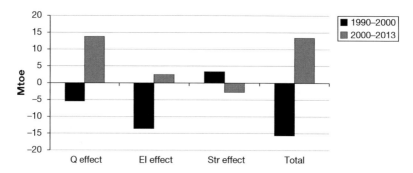

*Figure 5.25*  Decomposition analysis of changes in industrial energy demand in Kazakhstan

*Data source*: Author based on International Energy Agency (2015) and UNIDO (2016).

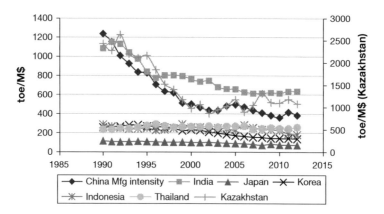

*Figure 5.26*  Energy intensity in the manufacturing industry of selected Asian countries

*Data source*: Author based on International Energy Agency (2015) and UNIDO (2016).

China had achieved a faster rate of improvement than India and by 2012, China's energy intensity in the manufacturing sector was almost one-half of that of India. However, China still consumes five times more energy than Japan for every dollar of output. The performance of South-East Asian countries like Indonesia and Thailand is relatively better compared to other regions but they can still learn from the experience of Japan or South Korea and reduce their energy requirement (See Figure 5.26).

# Conclusions

Industry plays an important role in Asia and accounted for about 38% of final energy demand in Asia in 2012. East Asia dominated industrial energy demand and consolidated its position over other sub-regions. There exists significant variation in the region in terms of industrial development, with strong sub-regional players. The review of industrial energy demand in the region shows that:

- The GDP elasticity of energy demand in developing Asia remains high and as a result, faster economic growth drives additional energy demand. Japan, however, has successfully decoupled energy demand from economic activities and achieved a low GDP elasticity, with a declining energy demand trend in its industries.
- Except in a few cases, the energy intensity in industry has shown a declining trend, but the rate of improvement varies significantly. Japan has achieved the best energy performance in the region despite relying on heavy industries. Technological change has played an important role in improving energy efficiency of the region's industries. For example, a gradual shift towards electric furnaces and recycled steel has reduced energy requirements in the steel industries. There is significant scope for the region's countries to learn from one another in order to improve energy efficiency in the industry.
- The composition of industry has evolved over time in the region. Countries which have successfully reduced energy demand in industry have managed a transition process to less energy-intensive activities. The recent focus on the green economy offers opportunities to integrate sustainable practices in the sector, which can contribute to energy demand reduction.
- Industries in most of the countries have suffered from the Asian financial crisis of 1997 and the global economic crisis of 2008. In general, the activity effect contributed to an energy demand increase except during the recession periods, while energy intensity has supported demand reduction. However, the drivers were not uniform for all countries and specific country conditions affected industrial demand growth.

Decoupling industrial energy demand from economic growth remains an important issue for Asia in its drive towards a sustainable future. Countries performing relatively better in this respect have taken initiatives to manage their industrial energy demand. The structural change in the industry is also an important element in this process and a transition to less energy-intensive activities and the promotion of green energies will improve sector performance.

As performance varies across the region, and, given that countries are at different stages of development, there are significant learning opportunities available from experiences across the region. Similarly, as Chinese economic growth slows down, it is possible that some industrial activities will relocate to other neighbouring countries in the region. Their energy demand will perhaps grow faster, but if they follow regional best practices in terms of technology, they may be able to better manage the industrialisation process.

## Notes

1  This chapter follows the same regional definition as indicated in Chapter 2.
2  Energy data used in this chapter comes from International Energy Agency (2015).

## References

Ang, B., 2005. The LMDI approach to decomposition analysis: a practical guide. *Energy Policy*, 33, pp. 867–871.

Chontanawat, J., Waiboonchutikula, P. & Buddhivanich, A., 2014. Decomposition analysis of the change of energy intensity of manufacturing industries in Thailand. *Energy*, 77, pp. 171–182.

Gomez, A., Dopazo, C. & Fueyo, N., 2014. The causes of the high energy intensity of the Kazakh economy: A characterization of its energy system. *Energy*, 71(July), pp. 556–568.

International Energy Agency, 2015. *World Energy Balance (via UK Data Service)*, Paris: International Energy Agency.

Ke, J. et al., 2012. China's industrial energy consumption trends and impacts of the Top-1000 Enterprises Energy-Saving Program and the Ten Key Energy-Saving Projects. *Energy Policy*, 50, pp. 562–569.

Luken, R. & Piras, S., 2011. A critical overview of industrial energy decoupling programs in six developing countries in Asia. *Energy Policy*, 39(6), pp. 3869–3872.

Oh, I., Wehrmeyer, W. & Mulugetta, Y., 2010. Decomposition analysis and mitigation strategies of CO2 emissions from energy consumption in South Korea. *Energy Policy*, 38(1), pp. 364–377.

Ouyang, X. & Lin, B., 2015. An analysis of the driving forces of energy-related carbon dioxide emissions in China's industrial sector. *Renewable and Sustainable Energy Reviews*, 45, pp. 838–849.

Price, L., Wang, X. & Yun, J., 2010. The challenge of reducing energy consumption of the Top-1000 largest industrial enterprises in China. *Energy Policy*, 38, pp. 6485–6498.

Reddy, B. & Ray, B., 2010. Decomposition of energy consumption and energy intensity in Indian manufacturing industries. *Energy for Sustainable Development*, 14(1), pp. 35–47.

UNIDO, 2016. *Industrial Development Report 2016*, Geneva: United Nations Industrial Development Organisation.

World Steel Association, 2014. *World Steel in Figures 2014*, Brussels: World Steel Association.

# 6

# TRANSPORTATION ENERGY DEMAND IN ASIA

## Status, trends, and drivers

*Govinda R. Timilsina and Ashish Shrestha*

### Introduction

Total final consumption of energy in Asia increased from 1,136 million tons of oil equivalent (Mtoe) in 1980 to 3,606 Mtoe in 2013, with a robust average annual growth rate of 3.6% (IEA, 2016a). The transport sector remains one of the most important energy consumption sectors in Asia, and its share of total regional energy demand has grown from about 11% to 16.5% over these 33 years, as shown in Figure 6.1.

This has happened despite the fact that China[1] and India, who account for almost 70% of the region's energy demand, feature a relatively low share of energy consumption from the transport sector (13.3 and 14.2%, respectively, in 2013) in their national energy demand, thus skewing the regional aggregates. Nevertheless, since rising incomes are associated with higher levels of car ownership and usage (Webster et al., 1986a, b) and greater trip rates and distances (Schäfer, 2000), transport activity and resulting energy demand could increase significantly in these countries along with economic growth and consumers' purchasing power. In most other Asian countries, the transport sector already accounts for a substantial share of total national energy demand (about 23% for Asia, other than China and India). Therefore, any attempt to address climate change in Asia must pay attention to transport sector energy demand. The identification of key factors driving transport energy demand and $CO_2$ emissions is essential for the formulation of effective climate change mitigation policies and strategies. One approach to accomplish this objective is to decompose the growth of energy demand into the possible affecting factors.

Most existing studies are focused on the decomposition of national $CO_2$ emissions and emission intensities. Examples include Wu et al. (2005) and Wang et al. (2005) for China, Kawase et al. (2006) for Japan, Rhee and Chung (2006) for Japan and South Korea; Lise (2006) for Turkey, Diakoulaki et al. (2006) for Greece, Saikku et al. (2008) for 27 EU member States, Lee and Oh (2006) for APEC countries, Luukkanen and Kaivo-oja (2002a) for ASEAN countries; Luukkanen and Kaivo-oja (2002b) for Scandinavian countries, Ebohon and Ikeme (2006) for sub-Saharan African countries, and Han and Chatterjee (1997) for nine developing countries (Brazil, Chile, Colombia, India, Korea, Mexico, Philippines, Thailand and Zambia). Some existing studies are focused on the decomposition of manufacturing and power sector $CO_2$ emissions or emission intensities (Liu et al. (2007),

81

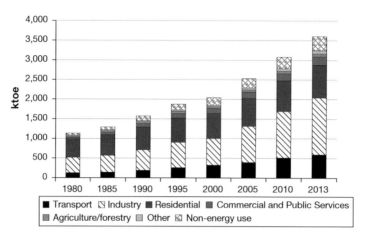

*Figure 6.1*  Total final consumption and energy sector demand growth in Asia

*Data source*: IEA (2016a).

Yabe (2004), Liaskas et al. (2000), Schipper et al. (2001), Chang and Lin (1998) and Bhattacharyya and Ussanarassamee (2004), Shrestha and Timilsina (1996), Nag and Kulshrestha (2000) and Shrestha and Marpuang (2006)). While the factors affecting $CO_2$ emissions and emission intensities of the industry and power sectors have been analysed in many countries, the transport sector has not been examined to the same extent, especially in developing countries.

Nevertheless, a few studies also examine factors affecting transport sector emissions growth. For example, Lakshmanan and Han (1997) attribute the change in transport sector $CO_2$ emissions in the US between 1970 and 1991 to growth in people's propensity to travel, population, and gross domestic product (GDP). Lu et al. (2007) decompose changes in $CO_2$ emissions from highway vehicles in Germany, Japan, South Korea and Taiwan during 1990–2002 into changes in emission coefficient, vehicle fuel intensity, vehicle ownership, population intensity and economic growth. Scholl et al. (1996) calculate how changes in transport activity, modal structure, $CO_2$ intensity, energy intensity and fuel mix affect $CO_2$ emissions from passenger transport in nine OECD countries between 1973 and 1992. Similarly, Schipper et al. (1997) identify the relative contribution of activity, modal structure, and energy intensity to changes in energy use and carbon emissions from freight transport in ten industrialized countries from 1973 to 1992. Schipper et al. (2000) attribute transport sector $CO_2$ emission growth to transportation activity, modal structure, modal energy intensity and fuel mix. Kveiborg and Fosgerau (2007) decompose the historical growth in national Danish road freight traffic using a Divisia index decomposition method. Wu et al. (2005) consider changes in transport energy intensity, average traveling distance, and number of vehicles (amongst numerous other factors) in their investigation of the underlying forces behind the stagnancy of China's energy-related $CO_2$ emissions from 1996 to 1999. Finally, Timilsina and Shrestha (2009) decompose $CO_2$ emissions growth in 20 Latin American and Caribbean countries into components associated with changes in fuel mix, modal mix, emission coefficients and transportation energy intensity, along with economic growth.

Understanding the factors affecting energy demand growth in the transport sector is critical because of its increasing prominence as a source of emissions in most countries and its relevance to the preparation of climate change mitigation strategies. Hence, this chapter aims to address this gap by executing a Divisia decomposition analysis of $CO_2$ emissions from the transport sector in 14 Asian countries during the 1980–2013 period. We attribute the growth of transport sector $CO_2$ emissions over these 33 years to five factors. These are: (i) fuel switching, (ii) modal shifting, (iii) sectoral energy intensity change, (iv) per capita economic growth and (v) population growth. Among these, three factors – change in transport sector energy intensity, and per capita GDP and population growth – are found primarily responsible for driving transport sector $CO_2$ emissions in Asia.

This chapter is organized as follows: the second section presents trends of transportation energy demand and that of potential factors driving the demand over the last 33 years. The roles of different factors are then quantified using an identity approach in the third section. The fourth section draws key conclusions.

## Potential factors driving the transport sector energy demand growth

Before discussing potential factors driving transport sector energy consumption growth, we first highlight the trend of energy consumption in selected Asian countries. This is followed by a discussion of direct factors, such as fuel switching, modal shifting and changes in transportation energy intensity. Moreover, we analyse some trends, such as population growth and urbanization, and economic growth and motorization, which provide further insights on the causes of transport sector energy consumption growth.

### *Energy demand*

Figure 6.2 presents the trend of transport sector energy demand in the 14 Asian countries that are responsible for more than 95% of the total energy demand in the region, and for which data is available over the entire study period. Aggregate transport sector energy demand at the regional level rose almost five-fold from 126 Mtoe in 1980 to 593 Mtoe in 2013, with a robust average annual growth rate of 4.8%, higher than the average growth in total final consumption over the same period.

Despite the increase in transport sector energy demand in absolute terms, the share of the sector in the national total in China and India are significantly smaller than in most other Asian countries. Figure 6.3 presents total national energy demand and the sectoral demand mix for the 14 Asian countries. The transport sector share of total national energy demand has historically been relatively high in Japan, Korea,[2] Malaysia and Thailand, and only modest growth is observed over the study period in those countries, along with Myanmar. The transport sector share of total national energy demand increased most substantially in Indonesia and the Philippines, exceeding even the sectoral shares in Japan, Korea and Thailand (but not Malaysia, which still features the highest transport share of energy demand in the region) by 2013. Although significant growth in transport energy demand is also found in all of the remaining countries, the transport sector in a number of countries, particularly Nepal and Myanmar (and also India and China), still does not account for a large share of the total final consumption relative to regional norms.

Because the transport, industry and residential sectors are the main contributors to national energy demand in almost all of the countries considered, changes in the magnitude of the demand from the other two sectors have a considerable impact on the transport sector's share of national energy demand. With a few exceptions, what has typically occurred is that the share of total final consumption from the residential sector has decreased while the share of the transport sector has increased, without much change in the low level of the other final consumption sectors. However, the commercial and public services sector in Japan, Korea, Malaysia and the

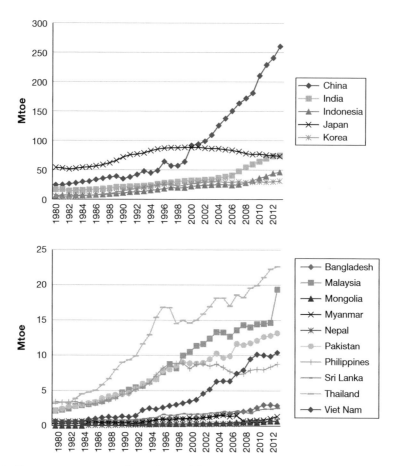

*Figure 6.2*   Transport sector energy demand growth by country

*Data source*: IEA (2016a).

Philippines has grown to become a key consumer of energy. Similarly, non-energy use, where fuels are used as raw materials, such as in the manufacture of bitumen, lubricants, paints, solvents, plastics, adhesives, and fertilizer, instead of being consumed as a source of energy, is an important final consumption sector in a few countries with strong manufacturing bases: Japan, Korea, Malaysia and Thailand.

## Modal mix in the transport sector

One potential factor driving transport sector energy demand growth could be modal shifting, from less energy intensive modes (in terms of ktoe per passenger/freight kilometre), such as railway and water transportation, to more energy intensive modes, such as commercial airplanes and private road vehicles. However, since modal mix data in terms of transportation services are available only for a few countries out of the 14 considered in this chapter, and the available data are not comparable across countries due to the lack of standardized reporting practices, we instead use energy consumption data as a proxy for transportation services[3] although passenger and freight kilometres (or any equivalent units) would be the desired measurement if available. Table 6.1 presents modal mix in terms of energy consumption.

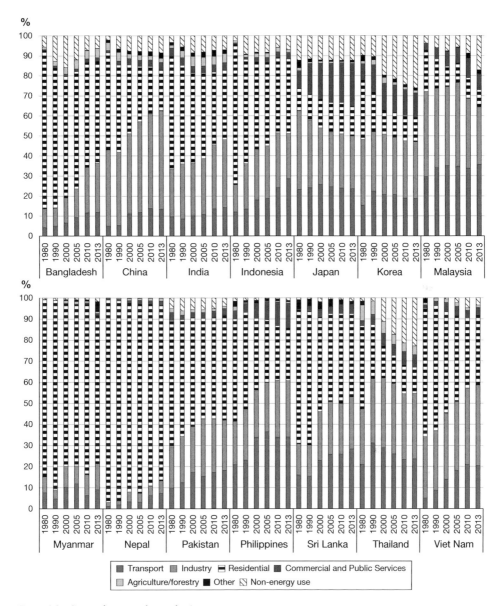

*Figure 6.3*   Sectoral energy demand mix

*Data source*: IEA (2016a).

Road was the predominant mode of transportation in all of the countries considered in 1980, especially in Malaysia, Thailand and Nepal, and the role of road transportation was even more prominent in 2013 as most countries have increased their reliance on road transportation since 1980, with only Bangladesh and Myanmar featuring modal shares of road transport under 70% in 2013. However, reliable data for fuel consumption in domestic aviation is not available for half the countries considered, likely resulting in a slight overstatement of the shares of the other modes of transport in those countries.

*Table 6.1* Modal mix in 1980 vs. 2013

| Country | 1980 | | | | | 2013 | | | | |
|---|---|---|---|---|---|---|---|---|---|---|
| | Total (Mtoe) | Air (%) | Water (%) | Rail (%) | Road (%) | Total (Mtoe) | Air (%) | Water (%) | Rail (%) | Road (%) |
| Bangladesh | 0.32 | n.a. | 29.5 | 13.2 | 57.3 | 2.89 | n.a. | 12.1 | 8.8 | 79.1 |
| China | 24.60 | 0.1 | 5.0 | 39.5 | 55.4 | 260.28 | 5.4 | 8.6 | 4.7 | 81.3 |
| India | 16.69 | 2.2 | 2.1 | 35.6 | 60.1 | 74.76 | 2.4 | 0.9 | 5.4 | 91.3 |
| Indonesia | 5.95 | 5.9 | 3.0 | 0.2 | 90.8 | 46.19 | 5.8 | 5.5 | 0.0 | 88.7 |
| Japan | 53.91 | 4.6 | 10.4 | 4.8 | 80.2 | 73.44 | 4.8 | 4.5 | 2.4 | 88.4 |
| Korea | 4.78 | 4.6 | 22.3 | 7.4 | 65.7 | 31.38 | 2.1 | 0.7 | 1.1 | 96.1 |
| Malaysia | 2.14 | n.a. | 0.0 | 0.0 | 100.0 | 19.36 | n.a. | 0.3 | 0.1 | 99.6 |
| Myanmar | 0.63 | 5.0 | 0.0 | 0.0 | 95.0 | 1.37 | 5.8 | 5.6 | 13.5 | 75.2 |
| Nepal | 0.05 | n.a. | 0.0 | 0.2 | 99.8 | 0.74 | n.a. | 0.0 | 0.1 | 99.9 |
| Pakistan | 2.21 | n.a. | 0.0 | 5.9 | 94.1 | 13.19 | 5.1 | 0.0 | 1.9 | 92.9 |
| Philippines | 3.46 | 0.2 | 5.2 | 0.0 | 94.6 | 8.78 | 4.9 | 7.3 | 0.1 | 87.7 |
| Sri Lanka | 0.68 | 18.7 | 0.6 | 5.1 | 75.5 | 2.55 | 0.1 | 1.4 | 1.4 | 97.1 |
| Thailand | 3.21 | n.a. | 0.1 | 0.0 | 99.9 | 22.63 | 2.5 | 0.7 | 0.4 | 96.4 |
| Vietnam | 0.65 | 15.1 | 0.0 | 9.7 | 75.2 | 10.41 | 1.7 | 0.4 | 0.0 | 97.9 |

*Data source*: IEA (2016b).

Only Sri Lanka and Vietnam utilized domestic air transport to a large extent in 1980, i.e., more than 10% of transport sector fuel consumption came from domestic air transport, but the modal share of domestic aviation had declined precipitously in both countries by 2013. Aside from Korea, the share of domestic aviation increased or remained steady in the other countries in 2013 compared with 1980, although domestic aviation generally accounts for a modest modal share (the highest share is 5.8% in Indonesia and Myanmar). Similarly, domestic navigation accounted for a very significant modal share in three countries, namely Bangladesh, Korea and Japan, in 1980, but this had declined significantly by 2013 in all three countries, most notably in Korea where the modal share declined from 22.3% to 0.7%. In 2013, only Bangladesh featured modal share of domestic navigation above 10%.

Rail transport was a critical mode of transportation in China and India, the two most populous countries in the world, in 1980, with modal shares in excess of 35%, and to a lesser extent in Bangladesh and Vietnam as well. All four countries experienced significant decline in the share of rail transport by 2013, especially China and India, where the modal share of rail had fallen to 4.7% and 5.4% respectively. This is reflective of the general drop in the modal share of rail transport across the region, with Myanmar the only country where rail gained a substantial share (going from nil to 13.5%) over this period.

Domestic navigation, which was a major transport mode in Bangladesh, Japan and Korea in 1980, was significantly less popular in 2013, particularly in Korea, where the modal share had fallen from 22.3% to 0.7%. Nevertheless, despite a considerable decline in modal share, domestic navigation remained an important transport mode in Bangladesh, which was the only country where its share was found to be above 10% in 2013. Most of the other countries witnessed a small increase in the modal share of domestic navigation over this period.

Overall, domestic aviation and domestic navigation tend to be less popular modes of transport as compared to rail and especially road, with modal shares for these two modes well under 10% in most countries and completely negligible in a few. The general trend is for modest growth in modal shares of domestic aviation and navigation, while the share of rail transport underwent significant decline in many countries, resulting in road transport entrenching its dominant position in terms of transport mode with significant growth in its modal share in many countries and a substantial reduction in only Myanmar and the Philippines.

## *Transport sector fuel mix*

Fuel substitution within a mode of transportation is another factor that explains transport sector energy demand growth, but fuel choices for transport remain limited, and gasoline and diesel are by far the most utilized fuel for transport across all of the countries considered. However, a few other fuels have gained market share in the transport sectors of several countries in recent years, leading to reductions in the share of diesel and gasoline in the fuel mix. A comparison of the fuel mix between 1980 and 2013 suggests significant substitution of gasoline with diesel in only one country – Vietnam, which provides sizable subsidies for diesel while taxing gasoline – whereas gasoline and diesel have been substituted as transport fuel in significant quantities by natural gas, liquefied petroleum gas (LPG) and biofuels in a number of countries (see Table 6.2).

Natural gas, which was scarcely utilized as a transport fuel in 1980, has emerged as vital fuel for transport in Bangladesh, Myanmar, Pakistan and Thailand, where compressed natural gas (CNG) is often used in lieu of gasoline by light duty vehicles, accounting for more than 10% of the fuel mix in all of these countries in 2013, and it is also gaining market share in China with 4.7% in 2013. It should be noted that the share of gasoline in Bangladesh is exceptionally low – only 10.3% – because the penetration of natural gas is so high. While the country has had significant natural gas production for quite some time, it only started to use natural gas as transport fuel in 2003, and since then the share of natural gas in the transport sector fuel mix has grown to 31.6% in 2013. The experience in Pakistan was similar, with natural gas adopted for transport fuel on a large scale in the late 1990s, and in 2013, it accounted for 16% of the transport fuel mix. Natural gas is considerably cheaper compared to gasoline and diesel in Pakistan, rendering it a popular choice for vehicle owners, as reflected in the fact that Pakistan leads the world in the number of natural gas vehicles (Khan and Yasmin, 2014).

LPG had also become a significant transport fuel by 2013, but only in two countries, namely Korea and Thailand. In Korea, where natural gas is used mainly in taxis, buses and trucks (Liu et al., 1997), the share of LPG in the transport fuel mix rose from 3.5% in 1980 to over 13% in 2013. Other than Korea, only Japan and Thailand used LPG as a transport fuel in 1980, it still accounts for a miniscule share in most countries as on 2013. Similarly, none of the countries used biofuels for transportation in 1980, but biofuels had become a notable transport fuel in Thailand and the Philippines in 2013, although its share was still relatively modest at 5.3% and 3.6% respectively.

China and India were both highly dependent on coal as fuel for rail transport in 1980 (coal comprised 38.6% and 29.2% of fuel consumption for transportation, respectively). By 2013, the share of coal in the transport fuel mix in China had declined to 1.1, while India completely phased out coal as fuel of rail transport in 1998. Between 1980 and 2013, considerable growth in the share of diesel in the fuel mix was observed in both countries, especially China, and some of this is the result of the direct substitution of coal with diesel in rail transport. However, since rail transportation itself has been significantly replaced with road transport, the substitution of coal with diesel only accounted for a small part of the gains in diesel consumption in China and India, whereas most of the growth in diesel demand can be attributed to the increase in road transportation. Vietnam, which had also relied on coal as fuel for rail transport in 1980, eliminated coal from its transport fuel mix by 1996. As in India and China, marginalization of rail transport is the main reason for the reduction in demand for coal as a transport fuel rather than substitution by diesel.

Table 6.2 Transport sector fuel mix (1980 vs. 2013)

| | | Bangladesh | China | India | Indonesia | Japan | Korea | Malaysia | Myanmar | Nepal | Pakistan | Philippines | Sri Lanka | Thailand | Vietnam |
|---|---|---|---|---|---|---|---|---|---|---|---|---|---|---|---|
| **1980** | | | | | | | | | | | | | | | |
| Total | Mtoe | 0.32 | 24.6 | 16.7 | 5.95 | 53.9 | 4.78 | 2.14 | 0.63 | 0.05 | 2.21 | 3.46 | 0.68 | 3.21 | 0.65 |
| Aviation Fuels | % | 0.0 | 0.1 | 2.2 | 5.9 | 4.6 | 4.6 | 0.0 | 5.0 | 0.0 | 0.0 | 0.2 | 18.7 | 0.0 | 15.1 |
| Electricity | % | 0.0 | 0.9 | 1.2 | 0.0 | 2.4 | 0.7 | 0.0 | 0.0 | 0.2 | 0.1 | 0.0 | 0.0 | 0.0 | 0.0 |
| Diesel | % | 73.3 | 15.8 | 55.7 | 42.5 | 32.1 | 60.5 | 39.5 | 53.2 | 82.5 | 71.4 | 59.9 | 64.4 | 52.1 | 15.5 |
| LPG | % | 0.0 | 0.0 | 0.0 | 0.0 | 3.1 | 3.5 | 0.0 | 0.0 | 0.0 | 0.0 | 0.0 | 0.0 | 1.2 | 0.0 |
| Biofuels | % | 0.0 | 0.0 | 0.0 | 0.0 | 0.0 | 0.0 | 0.0 | 0.0 | 0.0 | 0.0 | 0.0 | 0.0 | 0.0 | 0.0 |
| Gasoline | % | 19.2 | 44.3 | 9.8 | 50.3 | 50.3 | 17.8 | 60.5 | 41.7 | 17.4 | 26.2 | 38.6 | 16.9 | 46.6 | 59.7 |
| Natural Gas | % | 0.0 | 0.2 | 0.0 | 0.0 | 0.0 | 0.0 | 0.0 | 0.1 | 0.0 | 2.2 | 0.0 | 0.0 | 0.0 | 0.0 |
| Fuel Oil | % | 7.4 | 0.0 | 2.0 | 1.1 | 7.5 | 12.8 | 0.0 | 0.0 | 0.0 | 0.0 | 1.1 | 0.0 | 0.0 | 9.7 |
| Coal | % | 0.0 | 38.6 | 29.2 | 0.2 | 0.0 | 0.0 | 0.0 | 0.0 | 0.0 | 0.0 | 0.0 | 0.0 | 0.0 | 0.0 |
| Kerosene | % | 0.0 | 0.0 | 0.0 | 0.0 | 0.0 | 0.0 | 0.0 | 0.0 | 0.0 | 0.0 | 0.1 | 0.0 | 0.1 | 0.0 |
| **2013** | | | | | | | | | | | | | | | |
| Total | Mtoe | 2.89 | 260 | 74.8 | 46.2 | 73.4 | 31.4 | 19.4 | 1.37 | 0.74 | 13.2 | 8.78 | 2.55 | 22.6 | 10.4 |
| Aviation Fuels | % | 0.0 | 5.4 | 2.4 | 5.8 | 4.8 | 2.1 | 0.0 | 5.8 | 0.0 | 5.1 | 4.9 | 0.1 | 2.5 | 1.7 |
| Electricity | % | 0.0 | 1.9 | 1.8 | 0.0 | 2.1 | 0.6 | 0.1 | 0.0 | 0.1 | 0.0 | 0.1 | 0.0 | 0.1 | 0.0 |
| Diesel | % | 58.1 | 47.0 | 68.5 | 36.5 | 29.1 | 49.2 | 33.6 | 22.3 | 73.7 | 50.4 | 52.9 | 65.0 | 48.3 | 50.6 |
| LPG | % | 0.0 | 0.5 | 0.3 | 0.0 | 1.7 | 13.7 | 0.0 | 0.0 | 1.3 | 0.0 | 0.6 | 0.0 | 9.2 | 0.0 |
| Biofuels | % | 0.0 | 0.7 | 0.2 | 1.8 | 0.0 | 1.1 | 1.0 | 0.0 | 0.0 | 0.0 | 3.6 | 0.0 | 5.3 | 0.0 |
| Gasoline | % | 10.3 | 36.4 | 24.5 | 55.8 | 59.1 | 29.2 | 63.5 | 55.6 | 24.9 | 28.4 | 35.4 | 33.7 | 23.6 | 47.3 |
| Natural Gas | % | 31.6 | 4.7 | 2.0 | 0.1 | 0.1 | 3.8 | 1.5 | 14.0 | 0.0 | 16.0 | 0.0 | 0.0 | 11.0 | 0.0 |
| Fuel Oil | % | 0.0 | 2.3 | 0.4 | 0.0 | 3.1 | 0.3 | 0.3 | 2.3 | 0.0 | 0.0 | 2.4 | 1.2 | 0.0 | 0.4 |
| Coal | % | 0.0 | 1.1 | 0.0 | 0.0 | 0.0 | 0.0 | 0.0 | 0.0 | 0.0 | 0.0 | 0.0 | 0.0 | 0.0 | 0.0 |
| Kerosene | % | 0.0 | 0.0 | 0.0 | 0.0 | 0.0 | 0.0 | 0.0 | 0.0 | 0.0 | 0.0 | 0.0 | 0.0 | 0.0 | 0.0 |

Data source: IEA (2016b).

Coal was one of the most important fuels for transportation, primarily for rail, in China and India, and to a lesser extent, Vietnam, in 1980; but these countries managed to either dramatically reduce or eliminate entirely their reliance on coal by 2005. Mongolia, on the other hand, which did not use coal as a fuel for transportation at all in 1980, had incorporated it into its transport fuel mix by 2005. Finally, the use of kerosene as a share of total fuel consumption for transportation was and remains negligible, if not nil, in all countries considered.

The consumption of aviation fuels represented another notable source of energy demand from transportation in 1980, but only in a few countries such as Sri Lanka and Vietnam. Reliable aviation fuel consumption data is not available for all countries; however, the share of aviation fuel in total transport sector fuel consumption increased in the countries for which data is available, except for Korea, Sri Lanka and Vietnam. Utilization of electricity for transportation in 1980 was negligible in all countries except China, India, Japan and Korea, and while the share of electricity in the fuel mix has increased slightly in China and India, it remains a very minor source of energy for transport in all of the countries.

### Transportation energy intensity

Transportation energy intensity, which is the ratio of total fuel consumption for transportation in an economy to its gross domestic product, is displayed for all of the Asian countries considered over the period 1980–2013 in Figure 6.4. While transportation energy intensity varies significantly across countries and over time, some trends can be observed in Figure 6.4. Bangladesh and Nepal feature the lowest intensity by far over the study period, although it has increased rather significantly in Nepal since 2007, placing it at a similar level to India in recent year. Transportation energy intensity has grown over time in Malaysia, Thailand and Vietnam over the study period, with these three countries exhibiting the highest intensities in 2013. China and Myanmar, which had the highest transportation energy intensity in 1980, had both achieved drastic reductions by 2013, while India realized a more modest but steady decline. On the other

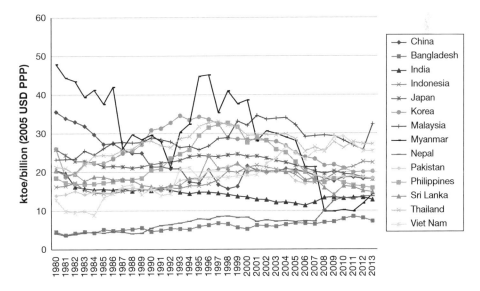

*Figure 6.4*   Transport energy intensity 1980–2013

*Data source*: IEA (2016a, 2016c).

*Table 6.3* Total and urban population size and growth rate

| Country | 1980 | | 1995 | | 2015 | | Average Annual Growth (1980–2015) | |
|---|---|---|---|---|---|---|---|---|
| | Population (millions) | Urban Share (%) | Population (millions) | Urban Share (%) | Population (millions) | Urban Share (%) | Total (%) | Urban (%) |
| Bangladesh | 81.4 | 14.9 | 118.4 | 21.7 | 161.0 | 34.3 | 2.0 | 4.4 |
| China | 981.2 | 19.4 | 1,204.9 | 31.0 | 1,371.2 | 55.6 | 1.0 | 4.1 |
| India | 697.2 | 23.1 | 960.9 | 26.6 | 1,311.1 | 32.7 | 1.8 | 2.8 |
| Indonesia | 147.5 | 22.1 | 197.0 | 36.1 | 257.6 | 53.7 | 1.6 | 4.2 |
| Japan | 116.8 | 76.2 | 125.4 | 78.0 | 127.0 | 93.5 | 0.2 | 0.8 |
| Korea | 38.1 | 56.7 | 45.1 | 78.2 | 50.6 | 82.5 | 0.8 | 1.9 |
| Malaysia | 13.8 | 42.0 | 20.7 | 55.7 | 30.3 | 74.7 | 2.3 | 4.0 |
| Myanmar | 34.5 | 24.0 | 44.7 | 25.5 | 53.9 | 34.1 | 1.3 | 2.3 |
| Nepal | 14.9 | 6.1 | 21.4 | 10.9 | 28.5 | 18.6 | 1.9 | 5.2 |
| Pakistan | 78.1 | 28.1 | 122.6 | 31.8 | 188.9 | 38.8 | 2.6 | 3.5 |
| Philippines | 47.4 | 37.5 | 69.8 | 48.3 | 100.7 | 44.4 | 2.2 | 2.7 |
| Sri Lanka | 14.7 | 18.8 | 18.1 | 18.5 | 21.0 | 18.4 | 1.0 | 0.9 |
| Thailand | 47.4 | 26.8 | 59.3 | 30.3 | 68.0 | 50.4 | 1.0 | 2.9 |
| Vietnam | 53.7 | 19.2 | 72.0 | 22.2 | 91.7 | 33.6 | 1.5 | 3.2 |

*Data source*: World Bank (2016).

*Note*: China does not include Hong Kong.

hand, transportation energy intensity in Korea and the Philippines deteriorated at first but improved from the mid to late 1990s so that it wasn't too much different in 2013 as compared to 1980. Therefore, a common trend cannot be found in terms of transportation energy intensity among the Asian countries considered.

## Population growth and urbanization

The Asian countries included in this chapter alone accounted for over half of the world's population in 2015, and these countries are also undergoing a period of rapid urbanization. Table 6.3 lists the population of Asian countries in 1990 and 2006, the percentage of urban dwellers, and the growth rates of total and urban populations. Urban population growth can be seen to be significantly higher than total population growth in all of the countries except Sri Lanka.

While developed and developing countries alike continue to urbanize, the rate of urbanization is especially swift in developing countries, where the majority of people have not been city-dwellers traditionally. The year 2007 was historic in that it commemorated the first time that half of the world's population lived in cities (ESCAP, 2007a), and this is only expected to increase given the higher growth rate of urban populations compared to total population. As Asian countries continue to grow and urbanize, increasing motorization can be expected to generate higher levels of energy consumption for transportation and place additional stresses on the transport infrastructure.

## Motorization[4] and economic growth

Growth in the rate of motorization is one of the key factors that explains the modal shift towards road transportation from other modes of transportation observed in Asia. Most Asian countries have experienced significant growth in their road transport fleets, particularly in urban areas, resulting in soaring transportation energy demand (ESCAP, 2007b). Table 6.4 presents the motorization rate (i.e., the number of passenger cars per 1000 persons) in 2000 and in 2011 for the 14 Asian countries considered, along with the average annual growth rate in motorization. Rapid growth in motorization can be seen in almost every country, including 20.7% growth in China, the most populous country in the world, and robust growth in two of the other most populous countries, India and Indonesia, albeit not at the tremendous pace as in China. It is only in the Philippines that a slight decline in motorization can be observed.

Table 6.4 also presents the per capita GDP and per capita GDP growth, along with motorization rate, and a positive relationship between higher levels of per capita GDP and increased motorization becomes apparent. The relatively more affluent economies of Japan, Korea and Malaysia feature much higher levels of motorization than the other Asian countries. The level of motorization in developing Asian countries lags well behind those in developed countries, but the expected expansion of developing Asian economies in the coming years means that the current levels of motorization are likely only a fraction of what they will be in a couple of decades. Excluding the mature economy of Japan, some of the remaining 13 countries have achieved impressive growth in per capita GDP from 2000 to 2011, for example, China with 9.8% growth, Myanmar with 9.1%, Vietnam with 5.7% and India with 5.6%. The demand for passenger cars in Asia is growing much faster than per capita income in every country considered, other than Myanmar and the Philippines, sometimes astonishingly so, as in the case

*Table 6.4* Motorization and GDP per capita (2005 USD using PPP)

| | 2000 | | 2011 | | Average annual growth rate | |
|---|---|---|---|---|---|---|
| | *Cars per 1,000 people* | *GDP per capita* | *Cars per 1,000 people* | *GDP per capita* | *Cars per 1,000 people (%)* | *GDP per capita (%)* |
| Bangladesh | 0.5 | 1,434 | 2.1[a] | 2,191[a] | 16.1 | 4.3 |
| China | 6.8 | 3,220 | 53.6 | 8,967 | 20.7 | 9.8 |
| India | 5.8 | 2,274 | 11.0[b] | 3,720[b] | 7.3 | 5.6 |
| Indonesia | 14.2 | 4,942 | 39.4 | 7,506 | 9.7 | 3.9 |
| Japan | 404.0 | 28,898 | 455.0 | 30,798 | 1.1 | 0.6 |
| Korea | 172.0 | 19,682 | 284.0 | 29,706 | 4.7 | 3.8 |
| Malaysia | 179.9 | 13,974 | 340.6 | 18,782 | 6.0 | 2.7 |
| Myanmar | 3.8 | 619 | 5.4 | 1,615 | 3.2 | 9.1 |
| Nepal | 2.0 | 1,404 | 4.0 | 1,820 | 6.5 | 2.4 |
| Pakistan | 7.7 | 2,993 | 16.0 | 3,798 | 6.9 | 2.2 |
| Philippines | 9.9 | 3,777 | 8.7[a] | 4,998[a] | −1.4 | 2.8 |
| Sri Lanka | 12.2 | 4,404 | 19.9[a] | 6,743[a] | 5.0 | 4.4 |
| Thailand | 34.7 | 7,959 | 73.5 | 11,394 | 7.1 | 3.3 |
| Vietnam | 0.6 | 2,359 | 14.0[c] | 3,479[c] | 56.8 | 5.7 |

*Data source*: ESCAP (2016) and IEA (2016c).

*Notes*: Road motor vehicles designed for the conveyance of passengers and seating not more than nine persons, including the driver. Taxis, jeep-type vehicles and station wagons are included. Special-purpose vehicles, such as two- or three-wheeled cycles or motorcycles, trams, trolley-buses, ambulances, hearses and military vehicles operated by police or other governmental security organizations, are excluded. China does not include Hong Kong.
a Data for 2010;
b data for 2009;
c data for 2007.

of China and Bangladesh. India still had a modest motorization rate of 11 passenger cars per 1,000 people in 2009, but its per capita GDP had only passed the $3,000 threshold a few years earlier, which has been called the tipping point past which vehicle ownership accelerates quickly (IEA, 2007).

# Quantification of factors driving transportation energy demand

## *Methodology*

In this section, we derive the methodology to decompose total transport sector energy consumption to its driving factors. Data needed to implement the methodology are also discussed here.

The total transport sector energy consumption in a country in year $t$ ($TE_t$) is the summation of energy consumption from all fuels used in all transport modes in that year, i.e.,

$$TE_t = \sum_{ij} TE_{ijt} \tag{6.1}$$

where subscripts $i, j$ and $t$ refer to fuel type (e.g., gasoline, diesel, electricity), transportation mode (e.g., road, rail, air and water) and year, respectively. In order to decompose the emission to the potential factors affecting it, Equation (6.1) can be expressed as

$$TE_t = \sum_{ij} \frac{FC_{ijt}}{FC_{jt}} \times \frac{FC_{jt}}{TS_{jt}} \times \frac{TS_{jt}}{TS_t} \times \frac{TS_t}{GDP_t} \times \frac{GDP_t}{POP_t} \times POP_t \tag{6.2}$$

where $FC$ refers to fuel consumption, $TS$ represents transport services (e.g., passenger kilometres, tons kilometres or any equivalent measurement representing transport services[5]) and GDP is used to measure economic output.

Unfortunately, data for transportation services are not available for the countries and for the time horizon considered in the study. We use an alternative approach as shown in Equation (6.3) to decompose the emission to the potential factors affecting it:

$$TE_t = \sum_{ij} \frac{FC_{ijt}}{FC_{jt}} \times \frac{FC_{jt}}{FC_t} \times \frac{FC_t}{GDP_t} \times \frac{GDP_t}{POP_t} \times POP_{tt} \tag{6.3}$$

As implied in Equation (6.3), we represent modal mix by energy consumption by mode instead of transportation services provided by the mode. Equation (3) can also be rewritten as

$$TE_t = \sum_{ij} FM_{ijt} \times MM_{jt} \times EI_t \times PC_t \times POP_t \tag{6.4}$$

where $FM$ refers to fuel mix (i.e., share of a fuel in a transportation mode), $MM$ represents modal mix (i.e., share of fuel consumption by a mode in total transport sector energy consumption); $EI$ is the transportation energy intensity (i.e., $FC/GDP$), $PC$ is economic activity as captured by per capita GDP, and POP is population.

The growth of emissions is often decomposed into the potential driving factors using different methods, such as the Laspeyres or Divisia methods. While studies such as Lin et al. (2008), Diakoulaki and Mandaraka (2007), Diakoulaki et al. (2006), and Ebohon and Ikeme (2006) use the refined Laspeyres techniques, studies such as Liu et al. (2007), Hatzigeorgiou et al. (2008) and Wang et al. (2005) use the Arithmetic Mean Divisia Index (AMDI) and the Logarithmic Mean

Divisia Index (LMDI) techniques. Like Timilsina and Shrestha (2009), this chapter follows the LMDI approach, which, unlike the AMDI approach, provides a residual-free decomposition and can accommodate the occurrence of zero values in the data set[6] (Ang, 2004). Although the refined Laspeyres methods also have these virtues, their formulae become increasingly complex when the number of factors exceeds three, and the linkages between the additive and multiplicative forms cannot be established easily.

Using LMDI (Ang, 2005), the additive decomposition of the change in transport sector energy demand from year $t-1$ to $t$ is expressed as

$$TE_t - TE_{t-1} = \sum_{ij} \sum_{ij} \tilde{w}_{ij} \ln \frac{FM_{ijt}}{FM_{ijt-1}} + \sum_{ij} \tilde{w}_{ij} \ln \frac{MM_{ijt}}{MM_{ijt-1}} + \sum_{ij} \tilde{w}_{ij} \ln \frac{EI_t}{EI_{t-1}}$$

$$+ \sum_{ij} \tilde{w}_{ij} \ln \frac{PC_t}{PC_{t-1}} + \sum_{ij} \tilde{w}_{ij} \ln \frac{POP_t}{POP_{t-1}} \tag{6.5}$$

where

$$\tilde{w}_{ijt} = \frac{TE_{ijt} - TE_{ijt-1}}{\ln TE_{ijt} - \ln TE_{ijt-1}} \qquad \text{for} \quad TE_{ijt} \neq TE_{ijt-1}$$

$$= TE_{ijt} \qquad \text{for} \quad TE_{ijt} = TE_{ijt-1} \tag{6.6}$$

Similarly, the multiplicative decomposition of the change in the transport sector energy demand from year $t-1$ to $t$ (again, following Ang (2005)) is given as

$$\frac{TE_t}{TE_{t-1}} = \exp \left[ \sum_{ij} \tilde{v}_{ij} \ln \frac{FM_{ijt}}{FM_{ijt-1}} \right] \times \exp \left[ \sum_{ij} \tilde{v}_{ij} \ln \frac{MM_{ijt}}{MM_{ijt-1}} \right] \times \exp \left[ \sum_{ij} \tilde{v}_{ij} \ln \frac{EI_t}{EI_{t-1}} \right]$$

$$\times \exp \left[ \sum_{ij} \tilde{v}_{ij} \ln \frac{PC_t}{PC_{t-1}} \right] \times \exp \left[ \sum_{ij} \tilde{v}_{ij} \ln \frac{POP_t}{POP_{t-1}} \right] \tag{6.7}$$

where

$$\tilde{v}_{ijt} = L(TE_{ijt}, TE_{ijt-1}) / L(TE_t, TE_{t-1}) \tag{6.8}$$

with

$$L(a, b) = \frac{a - b}{\ln a - \ln b} \qquad \text{for} \quad a \neq b$$

$$= a \qquad \text{for} \quad a = b \tag{6.9}$$

The first and second terms on the right-hand side of Equations (6.5) and (6.7) represent the fuel mix (FM) or fuel switching and the modal mix (MM) or modal shift effects, respectively. The third term represents the transportation energy intensity (EI) effect. And finally, the fourth and fifth terms represent the economic activity or per capita GDP (PC) effect and population (POP) effect, respectively.

We have carried out the decomposition analysis on an annual basis over the 33-year period between 1980 and 2013.

## Data

The study required a large set of data on energy consumption by fuel and by mode for the study period of 33 years. While national statistical agencies collect data at the level of detail needed in a few countries (e.g. Korea), they do not provide such data in most of the countries. Moreover, mixing data from different sources with different conventions and assumptions used for collection and aggregation would cause an artificial change in the trends. No source other than the IEA provides data at the required details and time series needed for the study. Therefore, we use transport sector energy consumption data by fuel type and mode from the International Energy Agency (IEA, 2016a, 2016b).

Fuels included are biofuel (i.e., ethanol and biodiesel), natural gas, liquefied petroleum gases (LPG), motor gasoline, aviation fuels (i.e., aviation gasoline, kerosene type jet fuel and gasoline type jet fuel), diesel oil, fuel oil, coal, kerosene and electricity. The modes of transportation considered are domestic aviation, road, rail and domestic navigation.[7] We have excluded energy consumption in oil and gas pipeline transport.

Data on gross domestic product (GDP), expressed in 2005 constant dollar measured at purchasing power parity, and population are also taken from the IEA (2016c). Data for China includes Hong Kong. Korea refers to the Republic of Korea.

## Results and discussion

All of the countries considered experienced significant growth in transportation sector energy demand during the 1980–2013 period. However, there remain meaningful differences in the magnitude of transport energy demand growth and the factors driving it. Table 6.5 summarizes the results of the additive decomposition of transport sector energy demand growth into fuel switching, modal shifting, transportation energy intensity, per capita GDP and population.

When a main influencing factor is considered as one that accounts for more than 25% of the change in energy demand (and since all of the countries experienced growth in transport energy demand, this can be taken to mean that the factor contributed at least 25% to the growth in transport energy demand), economic activity (i.e., per capita GDP growth) is a main influencing factor in all countries, as can be seen from Table 6.5. Population growth is a main influencing factor in India, Malaysia, Pakistan and the Philippines, and deterioration in transportation energy intensity is a main influencing factor in Nepal alone.

However, if we consider all the factors that have contributed to transport energy demand growth, regardless of the magnitude, then population growth must also be noted as an influencing factor in all of the countries. Similarly, deterioration in transportation energy intensity is found to contribute to emissions growth in Bangladesh, Indonesia, Malaysia, Pakistan, Thailand and Vietnam. Although fuel switching is not a main influencing factor in any of the countries considered, with the exception of India, Japan, Myanmar and Sri Lanka, it contributes directly to transport energy demand growth in all of the countries considered, but its impact is very modest (see Table 6.5). For example, in the Philippines and Korea, where the fuel switching effect makes the largest contribution, it still only accounts for 2.9% and 1.5%, respectively, of the average growth in transport energy demand.

While fuel switching appears to be a notable phenomenon in a few of the Asian countries during the 1980–2013 period (see Table 6.2), e.g., the penetration of natural gas as an important transport fuel in Myanmar, Pakistan, Thailand, and especially Bangladesh; the increased utilization of LPG in Korea and Thailand; and the increased reliance on biofuels in Thailand, the fuel switching effect is not found to play a major role in driving transport sector energy demand

Table 6.5 Average annual transport energy demand change and responsible factors (1980–2013)

| Country | Average energy demand change (ktoe) | Factors influencing the energy demand change | | | | | Influencing factors[a] | Main influencing factors[b] |
|---|---|---|---|---|---|---|---|---|
| | | Fuel mix (FM) | Modal mix (MM) | Transport energy intensity (EI) | Per capita GDP (PC) | Population (POP) | | |
| Bangladesh | 77.8 | 0.1 | 0.1 | 13.7 | 45.5 | 18.4 | FM, MM, EI, PC, POP | PC, |
| China | 7,188.8 | 37.3 | 11.0 | −747.9 | 7,252.9 | 635.5 | FM, MM, PC, POP | PC, |
| India | 1,759.5 | −0.5 | 1.0 | −284.0 | 1,528.1 | 515.0 | MM, PC, POP | PC, POP |
| Indonesia | 1,219.2 | 0.4 | 0.2 | 277.4 | 659.2 | 282.0 | FM, MM, EI, PC, POP | PC, |
| Japan | 591.8 | −0.1 | 0.6 | −744.1 | 1,167.3 | 168.1 | MM, PC, POP | PC, |
| Korea | 893.6 | 31.7 | −11.0 | −223.0 | 954.2 | 141.6 | FM, PC, POP | PC, |
| Malaysia | 521.6 | 0.3 | 0.1 | 77.6 | 266.7 | 176.9 | FM, MM, EI, PC, POP | PC, POP |
| Myanmar | 22.3 | −4.1 | −0.5 | −33.9 | 51.2 | 9.6 | PC, POP | PC, |
| Nepal | 20.8 | 0.0 | 0.0 | 10.4 | 6.3 | 4.1 | FM, MM, EI, PC, POP | EI, PC, |
| Pakistan | 332.5 | 0.8 | 0.5 | 27.9 | 140.0 | 163.3 | FM, MM, EI, PC, POP | PC, POP |
| Philippines | 161.3 | 4.7 | −4.2 | −79.9 | 106.8 | 133.9 | FM, PC, POP | PC, POP |
| Sri Lanka | 56.4 | 0.0 | 0.2 | −18.2 | 62.6 | 11.7 | MM, PC, POP | PC, |
| Thailand | 588.5 | 5.4 | −4.4 | 24.7 | 458.2 | 104.6 | FM, EI, PC, POP | PC, |
| Vietnam | 295.8 | 0.5 | 0.2 | 54.8 | 191.7 | 48.6 | FM, MM, EI, PC, POP | PC, |

Data source: Authors.

Note: The modal mix effect, as defined in this study, considers only four modes: road, rail, water and air. If necessary data is available to further disaggregate road transportation into auto, bus, etc., modal mix might be found to influence $CO_2$ emission growth.

a Factors in the same direction as average emission change.

b Factors that account for more than 25% of the energy demand change.

growth in these countries. Similarly, despite the significant decline in the modal share of domestic navigation in Bangladesh, Korea and Japan; of domestic aviation in Sri Lanka; rail transport in China, India and Vietnam; and the increase in modal share of rail transport in Myanmar, and of road transport in many countries, modal shift is not found to be a main influencing factor in the growth of transport energy demand in those countries. This is because the effects of fuel switching and modal shifting were eclipsed by the effects of the overwhelming growth in economic activity and population.

Nevertheless, despite the fact that fuel mix, modal mix, and transport energy intensity are not found to be main influencing factors in any of the countries (except for transport energy intensity in Nepal), they have been crucial in *restraining* the growth of transport energy demand in a few countries. For example, were it not for the ameliorating impact of fuel mix effect, the average transport energy demand would have been more than 18% higher in Myanmar and about 3% higher in the Philippines. The modal mix factor does not have such a large impact, but is seen to have reduced average transport energy demand growth in Myanmar and the Philippines by 2.3% and 2.6% respectively. When examining countervailing factors that have enabled countries to reduce their transport energy demand from what it would otherwise be, then the impact of improvements in transport energy intensity becomes clear. For example, if not for the reduction in transport energy intensity (i.e., decoupling of energy consumption from economic growth), the average transport energy demand would have been 152% higher in Myanmar and 126% higher in Japan over the study period. While not as effective as in Myanmar and Japan, improvement in transport energy intensity also significantly depressed transport energy demand in China, India, Korea, the Philippines and Sri Lanka.

Figure 6.5 displays indexed time-series charts of the multiplicative decomposition of transport energy and its driving factors in each of the 14 countries considered. A sharp decline in per capita GDP, reflecting the financial crisis in the region in 1997, can be observed for Indonesia, Korea, Malaysia and Thailand in 1997–1998, followed by renewed growth (see Figure 6.5d, f, g and m). This has a visible impact on transport energy demand, which follows the same trajectory.

Since Figure 6.5 offers a visual representation of this decomposition analysis, it reiterates much of the preceding discussion on the relative impacts of the various factors on transport energy demand. Most countries have experienced secular growth in economic activity and population, while the impact of fuel switching and modal shift has been negligible. This means that the year on year volatility of transport energy demand is largely explained by changes in transport energy intensity. One can see this effect most clearly in the case of Japan, where transport energy intensity rises over the early part of the study period and then declines from the late 1990s forward (see Figure 6.5e). On the other hand, in Nepal, deterioration in transport energy intensity clearly accounts for the drastic rise in energy demand from 2007. Finally, Myanmar is particularly notable for how it has managed to moderate transport energy demand growth through both significant improvements in transport energy intensity and fuel switching.

It could be surprising to note why the fuel-mix and modal-mix effects do not play a major role in driving the transport energy demand in most countries in the region. Fuel-mix is not expected to play a role because there are only two major fuels (i.e., gasoline and diesel) used for trans-portation[8] and energy contents of these two fuels are not much different. Modal-mix did not show much impact by the design of the decomposition technique in this study due to lack of data. In fact, the modes of transportation should have been measured in terms of passenger or tons kilometres; however, such data is not available mainly for road transportation, which is the main mode for transportation.

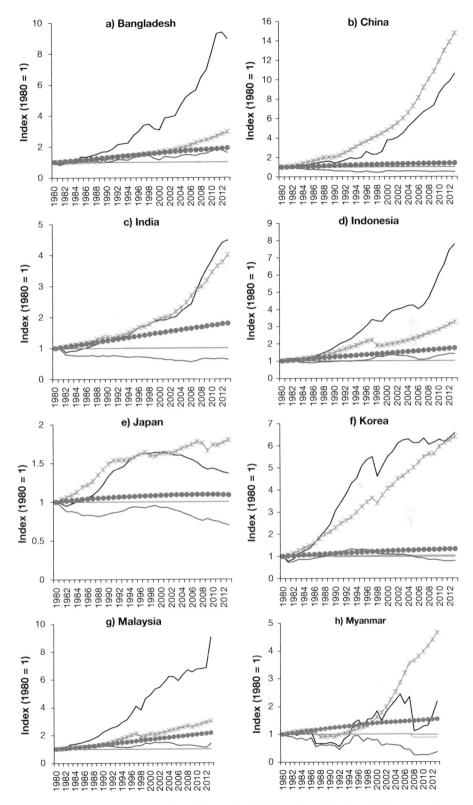

*Figure 6.5* Transport sector energy demand growth and driving factors in selected Asian countries

*Data source*: Authors.

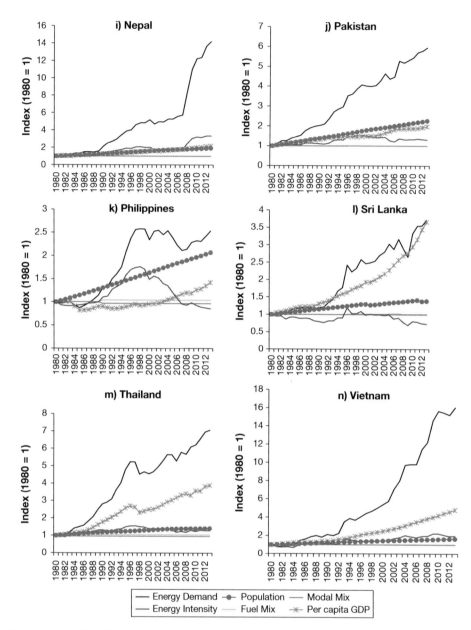

*Figure 6.5* (Continued)

## Conclusions

This chapter examines the growth of the transport sector energy demand and determines the underlying factors in 14 Asian countries over 33 years between 1980 and 2013. To identify the driving factors, we decompose the growth of transportation energy demand into income growth, population growth, changes in transportation energy intensity, change in fuel–mix and shifts in transportation modes. We used the LMDI approach for the decomposition analysis. The income (i.e., per capita GDP) growth effect and the population growth effect are found to be the principal drivers of transport energy demand growth in Asian countries over the study horizon, whereas fuel switching and modal shifting have not had a sizeable influence other than in Myanmar and the Philippines. While worsening of transport energy intensity was a major factor in the growth of energy demand in Nepal, in many countries, this one the main reason transport energy demand did not grow as much as it would have otherwise. To determine the reasons for transportation energy intensity change, this indicator itself can be decomposed into its driving factors, such as fuel efficiency of transportation by mode and transport service intensity of the economy, but this chapter could not explore those details due to lack of data.

The fuel switching and modal shifting effects are not found to have a sizeable influence on the growth of transport sector energy demand in any of the Asian countries studied. However, given the increased availability of transport fuels other than gasoline and diesel in many countries, along with rapid urbanization, which makes high-capacity public transport more viable, policy instruments to encourage these underutilized approaches could make a significant dent in transport energy demand growth if successfully implemented.

While this chapter does not detect modal shifting to be a main factor in the growth of transport sector energy demand, a major limitation of the analysis is that data specific to the various types of transport within road transport, such as cars, minibuses, buses and Bus Rapid Transit (BRT) systems, is not available. As road is by far the most popular mode of transport in Asia, disaggregated data on intra-modal shifting in road transport would enable a more accurate depiction of the impact of modal switching on transport sector energy demand. Furthermore, several large Asian cities have already constructed light rail (e.g. in China, India, Korea, Philippines, Malaysia and Thailand) and BRT systems (e.g. in Jakarta, Seoul and Beijing) to provide convenient and price competitive alternatives to private road vehicles, and more may elect to do so in response to urbanization in the region. While increases in economic activity and population may be beyond the mandate of transport planners, and transport energy demand growth associated with these factors may be difficult to control, improvements in transport energy intensity and modal mix could all be promoted to reduce transport sector energy demand.

## Acknowledgement and Disclaimer

The views expressed in this paper are those of the authors and do not necessarily represent the World Bank and its affiliated organizations.

## Notes

1 China refers to the People's Republic of China and Hong Kong unless otherwise specified.
2 Korea refers to the Republic of Korea.
3 In energy literature, this is a common practice to measure modal mix in the transportation sector (see e.g., EIA, 2007; IEA, 2004)
4 Motorization, as per the commonly utilized UN definition, is measured here as the number of private cars per 1000 people in a country.
5 Includes transport services provided to all sectors (e.g., households, industry, government).
6 In this approach zero values are replaced with a small positive constant.
7 In energy statistics, energy consumption by international aviation and maritime transportation are not considered part of national energy consumption. These are treated separately under their international conventions (i.e., International Civil Aviation Organization and International Maritime Organization).
8 In air and water modes, single fuel in each of these modes (jet fuel for air transportation and diesel for water transportation) are used.

# References

Ang, B.W., 2004. Decomposition analysis for policymaking in energy: which is the preferred method? *Energy Policy*, Volume 32 (9), 1131–1139.

Ang, B.W., 2005. The LMDI approach to decomposition analysis: a practical guide. *Energy Policy*, Volume 33 (7), 867–871.

Bhattacharyya, S.C. & Ussanarassamee, A., 2004. Decomposition of energy and $CO_2$ intensities of Thai industry between 1981 and 2000. *Energy Economics*, Volume 26 (5), 765–781.

Chang, Y.F. & Lin, S.J., 1998. Structural decomposition of industrial $CO_2$ emission in Taiwan: an input-output approach. *Energy Policy*, Volume 26 (1), 5–12.

Diakoulaki, D. & Mandaraka, M., 2007. Decomposition analysis for assessing the progress in decoupling industrial growth from $CO_2$ emissions in the EU manufacturing sector. *Energy Economics*, Volume 29 (4), 636–664 (Modeling of Industrial Energy Consumption).

Diakoulaki, D., Mavrotas, G., Orkopoulos, D. & Papayannakis, L., 2006. A bottom-up decomposition analysis of energy-related $CO_2$ emissions in Greece. *Energy*, Volume 31 (14), 2638–2651.

Ebohon, O.J. & Ikeme, A.J., 2006. Decomposition analysis of $CO_2$ emission intensity between oil-producing and non-oil-producing sub-Saharan African countries. *Energy Policy*, Volume 34 (18), 3599–3611.

EIA, 2007. International Energy Outlook 2007. Washington, DC: Energy Information Administration. US Department of Energy.

ESCAP, 2007a. Review of Developments in Transport in Asia and the Pacific 2007: Data and Trends. Thailand: United Nations Economic and Social Commission for Asia and the Pacific (ESCAP).

ESCAP, 2007b. Statistical Abstract of Transport in Asia and the Pacific. Thailand: United Nations Economic and Social Commission for Asia and the Pacific (ESCAP).

ESCAP, 2016. Statistical Yearbook for Asia and the Pacific 2015. Thailand: United Nations Economic and Social Commission for Asia and the Pacific (ESCAP).

Han, X. & Chatterjee, L., 1997. Impacts of growth and structural change on $CO_2$ emissions of developing countries. *World Development*, Volume 25 (3), 395–407.

Hatzigeorgiou, E., Polatidis, H. & Haralambopoulos, D., 2008. $CO_2$ emissions in Greece for 1990–2002: A decomposition analysis and comparison of results using the Arithmetic Mean Divisia Index and Logarithmic Mean Divisia Index techniques. *Energy*, Volume 33 (3), 492–499.

IEA, 2004. 30 Years of Energy Use in IEA Countries. Paris: International Energy Agency (IEA). p. 123.

IEA, 2007. World Energy Outlook 2007. Paris: International Energy Agency (IEA).

IEA, 2016a. "World energy balances", *IEA World Energy Statistics and Balances* (database). DOI: http://dx.doi.org.libproxy-wb.imf.org/10.1787/data-00512-en (Accessed on 18 July 2016).

IEA, 2016b. "Extended world energy balances", *IEA World Energy Statistics and Balances* (database). DOI: http://dx.doi.org.libproxy-wb.imf.org/10.1787/data-00513-en (Accessed on 15 August 2016).

IEA, 2016c. "World Indicators", *IEA World Energy Statistics and Balances* (database). DOI: http://dx.doi.org.libproxy-wb.imf.org/10.1787/data-00514-en (Accessed on 26 July 2016).

Kawase, R., Matsuoka, Y. & Fujino, J., 2006. Decomposition analysis of $CO_2$ emission in long-term climate stabilization scenarios. *Energy Policy*, Volume 34 (15), 2113–2122.

Khan, M.I. & Yasmin, T., 2014. Development of natural gas as a vehicular fuel in Pakistan: Issues and prospects. *Journal of Natural Gas Science and Engineering*, Volume 17, 99–109.

Kveiborg, O. & Fosgerau, M., 2007. Decomposing the decoupling of Danish road freight traffic growth and economic growth. *Transport Policy*, Volume 14, 39–48.

Lakshmanan, T. & Han, X., 1997. Factors underlying transportation $CO_2$ emissions in the USA: a decomposition analysis. *Transportation Research Part D*, 2(1), 1–15.

Lee, K. & Oh, W., 2006. Analysis of $CO_2$ emissions in APEC countries: A time-series and a cross-sectional decomposition using the log mean Divisia method. *Energy Policy*, Volume 34 (17), 2779–2787.

Liaskas, K., Mavrotas, G., Mandaraka, M. & Diakoulaki, D., 2000. Decomposition of industrial $CO_2$ emissions: The case of European Union. *Energy Economics*, Volume 22 (4), 383–394.

Lin, J., Zhou, N., Levine, M. & Fridley, D, 2008. Taking out 1 billion tons of $CO_2$: The magic of China's 11th Five-Year Plan. *Energy Policy*, Volume 36 (3), 954–970.

Lise, W., 2006. Decomposition of $CO_2$ emissions over 1980–2003 in Turkey. *Energy Policy*, Volume 34 (14), 1841–1852.

Liu, E., Yue, S.Y. & Lee, J., 1997. A study on LPG as a fuel for vehicles. Research and Library Services Division, Legislative Council Secretariat, Government of Hong Kong.

Liu, L.C., Fan, Y., Wu, G. & Wei, Y.M., 2007. Using LMDI method to analyze the change of China's industrial $CO_2$ emissions from final fuel use: An empirical analysis. *Energy Policy*, Volume 35 (11), 5892–5900.

Lu, I.J., Lin, S.J. & Lewis, C., 2007. Decomposition and decoupling effects of carbon dioxide emission from highway transportation in Taiwan, Germany, Japan and South Korea. *Energy Policy*, Volume 35 (6), 3226–3235.

Luukkanen, J. & Kaivo-oja, J., 2002a. ASEAN tigers and sustainability of energy use – decomposition analysis of energy and $CO_2$ efficiency dynamics. *Energy Policy*, Volume 30 (4), 281–292.

Luukkanen, J. & Kaivo-oja, J., 2002b. A comparison of Nordic energy and $CO_2$ intensity dynamics in the years 1960–1997. *Energy*, Volume 27 (2), 135–150.

Nag, B. & Kulshrestha, M., 2000. Carbon emission intensity of power consumption in India: A detailed study of its indicators. *Energy Sources*, Volume 22 (2), 157–166(10).

Rhee, H.C. & Chung, H.S., 2006. Change in $CO_2$ emission and its transmissions between Korea and Japan using international input-output analysis. *Ecological Economics*, Volume 58 (4), 788–800.

Saikku, L., Rautiainen, A. & Kauppi, P.E., 2008. The sustainability challenge of meeting carbon dioxide targets in Europe by 2020. *Energy Policy*, Volume 36 (2), 730–742.

Schäfer, A., 2000. Regularities in travel demand: an international perspective. *Journal of Transportation and Statistics*, Volume 3 (3), 1–31.

Schipper, L., Marie-Lilliu, C. & Gorham, R., 2000. Flexing the Link between Transport Greenhouse Gas Emissions: A Path for the World Bank. Paris: International Energy Agency.

Schipper, L., Murtishaw, S., Khrushch, M., Ting, M., Karbuz, S. & Unander, F., 2001. Carbon emissions from manufacturing energy use in 13 IEA countries: long-term trends through 1995. *Energy Policy*, Volume 29 (9), 667–688.

Schipper, L., Scholl, L. & Price, L., 1997. Energy use and carbon from freight in ten industrialized countries: an analysis of trends from 1973–1992. *Transportation Research Part D*, Volume 2 (1), 57–76.

Scholl, L., Schipper, L. & Kiang, N., 1996. $CO_2$ emissions from passenger transport: a comparison of international trends from 1973 to 1992. *Energy Policy*, Volume 24 (1), 17–30.

Shrestha, R.M. & Marpuang, C.O.P., 2006. Integrated resource planning in the power sector and economy-wide changes in environmental emissions. *Energy Policy*, Volume 34 (18), 3801–3811.

Shrestha, R.M. & Timilsina, G.R., 1996. Factors affecting $CO_2$ intensities of power sector in Asia: a Divisia decomposition analysis. *Energy Economics*, Volume 18 (4), 283–293.

Timilsina, G.R. & Shrestha, A., 2009. Factors affecting transport sector $CO_2$ emissions growth in Latin American and Caribbean countries: An LMDI decomposition analysis. *International Journal of Energy Research*, Volume 33 (4), 396–414.

Wang, C., Chen, J. & Zou, J., 2005. Decomposition of energy-related $CO_2$ emission in China: 1957–2000. *Energy*, Volume 30 (1), 73–83.

Webster, F.V., Bly, P.H., Johnson, R.H. & Dasgupta, M., 1986a. Part 1: urbanization, household travel, and car ownership. *Transport Reviews*, Volume 6 (1), 49–86.

Webster, F.V., Bly, P.H., Johnson, R.H. & Dasgupta, M., 1986b. Part 2: public transport and future patterns of travel. *Transport Reviews*, Volume 6 (2), 129–172.

World Bank, 2008. World development indicators. Washington, DC: World Bank. http://wdi.worldbank.org/tables (Accessed on August 12 2016).

Wu, L., Kaneko, S. & Matsuoka, S., 2005. Driving forces behind the stagnancy of China's energy-related $CO_2$ emissions from 1996 to 1999: the relative importance of structural change, intensity change and scale change. *Energy Policy*, Volume 33 (3), 319–335.

Yabe, N., 2004. An analysis of $CO_2$ emissions of Japanese industries during the period between 1985 and 1995. *Energy Policy*, Volume 32 (5), 595–610.

# 7

# RESIDENTIAL ENERGY USE IN ASIA

*Subhes C. Bhattacharyya*

## Regional overview

Residential energy demand is a major final energy consuming sector in many Asian countries.[1] As indicated in Chapter 2, this sector was the highest consumer of final energy in 1990 but has been displaced to the second place by industry, particularly due to strong industrialisation efforts in East Asia. Unlike other sectors which are dominated by modern commercial energies, residential energy demand in Asia is greatly influenced by traditional biomass. As indicated in Chapter 4, Asia has the largest number of people in the world using biomass and traditional energies for cooking purposes. As a consequence, the demand evolution varies depending on whether these traditional energies are included or excluded from the analysis. Figure 7.1 indicates a modest growth in total residential energy demand (i.e. including biomass) – just 1.6% per year on average for the region between 1990 and 2012. This is almost half the growth rate of total final energy use in the region for the same period. The demand has grown from just below 600 Mtoe to just above 800 Mtoe during this period. However, the use of modern fuel is growing in the sector and its share increased from about 31% in 1990 to about 44% in 2012 (International Energy Agency, 2015). The effect of Asian financial crisis in 1997 is clearly visible in the figure.

East Asia and South Asia greatly contribute to shape the overall residential demand in the region. Taken together, they accounted for 85% of the demand in 1990 with East Asia contributing almost 58% of the demand. The situation has improved marginally over the period as the overall share of these two sub-regions had fallen to 82.5% by 2012, with South Asia accounting for about 30% of the demand. West Asia has the lowest share (about 3% in 2012) in residential energy demand of the region whereas the remaining share comes from South East Asia (13.5% in 2012).

However, the picture changes quite significantly when the biomass energy is excluded. East Asia remains the dominant player but its share changed to 79% in 1990 and 68% in 2012 (see Figure 7.2). South Asia's share reduced to 12% in 1990 and 16% in 2012. South East Asia accounted for 6% and 8% respectively in 1990 and 2012. The rest mainly came from West Asia. The figure suggests an accelerated growth in demand in the new millennium, recording a growth rate of almost 4% per year between 2000 and 2012. The penetration of modern energy in the sector varies by country and sub-region. Dependence on biomass is practically non-

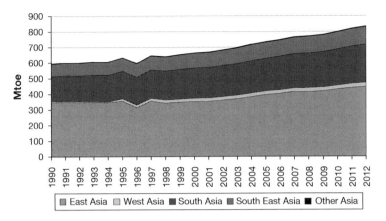

*Figure 7.1*   Evolution of total residential energy demand in Asia

*Data source*: International Energy Agency (2015).

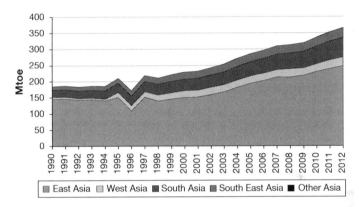

*Figure 7.2*   Trend of modern energy demand in the residential sector in Asia

*Data source*: International Energy Agency (2015).

existent in developed countries of the region (Japan, Singapore, South Korea) or in West Asian countries.

As population increased from three billion in 1990 to almost four billion by 2012 (UN DESA, 2015), the household energy consumption per person did not record any significant change. In 1990, the average consumption per person was 197 kgoe. This changed to 210 kgoe in 2012. However, there is a significant regional variation: West Asia has the highest level of consumption (close to 350 kgoe/person), followed by East Asia (above 250 kgoe/person), whereas south Asia has about 150 kgoe/person in 2012 (see Figure 7.3).

The energy mix of households has changed over this period of study. The biomass share reduced from 69% in 1990 to 58% in 2012. Coal has also lost its share from almost 16% in 1990 to 7% in 2012. Electricity share has gained rapidly: from 4.9% in 1990 to 16% in 2012. Natural gas has also recorded a significant gain in market share: from 2.7% in 1990 to almost 9% in 2012 (see Figure 7.4).

However, there is a significant regional variation in terms of energy mix (see Figure 7.5). As indicated earlier, West Asia has the lowest level of biomass dependence whereas South and

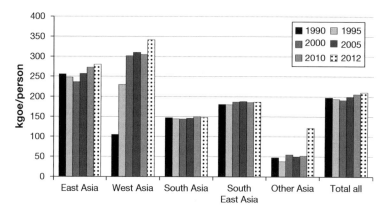

*Figure 7.3*  Household energy use per person in Asia (kgoe/person)

*Data source*: International Energy Agency (2015) and UN DESA (2015).

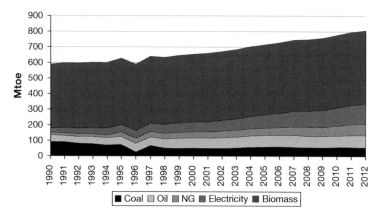

*Figure 7.4*  Fuel mix in the residential sector in Asia

*Data source*: International Energy Agency (2015).

South East Asia are still dependent on biomass for energy use. Natural gas has penetrated widely in West Asia and has secured its foothold in East Asia by 2012. The share of electricity, on the other hand, has grown in all regions – showing preference for convenient form of energy at the household level.

## Sub-regional pattern of residential energy use

Given the heterogeneity of regional demand pattern at the household level, a closer look at each sub-region is appropriate.

### Residential energy use in East Asia

Residential energy use in East Asia increased from 346 Mtoe in 1990 to 446 Mtoe in 2012, recording a modest growth on 1.1% per year. Out of this, about 200 ktoe of biomass was used, most of which was in China. The biomass use remained almost unchanged, thereby registering

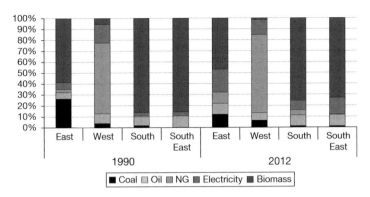

*Figure 7.5*   Regional variation in household energy mix

*Data source*: International Energy Agency (2015).

a decline in its share in the energy mix over this period. China's biomass dependence in the residential sector has fallen from 69% in 1990 to 53% in 2012. The growth in residential energy demand has been met by modern energies and the consumption has increased from 145 Mtoe in 1990 to 247 Mtoe in 2012, recording an average annual growth of 2.45%. Three countries, China, Japan and South Korea contributed more than 96% of the modern energy use in the sector of this region and are considered below.

The fuel mix in the region has evolved over time: coal use in the sector has declined from almost 91 Mtoe in 1990 to about 51 Mtoe in 2012. China and South Korea were the major coal users in 1980 but coal has been almost phased out in residential use in South Korea by late 1990s as natural gas started to penetrate the market. Coal is still used in the residential sector in China but is largely confined to rural areas. On the other hand, consumption of oil products, natural gas and electricity has increased manifolds: consumption of electricity and natural gas grew four times while that of oil products doubled (see Figure 7.6). However, the demand growth has been much faster in China compared to other countries. As a result, whereas Japan dominated the modern energy use in the region in this sector in 1990s, China has taken over the position in the last decade with a contribution of almost 60%. Japan still accounts for 20 to 30% of the sub-regional demand in the sector, with higher end ratios for oil and electricity and the lower end for natural gas.

Figure 7.6 also indicates high share of electricity in residential energy demand in all countries. Households, particularly urban households are benefitting from the convenience of electrical appliances. As the appliance stock improves, the share of electricity in the energy mix improves as well.

Despite growing residential energy use in China, its energy consumption per person is still low compared to Japan and South Korea. While Japan is showing a minor reduction in per capita consumption, it comes second to South Korea now, where on average 400 kgoe are being used per person. China, on the other hand, consumes almost a third of South Korean average household consumption (about 127 kgoe/person) (see Figure 7.7).

There is also significant urban–rural variation in residential energy use. According to Zheng et al. (2014), while urban residents live in smaller dwellings (96 m$^2$ compared to 135 m$^2$), they consume more energy (651 kgce/person compared to 445 kgce/person in rural areas). The fuel mix is also different: fuelwood still dominates the energy mix of rural areas whereas district heating accounts for 56% of energy used in urban households. Natural gas is almost non-existent in rural areas but 20% of urban residential energy comes from natural gas. The share of electricity

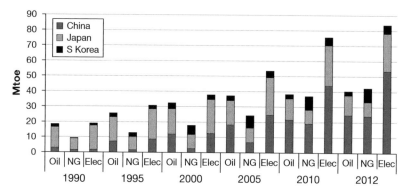

*Figure 7.6*   Evolution of modern energy use in the residential sector in East Asia

*Data source*: International Energy Agency (2015).

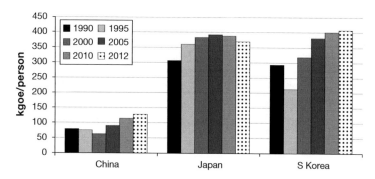

*Figure 7.7*   Comparison of residential energy use per capita (excluding biomass)

*Data source*: International Energy Agency (2015).

consumption in two areas is quite similar (about 15%) however. Space heating is the most energy consuming activity in urban areas whereas cooking consumes most energy in rural China.

However, China is urbanising rapidly: from 19% in 1980, the share of urban population increased to 56% in 2015 and by 2050, 77% of China's population will live in urban areas according to the United Nations (2014). Urbanisation has direct implications for residential energy use. The higher energy use per person in an urban household will exert pressure on energy demand and higher reliance on modern energies will reduce China's dependence on fuelwood. But high fossil fuel dependence will increase the carbon footprint, which suggests the importance of improving energy efficiency of appliances and building stocks. Urbanisation can facilitate improvements in energy usage efficiency due to higher concentration of users (Fan et al., 2017).

Gradual decline in residential energy use per person in Japan can be attributed to its population decline and intensification of efforts to improve energy efficiency. The Top Runner Programme, introduced in 1998, has set the energy efficiency targets for equipment and machinery and covers 70% of the household energy consumption (International Energy Agency, 2016). In addition, Japan has a relatively new building stock: 75% of its buildings are built after 1980 and only 2% of the buildings were built before 1950s (International Energy Agency, 2016). A high demolition rate and new building stock has helped Japan to achieve higher energy efficiency in buildings. Through these efforts Japan has managed to restrain its residential energy demand over time.

Increase in per capita residential energy use in South Korea, on the other hand, is driven by growing per capita income, rising appliance ownership and growth in household numbers. Between 1990 and 2012, per capita income (in constant 2005 $) increased from 8500 $/capita to $23,500 per person. The number of households in the country has also increased: in 1990, 11.1 million households lived in the country but by 2015, the number had increased to 19.5 million.[2] The housing availability also increased during the same period: from 7.4 million units in 1990 to 16.4 million in 2015. Rising income and household number also leads to a higher appliance stock. A recent study (Yoo & Kim, 2014) found that electrical appliances (such as fridge, TV, rice cooker, radiators, computers, washing machines, vacuum cleaners, etc.) largely contribute towards electricity consumption in households. The amount of electricity used increases as the family size increases but electricity consumption per person is higher for one-member families compared to a four-member family. The share of one-member household has increased over time and in 2015 represents the most common type of households in the country. With an aging population, it is likely that the share of one-member households will increase further in the future.

## West Asia

West Asia has seen a four-fold increase in residential energy use between 1990 and 2012 and as indicated in Figure 7.3, the sub-region has the highest per capita residential energy use in Asia. This region also has the lowest share of traditional energies in the residential fuel mix (just about 1%). The region also shows a dramatic change in energy use pattern in the 1990s when the countries became independent states subsequent to the collapse of the former Soviet Union. There is data uncertainty during the early period of 1990; accordingly the trend from 1995 is presented (see Figure 7.8). Natural gas dominates the residential energy mix in the region – with a 65% share in the fuel mix in 2012. Electricity and heat supplies constitute the other two major energy sources, while others (coal, oil and biomass) supply the balance. The effect of economic crisis of 2008 is also clearly visible on the residential energy use when the demand fell sharply but the turnaround was quick and the sector has returned to growth path again in 2011.

Three countries, namely Uzbekistan, Azerbaijan and Kazakhstan, dominate the regional scene of this sector, with Uzbekistan occupying a prominent place. 78% of natural gas consumption in the residential sector of this region originates from Uzbekistan. In fact, Uzbekistan meets the majority of its energy needs through natural gas. Likewise, Azerbaijan is also greatly dependent

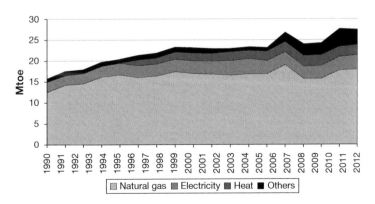

*Figure 7.8* Evolution of residential energy demand in West Asia

*Data source*: International Energy Agency (2015).

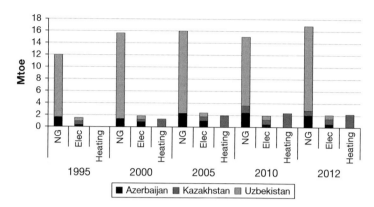

*Figure 7.9* Comparison of residential energy mix in three dominant West Asian countries

*Data source*: International Energy Agency (2015).

*Table 7.1* Per capita residential energy use (excluding biomass), kgoe/person

|  | *2000* | *2005* | *2010* | *2012* |
|---|---|---|---|---|
| Azerbaijan | 280.05 | 405.52 | 333.32 | 286.22 |
| Kazakhstan | 133.91 | 184.33 | 379.62 | 411.43 |
| Uzbekistan | 610.33 | 551.80 | 433.82 | 520.28 |

*Data source*: International Energy Agency (2015) and UN DESA (2015).

(80%) on natural gas for its residential energy needs. Kazakhstan, on the other hand, relies on a mix of fuels (coal, oil, natural gas, heat, and electricity), with district heating emerging as a major source of residential energy in recent times. The district heating system is very common in Kazakhstan that supplies steam, hot water and space heating to industrial, residential and commercial users (OECD–EAP Task Force, 2012).

Figure 7.9 presents the main residential energy use in three dominant countries of the region. Uzbekistan overshadows the other two very clearly in this figure. The consumption pattern is also relatively stable, with minor variation over the period. Notice also the reduction in electricity demand in the sector in Azerbaijan after 2005. This is attributed to an increase in electricity tariffs in the country (Energy Charter Secretariat, 2012).

The residential energy use per person varies quite significantly across these countries (see Table 7.1). Uzbekistan has the highest level of residential energy use per capita whereas Azerbaijan had the lowest in 2012. However, energy use per person in Kazakhstan has trebled between 2000 and 2012, thereby changing the order. The residential energy demand in Kazakhstan grew at 10% annually between 2000 and 2012, whereas electricity demand grew at 15% per year. This is driven by economic growth fuelled by oil and gas export revenue, which has resulted in a trebling of household income in the country (Kerimray et al., 2017). Consequently, the living space has expanded and appliance holding has increased, leading to higher energy demand.

A study (Kenisarin & Kenisarina, 2007) reported that the residential sector in Uzbekistan uses natural gas inefficiently for its heating needs due to poor metering, inappropriate tariff, poor insulation and inadequate regulatory requirements to save energy. The study suggested that almost 50% of the gas consumption could be reduced through a targeted programme. Similarly, specific heat consumption in Kazakh houses is reported to be very high (273 kWh/m2) compared to European countries (e.g. 130 kWh/m2 in England) (OECD–EAP Task Force, 2012). Residential consumption of heat is not metered and users pay for heat based on established norms. It is reported that the tariff does not reflect the true cost of energy use and does not provide any incentive to save energy. The district heating network is also outdated and is a source of heat loss (OECD–EAP Task Force, 2012).

Because of high dependence on fossil fuels and high per capita energy use, the residential sector of the region contributes significantly to carbon emissions. As the population of the region is likely to grow by 26% between 2015 and 2050 (UN DESA, 2015), continuing with the present trends will have significant implications for the future carbon emissions. However, all countries in the region are actively considering energy saving options through renovation, technological upgrading and better regulation.

## South Asia

The residential energy use in South Asia has increased by 50% between 1990 and 2012, recording an annual average growth rate of 1.9%. However, as biomass and traditional energies play a dominant role in the region and their size grew only 33% over this period, the modern energies recorded a faster growth in the region. Electricity use has recorded the fastest growth while coal use has gone out of fashion (see Figure 7.10). Natural gas is also gaining consumer support. In 2012, 75% of residential energy needs were satisfied by biomass and traditional energies, 10% by oil products, 9% by electricity, 5% by natural gas and 1% by coal.

The region remains highly dependent on biomass resources for its residential energy use (see Table 7.2). Although the dependence is declining, this is happening at a slow pace. Most of the countries rely on biomass for cooking and heating purposes, particularly in rural areas. Lack of affordable alternatives and lesser attention to cooking energy solutions are responsible for such a widespread reliance on traditional resources.

South Asian residential energy demand is highly influenced by India and Pakistan. Both of them accounted for almost 90% of the region's residential energy use in 2012. If Bangladesh is

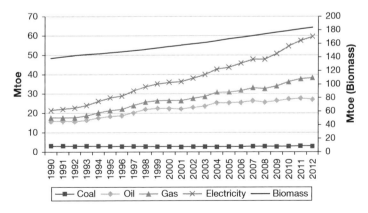

*Figure 7.10* Evolution of residential energy demand in South Asia

*Data source*: International Energy Agency (2015).

*Table 7.2* Biomass dependence in South Asia for residential energy needs (%)

|  | *1990* | *1995* | *2000* | *2005* | *2010* | *2012* |
|---|---|---|---|---|---|---|
| Bangladesh | 0.89 | 0.86 | 0.81 | 0.76 | 0.72 | 0.70 |
| India | 0.86 | 0.84 | 0.81 | 0.79 | 0.76 | 0.75 |
| Nepal | 0.98 | 0.97 | 0.96 | 0.96 | 0.97 | 0.97 |
| Sri Lanka | 0.97 | 0.95 | 0.93 | 0.91 | 0.88 | 0.88 |
| Pakistan | 0.82 | 0.80 | 0.78 | 0.77 | 0.74 | 0.72 |

*Data source*: International Energy Agency (2015).

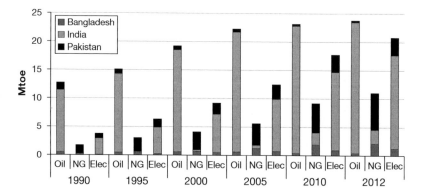

*Figure 7.11*    Evolution in commercial energy mix in major South Asian countries

*Data source*: International Energy Agency (2015).

included, the share increases to 95%. A focus on these three countries helps understand the regional picture. The evolution in the demand for major commercial energies is shown in Figure 7.11.

Oil products dominated the residential energy use in the 1990s but things have started to change in the new millennium when electricity has recorded a spectacular growth, particularly in India. As a result, in 2012, the gap between electricity and oil use has closed to a great extent. Natural gas demand has also grown substantially, particularly in Pakistan. The main petroleum products used by households are kerosene and LPG (liquefied petroleum gas). Kerosene is predominantly used in rural households whereas LPG is widely used by urban dwellers. As natural gas supply is contingent on the availability of distribution network, it is available in areas where the infrastructure exists or has been created.

South Asia remains the lowest residential energy user in Asia on a per capita basis. As Table 7.3 indicates the modern energy use per person remains modest. Pakistan leads the table with 56 kgoe per person per year, whereas on average households in Bangladesh use 24 kgoe per year. Population growth is one of the factors contributing to low energy use: between 1990 and 2012, more than 540 million people were added in these three countries.

India, being the most populous and largest country in this region, has contributed significantly to energy demand growth. Modern energy use in the sector has increased from about 17 Mtoe in 1990 to about 45 Mtoe in 2012, although traditional energies still support 75% of the overall energy needs of the sector. Oil products (mainly kerosene) remain the most important fuel, followed by electricity. Coal and natural gas account for the rest.

*Table 7.3* Per capita residential energy use in selected South Asian countries

| Per capita (kgoe) | 1990 | 1995 | 2000 | 2005 | 2010 | 2012 |
|---|---|---|---|---|---|---|
| Bangladesh | 7.64 | 9.60 | 13.39 | 18.59 | 22.40 | 24.42 |
| India | 20.18 | 23.35 | 27.34 | 30.31 | 35.48 | 35.87 |
| Pakistan | 32.70 | 37.85 | 41.55 | 45.44 | 49.67 | 56.04 |

*Data source*: International Energy Agency (2015) and UN DESA (2015).

India's residential energy demand has been influenced by a number of factors. Being the second most populous country in the world and having a quite young population, demography influences India's residential energy demand significantly. Between 1990 and 2012, India added 428 million people, more than double the Brazilian population or almost equal to the population of Middle East and North Africa. Moreover, the composition of the population is changing: the size of the working age population was 710 million in 2010, which contributed to the growth in labour supply and in turn to growth in energy demand. Alongside, the country has also seen significant urbanisation, economic growth and the rise of the middle income class. In the past, India's traditional rate of economic growth was less than 4% per year but in the new millennium, an accelerated growth was realised (7% or higher). This has helped the country to pull 137 million out of poverty between 2005 and 2012 (Reserve Bank of India, 2013). In addition, although officially 31% of India's population lived in urban areas in 2015, a World Bank Study suggested that 55% of the population lived in urban and urban-like areas in 2010 (Ellis & Roberts, 2016). 53% of India's GDP in 2012 originated in urban and urban-like areas (Brar et al., 2014) and the per capita GDP of urban districts is four to five times higher than the rural districts. As a result of urbanisation and economic growth, a sizeable middle income class has emerged in the country: according to Goldman Sachs (2010), the size of the middle income class has increased from 50 million households in 2002 to 100 million in 2010.

These changes have positively influenced residential energy demand. For example, based on National Sample Survey Office (2012) data for Indian households, there is a significant difference in energy use between urban and rural India (see Figures 7.12a and 7.12b).

While both urban and rural households use a combination of fuels, the rural households tend to rely more on traditional energies whereas urban households use modern energies to a greater degree. Moreover, the appliance stock changes dramatically with urbanisation. Rural households hardly use white goods such as refrigerators, air conditioners or washing machines but they are common in urban areas. Thus, urbanisation has shifted the pressure on traditional fuels to modern fuels and the energy use intensity has increased as well. This trend will continue in the future as the country becomes more urbanised (see Bhattacharyya (2015) for a more detailed analysis of this issue).

### South East Asia

The residential energy demand in this region has grown on average at 1.65% between 1990 and 2012. The fuel mix is highly dominated by biomass energy, accounting for 73% of the demand in 2012. Electricity accounted for 15% of the demand while oil products supported 10% of the demand. Coal and natural gas played a minor role in meeting the demand. Electricity demand has rapidly grown in the region – at 8.6% per year between 1990 and 2012 (see Figure 7.13). This is the fastest growing fuel in the region and as a result, electricity has established itself as the leader

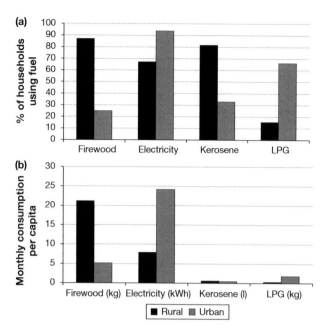

*Figure 7.12* (a) Urban–rural divide in energy use (b) Difference in fuel mix in urban and rural areas

*Data source*: National Sample Survey Office (2012).

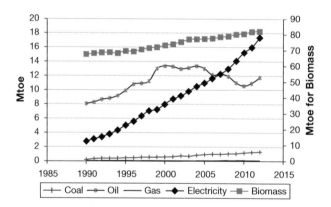

*Figure 7.13* Trend of fuel demand in the residential sector of South East Asia

*Data source*: International Energy Agency (2015).

in commercial energy category. On the other hand, oil products have recorded a declining trend, particularly between 2000 and 2010. A rapid slow-down in demand in Indonesia due to political instability, economic troubles and reduction in the subsidies was responsible for this trend.

The residential sector of the region depends on biomass energy to a great extent, except in Brunei and Singapore. A declining trend can be seen in biomass dependence over time in Malaysia, Thailand and Philippines (see Table 7.4). Myanmar, Cambodia, Indonesia and Vietnam still rely on biomass for more than two-thirds of their residential energy needs.

*Table 7.4* Biomass dependence in the residential sector of South East Asia

| Country | 1990 | 2000 | 2010 | 2012 |
|---|---|---|---|---|
| Brunei | 0 | 0 | 0 | 0 |
| Cambodia | na | 96% | 92% | 88% |
| Indonesia | 83% | 76% | 80% | 78% |
| Malaysia | 59% | 46% | 39% | 38% |
| Myanmar | 99% | 99% | 98% | 97% |
| Philippines | 68% | 69% | 58% | 57% |
| Singapore | 0% | 0% | 0% | 0% |
| Thailand | 65% | 66% | 60% | 53% |
| Vietnam | 83% | 84% | 70% | 68% |

*Data source*: International Energy Agency (2015).

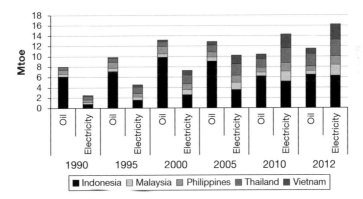

*Figure 7.14*   Changing fuel mix in the residential sector of South East Asia

*Data source*: International Energy Agency (2015).

In terms of modern energy use, while Indonesia dominates the energy scene of this region, four other countries, namely Vietnam, Thailand, Malaysia and the Philippines play an important role in the overall demand. The changing roles of oil and electricity can be clearly seen here (see Figure 7.14). Electricity has gradually taken over as the dominant modern residential fuel in the region. The region is urbanising rapidly – with more than 50% of the population living in urban areas by 2012. The preference for electricity in urban areas has driven the change. Moreover, IEA (2013) indicates the prevalence of widespread subsidy for petroleum products and electricity as a reason for demand growth (IEA, 2013).

The unit consumption (or per capita consumption) remains relatively low in the region (see Table 7.5). Vietnam has shown a spectacular growth (more than eight-fold increase between 1990 and 2012) in unit consumption in the residential sector. An average GDP growth of 7% between 1990 and 2012 and consequent three-fold increase in per capita income is the main reason. However, the income elasticity of demand appears to be much above one here, which may have happened due to the low starting point from where the demand is increasing in Vietnam. Thailand has also seen a four-fold increase in per capita modern energy use in this sector while in Malaysia the consumption per person has more than doubled during the same period. Indonesia and the Philippines, on the other hand, have recorded much lower

*Table 7.5* Trend of modern energy use per person in the residential sector of South East Asia

| Per capita energy use (ktoe) | 1990 | 1995 | 2000 | 2005 | 2010 | 2012 |
|---|---|---|---|---|---|---|
| Indonesia | 38.72 | 44.67 | 60.59 | 57.54 | 47.77 | 51.42 |
| Malaysia | 45.86 | 59.62 | 70.46 | 83.85 | 92.94 | 98.00 |
| Philippines | 20.50 | 27.70 | 32.90 | 28.17 | 27.61 | 27.02 |
| Thailand | 22.15 | 34.23 | 41.71 | 55.05 | 68.78 | 82.63 |
| Vietnam | 7.30 | 14.41 | 24.37 | 40.48 | 55.39 | 59.61 |

*Data source*: International Energy Agency (2015) and UN DESA (2015).

growth in unit consumption. High population growth and modest economic growth can explain this trend.

Based on the sub-regional analysis, it is found that sub-regional features get submerged at the regional analysis and the overall pictures hide the inner differences. While the biomass use remains high in the region, particularly for cooking energy purposes, the use of modern energies is increasing fast. Electricity use in increasing at a fast rate – countries with high economic growth rate and high urbanisation rate are demanding more modern energy. Electricity is emerging as the preferred fuel in the residential sector. There is significant variation across the region in terms per capita energy use but countries are catching up the high consumption pattern and over time, modern energy demand will only increase.

## Future outlook of residential energy use in Asia

The evolution of residential energy use depends on the main drivers of demand change and their interactions. As indicated before, demographic transition in the region has a significant influence. According to UN DESA (2015), Asia's population will increase by 530 million between 2015 and 2030 and another 344 million will be added between 2030 and 2050, making a total of 874 million additional people by 2050. Further, the working age population in most Asian countries will dominate the demography – thus the ratio of working to non-working population will still remain favourable, generating the population dividend. In addition, the size of urban population will increase: according to United Nations (2014), by 2050, 64% of Asian population will live in urban areas. As a result, the urban population will increase from 2.06 billion in 2014 to 3.31 billion in 2050. Moreover, urban agglomerations being centres of economic growth, a larger share of the economic output will come from urban areas and will fuel income growth as well. This will expand the size of the middle class in the region, who will have better buying power. This will be supported by the economic outlook for the region which remains promising and this will influence the residential energy use in the future. Further, the region is actively pursuing initiatives to improve the level of energy access, particularly in South and South East Asia. Improved access to sustainable energy is likely to reduce the dependence on biomass for cooking and the share of modern fuel use will increase in the region.

According to EIA (2016), China will experience an annual average increase of 2.4% in the residential energy demand between 2012 and 2040. But it projects a higher rate of demand growth (3.2% per year) for India and a slightly slower rate of 2.3% for other Asian countries. The total demand in developing Asia (excluding OECD Asia) reaches to 408 Mtoe in 2025 and increases to 580 Mtoe in 2040 (see Figure 7.15). Electricity becomes the most prominent fuel in

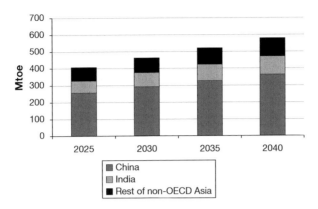

*Figure 7.15* Residential energy outlook in developing Asia (reference scenario)

*Data source*: EIA (2016).

the residential sector of the region but oil products continue to be used. Natural gas use increases in China but in the rest of Asia its role remains limited. Clearly, China's domination in the sectoral energy use continues and India's share rises to match the rest of developing Asia. However, the traditional energy use has not been included in this outlook.

In their forecast for the region, the Asian Development Bank includes residential energy sector in other sectors and suggests an average growth of 2.5% for the period up to 2035 (ADB, 2013). China, India and South East Asia will contribute to the growth in energy demand. The projected growth rates are higher than the historical growth noted earlier. The rise in population, sustained economic growth and better affordability will contribute to the growth in residential energy use.

## Conclusion

The residential sector in Asia is a major user of energy and its high dependence on biomass for cooking has remained an issue for most of the developing countries in the region. However, as economies develop, the preference for electricity as the versatile form of energy is becoming clear. Electricity demand has rapidly grown in the residential sector and the outlook suggests that this is likely to intensify in the future as the middle income class expands and the countries become more urbanised. However, as coal is the main fuel for electricity generation in the region, increased use of electricity adds to higher carbon emissions. The cooking energy will mainly shift towards liquid or gaseous petroleum fuels but it remains to be seen whether solid biomass is displaced in favour of modern, clean fuels. The regional demand is dominated by East Asia, followed by South Asia, and this pattern is likely to continue in the future. The region has significant potential for demand reduction through demand management and energy efficiency improvements. Appropriate pricing of energy holds the key but energy price rationalisation remains a politically sensitive issue in the region.

## Notes

1 This chapter follows the same regional coverage as in Chapter 2 and the sub-regional composition indicated in that chapter remains applicable here as well.
2 KOSTAT, Complete Enumeration Results of the 2015 Population and Housing Survey, http://kostat. go.kr/portal/eng/pressReleases/8/7/index.board

# References

ADB, 2013. *Energy Outlook for Asia and the Pacific*, Manila: Asian Development Bank.

Brar, J. et al., 2014. *India's Economic Geography in 2025: States, Clusters and Cities: Identifying the High Potential Markets of Tomorrow*, Mumbai: McKinsey & Co.

Bhattacharyya, S.C., 2015. Influence of India's transformation on residential energy demand, *Applied Energy*, Volume 143, pp. 228–237.

EIA, 2016. *International Energy Outlook 2016*, Washington, DC: US Energy Information Administration.

Ellis, P. & Roberts, M., 2016. *Leveraging Urbanisation in South Asia: Managing Spatial Transformation for Prosperity and Livability*, Washington, DC: World Bank.

Energy Charter Secretariat, 2012. *In-Depth Review of the Energy Efficiency Policy in Azerbaijan*, Brussels: The Energy Charter Secretariat.

Fan, J., Zhang, Y. & Wang, B., 2017. The impact of urbanization on residential energy consumption in China: An aggregated and disaggregated analysis. *Renewable and Sustainable Energy Reviews*, Volume 75, pp. 220–233.

Goldman Sachs, 2010. *India Revisited, White Paper*, s.l.: Goldman Sachs Asset Management.

International Energy Agency, 2013. *Southeast Asia Energy Outlook*, Paris: International Energy Agency.

International Energy Agency, 2015. *World Energy Balance (via UK Data Service)*, Paris: International Energy Agency.

International Energy Agency, 2016. *Energy Policies of IEA countries: Japan*, Paris: International Energy Agency.

Kenisarin, M. & Kenisarina, K., 2007. Energy saving potential in the residential sector of Uzbekistan. *Energy*, Volume 32(8), pp. 1319–1325.

Kerimray, A., Miglio, R., Rojas-Solorzano, L. & Gallachoir, B., 2017. Household energy consumption and energy poverty in Kazakhstan. *IAEE Energy Forum*, Q1, pp. 31–34.

National Sample Survey Office, 2012. *Household Consumption of Various Goods and Services in India, NSS 66th Round*, New Delhi: National Sample Survey Office, Government of India.

OECD–EAP Task Force, 2012. *Promoting Energy Efficiency in the Residential Sector in Kazakhstan: Designing a Public Investment Programme*, Paris: Organisation for Economic Cooperation and Development.

Reserve Bank of India, 2013. *Handbook of Statistics of the Indian Economy*, Mumbai: Reserve Bank of India.

UN DESA, 2015. *World Population Prospects: Key Findings & Advance Tables, 2015 Revision*, New York: Department of Economic and Social Affairs, United Nations.

United Nations, 2014. *World Urbanization Prospects: Highlights*, New York: United Nations.

Yoo, J. & Kim, K., 2014. Development of methodology for estimating electricity use in residential sectors using national statistics survey data from South Korea. *Energy and Buildings*, Volume 75, pp. 402–409.

Zheng, X. et al., 2014. Characteristics of residential energy consumption in China: Findings from a household survey. *Energy Policy*, 75, pp. 126–135.

# PART 2

# Energy supply in Asia

# 8

# OIL IN ASIA

*Tilak K. Doshi*

## Introduction

In the context of the global oil industry, the most useful geographical definition of Asia typically comprises countries east of, and including, Pakistan, tracing an arc around the Pacific Rim of the Eurasian landmass to Northeast Asia (China, Japan, South Korea and Taiwan). This geographical definition excludes countries in West Asia such as Afghanistan as well as the Central Asian republics of the former Soviet Union (FSU), all of which are minor oil consumers. In this definition, Asia includes South Asia (i.e. the countries of the Indian sub-continent), Southeast Asia, Oceania (i.e. Australia, New Zealand and the Pacific Islands) and Northeast Asia.

Given the vastness of this region and its heterogeneity across the domains of economics, politics, culture and demography, useful generalisations are difficult to make. In attempting to cover such an extensive subject area within the space of a book chapter, the practical approach would be to focus on key developments that highlight the most important aspects of the evolution of the Asian oil industry. In this chapter, we concentrate on three overarching themes: Asia's hitherto rapid oil demand growth, the trade and pricing nexus between the Middle East and Asia which covers the world's largest inter-regional crude oil flows, and the increasingly important role of Asian national oil companies (NOCs).

## Oil demand in Asia: rapid growth in net imports

The first factor in any narrative of 'oil in Asia' would be on the demand side. Even though Asia has had significant crude oil exporters such as Indonesia and Malaysia, the region has long been a net importer. The scale and speed of demand growth for oil among the Asian developing countries has been a dominating element in the evolution of the global oil industry over the past few decades. Popular references to an 'Asian century' preceding the onset of the twenty-first century reflected the shift in the centre of gravity of the global economy from North America and northwest Europe towards Asia. Arguably, this shift reflected the single most important economic development affecting the material welfare of the world's population in modern history. The lifting of the hundreds of millions of the poor in Asia has been seen as among the greatest achievements of humankind's experience since the Industrial Revolution.

119

*Table 8.1* GDP growth rates and share of world GDP for regions and select Asian countries

| Country name | Annual average GDP growth rates (%) | | | Percentage of world GDP |
|---|---|---|---|---|
| | 1990–2000 | 2000–2015 | 2015 | 2015 |
| East Asia and Pacific | 3.8 | 4.5 | 3.9 | 31.5 |
| East Asia and Pacific (excluding high income countries) | 8.0 | 8.4 | 6.4 | 21.9 |
| China | 9.9 | 9.5 | 6.9 | 17.2 |
| India | 5.6 | 7.0 | 7.6 | 7.0 |
| Japan | 1.5 | 0.8 | 0.5 | 4.2 |
| Korea, Republic of | 6.9 | 4.3 | 2.6 | 1.5 |
| Europe and Central Asia | 1.7 | 1.8 | 1.5 | 23.7 |
| North America | 3.2 | 2.0 | 2.2 | 17.2 |
| Middle East and North Africa | 4.5 | 4.3 | 2.9 | 7.0 |
| Latin America and the Caribbean | 2.9 | 2.9 | − 0.6 | 8.7 |
| Sub-Saharan Africa | 2.1 | 5.0 | 2.9 | 3.3 |
| World | 2.8 | 2.9 | 2.5 | 100.0 |

*Data source*: World Development Indicators Database, World Bank, accessed at http://data.worldbank.org/indicator/NY.GDP.MKTP.KD.ZG.

*Notes*: GDP growth rates are simple averages of annual percentage growth rate of GDP at market prices based on constant local currency; aggregates are based on constant 2010 US dollars; percentage share of 2015 world GDP measured in PPP current international dollars.

Indeed, the rapid ascent of the Asian emerging markets is the leading economics storyline of our age.

Table 8.1 indicates average GDP growth rates for select Asian countries as well as global regions over the periods 1990–2000 and 2000–2015 based on the World Bank's *WDI* database (World Bank, 2016). The World Bank's geographical delineation of the 'East Asia and Pacific' excludes South Asia and hence differs from the definition of Asia adopted in this chapter. The East Asia and Pacific region accounted for 31.5% of global GDP in 2015, and if India (the largest economy in South Asia by far) is added in, the region's share increases to 38.5%. The East Asia and Pacific region grew significantly faster since 1990 than all other regions save the far smaller region (by population and area) of the Middle East and North Africa. India experienced higher growth rates than the East Asia and Pacific region over the past quarter century. Indeed, the slowdown in China's growth rate in 2015 to less than 7% has made India the fastest growing large economy in the world (IMF, 2016). Developing East Asian countries were among the fastest growing economies in the region after China. The dynamism of developing Asia is apparent in the very high aggregate growth rates shown for the East Asia and Pacific region once the high income countries are excluded from the sample, with GDP growing by an annual average of 8–8.5% since 1990. This growth performance is exceeded only by China's stellar average of close to 10% through the period.

Commensurate with the rapid economic growth of non-OECD Asia, oil demand growth in the region also exceeded that of most other regions in the world with the exception of the Middle East. Table 8.2 shows estimated regional oil demand growth from BP's *Annual Review of World Energy Statistics*, which defines the Asia Pacific region coterminous to that of 'Asia' adopted in this chapter. From accounting for just under 15% of total global consumption of oil in 1970, Asia's

Table 8.2 Oil consumption growth by region

| | Thousand barrels daily | | | Share of total (%) | | | Cumulative average growth rate (%) | |
|---|---|---|---|---|---|---|---|---|
| | 1970 | 1990 | 2015 | 2015 | 1990 | 1970 | 1970–1990 | 1990–2015 |
| North America | 16,593 | 20,316 | 23,644 | 23.9 | 30.5 | 36.7 | 1.0 | 0.8 |
| South and Central America | 2,066 | 3,772 | 7,083 | 7.5 | 5.7 | 4.6 | 3.1 | 3.2 |
| Europe and Eurasia | 18,155 | 23,133 | 18,380 | 19.9 | 34.7 | 40.1 | 1.2 | − 1.1 |
| Middle East | 1,051 | 3,599 | 9,570 | 9.8 | 5.4 | 2.3 | 6.3 | 5.0 |
| Africa | 703 | 1,993 | 3,888 | 4.2 | 3.0 | 1.6 | 5.3 | 3.4 |
| Asia Pacific | 6,661 | 13,854 | 32,444 | 34.7 | 20.8 | 14.7 | 3.7 | 4.3 |
| Total world | 45,229 | 66,667 | 95,008 | 100.0 | 100.0 | 100.0 | 2.0 | 1.8 |

Data source: BP (2016).

Table 8.3 Incremental oil demand (thousand barrels per day)

| | 1970–1990 | 2000–2010 | 1970–2015 |
|---|---|---|---|
| China | 1,743 | 4,740 | 11,414 |
| India | 821 | 1,060 | 3,768 |
| Total Asia | 7,193 | 6,793 | 25,783 |
| Total world | 21,438 | 11,776 | 49,779 |
| China as % of Asia | 24 | 70 | 44 |
| India as % of Asia | 11 | 16 | 15 |

Data source: BP (2016).

share grew to more than 20% in 1990 and almost 35% in 2015 (BP, 2016). Asian oil consumption grew by a compound annual growth rate of 3.7% during 1970–1990 and by 4.3% during 1990–2015. This compares with the respective growth rates of 2.0% and 1.8% for global oil demand growth during the two periods. The rapid increase in Asia's oil demand is perhaps best measured by the region's share of global incremental demand. During the two decades 1970–1990, global demand grew by 21.4 million barrels per day (b/d), of which 7.1 million b/d were accounted for by Asia – approximately a third of the world total. In the past quarter century, Asian demand has increased by 18.5 million b/d, accounting for about two-thirds share of global demand growth.

China's role as the centrepiece of oil demand growth over the past two decades is remarkable (see Table 8.3). The country accounted for a little less than a quarter of Asia's increase in oil demand during 1970–1990. If we take a longer period of almost half a century from 1970 to 2015, China accounted for 44% of total Asian demand growth or almost a quarter of global demand growth. During the extraordinary commodity boom of the 2000s, China alone accounted for 70% of the total Asian oil demand increase or 40% of global demand growth. By any standard, China's weight on the demand side of global oil markets is notable. India is emerging as the other large non-OECD Asian oil consumer, and accounted for between 11% and 16% of incremental Asian demand during the three time spans examined in Table 8.3. In the past few years, India has emerged as an increasingly important oil consumer as its demand for oil has gained momentum,

*Table 8.4* US EIA forecasts for petroleum liquids consumption by region and country (million b/d)

|  | 2012 | 2040 | Growth rate (%) 2012–2040 |
|---|---|---|---|
| OECD Americas | 23.245 | 24.648 | 0.2 |
| OECD Europe | 14.052 | 13.975 | 0.0 |
| OECD Asia | 8.241 | 7.474 | − 0.3 |
| Total OECD | 45.539 | 46.097 | 0.0 |
| Non-OECD Asia | 21.502 | 38.909 | 2.1 |
| China | 10.175 | 16.358 | 1.7 |
| – India | 3.618 | 8.263 | 3.0 |
| – Other | 7.709 | 14.289 | 2.2 |
| Middle East | 7.705 | 13.232 | 2.0 |
| Africa | 3.609 | 6.929 | 2.4 |
| Central and South America | 6.68 | 9.602 | 1.3 |
| Total non-OECD | 44.765 | 74.792 | 1.9 |
| Total world | 90.304 | 120.889 | 1.0 |

*Data source*: US Energy Information Administration (EIA), 'World Liquids Consumption by Region', www.eia.gov/forecasts/aeo/data/browser/#/?id=5-IEO2016&sourcekey=0; accessed October 25, 2016.

although its weight in global incremental demand is far smaller than China's. For the most recent year, 2015, India accounted for almost a quarter (23.4%) of Asia's incremental demand (compared with 2014), while China contributed just under 58% (BP, 2016).

US Energy Information Administration long-range oil demand forecasts (EIA, 2016) – in common with other well-cited forecasts such as the IEA's World Energy Outlook and energy outlook reports by private sector companies such as BP and Exxon-Mobil – project stagnant OECD oil demand to 2040, with negative or nil growth in OECD Europe and OECD Asia, with North America growing by a tepid 0.2% per annum (see Table 8.4). Non-OECD Asia demand is forecast to grow annually by 2.1%, double the world average of 1.0%. China's share of future global incremental oil demand will likely diminish. Indeed, some observers have noted that were it not for China's aggressive agenda to fill its strategic petroleum reserves in the past few years, exceeding even its own ambitious reported plans for oil stockpiles, the country's imports would have been significantly lower in 2015 and 2016.[1] As China progressively shifts from export-led manufacturing to a services-based economy, the country's rates of oil demand growth will decelerate. While China's relative share in oil demand growth declines, other Asian developing countries in South and Southeast Asia, particularly India, will play an increasing role. Asia and the Middle East will continue to be the fastest demand growth regions, growing at almost double the world demand growth rate of 0.9%. From its low base, Africa will have the highest rate of demand growth of 2.4% after India's projected 3.0%.

## The Middle East: Asia oil trade and pricing nexus

With the rapid economic growth in developing Asia, there has been a corresponding eastward shift in the centre of gravity of the global oil trade. Asia's vast appetite for oil is accommodated by trade flows primarily out of the Middle East. For Middle Eastern oil exporters, Asia has been, and will continue to be for the foreseeable future, the largest growing importing region in the world and, given its relative geographical proximity, it enjoys logistical advantages. From an Asian perspective, the Middle East – endowed with the world's largest low cost reserves – is ideally positioned to meet Asia's growing demand. Thus, a majority of Middle Eastern exports head to Asia while, in turn, Asia hauls most of its crude oil imports from the Middle East. According to recent estimates of inter-regional crude flows for 2015, more than two-thirds of Asian oil imports were sourced from the Middle East; Asia, in turn, accounted for more than three-quarters of total Middle East oil exports (BP, 2016).

## Inter-area trade flows

Table 8.5 lists estimates of important inter-area oil flows (in million barrels per day, b/d) in 2011 and 2015, accounting for about 70% of global imports. Middle East crude oil exports to Asia constitute by far the single largest inter-area crude oil flow in the global oil trade. They accounted for 14.5 million b/d in 2011, and this increased by more than 3 million b/d in 2015, accounting for 41.2% of all the crude flows listed in the table. Exports to Europe from the former Soviet Union (primarily Russia) are the second largest inter-area flow, accounting for 15–16% of the flows listed or around 6 million b/d. The largest change in trade flows between 2011 and 2015 is the increase in Middle East exports to Asia, reflecting Asia's rapid growth in oil demand. The next major change is the sharp decreases in West African and Latin American oil exports to North America, which fell from 1.5 to 0.4 million b/d and from 2.3 to 1.8 million b/d respectively. Canadian exports to the US increased by 1 million b/d in the same period. Russia's exports to Asia increased by 0.6 million b/d. The other major oil-producing regions, West Africa and Latin America, also increased their exports to Asia, by 0.3 and 0.4 million b/d respectively.

These shifts in inter-area oil movements have been a result of two key factors. Major crude oil exporting regions are re-orienting towards Asia simply because this is where much of the global incremental demand is located, apart from the Middle East itself. In terms of demand growth,

*Table 8.5* Crude oil inter-area flows (million b/d)

| Regions | 2015 | | 2011 | | Change |
|---|---|---|---|---|---|
| | Million b/d | % | Million b/d | % | Million b/d |
| Middle East–Asia | 17.6 | 41.2 | 14.5 | 37.9 | 3.1 |
| FSU–Europe | 6.2 | 14.6 | 6.0 | 15.7 | 0.2 |
| Canada–US | 3.7 | 8.8 | 2.7 | 7.0 | 1.0 |
| Middle East–Europe | 2.6 | 6.2 | 2.5 | 6.5 | 0.1 |
| FSU–Asia | 2.3 | 5.3 | 1.7 | 4.4 | 0.6 |
| West Africa–Asia | 2.1 | 4.9 | 1.8 | 4.7 | 0.3 |
| Latin America–North America | 1.8 | 4.2 | 2.3 | 6.0 | − 0.5 |
| West Africa–Europe | 1.7 | 4.1 | 1.2 | 3.1 | 0.5 |
| Middle East–North America | 1.6 | 3.7 | 2.0 | 5.2 | − 0.4 |
| Latin America–Asia | 1.5 | 3.5 | 1.1 | 2.9 | 0.4 |
| North Africa–Europe | 1.1 | 2.6 | 1.0 | 2.6 | 0.1 |
| West Africa–North America | 0.4 | 0.8 | 1.5 | 3.9 | − 1.1 |
| Total | 42.6 | 100.0 | 38.3 | 100.0 | |
| Global imports | 60.7 | | 54.6 | | |
| Total listed flows as % of global imports | 70.2 | | 70.1 | | |

*Data source*: BP Statistical Review of World Energy, 2012 and BP Statistical Review of World Energy, 2016.

*Note*: Data for 2011 includes some refined product flows as there is no disaggregation between crude oil and refined products; 2015 data refers to crude oil flows only.

however, Asia has to be divided between the large but saturated oil markets of OECD Asia-Pacific, on the one hand (Japan and South Korea) – where import needs are expected to decline by the end of the century – and China and other non-OECD Asian importers, on the other hand, whose need for imported barrels will continue to grow. The second factor is the dramatic surge in unconventional production of liquids ('light tight oil') in the US and continued growth in heavy oil output from Canada which have displaced foreign imports into North America, especially from West Africa and Latin America.

The decision by OPEC in November 2014 to maintain output and market share despite sharp falls in oil prices in the latter half of that year has led to severe competitive pressures as major crude oil producers from outside the region compete aggressively in Asian oil markets. As a result of the shale oil boom in the US and sluggish demand in Europe, West African, Latin American and Russian Far East crude oil exports to Asian markets increased and challenged the traditional large exporters in the Middle East. For many of the large Asian crude oil importers who had been intent on diversifying regional sources of crude oil imports, the low oil price environment since mid-2014 has provided an increased momentum to Asian imports of non-Middle Eastern crude. This provides a perceived strategic diversification benefit for Asian importers from what was previously expected to be an inevitable growing dependence on Middle Eastern supplies.[2] Increasing volumes of crude oil formerly traded within the Atlantic Basin, such as those originating from West Africa, Latin America and even occasionally the North Sea, are now exported to Asia. The oil exporters of the Gulf Cooperation Council (GCC) such as Saudi Arabia and Kuwait are locked into a competitive struggle over Asian market share, not only with crude suppliers from other regions, but also with other major exporters within the Gulf OPEC region such as Iraq and Iran which had lost market share after years of wars and sanctions.

## *Pricing of Middle East crude oil in Asia*

Most primary commodity trade occurs in bilateral contracts which determine commodity quality specifications, time and place of delivery, and price. At one end of the spectrum, these contracts can be private and confidential between buyer and seller, and may be based on opaque arrangements, for instance when buyers have equity stakes in, or make long-term loans to, commodity suppliers in return for a long-term commodity supply contract at 'preferential' prices. At the other end, trade may be mediated by electronic commodity exchanges which allow trade of highly standardised commodity contracts with strictly specified volumes, qualities, delivery schedules, modes of allowed delivery (ex-ship or ex-storage facility, etc.), payment terms, and so on. In exchange-traded commodity contracts, the counter-party to every buyer and seller is the exchange's Clearing House. It should be noted that in most exchange-traded commodity contracts, physical delivery often plays a secondary role, as most futures contracts are liquidated via cash settlement, i.e. the purchase of an equal and opposite position on the contract before its physical delivery is due.

The group of commodities traded on modern electronic exchanges has expanded since the earliest ones were founded in the late nineteenth century in the US and the UK. The Goldman Sachs Commodity Index (GSCI) basket, for instance, includes 6 energy contracts,[3] 7 industrial and precious metals contracts, and 11 agriculture and livestock contracts. However, much of the physical trade still occurs outside of the commodity exchanges for reasons ranging from difficulty in grade standardisation and the role of dominant producers or buyers in preferring direct bilateral negotiations to government price or supply regulations. In highly concentrated industries such as iron ore, a few dominant sellers and buyers set market 'guideline' prices which are then used as a reference price for trading by numerous smaller buyers and sellers around the world.

Relatively few commodity prices are directly discovered by the formal commodity exchanges. Instead, there are a large class of commodities which are priced using assessments published by price reporting agencies (PRAs). Indeed, the largest (by value) internationally traded commodity categories – crude oil and refined petroleum products – are overwhelmingly transacted on the basis of price assessments published by private agencies such as Platts, ICIS and Argus Media. The influence of such PRAs extends across the commodities universe, including petrochemicals, coal and liquefied natural gas (LNG), metals, as well as other goods or services outside the traditional commodities space, such as oil freight rates and wholesale power tariffs. In the over-the-counter (OTC) oil derivatives market, an estimated 60–70 % of all contracts are priced or cash-settled by reference to quotes published by PRAs.[4] Among the PRAs, Platts plays a dominating role in oil-related price assessments. An estimated 80% of global crude oil and refined products trading, including almost all of the Middle East's sales of crude oil to its largest markets 'east of Suez', is transacted on the basis of Platts published price quotes.

Given that the Middle East–Asia oil trade is the largest among the world's inter-area oil flows by far, the pricing of Middle East crudes is a perennial and critical concern in the evolution of Asian oil markets. The pricing (or mispricing as some would have it) of crude oil traded in Asia is a key variable at the macro-level in determining financial flows and the balance of payments of the countries in Asian oil markets. And it is also central to the profit and loss accounts for the IOCs, NOCs and oil trading companies operating in Asia. Formula prices were first adopted by Mexico in early 1986 and widely accepted by Middle East oil exporters soon after. This was in the aftermath of the 1985/1986 collapse of the administered pricing system of the previous decade, where crude sales prices were directly set by OPEC oil ministers' meetings. The vast majority of Middle East crudes exported to Asia and elsewhere are based on long-term contracts (usually evergreen contracts which are renewable annually) on an FOB (free on board) Arabian Gulf basis. A small share of the crude exports is sold on a CIF (cost, insurance and freight) basis, with deliveries provided by the shipping arms of the Middle East NOCs. The pricing formula generally has four components: point of sale, a market-related base price, an adjustment factor that is reflective of crude oil quality, and a timing mechanism that stipulates when the value of the formula is to be calculated. For term buyers of Middle East crude oils marketed in Asia, the FOB Gulf crude oil price is linked to the monthly average spot (or 'front month') price of Oman and Dubai crude oils (O/D) during the month in which the crude is loaded at an Arabian Gulf port for delivery. The base price for crude lifted in the Gulf is then adjusted by adding or subtracting a monthly announced 'offset' or adjustment factor. The adjustment factor takes into account the quality differential between the given Middle East crude grade and the reference crude it is being priced off.[5]

Unlike the Atlantic Basin markets where crude reference prices (Brent Blend and West Texas Intermediate) are discovered in liquid futures markets such as the InterContinental Exchange (ICE) and the New York Mercantile Exchange, the reference or 'marker' crude price for Asian markets is assessed by price reporting agencies. The O/D base price referenced in typical Asian crude oil sales agreements refers to price assessments published by Platts, a leading PRA and a division of Standard and Poors.[6] The problems with crude oil price discovery in Asia, and specifically with the assessment methodology related to Platts O/D price quotes, have long been fodder for editorials in industry journals and the trade press over the years. It has also been subject to some academic attention (see for example, Fattouh (2011) and Imsirovic (2014)). There have also been presentations by government-funded research institutes in Northeast Asian countries which argued that Middle Eastern crude oil sellers charged an 'Asian premium' to their customers in Asia (see Ogawa (2003) and Moon & Lee (2003) for the concept and see Doshi & D'Souza (2011) for a critique).

## Alternatives to PRA assessments

Three potential alternatives to the PRA's O/D assessments have been identified in the trade press on Asian crude oil benchmark pricing: the Oman futures contract traded on the Dubai Mercantile Exchange (DME); ESPO crude in Russia's Far East; and the proposed sour crude futures contract on the Shanghai International Energy Exchange (INE).

### DME

The DME launched the Oman futures contract in June 2007, and since then it has established itself as the key arena for physical Oman crude oil delivery. However, its estimated average daily traded volumes of 5,000–6,000 contracts pale in comparison with the daily volume of more than 800,000 Brent futures contracts traded on the InterContinental Exchange (ICE) in the month of September 2016.[7] The emergence of the DME Oman futures contract as a viable instrument for establishing a reference price for Middle East crude oil exports to Asia is contingent on whether key market participants support the use of that instrument as a mechanism for price discovery. Until such time as a major Middle East national oil company or a major importing country such as China elects to use the DME Oman futures contract price as a price benchmark, the contract will continue being traded as a tool for effecting physical delivery of Oman crude (see Fattouh (2008) for a detailed analysis of DME).

DME's ambitions for the contract's wider role as a pricing reference and risk management instrument for Middle East crudes sold in Asia will likely remain out of reach until a major stakeholder or group of stakeholders finds the existing PRA assessments of oil benchmark prices too dysfunctional and unilaterally opts for an alternative. In 2008 and 2009, when the Platts' benchmark WTI assessments for delivery at Cushing (Oklahoma) were used in Saudi Aramco's pricing formula for crude oil sales to the US, WTI crude was often 'disconnected' and sold at steep discounts to other global benchmark crude prices as a result of logistical bottlenecks at the Cushing delivery point. In the event, Saudi Aramco announced a switch in its price reference from January 2010 to a competing PRA's (Argus Media) price assessments of an alternative sour crude index (Fletcher, 2009). The latter, known as the Argus Sour Crude Index (ASCI), is a volume-weighted average of daily spot sales of the three US Gulf Coast medium sour crudes, Mars, Poseidon and Southern Green Canyon.

### ESPO

The completion of the ESPO (East Siberia–Pacific Ocean) oil pipeline allowed crude oil cargoes to be loaded out of the port of Kozmino in Russia's Far East. Kozmino's proximity to the oil refineries of Northeast Asia, within three to five days' sailing time from markets in China, South Korea and Japan (which account for more than half of total Asian demand for crude oil), confers significant locational rents to ESPO Blend crude oil relative to similar quality crudes which need to be imported from much further distances in the Middle East, West Africa and Latin America. It can take anywhere from two to three weeks to ship oil cargoes from these latter locations to Northeast Asian destination ports. The beginning of ESPO Blend exports from Kozmino in 2010 led several market observers to suggest that the new crude marketed into Asia had attributes that could lead it to serve as a new pricing benchmark (see Hall, 2011). Among such attributes would be that the crude oil should be freely tradable (i.e. with no restrictions on its re-sale), its spot trade should have adequate liquidity without dominant buyers or sellers, there should be adequate loading facilities with transparently set loading schedules agreed between buyers and sellers, and there should be a stable regulatory environment for trading the crude oil (Horsnell, 1997).

While deliveries of ESPO crude at Kozmino are significant in volume (estimated to be more than 500,000 b/d in 2014–2015), various diversions of the crude away from spot trade in favour of long-term supply commitments and sales via tender to invited participants have increasingly constrained spot market liquidity (Weber, 2015). Concerns about concentration on the supply side, with two companies – Rosneft and Surgutneftegaz – accounting for almost three-quarters of ESPO production, also militates against the chances of ESPO spot trade leading to independent price discovery. On the demand side, when the ESPO trade out of Kozmino gained momentum from 2010 onward, it drew a wide range of customers including Singapore, Malaysia, Australia and the US outside the core markets of Northeast Asia (China, Japan and South Korea). In the past two years, however, the list of buyers has narrowed considerably. Effectively customers from only two countries are left – China and Japan (Yagova, 2015). As the ESPO Blend draws new supply from different oil fields in Eastern Siberia, there are also concerns about the stability of crude oil quality over the long term. For these reasons as well as uncertainty over government policy and perceptions that the ESPO market could be influenced by political exigencies of Rosneft, a state-owned company, suggest that the spot trade in ESPO is unlikely to lead to independent price discovery. ESPO crude will most likely continue to be priced off Dubai price assessments.

## INE

In 2012, the Shanghai Futures Exchange (SHFE) announced its plan to launch a crude oil contract based on a medium sour crude oil with specific gravity ranging from 30 to 34 degrees API and a maximum sulphur content of 2%. The proposed contract would include the commonly spot traded ME crudes such as Oman, Dubai, Basrah Light, Upper Zakum and Qatar Marine as well as Shengli, a domestic crude, delivered to specified locations in China (Platts, 2016). The Shanghai crude oil futures contract was initially planned to start trading on the SHFE. Since its announcement, the launch of the contract has been continually delayed. In 2013 the planned launch of the futures contract was moved to a new exchange, the Shanghai International Energy Exchange (INE) located in Shanghai's new free trade zone. The contract was set to be the first Chinese commodities contract to be fully open to foreign investors, a landmark in the opening of China's financial markets with tax incentives and promise of full convertibility of the yuan. In 2015, expectations rose that China's first internationally traded crude futures would be ready to begin trading later in the year. According to the latest reports, however, the INE preparations may not be completed until year-end (2016), essentially pushing the launch date into the next year at the earliest.

There is a popular view that what Asia needs is a 'genuinely Asian marker', that a shift of crude pricing benchmarks 'eastward' is a natural move given the shift in the centre of gravity in crude oil trading to Asia. A variant of this argument is that the sheer size of China's oil market 'is enough to justify its own pricing benchmark'. The scale of Chinese demand in global commodity markets can lead to rapid growth in domestic liquidity on the commodity exchanges; for instance, the Dalian Commodities Exchange is home to the world's first and third most actively traded commodity contracts (steel reinforcement bars and iron ore) (Sanderson, 2016). China's efforts in launching a crude oil futures contract seem to be geared towards having their commodity imports 'priced as much as possible off of Chinese reference contracts whenever they can', according to one close observer of Asian crude oil markets cited in the media.[8] However, the sharp sell-off in China's stock market in mid-2015, followed by the government's rushed regulatory and policy shifts to re-assert control over the market, has dampened foreign enthusiasm for the INE sour crude contract (Hsu, 2015). More recently, actions by the country's National Development and Reform Commission (NDRC) to rein in surging coal prices by administrative fiat again brought attention to the government's reluctance to allow commodity markets to trade freely and openly (Meng & Mason, 2016).

## Price discovery and government regulations

In the wake of the Libor scandal in 2012, the high degree of reliance on PRAs in crude oil price discovery and the potential for price manipulation by market actors gained attention from regulatory authorities in the US and Europe (see Kwiatkowski & Zhu, 2013; Campbell, 2013). One critique was that the oil PRAs have 'too much power' and that they use that power 'to determine the contractual framework in which trades take place rather than simply reporting trades'.[9] However, as the use of price assessments published by PRAs is solely at the discretion of buyer and seller for effecting physical or derivative transactions, it is not clear just what 'power' the PRA possesses in the market place beyond that of delivering its service to willing buyers: that is, deriving price assessments within certain methodological and editorial guidelines that are made known to all participants.

Some observers have charged that PRAs effectively engage in 'selective reporting', since they do not require oil market participants to submit data on *all* their trades. A further concern is that a narrow pool of market participants could 'game' or collude in price assessments for their own benefit at the cost of consumer welfare. In this view, the PRA price assessment processes reflect a market for 'big boys', described by one oil trader as an 'oligopoly of large "too big to fail" companies with close links to governments and regulators with a vested interest to keep the "status quo" which they find so profitable' (Imsirovic, 2013). Yet, so long as there are no barriers to entry for oil traders willing and able to participate in the price assessment process, any 'selective' reporting can always be challenged by competitors in the market place who can engage in counter bids or offers to reflect their opposing (buy or sell) side of the market. PRAs, after all, have every incentive to produce accurate data since the parties on both sides of any transaction rely on them. Any PRA that loses the trust of market participants would lose its revenues from subscription services for price assessments.

Attempts to regulate PRAs would lead oil market participants to refrain from reporting transactions if such information could be used by regulatory agencies to impose liabilities. If regulations were imposed that would make both traders that submit quotes and the reporters that publish them legally liable for any losses in oil markets resulting from 'incorrect' prices, this would naturally lead oil traders to withdraw from making any information on oil transactions publicly available. This in turn would lead to opacity in market activity and a reversal from open and freely traded oil markets. It could also constrain PRAs from performing their price assessment functions as government officials determine how PRA reporters and editors assess prices (Verleger, 2012).

Indeed, intrusive regulation of PRAs could lead oil market pricing back to the earlier system when the oil industry was largely based on global vertically integrated oligopolistic firms which controlled markets from upstream exploration and production to downstream oil refining and marketing and erected formidable barriers to entry for smaller upstart firms. Indeed, markets hardly existed as most flows from 'source rock to petrol pump' were transactions that were carried out within the globally integrated oil firms at opaque internal transfer prices established by the major oil companies. After the dislocations and nationalisation of oil resources in the 1970s, global oil markets have become far more fungible and liquid with a much larger role for market forces and much more flexible as a result. By the 1980s, spot physical and derivative markets including futures and swaps helped establish competitive price signals for global trade in crude oils and refined oil products. PRAs played a critical role in this process of increased price transparency and competitive markets. The fact remains that the world's largest inter-regional flows of crude oil – including the flow from the Middle East to Asia, amounting to over 17 mmbd in 2015 – are priced off PRA assessments. To date, the Saudi, Kuwaiti, Iranian and other Middle East OSPs for Asia-destined long-term crude oil sales are based on PRA assessments, and there is no indication that this will change anytime soon.[10] No official announcements have been made by the region's NOCs or their governing ministries regarding intentions to adopt alternative benchmarks as their pricing basis in Asian sales.

## Growing importance of Asian national oil companies

The first national oil companies (NOCs) were established in the early years of independence from the ex-colonial powers, primarily representing the large oil exporting countries of OPEC. The NOCs of the large Asian net oil importers are relative new-comers.[11] However, Asian NOCs from China, Japan, South Korea and India have become important players in the acquisition of upstream assets in Africa, the Middle East, Central Asia, Latin America and Southeast Asia/Australasia. In the past decade, the Asian NOCs have become increasingly internationalised in their upstream asset portfolios. They compete as well as form joint ventures with the international oil companies (IOCs) for access to oil and gas deposits around the world. Among the larger and more active Asian NOCs with strong state mandates for international upstream exploration and production (E&P) activity are China National Offshore Oil Corporation (CNOOC), China National Petroleum Corporation (CNPC) and China Petrochemical Corporation (Sinopec) of China, Oil and Natural Gas Corporation (ONGC) of India, Korea National Oil Corporation (KNOC), Petroliam Nasional Berhad (Petronas) of Malaysia and PTT Public Co. Ltd (formerly Petroleum Authority of Thailand). The indebted Japanese National Oil Company (JNOC) was discontinued in 2001, and was replaced by publicly listed exploration and production companies that are expected to reflect national and strategic interests of the country while operating as independent E&P companies accountable to their shareholders.[12]

The Chinese NOCs in particular have gained increasing attention from the industry in recent years by the scale of their investments in foreign oil and gas assets (Xu, 2013). According to the IEA, China spent $47.6 billion to acquire oil and gas assets worldwide, accounting for 61% of total acquisition value by all NOCs in 2009 (IEA, 2011). From 2009 to 2011, the worldwide net investment by the top 50 oil companies ranked by the *Petroleum Intelligence Weekly* was $107 billion; the Chinese NOCs alone represented more than half of the total ($54.1 billion) (PIW, 2011). According to another estimate, Chinese NOCs spent some $187.3 billion in overseas acquisition of oil and gas assets between 2005 and the third quarter of 2012 (Deloitte, 2013).

Asian NOCs vary considerably in their corporate governance structures and in their economic performance, reflecting the considerable heterogeneity in historical circumstance, resource endowments and government policy in their respective countries. A key motivation for Asian NOCs of the large net oil importers has been to acquire upstream assets around the world. Given the relentless increases in import demand faced by the large Asian energy consumers with inadequate (China, India) or practically non-existent (Japan, South Korea) domestic hydrocarbon reserves, the perceived need to gain 'control' of larger levels of oil and gas reserves across the globe has been among the key strategies pursued by Asian governments to enhance energy security.[13] With the steep increase in energy prices during the commodity boom of 2002–2008, and with continued unrest and political instability in the Middle East, energy security has been a high priority for Asian governments. Oil in particular is seen as a strategic commodity 'too vital' to be left entirely to international markets. The cited investment objectives of NOCs include the need to expand oil and gas production, diversify supplies geographically, acquire technological and managerial expertise, and become an 'internationalised company' (see for instance Xu (2007) and IEA (2011)).

Asian NOCs have often been cast as agents charged to advance their government's strategic and political initiatives across the globe. Thus, the attempt by China's CNOOC to acquire Unocal, a US oil company, in 2005 quickly became politicised and was subject to intense national security debates within US government policy circles (see for instance, Pottinger et al. (2005)). Asian NOCs have been perceived in conventional wisdom as opaque and adept at using oil and gas acquisitions as a tool to advance their government's strategic and political initiatives across the globe. This state-directed role of Asian NOCs has been termed 'neo-mercantilist' in that while NOCs participate in the global market for upstream acquisitions in competition with other industry players, governments seek to gain advantage for their NOCs through financial and diplomatic support (Ziegler, 2008). Investments by Chinese state-owned enterprises across the natural resource extractive sector in sub-Saharan Africa, for example, have been the staple of a large literature on Chinese geopolitics and international relations, although Africa accounts for only a tiny percentage of China's overall foreign economic activity (Sun, 2014). Asian NOCs exhibit higher risk tolerances than their IOC counterparts, investing in countries such as Myanmar, Chad, Sudan, Central Asia and Venezuela. Indeed, some Asian NOCs might have found it initially to their strategic advantage to operate in regions which had troubled relations with Western countries and where the IOCs did not dominate oil and gas exploration and production opportunities. Nevertheless, and not surprisingly, Asian NOCs have not been protected from poor investment decisions and the political and fiscal risks of host countries either.[14]

There is a vast gulf between the best and the worst performers among NOCs.[15] At one end are examples of organisations that are mere instruments of plunder for favoured or powerful constituencies. Among the well-known examples of NOCs which have perennial problems of corruption, poor decision-making and the squandering of resources are those of Nigeria and Venezuela. At the other end of the spectrum are NOCs such as Statoil of Norway or PTT of Thailand which are practically indistinguishable from commercial firms maximising shareholder profits. Most Asian NOCs lie somewhere between these extremes. For the large NOCs, governmental oversight and control is often compromised by 'principal–agent' problems, i.e. where the principal cannot directly ensure that the agent is always acting in its (the principal's) best interests since the agent's actions are either not observable or costly for the principal to observe.[16]

Some state-owned enterprises evolved into behemoths loosely constrained by accountability except to the highest political authorities, with equivalent rank to central government ministries. At the limit, ostensible oversight ministries can become mere appendages of the NOC rather than exercising overall control as independent regulators. Thus, in describing the relative autonomy of the Chinese NOCs, one analyst remarked that the

> regulatory power over the petroleum industry at the central level is fragmented among different bureaucracies, who often lack the clout or staff necessary to regulate the petroleum industry. As a result, the regulatory agencies often defer to the three NOCs over important policy problems and increasingly have relied on them to identify policy problems, formulate corresponding responses, and implement policies.
>
> (*Zhang, 2014*)

The image of NOCs as mere puppets of their governments also obscures the material incentives which give momentum to the global expansion of Asia's NOCs. As a manifestation of the principal–agent problem, NOC managers have strong incentives to use surplus financial reserves to aggressively fund international asset acquisitions. This increases the size and growth of company assets which often determine the rank and compensation of management personnel in NOCs. As dividends, profits and real rates of return to capital employed are not typically used as key evaluative metrics in state-owned enterprises, NOC managers often have a greater interest in acquiring assets than in generating higher returns to existing assets. Asset accumulation can be a more important objective than asset returns when national image and 'energy security' claims trump profits as a measure of NOC corporate performance in international E&P ventures.

Do Asian NOC acquisitions of upstream assets enhance their energy supplies at the expense of other energy consumers? Do Asian NOCs 'lock-up' fuel supplies and distort the competitive landscape with preferential access to such resources? If the NOC by acquiring a company invests in expanding the acquired company's production capacity, then there is no necessary adverse impact on other buyers and consumers of oil and gas. By definition, such investments lead to larger and more diversified oil supplies, benefiting all users. One empirical study of China's acquisitions, loans and long-term procurement deals by state-owned enterprises draws the distinction between those deals that work towards 'consolidating an existing structure of production for a given resource base', as opposed to 'multiplying and diversifying sources of supply while adding new output at the margin faster than the growth in world demand' (Moran, 2010). The study found that the latter outcome was more common. While the largest Chinese natural resource procurement arrangements show 'a few instances in which Chinese natural resource companies take an equity stake to create a "special relationship" with a major producer . . . the predominant pattern is to take equity stakes and/or write long-term procurement contracts with the competitive fringe' (Moran, 2010).

While only the largest Asian NOCs possess the financial capacity to be significant players in the acquisition of upstream oil and gas assets internationally, they operate the vast majority of the region's refining capacity, with the rest owned by the IOCs in open refining markets such as Singapore or run as joint ventures between domestic NOCs and either IOCs or other NOCs from the oil-exporting countries in the Middle East. According to BP data, Asia accounted for just over 32 million barrels per day or about a third of global refining capacity in 2015, roughly similar to its share of global crude oil consumption (BP, 2016). Asia is set for the biggest refining capacity expansion after the past three years, with new and expanded refineries coming on-stream in 2017 primarily in India and China (Tan, 2016). Combined with planned refinery closures in Japan (which has continued to have a surfeit of capacity over the past decade), a net capacity expansion of half a million barrels per day of capacity will be added in 2017.

Perhaps the most notable development in the Asian refining industry in recent years has been the emergence of so-called 'teapot' refiners in China which are smaller establishments operating simpler but more flexible plants compared with the giant NOCs such as China Petroleum & Chemical Corp., or Sinopec, PetroChina Co., and China National Offshore Oil Corp. (CNOOC). These smaller refiners typically bought crude oil from the large Chinese NOCs which had licences to import crude oil. They then sold much of their output back to the big state-owned companies which had vast refined product marketing arms throughout the country. However, since July 2015, the government began granting the 'teapots' crude oil import licences to promote competition in the sector. Since then, they have emerged as key players in China and the greater Asia region, accounting for about 20% of China's total refining capacity. With their more flexible and market-oriented approach to doing business, these refiners operate at higher capacity rates with lower operating costs and have contributed to a surge in refined product exports from China. As China's domestic demand for oil products has slowed, particularly for diesel, the 'teapots' have become a major factor in Asian oil-refining dynamics, both as aggressive buyers in the spot crude oil market and as exporters of refined products.

In the context of the low oil prices that followed the second half of 2014 price collapse, there has been increased competition in the Asian oil market. Asian refiners have added greater weight to spot purchases in their crude oil buying portfolios. This development has led even the most conservative Middle East NOCs to engage in spot market sales, albeit intermittently. The Middle East NOCs which had long preferred to sell their oil under long-term crude oil sales agreements (COSAs) have begun to engage in opportunistic spot market sales in order to protect or grow their market share, while minimising price discounts in their formula pricing for their long-term clients. This is a radical departure from the crude oil-exporting NOCs' preference for term contracts. These one-off sales are being experimented with by Middle East exporters because customers in the Far East are increasingly willing to consider purchasing cargoes in the spot market. China's teapot refineries, located mainly in Shandong province, are a good example of commercial entities which are new to the global oil market and which view spot purchases as a way to 'test' the market values of crudes and compare the reliability of competing exporters. Middle East NOCs can most easily reach these new customers by selling them spot cargoes from their leased storage facilities in Asia. Saudi Aramco, for example, sold a cargo of crude from its leased Okinawa storage to a Chinese independent refiner on a spot basis in 2016. Saudi Aramco reportedly priced this cargo on a FOB Okinawa basis, instead of quoting an official selling price (Rapoza, 2016).

## Concluding remarks

As the largest oil-consuming region in the world, Asia occupies a central place in the planning and analysis departments of most corporations in the oil industry. Given that the OECD countries are already into, or facing impending, 'peak oil demand' in their energy outlooks, Asia is commonly seen as the major demand growth region for oil in the coming decades. In a context of low oil prices and robust production of unconventional oil in North America, the major crude oil producers in West Africa, Latin America and Russia are aggressively competing with the Middle East exporters for market share in Asia. Asian consumers, spoilt for choice in a buyers' market after almost a decade of high oil prices, now actively look at spot purchases of crude oil from Latin America and even the North Sea, apart from their regular supplies from the Middle East. West African oil, having dramatically lost market share in the US with the surge of light tight oil as a result of the 'shale revolution', now faces static demand in the European region (which is still facing an uncertain recovery from the Great Recession of 2008–2009) and an imperative to compete in Asia for incremental demand alongside other crude oil producers.

Unlike the Atlantic Basin where crude reference prices (Brent and West Texas Intermediate) are discovered in liquid futures exchanges such as ICE and NYMEX, the Asian market does not have any traded futures contract for crude oil which serves as widely used pricing benchmark for sour crude. The reference Dubai crude price is, as already noted, discovered by PRAs such as Platts and Argus Media. While the role of PRAs in oil price discovery has been the subject of considerable debate and controversy, there seem to be no plausible alternatives. The current system of voluntary reporting of trades, bids and offers to PRAs, evolved since the mid-1980s at the end of the OPEC-administered pricing system in place previously, has proved resilient despite the many deficiencies emphasised by market observers. Reflecting Asia's dominating role in global incremental demand for oil, Asian NOCs have emerged as important players in the international market for upstream oil and gas acquisitions. Concerns about whether Asian NOCs threaten the welfare of other consuming countries by acquiring and 'locking up' oil and gas assets from competitors seem overblown. Evidence suggests that most acquisitions by Asian NOCs lead to investments that open up new or increased sources of oil supplies, and on balance, are of net benefit to the global oil industry. While it is apparent with hindsight that Asian NOCs have often over-paid for access to international upstream assets in the aftermath of the oil price collapse in the second half of 2014, the resulting steep cutbacks in upstream investments are traits they share in common with their IOC counterparts.

## Notes

1 *Bloomberg News*, 'Oil Bulls Beware Because China's Almost Done Amassing Crude', 30 June 2016.
2 For instance, forecasts in the 1990s suggested that up to 95% of Asian imports of crude oil would be sourced from the Middle East by 2010 (Fehsaraki et al., 1995).
3 Brent Blend crude oil, West Texas Intermediate (WTI) crude oil, US gasoline, US heating oil (gasoil) and US natural gas.
4 Liz Bossley cited in Tett (2014); Verleger (2012).
5 The quality differential is measured as the difference in Gross Product Worth (GPW) of the particular crude relative to the reference crude. The GPW measures the total value of all the refined product processed from the crude and determines the crude oil's refining value. In Asia, refined product prices quoted at FOB Singapore are taken as the reference prices for calculating the GPW of the crude oil.
6 For a full description of how Platts assesses Dubai and Oman physical crude oil prices, see Platts website at www.platts.com/IM.Platts.Content/MethodologyReferences/MethodologySpecs/Crude-oil-methodology.pdf. This is unlike the situation in the Atlantic markets of Europe and North America, which have liquid exchange-traded futures in West Texas Intermediate and Brent contracts that serve as the reference prices for Middle East producers which export to the two major regions. (It should be noted that Middle East exports to the US have been based on PRA assessments of WTI prices, not on exchange-traded prices.)
7 Data from ICE website for historical monthly volumes of contracts traded on the Exchange, accessed at www.theice.com/marketdata/reports/. It is assumed that a month has 22 working days on average.
8 Dave Ernsberger, quoted in Raval. & Sheppard (2015).
9 Liz Bossley, quoted in Lawler & Mably (2012).
10 Among the few Gulf crudes sold on the 'spot' market (i.e. not based on term contracts with end-user and re-sale restrictions) are Oman and Dubai. For a full if dated description of Middle East crude exports and pricing in Asia, see Horsnell (1997).
11 The NOCs of Asian oil producers Indonesia and Malaysia, Pertamina and Petronas, were founded in 1968 and 1974 respectively. Indonesia joined OPEC initially in 1962 and suspended its membership as of 1 January 2009, but this was reactivated as of 1 January 2016.
12 Japan Oil, Gas and Metals National Corporation (JOGMEC) is a government 'Administrative Institution' was created in 2004 when the former Japan National Oil Corporation merged with the former Metal Mining Agency of Japan.
13 China's 'White Paper on Energy' states that Chinese policies 'will, step by step, change the current situation of relying too heavily on spot trading of crude oil, encourage the signing of long-term supply contracts, and promote the diversification of trading channels.' See China State Council Information Office (2007).
14 Xu (2013) refers to a 2010 report by the China University of Petroleum that found that many foreign investments by Chinese NOCs were loss-making or abandoned, due to poor assessment of risks.
15 For an empirical study of comparative NOC operational performance, see Hartley & Medlock (2013).
16 For exposition of the principal–agent problem, see Grossman & Hart (1983).

## References

BP, 2016. *Statistical Review of World Energy 2016*, London: British Petroleum.
Campbell, R., 2013. *How to Manipulate Oil Price Assessments*, 15 May, s.l.: Reuters.
China State Council Information Office, 2007. *China's Energy Conditions and Policies (White Paper on Energy)*, Beijing: China State Council Information Office.

Deloitte, 2013. *The Resurgent Dragon: Searching for Value in Troubled Times*, Hong Kong: Deloitte.

Doshi, T. K. & D'Souza, N. S., 2011. The 'Asia Premium' in Crude Oil Markets and Energy Market Integration. In: F. Kimura & X. Shi, eds. *Deepen Understanding and Move Forward: Energy Market Integration in East Asia*. Jakarta: Energy Research Institute of the ASEAN, pp. 152–190.

EIA, 2016. *International Energy Outlook 2016*, Washington, DC: US Energy Information Administration.

Fattouh, B., 2008. *Oxford Energy Comment (March)*. [Online] Available at: www.oxfordenergy.org/wpcms/wp-content/uploads/2011/01/March2008-DMECrudeOil-BassamFattouh.pdf [Accessed 27 March 2017].

Fattouh, B., 2011. *An Anatomy of the Crude Oil Pricing System*, Oxford: Oxford Institute of Energy Studies.

Fehsaraki, F., Clark, A. & Intarapravich, D., 1995. *Pacific Energy Outlook: Strategies and Policy Initiatives to 2010*, Honolulu, Hawaii: East West Centre Occasional Paper, March.

Fletcher, S., 2009. Aramco Switches from WTI Benchmark. *Oil and Gas Journal*, 11 September.

Grossman, S. & Hart, O., 1983. An Analysis of the Principal–Agent Problem. *Econometrica*, 51(1), pp. 7–45.

Hall, S., 2011. *IEA: Russia's ESPO Crude May Become an Asian Benchmark*, 18 January, New York: Dow Jones Newswires.

Hartley, P. & Medlock III, K., 2013. Changes in the Operational Efficiency of National Oil Companies. *The Energy Journal*, 34(2), pp. 21–56.

Horsnell, P., 1997. *Oil in Asia: Markets, Trading, Refining and Deregulation*, Oxford: Oxford University Press.

Hsu, J., 2015. The Long Wait for Chinese Oil Futures Continues. *Wall Street Journal*, 11 November.

IEA, 2011. *Overseas Investments by Chinese National Oil Companies: Assessing the Drivers and the Impacts*, Information Paper, Paris: International Energy Agency.

IMF, 2016. *World Economic Outlook 2016*, Washington, DC: International Monetary Fund.

Imsirovic, A., 2014. *Oil Markets in Transition and the Dubai Crude Oil Benchmark*, Oxford: Oxford Institute of Energy Studies.

Imsirovic, A., 2013. Don't Blame PRAs for Oil Industry's Structural Failures, *Letters, Financial Times*, Tuesday, 21 May.

Kwiatkowski, A. & Zhu, W., 2013. *EU Oil Manipulation Probes Shines Light on Platts Pricing*, 15 May, New York: Bloomberg.

Lawler, A. & Mably, R., 2012. Oil Price Agency Platts too Powerful, Regulator Told, *Reuters*, 5 April.

Meng, M. & Mason, J., 2016. *China Coal Fumble Casts Doubt on Its Global Commodities Pricing Goal*, 11 November, London: Reuters.

Moon, Y.-S. & Lee, D.-S., 2003. Asian Premium of Crude Oil, paper prepared for international workshop on Cooperative Measures in Northeast Asian Petroleum Sector: Focusing on Asian Premium Issue. Seoul: Korea Energy Economics Institute. Available at: www.keei.re.kr/web_keei/en_publish.nsf/0/EFD85D3DA10B33A549256E360027F9D2/$file/YSMoon.pdf.

Moran, T., 2010. *Chinese Strategy to Secure Natural Resources: Risks, Dangers, Opportunities*, New York: Peterson Institute of International Affairs, Columbia University Press.

Ogawa, Y., 2003. Asian Premium of Crude Oil and Importance of Development of Oil Market in Northeast Asia, paper prepared for international workshop on Cooperative Measures in Northeast Asian Petroleum Sector: Focusing on Asian Premium Issue. Seoul: Korea Energy Economics Institute. Available at: www.keei.re.kr/web_keei/en_publish.nsf/0/EFD85D3DA10B33A549256E360027F9D2/$file/YOgawa.pdf.

PIW, 2011. *PIW Ranks the World's Top 50 Oil Companies*, New York: Petroleum Intelligence Weekly, Energy Intelligence.

Platts, 2016. *Launch of China's First Crude Futures Contract Pushed to Late 2016*, 25 September, London: Platts.

Pottinger, M., Gold, R., Phillips, M. & Linebaugh, K., 2005. Cnooc Drops Offer for Unocal, Exposing US–Chinese Tensions. *Wall Street Journal*, 3 August.

Rapoza, K., 2016. Saudi Russia Fight for China Market Make Oil Price a Sham. *Forbes*, 1 May.

Raval, A. & Sheppard, D., 2015. Oil Futures Plan Fuels up China's Ambitions, *Financial Times*, 17 September.

Sanderson, H., 2016. Speculators March into China Commodities. *Financial Times*, 27 April.

Sun, Y., 2014. *Africa in China's Foreign Policy*, Washington, DC: Brookings Paper, Brookings.

Tan, F., 2016. *Asia Set for Biggest Refining Capacity Jump in Three Years*, s.l.: Reuters.

Tett, G., 2014. Oil Markets Should Heed Libor Lessons: Setting of Oil Prices May Come Under Similar Scrutiny, *Financial Times*, 13 April.

Verleger, P., 2012. *Regulating Oil Prices to Infinity*, PK Verleger LLC. Available at: www.pkverlegerllc.com/assets/documents/120812_Regulating_Oil_Prices_to_Infinity.pdf.

Weber, F., 2015. *Eastward Shifting Oil Markets and the Future of Middle Eastern Benchmarks*, Oxford: Oxford Energy Comments, Oxford Institute of Energy Studies.

World Bank, 2016. *World Development Indicators*, Washington, DC: World Bank.

Xu, C., 2013. Chinese NOC's Expansion. *Oil and Gas Journal*, 22 April.

Xu, X., 2007. *Chinese NOCs' Overseas Strategy, Background, Comparison and Remarks*, Houston: James A. Baker III Institute of Public Policy, Rice University.

Yagova, O., 2015. *Russia's ESPO Blend Crude Still Struggles in Asia-Pacific*, 30 December, s.l.: Reuters.

Zhang, A., 2014. Foreign Direct Investment from China: Sense and Sensibility. *Northwestern Journal of International Law & Business*, 34(3), p. 394.

Ziegler, C., 2008. Competing for Markets and Influence: Asian National Oil Companies in Eurasia. *Asian Perspective*, 32(1), pp. 129–163.

# 9

# NATURAL GAS TRADE AND MARKETS IN ASIA

*Ronald D. Ripple*

## Introduction

This chapter aims to present a historical overview of the evolution of the natural gas markets in the Asian region and to provide a view to the future evolution given the expected changes in both demand and supply within and from outside the region.

The twenty-first century is frequently cast as the Asian Century. This is meant typically to address the region's rise in significance in geopolitical, economic, and military realms. However, such a label is also applicable to the expectations of the region's role, engagement, and influence on energy. And, within the broad definition of energy, the region's role in the developments of natural gas production, consumption, and trade will be no less influential.

Natural gas has played a relatively small role in Asia. Approximately 24 percent of the world primary energy mix is accounted for by natural gas, while the Asia Pacific region used natural gas to provide for only 11 percent of its energy needs in 2015. This compares to 32 percent for both North America and Europe/Eurasia. The Asia Pacific region relied on coal for 50 percent of its primary energy needs in 2015. BP and the International Energy Agency (IEA)[1] expect the share of natural gas to grow to 13 percent for the Asia Pacific region by 2035.

Asia shifted from near production-consumption balance for natural gas in the early 1980s to an imbalance of 144.5 Bcm (14 Bcf/d)[2] by 2015. The imbalance is expected to expand in the future, and this implies that the role of inter-regional trade in natural gas is expected to grow. It seems likely that most of the growth in inter-regional natural gas trade will be met via liquefied natural gas (LNG)[3] trade.

Asia is expected to continue its role as a key global driver of energy demand over the next several decades. The regional character of this demand is expected to change with China's role, while continuing to increase in absolute terms, diminishing relative to India's role. Much of this growth across the region is expected to focus on natural gas as a cleaner primary energy source and as a good substitute for coal in power generation and backup for intermittent renewables.

Numerous energy outlooks suggest that the Asian region will continue to rely on cross-border and inter-regional trade to meet demand growth, as internal supply growth, even from unconventional (e.g. shale and coal seam) natural gas, will be insufficient. Several natural gas pipelines are currently proposed, but the geography and geopolitics of the region make cross-border pipelines challenging. Hence, reliance on LNG-sourced gas trade is expected to continue to expand. Currently, 73 percent of global trade in LNG-sourced gas flows to Asian countries (with nearly half of this flowing to Japan), and only about 41 percent of these supplies originate within the region. Australia is expected to become the largest global exporter of LNG-sourced natural gas by 2018, but new players will also be entering the regional trade with natural gas sourced from the United States, as well as the potential for volumes from East Africa, Canada, and others. Asia is likely to be a prime target of nearly all of this capacity, especially if historical price differentials are a reasonable expectation for the future. The new entrants and volumes will change the regional and sub-regional market dynamics and potentially the underlying pricing structures that have persisted within the region for decades. All of this, as well as the potential complications of a persistence of the current low-price environment being faced around the globe will be addressed in the chapter.

## The region

Asia is a very large and diverse region, and there are more than a few definitions of what constitutes the Asian region. Therefore, the region that is discussed in this chapter and the countries therein contained must be defined. If Asia is defined as continental Asia, it will include some 50 countries including those of the Middle East, the Russian Federation east of the Urals, the Caucasus, and Central Asia, along with all of the countries stretching across from Afghanistan and Pakistan south of the Himalayas to China and North and South Korea. The various sources of data and outlooks for energy supplies and demands employ a range of different country groupings, but none employs the continental definition with 50 countries. The discussion that follows will employ the regional definition used by BP in its Annual Statistical Review of World Energy and its Outlooks that is referred to as the Asia Pacific; this definition is quite consistent with the IEA when combining Organisation for Economic Co-operation and Development (OECD) Asia Oceania and Non-OECD Asia.

## Reserves, production, and consumption of the region

The countries included in this regional definition that contain individually identified proved reserves of natural gas are Australia, Bangladesh, Brunei, China, India, Indonesia, Malaysia, Myanmar, Pakistan, Papua New Guinea, Thailand, and Vietnam; additional countries in the region contain proved reserves, but these are deemed too small to report separately; see Table 9.1 for reserves and Table 9.2 for production.

The largest proved reserves are in China, followed closely by Australia. Japan, South Korea, and Taiwan, traditional LNG-sourced natural gas importers, have no domestic natural gas resources. The largest producer of natural gas is China, and it has been since 2007. The reserves of the Asia Pacific region are 8.4 percent of the world total, and the region's production in 2015 represented 15.7 percent of global production. The driver for inter-regional trade is the fact that in 2015, the region's consumption accounted for 20.1 percent of global consumption.

Table 9.3 shows that China is the largest consumer of natural gas in the region, followed by Japan. Similar to its position with coal, China has the largest reserves, the highest production, and the largest consumption. However, since Japan must import all of its gas, China falls second to Japan for importing volumes; in 2015 China imported 59.8 Bcm via pipe and LNG, while Japan imported 118 Bcm via LNG.

Consumers of natural gas in the region, in addition to the 12 above, include Japan, New Zealand (which has its own reserves but too small to be separated out in the BP statistics), the Philippines (same situation as New Zealand), Singapore, South Korea, and Taiwan; see Table 9.3.

*Table 9.1* Asia Pacific natural gas proved reserves

| Country | Proved reserves (Bcm) | | | | | | | |
|---|---|---|---|---|---|---|---|---|
| | 1980 | 1985 | 1990 | 1995 | 2000 | 2005 | 2010 | 2015 |
| Australia | 172.2 | 673.6 | 880.8 | 1200.9 | 2093.1 | 2236.2 | 3483.3 | 3471.4 |
| Bangladesh | 286.0 | 353.0 | 725.0 | 272.0 | 306.3 | 407.0 | 354.0 | 232.2 |
| Brunei | 207.0 | 238.0 | 331.0 | 400.0 | 366.0 | 340.0 | 301.2 | 276.0 |
| China | 726.1 | 900.9 | 1032.1 | 1725.1 | 1412.2 | 1585.5 | 2816.1 | 3841.3 |
| India | 343.0 | 480.0 | 700.0 | 676.1 | 759.8 | 1101.0 | 1148.6 | 1488.5 |
| Indonesia | 822.0 | 1982.2 | 2863.6 | 1950.0 | 2682.0 | 2478.0 | 2965.1 | 2839.0 |
| Malaysia | 850.0 | 1494.0 | 1640.0 | 2271.0 | 2337.0 | 2480.0 | 1081.6 | 1169.3 |
| Myanmar | 93.0 | 268.0 | 265.0 | 268.0 | 287.0 | 538.0 | 221.1 | 528.4 |
| Pakistan | 450.0 | 623.0 | 642.0 | 596.0 | 677.0 | 852.0 | 657.4 | 542.6 |
| Papua New Guinea | – | 0.1 | 1.7 | 3.0 | 3.0 | 2.4 | 155.3 | 141.3 |
| Thailand | 292.0 | 218.0 | 224.0 | 176.0 | 360.0 | 304.0 | 299.6 | 219.5 |
| Vietnam | – | – | 15.0 | 147.0 | 170.0 | 220.0 | 617.1 | 617.1 |
| Other Asia Pacific | 243.0 | 254.0 | 290.0 | 406.0 | 338.0 | 413.0 | 287.8 | 281.7 |
| Total Asia Pacific | 4484.3 | 7484.9 | 9610.2 | 10,091.2 | 11,791.2 | 12,957.1 | 14,388.0 | 15,648.1 |

*Data source*: BP (2016b).

*Table 9.2* Asia Pacific natural gas production

| Country | Natural gas production (Bcm) | | | | | | | |
|---|---|---|---|---|---|---|---|---|
| | 1980 | 1985 | 1990 | 1995 | 2000 | 2005 | 2010 | 2015 |
| Australia | 11.1 | 13.0 | 19.7 | 28.3 | 32.1 | 39.2 | 52.6 | 67.1 |
| Bangladesh | 1.3 | 2.7 | 4.8 | 7.0 | 9.4 | 13.8 | 20.0 | 26.8 |
| Brunei | 8.6 | 8.6 | 8.9 | 11.8 | 11.3 | 12.0 | 12.3 | 12.7 |
| China | 14.7 | 13.4 | 15.8 | 18.5 | 28.1 | 51.0 | 99.1 | 138.0 |
| India | 1.2 | 4.5 | 12.0 | 18.8 | 26.4 | 29.6 | 49.3 | 29.2 |
| Indonesia | 18.5 | 32.3 | 43.9 | 60.0 | 69.6 | 75.1 | 85.7 | 75.0 |
| Malaysia | 2.5 | 10.7 | 17.2 | 26.8 | 46.6 | 63.8 | 60.9 | 68.2 |
| Myanmar | 0.4 | 0.9 | 0.9 | 1.6 | 3.4 | 12.2 | 12.4 | 19.6 |
| Pakistan | 7.2 | 8.8 | 12.2 | 15.6 | 21.5 | 39.1 | 42.3 | 41.9 |
| Thailand | – | 3.1 | 6.5 | 11.4 | 20.2 | 23.7 | 36.2 | 39.8 |
| Vietnam | – | ^ | ^ | 0.1 | 1.6 | 6.4 | 9.4 | 10.7 |
| Other Asia Pacific | 7.6 | 9.6 | 7.6 | 7.5 | 8.9 | 11.1 | 17.6 | 27.7 |
| Total Asia Pacific | 73.2 | 107.6 | 149.5 | 208.2 | 279.2 | 377.0 | 497.8 | 556.7 |

*Data source*: BP (2016b).

*Note*: ^ means less than 0.05 BCM. Totals may differ from column sums due to rounding.

Consumption of natural gas in the region has increased by 858 percent between 1980 and 2015, while total global consumption has increased by 142 percent.

The Asia Pacific region is expected to increase consumption by about 76 percent from its 2014 level. This will exceed that of all other regions, with the exception of Africa. However, Africa's growth is from a much lower initial base, and its 2035 consumption level is expected to reach 211 Bcm compared with 1073 Bcm for the Asia Pacific, placing it second only to North America.

*Table 9.3* Asia Pacific natural gas consumption

| Country | Natural gas consumption (Bcm) | | | | | | | |
|---|---|---|---|---|---|---|---|---|
| | *1980* | *1985* | *1990* | *1995* | *2000* | *2005* | *2010* | *2015* |
| Australia | 10.4 | 13.0 | 16.0 | 18.5 | 22.0 | 24.9 | 33.2 | 34.3 |
| Bangladesh | 1.3 | 2.7 | 4.8 | 7.0 | 9.4 | 13.8 | 20.0 | 26.8 |
| China | 14.7 | 13.4 | 15.8 | 18.3 | 25.3 | 48.2 | 111.2 | 197.3 |
| China Hong Kong SAR | – | – | – | ^ | 3.0 | 2.7 | 3.8 | 3.2 |
| India | 1.2 | 4.5 | 12.0 | 18.8 | 26.4 | 35.7 | 61.5 | 50.6 |
| Indonesia | 7.0 | 12.4 | 16.9 | 28.1 | 32.5 | 35.9 | 43.4 | 39.7 |
| Japan | 24.1 | 38.3 | 48.1 | 57.9 | 72.3 | 78.6 | 94.5 | 113.4 |
| Malaysia | 2.5 | 4.4 | 7.6 | 12.9 | 26.6 | 34.9 | 34.5 | 39.8 |
| New Zealand | 0.9 | 3.3 | 4.3 | 4.3 | 5.6 | 3.6 | 4.3 | 4.5 |
| Pakistan | 7.2 | 8.8 | 12.2 | 15.6 | 21.5 | 39.1 | 42.3 | 43.4 |
| Philippines | – | – | – | ^ | ^ | 3.1 | 3.5 | 3.3 |
| Singapore | – | – | – | 1.5 | 1.7 | 6.5 | 8.8 | 11.3 |
| South Korea | – | – | 3.0 | 9.2 | 18.9 | 30.4 | 43.0 | 43.6 |
| Taiwan | 1.9 | 1.1 | 1.9 | 4.0 | 6.2 | 9.4 | 14.1 | 18.4 |
| Thailand | – | 3.1 | 6.5 | 11.4 | 21.9 | 32.5 | 45.1 | 52.9 |
| Vietnam | – | ^ | ^ | 0.1 | 1.6 | 6.4 | 9.4 | 10.7 |
| Other Asia Pacific | 1.9 | 2.9 | 2.5 | 3.4 | 3.5 | 5.2 | 5.8 | 7.8 |
| Total Asia Pacific | 73.2 | 107.9 | 151.6 | 211.0 | 298.5 | 410.8 | 578.4 | 701.1 |

*Data source*: BP (2016b).

*Note*: ^ means less than 0.05 BCM. Totals may differ from column sums due to rounding.

Figure 9.1 provides a graphical view of the regional growth in natural gas consumption since 2000. The three biggest players have been China, Japan, and India. By 2015 they accounted for over 50 percent of the region's total natural gas consumption, and they are expected to expand this role within the region.

## Natural gas pricing and LNG shipping costs

Historically, natural gas pricing in the Asia region has been linked to the price of crude oil. Specifically, LNG-sourced natural gas has traded under long-term contracts with pricing clauses that link the price for the natural gas, in terms of MMBtu, to the price of a barrel of crude oil imported into Japan. This is the so-called JCC[4] pricing link.

Prior to the recent downturn in the price of crude oil, being mid-year 2014, the price of natural gas in Asia had been at a significant premium to that in the USA and Europe, reaching prices of US$18–US$19 per MMBtu, while prices in Europe were in the US$8–US$10 per MMBtu range and in the USA prices were in the US$2–US$4 per MMBtu range. Indeed, it was this large differential that motivated the large number of proposed natural gas export projects in the USA. The expectation was that such margins would lead to significant profits by liquefying USA natural gas and exporting it primarily to Asia. With the collapse of the crude oil price, the slowdown in the growth of energy demand, and the introduction of new export volumes from new LNG projects, the price of natural gas in Asia has fallen significantly leaving little if any margin for LNG shipments from the USA. Even the completion of the Panama Canal expansion,

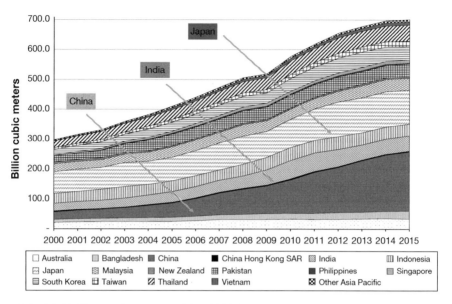

*Figure 9.1* Asia Pacific natural gas consumption, 2000–2015

*Data source*: BP (2016b).

*Table 9.4* LNG carrier shipping cost comparison

| Port-to-port | | Approximate distance (nautical miles) | Fuel cost (US$) | 18 knots | | Day rate (US$40,000) | Cost/MMBtu |
|---|---|---|---|---|---|---|---|
| | | | | Days | Hours | 18 knots | US$ |
| Sabine Pass, USA | Zeebrugge | 4861 | 340,248 | 13 | 6 | 1,060,000 | 0.46 |
| | Tokyo (via South Africa) | 15,825 | 1,107,755 | 36 | 12 | 2,920,000 | 1.34 |
| | Tokyo (via Panama) | 9149 | 640,440 | 21 | 8 | 1,706,667 | 0.96 |
| Dampier, Australia | Tokyo | 3762 | 263,319 | 8 | 12 | 680,000 | 0.31 |

*Data source*: Author's calculations.

*Notes*:

Assumptions: 160,000 m$^3$ tanker ⇒ ~3,500,000 MMBtu;

Accounts for round trip, includes two additional days for loading and unloading, US$35/nm fuel cost and US$0.18/MMBtu for Panama.

which shortens the voyage time between the USA and Japan by about 11 days, is not sufficient to produce the necessary margins to justify the construction of any additional capacity in the USA not already under contract and construction (see Table 9.4).

The delivered prices into Japan during 2016 have remained at or below US$7 per MMBtu, falling as low as US$4.1 in May. During this 2016 period, the price of natural gas in the USA, priced at Henry Hub, has been near US$3 per MMBtu, ranging a bit above and a bit below. When accounting for natural gas acquisition, liquefaction, and shipping costs, the natural gas from the USA cannot be delivered to Japan and produce a positive margin. To understand this, it is first necessary to understand the character of the business model for exports processed through the Cheniere Sabine Pass LNG project, based in Louisiana. An example follows.

A simple statement of the Cheniere business model is that Cheniere acquires the natural gas, processes it into LNG, and then delivers it to the contractual off-taker at shipside. Cheniere charges the Henry Hub price plus 15 percent for the gas, and adds the cost of processing that has been contractually agreed to with the off-takers,[5] which range from US$2.25 to US$3.00 per MMBtu. In September 2016, a typical price for natural gas at Henry Hub was US$3.12 per MMBtu. If we use BG's lowest cost processing agreement for our example, we find that the shipside delivered price, for full-cost recovery would have been US$5.84 per MMBtu. The September delivered price for spot trade into Japan was US$5.70 per MMBtu. So, even before accounting for shipping costs a negative margin will apply.

Table 9.4 provides an example of the shipping costs associated with transporting natural gas in the form of LNG from Sabine Pass to various delivery ports, with a comparison to the cost of shipping from the Northwest Shelf Project in Western Australia to Tokyo. For this example, the focus is on shipping via the expanded Panama Canal from Sabine Pass to Tokyo, and the spot day rate for the LNG tanker is set at US$40,000 per day.[6] The cost to ship the LNG in a 160,000-cubic meter tanker for the 9149 nautical mile voyage via Panama (and the return voyage for the ship) would be US$0.96 per MMBtu. When the natural gas cost, liquefaction fee, and shipping cost are totaled, the full-cost recovery delivered cost will be US$6.80 per MMBtu. Given the spot delivered price of US$5.70, this implies a loss of US$1.10 per MMBtu.

This suggests that in the current low-price environment there is little incentive to construct new LNG export projects in the USA to meet any shortfall in Asia. However, this does not mean that no volumes will flow from the USA Gulf Coast to Asia.[7]

## Natural gas trade

Trade in natural gas has grown from 106.45 Bcm in 2001 (consisting of 4.25 Bcm imported via pipeline from within the region into Singapore and Thailand, and 102.2 Bcm imported in the form of LNG into Japan, South Korea, and Taiwan from both within and outside the region) to 299.8 Bcm in 2015 (consisting of 61.2 Bcm of pipeline trade and 238.6 Bcm of LNG-sourced gas trade). This represents a 182 percent increase.

### Intra- and inter-regional trade

The trade of natural gas began within the region with LNG-sourced flows from Brunei in 1973. These followed the initial LNG-sourced trade between the USA (from the State of Alaska) and Japan beginning in 1969. Indonesia (1977) and Malaysia (1983) followed, and Australia joined the trade in 1989. With the exception of Brunei, each of these players expanded their export capacity, and from within the region were joined by Papua New Guinea in 2014. And while not included in the Asia Pacific region definition, Russia's Sakhalin 2 entered the LNG-sourced trade in 2009.

Pipeline trade within the region began with flows to Singapore from Malaysia (1992). Myanmar/Burma began pipeline exports to Thailand in 1998, and Indonesia began exports to Singapore in 2001. This form of movement within the region was further expanded with the completion of the Sino-Myanmar pipeline project in 2015 to move natural gas from offshore Myanmar to China.

China initiated pipeline imports from outside the region with the completion of the second west-to-east pipeline project in 2010, which carries natural gas sourced from Turkmenistan; the first west-to-east pipeline carried domestically produced gas from the Tarim Basin eastward to the major domestic consuming regions. The west-to-east capacity has been further expanded with the third west-to-east pipeline, and imports now come from Turkmenistan, Kazakhstan, and Uzbekistan. The second and third west-to-east pipelines each have a capacity of 30 Bcm per year.

The largest form of natural gas importation in the region is by way of LNG-sourced movements. This trade has grown from the initial small volumes moving from Alaska to Japan to 10 percent of total world consumption in 2015 [total world inter-regional trade accounted for 1042.4 Bcm, or 30 percent of total world consumption]. Total imports into the Asian region in 2015 amounted to just under 300 Bcm, which represented 29 percent of the world inter-regional trade flows. However, the Asian region accounted for 70 percent of the world trade in the form of LNG, including flows both within and from outside the region.

Table 9.5 shows the imbalance in production and consumption for the region since 1980. From near production-consumption balance, the region transitioned to a shortfall of 144.5 Bcm in 2015. The majority of this shortfall was met by LNG-sourced imports amounting to 125.4 Bcm from outside the region; the total LNG-sourced volumes for Asia Pacific amounted to 238.6 Bcm, with 47 percent of this produced within the region.

Japan, South Korea, and Taiwan have no domestic natural gas production. They also have no pipeline interconnections with any other country, so all of their natural gas is supplied via LNG-sourced imports. In the early period of LNG-sourced natural gas imports into the region only Japan, South Korea, and Taiwan imported. By 2015, they were joined by China, India, Malaysia, Pakistan, Singapore, and Thailand. Malaysia is actually an interesting example of energy economics at work. While Malaysia is now an importer of natural gas in the form of LNG (and via pipeline) it is also still an exporter of natural gas via LNG shipments. This reflects the fact that it is more economic for Malaysia to import to meet natural gas demand in parts of the country that are not nearby its natural gas production regions; it also has long-term contractual obligations that must be met from its existing production and liquefaction operations.

Tables 9.6a and 9.6b provide a view to the future while putting the Asia Pacific region imbalance into a global context; Table 9.6a is in Bcm and Table 9.6b is converted to Mtpa since much of the shortfall will be met by LNG-sourced trade, which will require both additional liquefaction and regasification capacity, which is typically stated in Mtpa. The projections are based on BP's Outlook to 2035, published in 2016 (BP, 2016a).

For this chapter, the key focus is on the row for Asia Pacific, which reports only negative values that are growing throughout the period from 1990 through 2035. Thus, as a region,

*Table 9.5* Asia Pacific production: consumption imbalance (Bcm)

| Items | 1980 | 1985 | 1990 | 1995 | 2000 | 2005 | 2010 | 2015 |
|---|---|---|---|---|---|---|---|---|
| Production | 73.2 | 107.6 | 149.5 | 208.2 | 279.2 | 377.0 | 497.8 | 556.7 |
| Consumption | 73.2 | 107.9 | 151.6 | 211.0 | 298.5 | 410.8 | 578.4 | 701.1 |
| Difference | 0.01 | (0.32) | (2.10) | (2.78) | (19.32) | (33.79) | (80.56) | (144.48) |
| BCF/day | 0.00 | (0.03) | (0.20) | (0.27) | (1.87) | (3.27) | (7.79) | (13.97) |

*Data source*: BP (2016b).

*Table 9.6a* Regional imbalance (production minus consumption) in Bcm

| Region | 1990 | 1995 | 2000 | 2005 | 2010 | 2014 | 2015 | 2020 | 2025 | 2030 | 2035 |
|---|---|---|---|---|---|---|---|---|---|---|---|
| North America | 5.50 | (24.62) | (29.56) | (31.63) | (27.56) | (0.03) | 2.04 | 97.75 | 100.00 | 189.54 | 188.11 |
| South and Central America | 0.31 | 0.53 | 6.13 | 16.64 | 14.60 | 4.92 | 1.80 | (7.48) | (16.26) | (29.87) | (40.58) |
| Europe and Eurasia | (12.39) | (37.74) | (50.86) | (69.68) | (99.66) | (7.24) | (30.21) | (24.98) | 6.02 | (0.10) | (14.79) |
| Middle East | 7.24 | 7.03 | 20.70 | 41.73 | 93.16 | 135.91 | 126.64 | 123.45 | 121.36 | 115.81 | 119.63 |
| Africa | 29.14 | 37.81 | 71.86 | 91.79 | 106.17 | 82.57 | 75.21 | 72.03 | 61.70 | 81.38 | 125.52 |
| Asia Pacific | (1.94) | (2.33) | (18.87) | (34.77) | (76.86) | (147.47) | (144.85) | (172.63) | (289.19) | (353.36) | (389.73) |
| Total natural gas imbalance | 27.87 | (19.32) | (0.58) | 14.08 | 9.86 | 68.65 | 30.63 | 88.13 | (16.37) | 3.40 | (11.83) |

*Data source:* Author's calculations based on BP (2016b). Totals may differ from column sums due to rounding.

*Table 9.6b* Regional imbalance (production minus consumption) – Mtpa

| Region | 1990 | 1995 | 2000 | 2005 | 2010 | 2014 | 2015 | 2020 | 2025 | 2030 | 2035 |
|---|---|---|---|---|---|---|---|---|---|---|---|
| North America | 4.06 | (18.16) | (21.80) | (23.33) | (20.33) | (0.02) | 1.50 | 72.11 | 73.76 | 139.82 | 138.77 |
| South and Central America | 0.23 | 0.39 | 4.52 | 12.27 | 10.77 | 3.63 | 1.33 | (5.52) | (11.99) | (22.04) | (29.93) |
| Europe and Eurasia | (9.14) | (27.84) | (37.52) | (51.40) | (73.52) | (5.34) | (22.28) | (18.43) | 4.44 | (0.07) | (10.91) |
| Middle East | 5.34 | 5.19 | 15.27 | 30.78 | 68.72 | 100.26 | 93.42 | 91.06 | 89.52 | 85.43 | 88.25 |
| Africa | 21.50 | 27.89 | 53.01 | 67.71 | 78.32 | 60.91 | 55.48 | 53.13 | 45.52 | 60.03 | 92.59 |
| Asia Pacific | (1.43) | (1.72) | (13.92) | (25.65) | (56.69) | (108.79) | (106.85) | (127.34) | (213.33) | (260.66) | (287.49) |
| Total natural gas imbalance | 20.56 | (14.25) | (0.43) | 10.39 | 7.27 | 50.64 | 22.59 | 65.01 | (12.08) | 2.51 | (8.73) |

*Data source:* Author's calculations based on BP (2016b). Totals may differ from column sums due to rounding.

Asia Pacific has been and will continue to be net importer of natural gas to be able to meet the projected regional consumption given expected regional productive capacity.

## The big players in demand and supply

Japan, China, and India are seen as continuing their role as the major consumers of natural gas in the region, and Australia is seen to be the largest player in terms trade in natural gas into the future.

The BP Outlook does not provide a country-by-country projection by fuel, so we will turn to the World Energy Outlook of the IEA. There is not complete agreement between the two Outlooks, but they are close enough to be able to make useful observations.

The Asia Pacific consumption of natural gas in 2013 equaled 630 Bcm,[8] which constituted 19.5 percent of total world consumption. According to BP statistics, the share of world natural gas consumption grew to just over 20 percent by 2015, at 701.1 Bcm. The BP Outlook 2016 has Asia Pacific natural gas consumption growing by 75 percent between 2014 and 2035, while the IEA has it growing by 84 percent between 2013 and 2035. BP's 2035 consumption level is 1073.3 Mtoe, whereas the IEA has it at 1046 Mtoe. If we adjust the IEA value for consumption in to 2013 to reflect the update in BP's Statistical Review, the projected IEA growth is the about 72 percent, so the two outlooks are in relatively close agreement as to both the future levels of consumption demand for natural gas in Asia and the growth from earlier observed consumption levels.

BP and the IEA see the future growth for natural gas in the region at similar levels. Per Table 9.7, the IEA projects 2035 natural gas component of TPED to be 1162.2 Bcm, while the BP Outlook to 2035 (2016a) has it at 1193 Bcm. The IEA projection has natural gas consumption doubling between 2013 and 2040. And over this period the share of the region accounted for by Japan, China, and India expands from 51 percent to 60 percent.

The share of natural gas in the energy mix is expected to increase significantly for the non-OECD Asian countries, and this can be seen for China (lifting from 5 percent to 11 percent) and India (lifting from 6 percent to 8 percent) by 2035.

*Table 9.7* Natural gas total primary energy demand

| | Natural gas demand (Bcm) | | | | | | Growth (%) 2013–2040 | Share of mix | |
|---|---|---|---|---|---|---|---|---|---|
| | *2013* | *2020* | *2025* | *2030* | *2035* | *2040* | | *2013* | *2040* |
| OECD Asia Oceania | 210.0 | 196.7 | 198.9 | 206.7 | 208.9 | 206.7 | −1.6 | 22 | 21 |
| Non–OECD Asia | 420.0 | 588.9 | 710.0 | 830.0 | 953.3 | 1064.4 | 153.4 | 8 | 13 |
| Japan | 117.8 | 95.6 | 93.3 | 95.6 | 96.7 | 95.6 | −18.9 | 23 | 22 |
| China | 157.8 | 280.0 | 352.2 | 416.7 | 468.9 | 506.7 | 221.1 | 5 | 11 |
| India | 50.0 | 64.4 | 90.0 | 114.4 | 140.0 | 165.6 | 231.1 | 6 | 8 |
| OECD – Oceania less Japan | 92.2 | 101.1 | 105.6 | 111.1 | 112.2 | 111.1 | 20.5 | | |
| Asia Pacific total | 630.0 | 785.5 | 908.9 | 1036.7 | 1162.7 | 1271.1 | 101.8 | | |
| Japan's share (%) | 18.7 | 12.2 | 10.3 | 9.2 | 8.3 | 7.5 | − 59.8 | | |
| China's share (%) | 25.0 | 35.6 | 38.8 | 40.2 | 40.3 | 39.9 | 59.2 | | |
| India's share (%) | 7.9 | 8.2 | 9.9 | 11.0 | 12.0 | 13.0 | 64.1 | | |
| Total of Japan-China-India share | 51.7 | 56.0 | 58.9 | 60.5 | 60.7 | 60.4 | 16.9 | | |

*Data source*: BP (2016a).

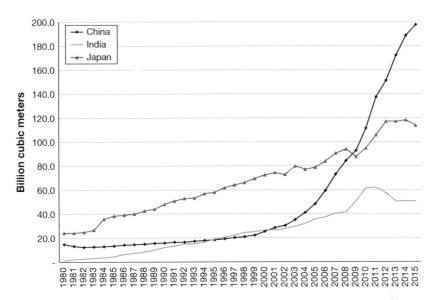

*Figure 9.2*   Natural gas consumption: Japan, China, and India

*Data source*: BP (2016b).

Figure 9.2 shows the dramatic increase in natural gas consumption in China, and the abrupt increase for Japan following the Fukushima Daiichi disaster.

The combination of Japan, China, and India accounts for 51 percent of the region's consumption in 2015. While the entire region is expected to continue to increase the absolute volumes of natural gas going forward, these three are expected to continue to dominate the region's consumption requirements, reaching 60 percent of the region by 2030 according to the IEA's New Policies Scenario projections. Growth in demand for natural gas will get a strong boost from China and India as they are expected to account for nearly half of the world's GDP growth through 2035, according to BP's Outlook to 2035.

## India

India has announced its plan to expand its LNG import capacity to 50 Mtpa, up from 21 Mtpa currently. The timeline has not been specified, but given current expansions and new projects under development in 2015–2016, this target seems reasonable. In addition to the existing 21 Mtpa of regasification capacity an additional 20 Mtpa is under construction at four different locations (Dahej, Mundra, Kakinada, and Ennore, according to GIIGNL (2017)). All are expected to be operational by 2020. The additional 10 Mtpa needed to meet the 50 Mtpa target are not yet specified.

India has been in negotiations for many years to access Iranian natural gas via pipeline. However, this pipeline would have to transit Pakistan, and agreement has not been successfully brought to the final investment decision nor to the beginning of construction. As a result, India has turned its efforts in the direction of developing domestic resources and investing in LNG regasification terminals to meet its needs.

Like China, India is heavily reliant on coal for its energy requirements. It too would like to wean itself from coal to help satisfy international emissions reduction commitments and to clean its environment. The expectation of a less dramatic increase in the share of natural gas in the total country energy mix suggests that India's reliance on coal will continue more strongly into the future compared to China. Nevertheless, the IEA projection foresees a 231 percent increase in consumption, which will require substantial investment in natural gas import facilities and domestic transportation and distribution infrastructure. And this projected growth in natural gas consumption will propel India past Japan by 2030, but India will still be a distant second to China's natural gas consumption level.

## Japan

Japan's natural gas consumption is not expected to increase from current levels. This is because it is expected that Japan will be bringing some, perhaps much, of its nuclear generation fleet back on line, and this will decrease the need for natural gas. While Japan's natural gas consumption had been on the rise throughout the 2000s, it showed a significant jump following the Fukushima Daiichi disaster in March 2011. Moving into 2012 Japan shut down all of it nuclear power plants, and natural gas was one of the main replacement fuels to help meet electricity generation needs. That surge in demand was a key driver behind the spike in regional natural gas prices to the US\$18–US\$19 per MMBtu range, with annual average prices nearing US\$17 per MMBtu.

If we assume that simple cycle natural gas generation plants will be the first to shut down when nuclear power plants are brought back online, the reduction in demand for natural gas imports could be quite significant. For each 1000 MW nuclear power plant that is restarted, natural gas demand tied to simple cycle power plants could fall by as much as 1.5 Mtpa of LNG-sourced imports.[9]

Japan has 44 GWe of nuclear power capacity, comprising 42 reactors.[10] Of the 42, 24 reactors were in the process of restart approval in 2016, with two having restarted in the second half of 2015. Each reactor is roughly equivalent to the hypothetical 1000 MW plant in this example. So, if all 26 plants are operating by 2020, that could lead to a decrease in Japan's natural gas import demand by 39 Mtpa, or 43 Bcm. The IEA projections reported in Table 9.7 suggest a decrease of 22.2 Bcm between 2013 and 2020. The difference between these two may be due to the IEA assuming fewer nuclear reactors return to service or that the natural gas generation fleet in use is more thermally efficient, or both, and it may also assume that growth in other non-power generation uses for natural gas will offset some of the decline from the power sector. The decrease in demand of 22.2 Bcm implies 20 Mtpa, so there may be a range of 20–39 Mtpa of uncertainty for the demand for natural gas imports from Japan over just the next five or so years.

## China

China's 13th Five Year Plan aims to have natural gas account for 10 percent of primary energy by 2020. This will be an increase from 5.9 percent in 2015. The IEA projection puts natural gas at 11 percent of primary energy for China by 2040; the IEA projection pre-dates the 13th Five Year Plan. BP expects China to contribute significantly to natural gas production, especially toward the end of its outlook to 2035. BP projects that China will contribute more to shale gas production growth than any other country in the world.

China appears to have significant shale resources that are deemed to be technically recoverable according to the 2013 study done for the IEA by Advanced Resources International. China's estimated technically recoverable shale gas resources amount to 1115 trillion cubic feet (31.6 trillion cubic meters), which is nearly double the estimated technically recoverable shale gas resources for the USA. However, there are significant questions about the commerciality of these technically recoverable resources. China has missed its production targets for coal seam gas consistently over the past five or so years, and many question the near future potential for China's large shale gas resource. This is further exasperated by the current low-price environment affecting investment viability around the world. This situation was exacerbated by the late-2014 decision of OPEC to maintain crude oil production levels to attempt to regain its market share. The decision sent negative signals across all energy sectors regarding the uncertainty of prices that may be earned on virtually any form of energy production.[11]

Nevertheless, BP expects China to import 40 percent of its natural gas requirements by 2035. This compares to their imports of 30 percent of consumption in 2015. If we accept the IEA projection for China's natural gas demand in 2035 at 468.9 Bcm, BP's expectation implies China's imports will reach 187 Bcm, compared to 59.8 Bcm in 2015. At the end of 2015, China had two import pipelines with combined capacity of 60 Bcm and 13 LNG regasification import facilities with combined capacity of about 46 Bcm (41.3 Mtpa). According to GIIGNL, there are eight additional regasification projects under construction with import capacity of 26 Bcm (24 Mtpa). This implies that existing and under-construction import capacity equals 132 Bcm. If we accept the BP–IEA-based projection of 187 Bcm imports for China in 2035, there must be additional investment in, and expansion of, import capacity equaling 55 Bcm. If all of this were to come in the form of LNG regasification, it would add just under 50 Mtpa of import capacity and increased demand.

There is much talk about pipeline imports to China from Russia, and there are two agreements between the countries and their respective national energy companies. Nevertheless, it is not likely that these projects will go forward any time soon. This is due at least in part to the current relatively low oil and natural gas prices and the negative effect this has had on Russia's export revenues that could be applied to natural gas field development and pipeline construction.

## *Australia*

Australia has been part of the regional trade in natural gas since the Northwest Shelf Project (NWS) came online in 1989. The NWS is located in Western Australia about 1500 km north of Perth and has been expanded to five trains with a capacity of 16.3 Mtpa. The NWS project was followed by Darwin LNG, in the Northern Territory, and Pluto LNG, also in Western Australia, before several project consortia took final investment decision on seven new LNG export projects. The seven projects include Gorgon (15.6 Mtpa), Queensland Curtis LNG (QCLNG – 8.5 Mtpa), Gladstone LNG (GLNG – 7.8 Mtpa), Asia Pacific LNG (APLNG – 9.0 Mtpa), Wheatstone LNG (8.9 Mtpa), Ichthys LNG (8.4 Mtpa), and Prelude LNG (3.6 Mtpa).[12] When all projects are completed to planned full capacity, the ten LNG projects will constitute 85 Mtpa of export capacity. It was planned that all projects would be developed to full capacity by 2018, but several projects have announced slow-downs in the further development of trains in response to the low-price environment. Nevertheless, when all projects reach full capacity, Australia's LNG export capacity will exceed that for Qatar, the current largest exporter with 77 Mtpa.

Of the "new" seven, QCLNG, GLNG, APLNG, and Gorgon have all begun operation. QCLNG and GLNG began exports in 2015, and Gorgon and APLNG started operations in 2016. The three Queensland projects (QCLNG, GLNG, and APLNG) are each supplied with natural gas sourced from coal seams. These are the first LNG projects in the world to employ coal seam gas (also referred to as coal bed methane) as the feedstock for an LNG project. The Gorgon project, based in Western Australia is the largest LNG project in Australia history, with an estimated cost of US$54 billion.

The Prelude LNG project is led by Shell and will be a floating LNG (FLNG) liquefaction facility located offshore to the north of Western Australia. The floater is being constructed in South Korea.

Australia is already the largest exporter of natural gas within the Asia Pacific region, exporting 39.8 Bcm in 2015, all in the form of LNG. This represented nearly 60 percent of its production, and the share of production to be exported in the future is expected to increase. This is made possible by the relative size of its proved reserves to its domestic consumption. With a population of only about 24 million, there is not a large domestic draw on its resource base. From Tables 9.1 and 9.3, it is observed that Australia has a proved reserve base of natural gas equal to 3471.4 Bcm in 2015, with domestic consumption in that year of just 34.3 Bcm.

It is important to keep in mind, however, that Australia is a member of the Asia Pacific region being discussed. Thus, the imbalance identified between production and consumption in the region already includes all of these projects and their associated production, and it may include the assessment of BP and the IEA of potential expansions of existing projects and additions of new ones. So, while Australia will become the largest exporter of natural gas in the form of LNG, and most of the exports will be delivered within the Asia Pacific region, the shortfall identified will still have to be from outside the region via inter-regional trade.

## Conclusion

The Asia Pacific region is a growing force in the markets for natural gas on a global scale. In the twenty-first century, it is expected to exhibit consumption growth at a higher rate than any other global region, with the exception of Africa. Asia Pacific's consumption will be second only to North America by then.

To meet this growth there will be a need for significant infrastructure investment both within and outside the region. This is because total global consumption of natural gas is expected to grow beyond just that in the Asia Pacific, and because the region is not capable of satisfying its own consumption demands with its own production. While the region is deemed to hold significant natural gas resources, which are identified as technically recoverable, there are concerns about the ability to produce these resources commercially.

The twenty-first century of natural gas is also likely to be the Asia Century. While virtually all member countries in the region will see substantial growth in consumption, and some with growth in production, the key players to watch appear to be Japan, China, and India for consumption, and Australia for production and exports.

## Notes

1 Throughout this chapter the IEA projections discussed are drawn from the 2015 edition of the IEA's World Energy Outlook (International Energy Agency, 2015), and specifically the New Policies Scenario (NPS) outlook. The NPS projects are based on the assumption that all proposed energy policies will be implemented (albeit conservatively) and added to those policies officially in place as of mid-year 2015. BP also makes underlying assumption about the implementation of proposed energy policies beyond just those currently enacted in laws. These differ from outlooks by the Energy Information Administration in the USA, which is prohibited by law from assuming anything other than policies already enacted.
2 There are 35.3 cubic feet per cubic meter.
3 LNG (liquefied natural gas) is simply a transport phase for natural gas, and does not constitute a distinct commodity class. Natural gas (primarily methane) is cooled to −161 degrees Celsius, at which point it becomes a liquid. When liquefied in this manner the space required to contain the natural gas shrinks roughly 600 times while retaining it heat/energy content. It is this significant reduction in required space

that makes it economic to move natural gas in this form. Nevertheless, upon delivery the LNG is always re-gasified prior to use or injection into a pipeline system where it is comingled with other natural gas.

4 JCC is colloquially referred to as the Japanese Crude Cocktail, but it is actually Japanese Customs Cleared. The price of natural gas is linked to relative heat/energy content of natural gas to crude oil. An exact heat content linkage depends on the heat content of the average barrel of crude oil being imported into Japan. However, it is typically accepted that there are 5.8 million Btu (MMBtu) per barrel of crude oil. In this case, an exact conversion for the natural gas would call for multiplying the appropriate JCC price of crude oil by 0.1724 to arrive at the natural gas price per MMBtu. This multiple (as referred to as the slope factor) is an actively negotiated element of the long-term contracts, and it is typically assumed to be closer to 0.15, with some level of constant dollars and cents added. A version of this pricing formula that I estimated in the past is JCC × 0.1485 + 0.8, based on a regression analysis of Japanese crude oil prices and reported LNG delivered prices.

5 The contractual off-takers for the Sabine Pass LNG project are BG, GNF, GAIL, KOGAS, TOTAL, and Centrica. BG was first in the door, and has the lowest processing cost at US$2.25 per MMBtu.

6 The spot day rates vary according to the supply and demand for LNG tankers. The US$40,000 per day was in effect during September 2016. However, during a roughly five-year period the rate had been as high as US$130,000 and as low as US$30,000. At US$130,000 the shipping cost per MMBtu would be US$2.23 to Tokyo via Panama.

7 It is worth noting, however, that on a marginal cost recovery basis, which is relevant for already constructed and contracted facilities, there is still an incentive for off-takers to take delivery from Cheniere and ship even to Asia. In the example it is possible that the only marginal cost faced by, say BG, is the natural gas acquisition cost, which was US$3.59 per MMBtu. This is because the liquefaction fee is a demand charge (take-or-pay), which is a fixed cost. It is also potentially the case that the LNG tankers to be used will be on long-term contract, rather than spot terms, which is again a fixed cost. Under these conditions, there would then be a margin of US$2.11 per MMBtu, and this could then be allocated to cover at least a portion of the fixed costs.

8 The IEA World Energy Outlook 2015 reports TPED of natural gas as 567 Mtoe for the combined OECD Asia Oceania and Non-OECD Asia (International Energy Agency, 2015). Using BP conversion of 1.11 × Mtoe = Bcm results in 630 Bcm. The BP Statistical Review of World Energy for 2014 reports Asia Pacific consumption for 2013 at 575.2 Mtoe and 639.2 Bcm. However, BP updates these values in the 2016 Stat Review (BP, 2016b) to 610.5 Mtoe and 678.4 Bcm.

9 This assumes the 1000 MW nuclear plant runs at 90 percent of capacity over the year, and that simple cycle natural gas power plants are 38 percent thermally efficient. The 1.5 Mtpa of LNG equates to the 70,790,021 MMBtu of natural gas required to produce the comparable flow of electricity using simple cycle power plant.

10 www.world-nuclear.org/information-library/country-profiles/countries-g-n/japan-nuclear-power.aspx.

11 The more recent OPEC decision to constrain production is yet to send strong new investment signals to most of the world, so it is not clear how this may affect China's investment decisions regarding natural gas developments.

12 The capacities shown are planned full project capacities. Most of the projects are bringing on one a train at a time, and due to the relatively low-price environment some of the following trains are being delayed.

# References

BP, 2016a. *BP Energy Outlook, 2016*, London: British Petroleum.

BP, 2016b. *Statistical Review of World Energy 2016*, London: British Petroleum.

GIIGNL, 2017. *The LNG Industry—GIIGNL Annual Report 2017*, Neuilly-sur-Seine, France: International Group of Liquified Natural Gas Importers.

International Energy Agency, 2015. *World Energy Outlook 2015*, Paris: International Energy Agency.

# 10

# THE ROLE OF COAL IN ASIA

*Subhes C. Bhattacharyya*

## Introduction

Coal plays an important role in Asian energy scene. The region has been relying on coal to fuel its growing energy needs to support faster economic growth. The region is well endowed with coal resources and it is a comparatively cheaper fuel. However, high reliance on coal brings local environmental issues and it has significant climate change implications. With rising local and global environmental concerns, the region is committing to cleaner alternatives to support climate protection and sustainable development initiatives.

This chapter provides a brief review of coal supply and use in Asia and discusses how the region is trying to support economic growth while reducing the environmental damages. The chapter is organised as follows: the second section presents an overview of coal supply in Asia, which is followed by a section on coal use. A brief review of the coal industry in selected countries is presented in the fourth section, while the outlook for coal is presented in the fifth section. Some concluding remarks are given at the end of the chapter.

## Overview of coal supply in Asia

As presented in Chapter 2, coal is the dominant form of energy in Asia, accounting for 50% of the primary energy supply in 2012. Between 1990 and 2012, the primary coal supply in the region has grown from 844 Mtoe in 1990 to 2710 Mtoe in 2012 (International Energy Agency, 2015b), representing an annual average growth rate of 5.4%. However, the use of coal has grown even faster in the new millennium, reaching an average growth rate of 7.8% per year. As indicated in Figure 10.1, most of the regional supply came from within the region. Domestic production by member countries of the region accounted for more than 93% of the supply, whereas coal trade (i.e. imports and exports) and stock change account for the balance. The share of imports in the supply has grown in recent times, reaching almost 19% of the primary supply in 2012 but the region has also exported about 10% of its production.

East Asia dominates the coal supply and use in Asia. Almost 80–82% of primary coal demand[1] of the region comes from East Asia; 89% of East Asian coal demand came from China in 2012, playing the most important role. In fact, China accounts for half of global coal demand

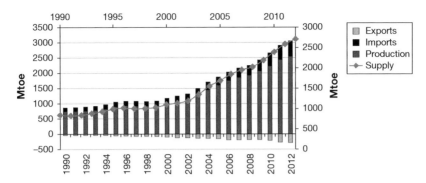

*Figure 10.1*    Trend of Asian coal supply

*Data source*: International Energy Agency (2015b).

(Cornot-Gandolphe, 2014). South Asia accounts for about 13% of the demand, while the rest comes from West and South East Asia (see Table 10.1). India accounted for 98.5% of South Asian coal demand in 2012. These shares have remained practically unchanged over the past two decades, with the exception that the share of coal use in West Asia has declined from 6% in 1990 to 1% in 2012, whereas South East Asia has recorded a corresponding gain in the share.

In terms of coal production in the region, seven countries account for 99% of the output (see Table 10.2). Seventy-five per cent of the regional production comes from China, while India and Indonesia account for 10% each. Kazakhstan produces 2% of the Asian coal while Vietnam, Mongolia and DR Korea produce 1% each. China is the largest producer of coal in the region and in fact in the world. The resource is distributed over an area of 600 thousand square kilometres (Wu, et al., 2017). The Chinese coal output (in tonnes) is about six times more than that of India and four time that of the USA. Only recently when China's economy has started to slow down, has its coal output has started to fall. A recent study (Wu, et al., 2017) indicates that the year-on-year coal production fell by 2.5% in 2014 and 4.1% in 2015.

Despite producing significant amounts of coal, the region has to import some coal for various reasons. Some countries (like Japan and Korea) have limited resource endowments and their domestic production is insufficient to meet their needs. The quality of coal is an issue and even large producers like China and India are importing better quality coal and coking coal. Moreover, the geographical location of the deposits often makes domestic transport of coal difficult, particularly to distant areas due to inadequate transport facilities. For example, coal is concentrated in the eastern and northern parts of China and in the eastern and central parts of India. In such cases, importing coal can be a cheaper option, particularly for the coastal locations. China accounted for 29% of the coal imports in the region in 2012, followed by Japan at 22%. India and Korea are two other major importers of coal in Asia. Four of them accounted for 83% of coal imports in 2012. According to Cornot-Gandolphe (2014), China is the 'swing buyer' of the coal market – imports coal when international prices are lower than domestic prices and vice-versa. On the other hand, Indonesia is the leader in coal export in the region, accounting for 79% of export from Asia in 2012. Mongolia and Kazakhstan are also main exporters. These three countries account for 90% of coal export from the region.

With 114.5 billion tonnes of proved reserves at the end of 2015, China tops the coal resource endowment in the region (BP, 2016). However, with current level of production, its proven reserves will last for about 31 years. On the other hand, India and Indonesia, despite having much

*Table 10.1* Distribution of coal use in Asia by region (%)

| Region | 1990 | 2000 | 2010 | 2012 |
|---|---|---|---|---|
| East | 80 | 80 | 82 | 82 |
| West | 6 | 2 | 2 | 1 |
| South | 13 | 15 | 13 | 13 |
| South East | 1 | 3 | 4 | 3 |

*Data source*: International Energy Agency (2015b).

*Table 10.2* Major coal producers, importers and exporters in Asia

| Major producers (share in total output, %) | | Major importers (share in total import %) | | Major exporters (share in total export, %) | |
|---|---|---|---|---|---|
| China | 75 | China | 29 | Indonesia | 79 |
| India | 10 | Japan | 22 | Mongolia | 6 |
| Indonesia | 10 | India | 17 | Kazakhstan | 5 |
| Kazakhstan | 2 | Korea | 15 | Vietnam | 3 |
| Vietnam | 1 | Chinese Taipei | 8 | China | 3 |
| DR Korea | 1 | Malaysia | 3 | DR Korea | 3 |
| Mongolia | 1 | Thailand | 2 | India | 1 |
| | | Hong Kong | 1 | Philippines | 1 |
| | | Philippines | 1 | | |

*Data source*: International Energy Agency (2015b).

lower reserves (with 61 and 28 billion tonnes of proved reserves respectively at the end of 2015), have a higher reserve to production ratio (89 years for India and 71 years for Indonesia) (BP, 2016). However, the quality of output can be an issue as more coal is extracted and the reserve figures may prove illusory in such cases.

## Coal use in Asia

Although energy balances make distinction between energy used for transformation and that for final use, in this section, the information is combined to present coal use clearly. Coal is mainly used for electricity generation either in central utility power stations or by users themselves using captive power plants. In Asia, many large users (like cement plants, steel plants and the like) generate their own power to ensure a reliable supply or to supplement power from the grid. Coal used for both these categories of electricity generation is included under electricity. Some amount of coal is also used for steel making and other secondary energy transformation (such as liquid fuels using coal to liquid fuel technologies). Generally, coking coal is used for steel production and thermal coal is used for other transformation activities. Coal used for steelmaking also includes any consumption for other transformative activities. In the final use category, industrial use of coal remains important while some coal is used in rail transport and in other sectors (residential, commercial and non-energy use). Other use category includes coal used in transport, other sectors and in the energy industry (own-use). Accordingly, the shares shown here will differ from conventional final energy use data reported generally.

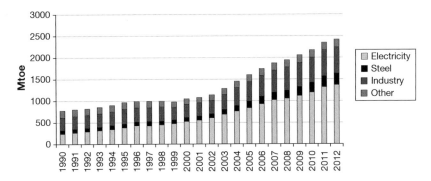

*Figure 10.2*   Uses of coal in Asia

*Data source*: International Energy Agency (2015b).

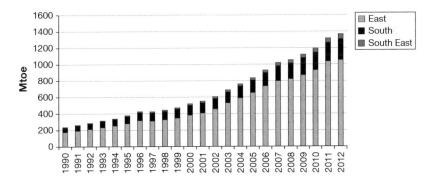

*Figure 10.3*   Coal demand for electricity generation in Asia

*Data source*: International Energy Agency (2015b).

Electricity generation drives the demand for coal in Asia (see Figure 10.2). This sector accounted for 31% of coal demand in the region in 1990, but the share has gradually increased to 57% in 2012. Industry remains the second most important user of coal: its share, however, has fallen from 37% in 1990 to 25% in 2012. Miscellaneous use of coal was the third main user in 1990s, but this has been pushed to the fourth position by steelmaking in recent times. As noted earlier, the coal demand has seen a rapid growth in the new millennium and the demand for coal from electricity generation has been responsible for this dramatic growth.

The regional break down of coal demand for electricity generation clearly shows the domination of East Asia. In 2012, 77% of coal used for electricity generation was in East Asia (see Figure 10.3). Eighty-seven per cent of East Asian coal demand originated from China, while 6% came from Japan and 5% came from Korea. South Asia accounted for 18% of coal demand for electricity generation and the remaining 5% came from South East Asia. India is the principal coal user in South Asia for electricity generation, while in South East Asia, the demand is shared between Indonesia (43%), Malaysia (23%), Thailand (13%), Philippines (11%) and Vietnam (9%).

The regional electricity output from coal-fired plants follows the same trend as coal use. East Asia, with an 80% share in 2012, dominates the coal-based electricity production in the region. Eighty-four per cent of the East Asian electricity output from coal-based plants came from China in 2012, while Japan and Korea accounted for 7% and 5% of the outputs, respectively.

South Asia accounts for only 14% of the coal-based electricity output at the regional level, while SE Asia and West Asia accounted for 4% and 1%, respectively (see Figure 10.4). Note that South East Asia holds the highest growth rate in coal-based electricity generation (10.2% per year compared to 9.2% for East Asia and 6.7% for South Asia) for the period between 1990 and 2012. South East Asia is starting from a very low base, which explains their faster growth in this area.

In terms of importance, China's coal-based electricity output (in TWh) was 4.7 times higher than that of India in 2012. India's output was 2.6 times more than that of Japan in the same year, whereas Korea's output was about 79% of Japan's. These four countries accounted for 90% of the coal-based electricity output in the region in 2012 (see Table 10.3) and taken with Indonesia, Malaysia, Kazakhstan, Thailand and Philippines, they cover 99% of coal-thermal electricity output of the region.

Coal use in industry is the second main source of demand. As before, East Asia accounted for 80% of the demand for this end use in 2012, while South Asia accounted for 13% and South East Asia 4% (see Figure 10.5). However, in terms of demand growth rates, South East Asia recorded an average growth of 7% per year in coal demand for this sector between 1990 and 2012, while

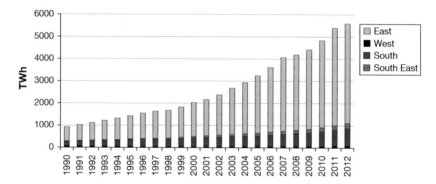

*Figure 10.4*   Electricity output from coal-based power plants in Asia

*Data source*: International Energy Agency (2015b).

*Table 10.3* Share of coal demand in different uses by country in 2012

| Country | Share in regional coal-based electricity output (%) | Share in regional industrial coal demand (%) |
| --- | --- | --- |
| China | 68 | 72 |
| India | 14 | 13 |
| Japan | 5 | 4 |
| Korea | 4 | 1 |
| Chinese Taipei | 2 | 1 |
| Indonesia | 2 | 1 |
| Kazakhstan | 1 | 2 |
| Malaysia | 1 | 0 |
| Thailand | 1 | 1 |
| Philippines | 1 | 0 |
| Vietnam | 0 | 2 |

*Data source*: International Energy Agency (2015b).

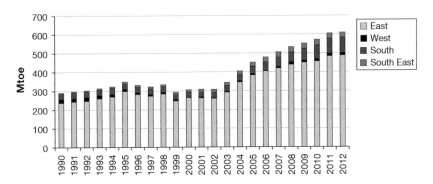

*Figure 10.5*   Coal use in industry in Asia

*Data source*: International Energy Agency (2015b).

South Asia recorded a 5% annual growth over the same period. East Asia, on the other hand, posted a growth rate of 3% per year. A slowdown in coal use in industry after 1998 is clearly visible – which reflects the post-1997 financial crisis. The demand, however, bounced back in 2003 and a rapid growth in coal demand in industry in the subsequent years is visible as well. Only recently (in 2015 – not shown in the figure), China's coal demand has started to slow down as the economy adjusts to a new equilibrium of moderate growth.

As indicated in Table 10.3, China has a 72% share in the regional coal demand in the industrial sector. India and Japan account for 13% and 4%, respectively, while Kazakhstan and Vietnam account for 2% each.

## Coal industry development in selected Asian countries

Although the coal industry has a long history, it has come into prominence in recent times. This section provides a brief review of coal industry in selected Asian countries.

### *Coal industry in China*

China's coal endowment is spread over 6% of its land area and is concentrated in five coal endowment districts, namely northeast, north, south, northwest, and Tibet and Yunnan (Tu, 2011). The resources are unevenly distributed and more than 90% of the reserves are located in less-developed, environmentally vulnerable areas. China's coal industry has a long history – in fact, the organised production and consumption of coal started long (by the year 1000) before other countries and China remained the world's largest producer of coal until steam engines came into existence (Thomson, 2003). But as the new steam engine technology did not reach China until the mid to late 1800s (Thomson, 2003), the coal mining remained small scale and the industry was somewhat insulated from the rest of the world.

However, as international trade started to expand with the arrival of steam-powered ships and railways, there was a growing foreign interest in the Chinese coal industry. Large-scale mine development started under foreign investment and takeover of Chinese mines in the latter half of the 19th century and early 20th century. The industry grew rapidly under Japanese domination in the 1930s. Alongside large mines developed through foreign investment, there were many small local mines, often working in winter and in many cases illegally in concession areas held by foreign companies. Output grew rapidly and by 1949, China produced 32 million tonnes of coal.

However, the Sino-Japanese war that broke out between 1937 and 1945, and the civil war that followed in 1947–1948, damaged the industry considerably and at the time of creation of the Peoples Republic of China, the country inherited a fragmented, poorly equipped industry.

The industry recorded modest growth during the Maoist era of central planning. Coal mines were organised under three categories, namely those under the Central Mining Administrations, the local state mines and the local non-state mines. Under the Soviet-style planning model, the industry worked to meet a set production target and depended on Russian technology. Non-state sector small coal mines were promoted during this period as a biomass conservation effort in rural areas through a government directive in 1957. Small coal mines started to emerge in rural areas and the production started to increase, recording a threefold increase between 1957 and 1960 and a fourfold increase between 1960 and 1965 (Bhattacharyya & Ohiare, 2012). However, the state sector coal mining was hard hit under Mao's the Great Leap Forward and the Cultural Revolution, and the strained Sino-Soviet relations in the 1960s. With a shift in the energy policy, the better northern reserves were abandoned and exploration was started in south and east. Production fell as a result, leading to energy shortages during the period (Arruda & Li, 2003).

The coal industry started to grow rapidly during the liberalisation period. To support economic growth, major emphasis was laid on coal industry development and expansion. Alongside development of state-owned enterprises (SOEs), township and village enterprises (TVEs) were allowed to expand to support local economic activities. The three-tier production system continued during this period, namely the large SOEs, local SOEs and finally a large number of small, decentralised TVEs (Tu, 2011; Cornot-Gandolphe, 2014). Large SOEs are governed by provincial governments, whereas small SOEs are local in nature. Agricultural collectives invest and control TVEs but because of their artisanal approach to mining, they tend to be inefficient, accident prone and less compliant with health and safety as well as environment regulations. According to Umbach & Wu (2016), 65% of coal production in 2012 came from state companies. It is reported that at the end of 2015, China has about 11,000 coal mines, of which 1050 mines are large ones having annual output of over 1.2 million tonnes. On the other hand, there were more than 7000 small mines with less than 0.3 million tonnes of output (Wu, et al., 2017). The rapid growth in coal production in the new millennium is partly attributed to outputs from these small mines.

The recent slowdown in Chinese coal demand has led to the suggestion that coal demand in the country has peaked. The government has initiated actions to combat air pollution by capping coal use in eastern demand centres, promoting non-fossil fuel energies and improving energy efficiency. Simultaneously, China has started to close down smaller mines and is restructuring the coal industry by consolidating the mining activities. The power plant technology has moved towards more efficient supercritical and ultra-supercritical plants, while the smaller, inefficient plants have been shut down, thereby improving the overall efficiency of power generation.

However, the coal industry faces important issues. The cost of production tends to be higher in SOEs due to higher level of mechanisation, better compliance with safety and environmental standards and a bloated workforce, a legacy from the past era. TVEs, on the other hand, pay low wages, are poorly mechanised and have poor compliance with standards (Tu, 2011). The cost difference creates competitive pressures and leads to overproduction of coal. The oversupply and the resultant low price have affected the industry financially. It is reported that 70% of the coal companies are incurring losses as a result (Cornot-Gandolphe, 2014).

High reliance on coal has also caused severe environmental and health issues in China. Air pollution is causing 1.2 million premature deaths and costing 3.5% of national GDP (Umbach & Wu, 2016). Due to environmental concerns and health and safety issues of small mining activities in China, the government has been trying to reorganise the activities of the coal industry over the past two decades. The government was encouraging mergers and acquisitions to create ten large coal companies and to bring the overall number to 4000 mines by 2015. However, these efforts have been less successful due to strong opposition from TVEs (Umbach & Wu, 2016). The rationalisation and consolidation of the industry is essential for the long-term viability of the sector.

## Coal industry in India

India has fifth largest recoverable coal reserves in the world, but 78% of its total proved reserves are located in the eastern states of West Bengal, Odisha, Jharkhand and Chhattisgarh. Sixty-five per cent of its coal production comes from these states, while another 23% comes from Andhra Pradesh and Madhya Pradesh (Office of the Chief Economist, 2015). Indian coal has its own specific features: it has low energy content and high ash content. This makes Indian coal less suitable for export and larger quantities of coal are required to meet any specific needs. Often better quality imported coal is mixed with local coal to achieve improved performance, and coal washing (or beneficiation) has been identified as an option to improve effectiveness of coal and to reduce pollution, but so far the capacity of such treatment plants is limited in the country.

Unlike China, the Indian coal industry is relatively new and has a history of about two centuries. Coal mining started Bengal (Raniganj) in 1774 during the time of Warren Hastings under the British rule (Gee, 1940). But the progress was slow and limited for almost one century due to lack of transportation infrastructure to ship coal to the demand centre in Calcutta. The activities started to gain momentum in the mid-19th century when the railway link between Calcutta and the mining fields in Eastern India came into existence. Three main groups of players are identified in the literature: the landowners, the operating houses and the Rail Board (Lahiri-Dutt, 2016). As the government did not claim rights to the mineral wealth of the area, the companies had to negotiate royalties and mining rights with local land owners, which proved to be very complex and litigation prone, leading to failed initiatives (Gee, 1940). But during the colonial period, the British controlled agency houses and the Rail Board managed to generate profits by controlling trade and the market (Lahiri-Dutt, 2016). By 1890, coal production reached 6.2 million tonnes (Mt) and the production rose rapidly during the First World War. By 1950, India had very similar levels of production as China (around 33 Mt) (PricewaterhouseCoopers, 2016a).

Post-independence, the industry operated mainly under private ownership until its nationalisation. Two exceptions were the creation of the National Coal Development Corporation in 1956 by the Government of India to take control of mines under the railways and the takeover of Singareni Collieries Company Limited (SCCL), by the state of Andhra Pradesh in 1956. During this period, the sector did not grow fast due to lack of investment in the sector because of low coal prices and strong trade union influence on the labour force. However, the oil price shock in the early 1970s led to the realisation of strategic importance of coal in the country and the coal industry was progressively nationalised between 1972 and 1973. All coking coal mines except those operated by Indian Iron and Steel Company, Tata Iron and Steel Company and Damodar Valley Corporation were nationalised in 1972. In the following year, all non-coking coal mines were nationalised and a national coal company, Coal India Limited (CIL), was established in 1975 (PricewaterhouseCoopers, 2016a).

The coal industry is still dominated by the public sector entities, CIL and SCCL. Their share in coal production was around 95%, while the private sector accounted for only 5%. Most of the coal produced in India is non-coking in nature (90%). In terms of mining technology, about 93% of the output comes from open-cut mines and the rest (7%) comes from underground mining

(Government of India, 2016). Although India and China started from a similar production level in 1950, two countries have followed very different trajectories and India has been left far behind over the past decades. India has now initiated the process of involving private sector through allocation of coal blocks but the initial allocation process was turned down by the Supreme Court in 2014. A competitive bidding based process is now being followed and the benefits of these investments will emerge only in due course.

## Coal industry in Indonesia

Indonesia is the world's largest coal exporter since 2011 and the most important coal producer in South East Asia. Based on 2015 estimates, the country has 127 Gt of coal resources and 32 Gt of reserves with 8 Gt of proven reserves (Cornot-Gandolphe, 2017). Eighty-four per cent of the reserves are located in two geographical areas, namely East Kalimantan (accounting for 45% of the proved reserves) and South Sumatra (accounting for 39% of the proved reserves). Coal output has expanded rapidly over the past decade recording an annual average growth rate of 15% (International Energy Agency, 2015a). The island of Kalimantan accounts for almost 90% of the coal produced in the country and only 8% of the production comes from Sumatra (Cornot-Gandolphe, 2017). Indonesian coal enjoys competitive advantages, such as low ash content, low sulphur content, proximity to coastal areas, proximity to major demand centres in China and India, and low cost of production due to surface mining (International Energy Agency, 2015a).

The coal industry in Indonesia is relatively new, with its formative period starting in late 1960s. Prior to 1990, the mining activity was very limited and the activity was concentrated in Sumatra. In the 1990s, mining activities started to pick up, particularly in Kalimantan island and the production started to pick up from around 2005 to reach a high level of 490 Mt in 2013 (Cornot-Gandolphe, 2017). The mining activity in Indonesia was previously governed under a regulatory system known as Contracts of Work that was introduced in 1967. For the coal industry, the relevant regulatory arrangements were Coal Cooperation Agreements, Coal Contracts of Work and Mining Rights (PricewaterhouseCoopers, 2016b). Most of the coal output came from Coal Contracts of Work. But in 2009, a new Mining Law was introduced and the government intended to bring all previous arrangements in line with the new regulatory provisions that are more transparent and which introduced standard tendering and licenses for coal mining. The Indonesian coal industry is open and competitive, with only 5% of the assets under direct state control. This is in contrast to other countries in the region where state ownership still plays an important role.

Indonesia has launched an ambitious coal-based electricity generation programme where the country intends to add 35 GW of electricity capacity by 2019. This programme, if successful, will increase the domestic coal demand dramatically. This may offset the oversupply issue arising from slowdown in import demand from China, but a domestic demand growth may also reduce the profitability of coal mining in the country, particularly if there is a domestic obligation to supply coal at below market prices (International Energy Agency, 2015a). Indonesia has relied on brown field expansion so far for its incremental coal supply but there is a large green field expansion programme in Sumatra. This is expected to change the share of coal supply from two main geographical areas in the future (International Energy Agency, 2015a).

## Outlook for coal in Asia

Various studies carried out by international organisations maintain that coal will maintain its dominant role in the primary energy supply in the future. ADB (2013) suggests that 42% of primary energy demand in Asia and the Pacific will be met by coal in 2035 in the business as usual case. On the other hand, in the developing member countries, the share of coal reaches 43% by the same year under the same scenario. A forecast by the Energy Information Administration suggests that 37% of Asia's primary energy demand in 2035 will be met by coal but the share of coal in the primary energy mix of developing countries in Asia will be almost 40% (EIA, 2016). India and China will account for 74% of the regional coal demand in 2035 according to this study. India's coal demand is likely to grow at a faster rate, whereas China will see a very modest growth in coal demand. Although both the studies suggest a reduction in the coal share in the primary energy mix, the quantity of coal use will increase due to bigger size of the pie. However, compared to the share of coal in 2012, the region will see a gradual decline in the role of coal during this period (EIA, 2016), but in South East Asia the coal share will increase in the primary energy mix in the future – from 15% in 2013 coal share to 27% in 2035 and 29% by 2040 (International Energy Agency, 2015a, 2015b).

The power sector will remain the mainstay of coal demand in the region, particularly for East Asia, South Asia and South East Asia. Coal will continue as the main fuel for power generation in these regions in 2035 under the business as usual scenario. Industry will remain the next major use of coal in the region. The cost advantage arising from the low coal price and local availability drives coal use, but the commitment of Asian countries towards sustainable development and climate mitigation will put pressure on adopting more efficient technologies for power generation and for using clean coal technologies. China, for example, has adopted clean coal technologies as a strategy, which covers the entire value chain, namely mining and production, coal conversion into value added products (such as coal to liquids) and waste disposal (carbon capture and storage) (Cornot-Gandolphe, 2017). As Asian countries are unlikely to abandon coal, it is important to support such strategies to reduce the adverse impacts on the environment and the society.

## Conclusions

Coal has aptly supported Asia's economic development over the past few decades, particularly in the aftermath of the first oil crisis and has helped the region to emerge as a major player in the global energy scene. The price advantage of coal and its abundant supply has ensured its dominant position in the primary energy mix of the region. Despite slowdown in the rate of growth in the future, the 'black diamond' is expected to maintain its dominant position in the region but the consequent local environmental challenges and the climate change implications will require adoption of best practice technologies and processes to reduce the possible impacts.

## Note

1  As indicated in Chapter 2, the terms primary energy supply and primary energy demand have been used interchangeably in this chapter. In the energy balance framework, they are equivalent as supply equals demand.

## References

ADB, 2013. *Energy Outlook for Asia and the Pacific*, Manila: Asian Development Bank.

Arruda, M. & Li, K., 2003. China's energy sector: Development, Structure and Future. *China Law & Practice*, 17(9), pp. 12–17.

Bhattacharyya, S. & Ohiare, S., 2012. The Chinese electricity access model for rural electrification: Approach, experience and lessons for others. *Energy Policy*, 49, pp. 676–687.

BP, 2016. *Statistical Review of World Energy 2016*, London: British Petroleum.

Cornot-Gandolphe, S., 2014. *China's Coal Market: Can Beijing Tame 'King Coal'?*, Oxford: Oxford Institute of Energy Studies.

Cornot-Gandolphe, S., 2017. *Indonesia's Electricity Demand and the Coal Sector: Export or Meet the Domestic Demand? OIES Paper CL-05*, Oxford: Oxford Institute of Energy Studies.

EIA, 2016. *International Energy Outlook 2016*, Washington, DC: US Energy Information Administration.

Gee, E., 1940. *History of coal mining in India*. [Online] Available at: www.insa.nic.in/writereaddata/UpLoadedFiles/PINSA/Vol06_1940_3_Art03.pdf [Accessed 31 March 2017].

Government of India, 2016. *Provisional Coal Statistics, 2015–16*, Kolkata: Coal Controller's Organisation, Ministry of Coal, Government of India.

International Energy Agency, 2015a. *South East Asia Energy Outlook 2015, World Energy Outlook Special Report*, Paris: International Energy Agency.

International Energy Agency, 2015b. *World Energy Balance (via UK Data Service)*, Paris: International Energy Agency.

Lahiri-Dutt, K., 2016. The diverse worlds of coal in India: Energising the nation, energising livelihoods. *Energy Policy*, 99, pp. 203–213.

Office of the Chief Economist, 2015. *Coal in India*, Canberra: Department of Industry and Science, Austalian Government, Commonwealth of Australia.

PricewaterhouseCoopers, 2016a. *Bridging the Gap: Increasing Coal Production and Sector Augmentation*, New Delhi: PricewaterhouseCoopers.

PricewaterhouseCoopers, 2016b. *Mining in Indonesia; Investment and Taxation Guide*, Jakarta: PricewaterhouseCoopers (Indonesia).

Thomson, E., 2003. *The Chinese Coal Industry: An Economic History*, London: Routledge.

Tu, J., 2011. *Industrial Organisation of the Chinese Coal Industry, Working Paper 103*, Stanford: Program on Energy and Sustainable Development, Stanford University.

Umbach, F. & Wu, K., 2016. *China's Expanding Overseas Coal Power Industry: New Strategic Opportunities, Commercial Risks, Climate Challenges and Geopolitical Implications*, London: King's College London.

Wu, Y., Xiao, X. & Song, Z., 2017. Competitiveness analysis of coal industry in China: A diamond model study. *Resources Policy*, 52, pp. 39–53.

# 11

# REVIEW OF ELECTRICITY SUPPLY IN ASIA

*Subhes C. Bhattacharyya*

## Introduction

This chapter provides a brief review of the electricity sector in Asia and presents the current status of electricity demand in the region, supply mix and the future outlook. As electricity is covered in other chapters (particularly in Chapters 15 and 16 on electricity sector reform but also in other chapters on renewable energy, energy access, and climate change policies), this chapter does not cover sector structure or low-carbon transition issues. References to other chapters are indicated throughout to direct the readers to the relevant chapters.

Asian electricity demand is growing rapidly:[1] between 1990 and 2013, the demand has grown at 6.1% per year on average, which is double the growth rate of world electricity demand for the same period. The demand growth has accelerated in the new millennium when the demand has grown at above 7% per year. Consequently, the Asian share in the global electricity demand has almost doubled between 1990 and 2013: in 1990, the share was about 21% but in 2013, this has exceeded 41% (see Figure 11.1). In volume terms, electricity supply in Asia has increased from 2331.5 TWh in 1990 to 9135.7 TWh in 2013, recording a fourfold increase in supply (IEA, 2015b). In fact, electricity supply has grown faster than population growth and the region has recorded about a threefold increase in per capita electricity supply from 773 kWh per person in 1990 to 2169 kWh per person in 2012 (see Table 11.1).

As with other types of energy discussed in previous chapters, the domination of East Asia (particularly China) in Asian electricity supply is clearly visible (see Figure 11.2). East Asia accounted for about 75% of electricity supply in the region in 2013 followed by South Asia with a share of 14%, and South East Asia with a share of 8.5%. The balance comes from West Asia, with rest of Asia contributing a very small share. China accounted for 73% of the electricity in East Asia, whereas India accounted for 87% of the electricity supply in South Asia in 2013. Japan and Korea are two other major electricity users in East Asia – accounting for 14.5% and 7.5% of electricity supply in East Asia in 2013, respectively.

The regional disparity in electricity use is further evident in terms of the rate of electrification, per capita electricity use and intensity of electricity use. As discussed in Chapter 4, in 2013 developing Asia has an electrification rate of 86% and 526 million people lacked access to electricity (IEA, 2015a). Ninety-six per cent of the urban population is electrified but only 78% of the

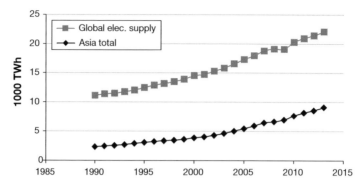

*Figure 11.1* Evolution of Asian electricity supply

*Data source*: IEA (2015b).

*Table 11.1* Salient electricity information for Asia

| Description | 1990 | 2012 |
|---|---|---|
| *Electricity supply (TWh)* | | |
| East Asia | 1615.9 | 6845.8 |
| West Asia | 242.7 | 218.3 |
| South Asia | 319.3 | 1278.6 |
| South East Asia | 147.2 | 770.9 |
| Other Asia | 6.4 | 22.1 |
| Total Asia | 2331.6 | 9135.7 |
| *Per capita electricity supply (kWh/person)* | | |
| East Asia | 1190.0 | 4305.5 |
| West Asia | 3666.3 | 2709.0 |
| South Asia | 293.1 | 777.7 |
| South East Asia | 339.1 | 1277.4 |
| Other Asia | 95.4 | 540.0 |
| Total Asia | 773.3 | 2168.6 |
| *Electricity intensity (kWh/$ constant 2005)* | | |
| East Asia | 310.9 | 587.5 |
| West Asia | 2197.7 | 1129.6 |
| South Asia | 337.6 | 771.8 |
| South East Asia | 201.7 | 563.3 |
| Total Asia (excluding other Asia) | 333.0 | 612.8 |

*Data source*: IEA (2015b) and International Monetary Fund (2015).

rural population has access to electricity. South Asia accounts for the largest share of non-electrified population in the region. As a result, South Asia has lowest electricity use per person compared to other areas in Asia, except rest of Asia. On the other hand, the faster growth in electricity supply in East Asia has resulted in a rapid growth in per capita electricity use and with above 4300 kWh per person electricity use, East Asia has emerged as the highest user of electricity per person in Asia. Apart from West Asia where the consumption level per person has fallen by

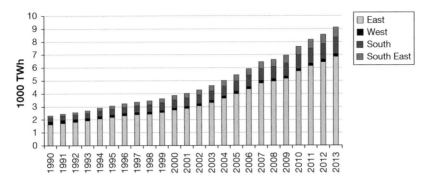

*Figure 11.2*   Regional distribution of electricity supply in Asia

*Data source*: IEA (2015b).

more than 25% between 1990 and 2012, all regions of Asia have shown significant growth in per capita electricity use (see Table 11.1). Yet, electricity use per person in all regions of Asia, except East Asia, remained below the world average of 2972 kWh in 2012.

Electricity intensity (i.e. electricity use per unit of gross domestic product) has also increased over the same period, except in West Asia. South East Asia has recorded the fastest growth in electricity intensity between 1990 and 2012, at an average rate of 4.8% per year. South Asia comes next with an average annual intensity growth of 3.8%. West Asia has recorded a 50% reduction in electricity intensity over the same period, which was related to the economic shock arising from the collapse of the former Soviet Union and the consequent economic and social adjustments.

The rapid growth in electricity use and its increasing intensity in Asia is a cause for concern from an environmental point of view. High dependence on coal for electricity generation, as indicated in Chapter 2, is the main source of concern for local as well as global environment. In addition, the electricity system performance remains an issue in many countries, which results in a higher losses and poor services.

## Electricity profiles of Asian countries

Asian countries covered in this chapter had an installed capacity of 2442 GW in 2014 (UN, 2016), of which 1900 GW was in East Asia, 356 GW in South Asia, 118 GW in South East Asia and 66 GW in West Asia (see Table 11.2 for details). At the country level, China is by far the leader in electricity capacity with 1462 GW in installed capacity in 2014. India, with an installed capacity of 316 GW, and Japan, with an installed capacity of 315 GW, are distant second and third ranking countries, respectively, in the region in terms of capacity. The capacity indicated above includes main utility capacity and auto-producer capacity.

Table 11.2 also indicates that about 70% of the installed capacity uses combustible fuels and are thermal power plants. The remaining 30% is composed of hydropower, nuclear power and other renewable energies. However, there is some variation across the regions. For example, 90% of electricity capacity in Thailand uses combustible fuels, whereas Nepal and Tajikistan have more hydro-dominated power systems (only 6% coming from thermal power stations). Both China and India depend on coal to a great extent, whereas Japan and Korea rely on natural gas (imported as liquefied natural gas) and nuclear power for their electricity generation.

Due to differences in the technology, the age of the fleet and the operational characteristics of the plants, the conversion efficiency of combustible fuel-burning power plants varies across

Table 11.2 Electricity profiles of selected Asian countries

| Country | Installed capacity (MW) in 2014 | Generation in 2014 (GWh) | Installed combustible fuel capacity (MW) in 2014 | Thermal plant efficiency in 2014 (%) | Distribution losses in 2013 (%) |
|---|---|---|---|---|---|
| China | 1,461,939 | 5,649,583 | 1,005,720 | 39 | 6.3 |
| Hong Kong | 12,625 | 39,803 | 12,625 | 34 | 11.5 |
| Korea S | 99,830 | 55,0933 | 69,122 | 40 | 3.5 |
| DR Korea | 9500 | 17,909 | 4500 | 26 | 17.5 |
| Mongolia | 1370 | 6725 | 1370 | 20 | 13.5 |
| Japan | 315,318 | 1,040,676 | 194,857 | 43 | 4.8 |
| East Asia | 15,090,582 | 7,305,629 | 1,288,194 | | |
| Armenia | 4091 | 7750 | 2390 | 41 | 15.3 |
| Azerbaijan | 7355 | 24,728 | 6270 | 39 | 14.9 |
| Georgia | 4350 | 10,370 | 1688 | 33 | 8.1 |
| Kazakhstan | 25,011 | 105,068 | 22,500 | 31 | 13.2 |
| Kyrgyzstan | 3864 | 14,572 | 800 | 21 | 21.1 |
| Tajikistan | 5190 | 16,472 | 318 | 47 | 15.7 |
| Turkmenistan | 4001 | 20,400 | 4000 | 20 | 16.3 |
| Uzbekistan | 12,722 | 55,400 | 10,867 | 25 | 9.3 |
| Total West Asia | 66,584 | 254,760 | 48,833 | | |
| Bangladesh | 11,557 | 55,845 | 11,302 | 33 | 14.0 |
| India | 316,379 | 1,308,873 | 241,396 | 39 | 19.7 |
| Nepal | 829 | 3508 | 53 | 39 | 24.5 |
| Sri Lanka | 4058 | 12,832 | 2243 | 34 | 10.2 |
| Pakistan | 23,686 | 105,305 | 15,888 | 36 | 17.6 |
| Total | 356,509 | 1,486,363 | 270,882 | | |
| Cambodia | 1513 | 3059 | 582 | 32 | 13.1 |
| Indonesia | 60,588 | 22,0794 | 53,692 | 26 | 9.8 |
| Malaysia | 29,974 | 147,461 | 25,040 | 38 | 4.2 |
| Myanmar | 4805 | 14,156 | 1620 | 24 | 11.2 |
| Philippines | 17,970 | 77,295 | 12,056 | 38 | 11.2 |
| Singapore | 13,047 | 49,310 | 13,047 | 42 | 0.5 |
| Thailand | 53,472 | 180,862 | 48,238 | 33 | 6.1 |
| Vietnam | 34,080 | 145,730 | 18,368 | 39 | 9.1 |
| Total SE | 215,449 | 838,667 | 172,643 | | |
| Asia total | 2,539,124 | 9,885,419 | 1,780,552 | | |

Data source: UN (2016) and IEA (2015b).

countries in the region. Plants in Japan, Singapore and Korea have achieved an efficiency of 40–43% in 2014. China, India, Malaysia and Vietnam have achieved an efficiency of about 38–39%. On the other hand, the conversion efficiency is low in Indonesia (26%) and some other countries. However, the average efficiencies often mask the plant level picture. For example, China has been aggressively expanding its power plant capacity over the past decade and as a result the country has a young generation capacity that is more efficient. Eighty-nine per cent of its coal fleet is less than 20 years old and 69% is younger than 10 years. In addition, China has installed supercritical and ultra-supercritical thermal power plants that achieve efficiencies as high as 46%. Simultaneously, China has shut-down its small, inefficient power plants, which has helped improve the overall efficiency of power generation in the country (Cornot-Gandolphe, 2014).

India, on the other hand, still relies on sub-critical thermal power plants to a large extent. Only 15% of capacity relies on supercritical technology and there is no ultra-supercritical power plant in India (Kanchan & Kumarankandath, 2015). Similarly, India's coal-fired plant fleet is relatively old: only 18% of the fleet is less than 10 years old whereas 28% of the plants have completed their design life of 25 years (Kanchan & Kumarankandath, 2015). The plant size is comparatively small as well: the standard unit sizes were 210 MW and 500 MW and only recently bigger units of 660 or 800 MW are being introduced. As the performance of plants deteriorates with age, India needs to phase out older plants and introduce more efficient state-of-the art technologies following the Chinese example. There are plans to introduce ultra-supercritical plants (Henderson, 2015) but it may be a slow process.

Another important feature of the power sector in developing Asia is its high transmission and distribution (T&D) losses. Whereas countries like Korea and Japan record a T&D loss of 3–5% of electricity supply, the developing countries have much higher losses often above 10% and in some case beyond 20% (see Table 11.2). India is a case in point. T&D losses have remained a major issue in the sector over the past three decades. This has partly to do with the practice of unmetered supply to agricultural consumers and poor residential consumers who were given a single point connection. Utility companies tried to minimise their investment cost for these low income-generating consumers (due to highly subsidised tariffs), but this practice created problems for energy accounting. Only in the 1990s, when the sector reform initiatives were taken (see Chapter 15 for reform details), did the issue of losses received regulatory attention. The country now makes a distinction between technical (T&D) and commercial losses and reports an aggregate technical and commercial (AT&C) loss figure. The commercial loss covers theft of power and loss of supply arising from inefficient billing, metering and revenue collection practices. While the technical component of the loss is close to 20% (see Table 11.2), the combined technical and commercial loss at the national level is estimated at 24.6% for 2014–2015 (PFC, 2016). Eastern and North-Eastern states reported a loss of 35–40%, whereas southern and western states reported a loss between 18% and 21% (PFC, 2016).

Clearly, any technical loss in the network increases the amount of generation required to meet a given demand. This also increases the combustible fuel requirement due to conversion losses and compounds the environmental effects. Improving the distribution system and reducing the losses can therefore benefit the electricity system and the global climate.

## Regional power markets

Given the specific features of each sub-region, this section briefly discusses the electricity market at the regional level. Four regions, namely East, West, South and South East are considered in turn.

### *East Asian power market*

East Asia represents the largest electricity market in Asia with an installed capacity of 1900 GW and generation of 7305 TWh in 2014 (see Table 11.2). China is the largest market accounting for 77% of the capacity share, but Japan and Korea are two other major players in the region with 1041 TWh and 551 TWh of electricity generation, respectively. The trend of electricity production clearly indicates an exponential growth pattern for China, taking off in the new millennium (see Figure 11.3). Korea also shows a relatively high growth of 7.5% per year between

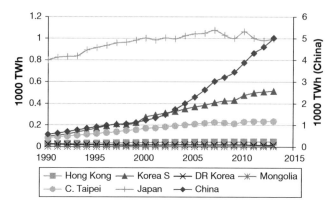

*Figure 11.3* Electricity generation trend in East Asia

*Data source*: IEA (2015b).

1990 and 2013. On the other hand, Japan shows practically no change in its supply during this period. The electricity industry in China is covered in Chapter 16 and is not discussed here.

Unlike China which relies on indigenous coal for its electricity production, resource-constrained Korea and Japan are more reliant on imported fuels (mainly natural gas in the form of liquefied natural gas but some coal and oil) as well as nuclear power. Japan has a well-diversified fuel mix with no single fuel dominating its electricity sector. However, the country has undergone a significant evolution in its fuel mix since the first oil shock in 1973 when it was highly dependent on oil for its electricity generation. By 1990, the share of oil in the power generation technology was drastically reduced to 32.5% and the country moved towards nuclear power for energy security purposes (see Table 11.3). The process of oil substitution continued over the past two decades when coal and natural gas have gained importance alongside nuclear power. However, since then the nuclear accident in 2011 has forced the government to reduce its nuclear generation. Renewable energies are being promoted vigorously to support a low-carbon transition. Meanwhile, natural gas and coal are continuing as major fuels for electricity supply (IEA, 2016a), making the country highly dependent on fossil fuels for its electricity generation.

Similarly, Korea has also recorded an intensification of coal use in its power sector at the cost of oil and nuclear power. In 2015, 43% of its electricity generation came from coal, 30% nuclear and 22% from natural gas, with oil accounting for only 1% (IEA, 2016b). But, in 2000, oil accounted for 12% of electricity generation, while natural gas, coal and nuclear accounted for 10%, 39% and 38% of supply, respectively (IEA, 2012). The country plans to continue with coal, natural gas and nuclear as the main source of electricity, although the government is aiming at increasing the renewable energy share through a renewable energy portfolio standard that replaced the feed-in tariff policy in 2012 (Hong, et al., 2014).

The power systems in Japan and Korea are national systems that are isolated from the rest of the world and accordingly there is no power trading taking place at the moment. Power trading in East Asia takes place mainly with China but trading still plays a minor role in the region.

Industry is the major user of electricity in China: 76% of electricity was used in industry in 1990 but this has reduced to 68% in 2012. Industry is major user of electricity in Korea as well (with a 52% share in 2012), but the commercial sector also contributes significantly (32% share in 2012). On the other hand, Japan has a well-distributed demand across industry, residential use and

*Table 11.3* Evolution of fuel mix for power generation in Japan

| Fuel | 1973 | 1990 | 2000 | 2014 |
|---|---|---|---|---|
| Coal (%) | 8.0 | 13.5 | 21.5 | 33.7 |
| Oil (%) | 73.2 | 32.5 | 16.5 | 11.2 |
| Natural gas (%) | 2.3 | 19.6 | 23.3 | 40.6 |
| Waste and biomass (%) | | 1.1 | 0.9 | 3.4 |
| Nuclear (%) | 2.1 | 23.2 | 29.6 | 0 |
| Hydro (%) | 14.3 | 10.0 | 7.8 | 7.9 |
| Other renewable energies (%) | 0.1 | 0.1 | 0.4 | 3.2 |
| Total (%) | 100 | 100 | 100 | 100 |

*Data source*: IEA (2016a).

commercial use. The commercial sector has emerged as the most important user with a 36% share in 2012, but residential and industrial use closely follows at 31 and 30%, respectively.

Countries in the region face different issues related to their power sector. Japan is trying to adjust to the post-nuclear accident situation while trying to contain the impact on climate change. In July 2012, Japan has introduced its feed-in tariff for renewable electricity and offered an attractive tariff. This has resulted in a rapid growth in renewable energy capacity, particularly solar photovoltaic capacity which has risen from 7 GW in 2012 to 32 GW in 2015).[2] However, the high cost of renewable energy will have an impact on the electricity tariff paid by the consumers. Even with rapid renewable capacity addition, the share of non-fossil fuel electricity remains low and the country will have to rely on fossil fuels in the foreseeable future to meet its electricity needs as it pursues its nuclear phase-out policy. The challenge of containing greenhouse gas emissions will remain the most important issue for the country.

Rapid growth in electricity demand in Korea, on the other hand, is attributed to relatively low electricity tariff compared to other OECD countries (IEA, 2012). The average tariff is below the production cost and this encourages higher consumption. Moreover, the concentration of its generation facilities far away from its demand centres creates the transmission and power evacuation challenge. In addition, the reliance on fossil fuels remains high and the transition to a low-carbon electricity system remains a major challenge.

## West Asian power market

The West Asian power system has an installed capacity of 66 GW in 2014 and it produced about 255 TWh of electricity. This is the smallest system in Asia and it did not record practically any supply growth between 1990 and 2013 (see Figure 11.4). The sector is dominated by two main players, Kazakhstan and Uzbekistan. Kazakhstan has seen a rapid fall in its power supply between 1990 and 2000 and a turn-around from 2004, but by 2013 the supply had not reached the level seen in 1990. Uzbekistan and most of the other countries of the region have maintained their 1990 level of supply in 2013. This represents a strong departure from the rest of Asia.

West Asia has vast energy resources: Kazakhstan is rich in oil, coal and nuclear fuels; and Kyrgyzstan and Tajikistan are rich in water resources and provide hydroelectricity for the region. There is an interconnected power grid via 500 kV transmission network that connects five states, namely Kyrgyzstan, Tajikistan, Uzbekistan, Turkmenistan and south of Kazakhstan. This was developed in the Soviet era but post-independence the coordinated management of the network has emerged as an issue due to political, economic and priority differences among the states.

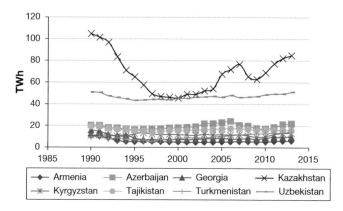

*Figure 11.4* Evolution of electricity supply in West Asia

*Data source*: IEA (2015b).

*Table 11.4* Aging power generation fleet in West Asia

| Age range | Kazakhstan (%) | Kyrgyzstan (%) | Tajikistan (%) | Uzbekistan (%) |
|-----------|----------------|----------------|----------------|----------------|
| <10 years | 11 | 4 | 14 | 7 |
| 11–20 years | 11 | 9 | 0 | 5 |
| 21–30 years | 33 | 23 | 12 | 13 |
| >30 years | 44 | 64 | 74 | 75 |

*Data source*: Boute (2015).

Consequently, the cross-border trade within the region has suffered and the volume of transaction has fallen to 10% of the level achieved in the 1990s. But countries are looking outwards either to connect with countries outside the region (e.g. Iran and Afghanistan) or with Russia. As a result of poor coordination, the hydro-based systems are suffering from seasonal imbalances (i.e. shortage in winter and surplus in summer), resulting in inefficient use of valuable water resources. The outcome is costly for thermal-based systems as well, as some are not able to meet their consumer demand in the absence of imports, which in turn is increasing the level of unserved energy in these countries.

Aging infrastructure of the region is another issue. Most of the generating plants have completed their planned life (see Table 11.4), but due to lack of funds the utilities are unable to replace and retire these capacities. As a result, the generating plants operate in a 'state of crisis' (Boute, 2015). The region also has a high electricity intensity compared to the rest of Asia (see Table 11.1). The hydro-dominant countries of the region, Kyrgyzstan and Tajikistan, use electricity for space heating in winter. But energy intensive industrial activities such as aluminium factories in Tajikistan and steel making in Kazakhstan also contribute to high electricity intensity (Boute, 2015). Accordingly, industrial energy use accounted for 72% of electricity final use in 2012 in Kazakhstan, whereas in Uzbekistan, industry had a 38% share, residential and commercial sectors had 25% share and agricultural activities accounted for 32% in 2012.

Aging infrastructure also contributes to high T&D losses. The region exhibits a wide range of T&D losses – 8% in Georgia and 21% in Kyrgyzstan, with most countries around 15% losses (see Table 11.2). Investment in electricity infrastructure for system modernisation and integrated operation of the power system for mutual benefits are two main issues of the sector.

## South Asian power market

South Asia is the second largest power system in Asia after East Asia. It has an installed capacity of 356 GW and an annual production of 1486 TWh in 2014 (see Table 11.2). India, accounting for 88% of the output, dominates the South Asian system. Combustible fuel (mainly coal) based power generation accounts for 76% of the capacity, and hydropower contributing a major part of the balance capacity. While the thermal power generation efficiency is reasonably good, the region suffers from high T&D losses. Most of the countries in the region have high losses. Industry accounted for 44% of final electricity consumption in India in 2012. Residential and commercial activities and agriculture accounted for another 31% and 18%, respectively, in 2012. Chapter 15 provides further discussions on South Asian power market.

## South East power market

The power market in South East Asia has an installed capacity of 215 GW and a supply of about 10,000 TWh (see Table 11.2 for details). Indonesia and Thailand are two big players in the region and account for 52% of the regional installed capacity. Malaysia and Vietnam are the other important players, but Vietnam is growing fast and has ambitious plans. Electricity supply in this area is growing at a fast rate – averaging above 7% between 1990 and 2013. Vietnam is the fastest growing electricity market in the region, but Indonesia, Malaysia and Myanmar have also recorded impressive growth rates of above 8% per year (see Figure 11.5).

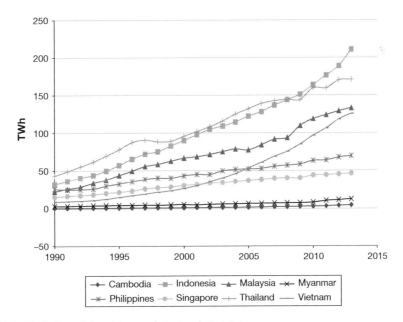

*Figure 11.5*  Evolution of electricity supply in South East Asia

*Data source*: IEA (2015b).

The electricity sector is still dominated by state utilities in Indonesia, Thailand, Malaysia and Vietnam, but they allow various degrees of private sector participation, mainly following the single buyer market model. The Philippines, on the other hand, has fully reformed its sector and has introduced retail competition. Cambodia, Malaysia, Thailand and Philippines have independent electricity regulators while the ministries in other countries are responsible for sector governance (IEA, 2015c).

There is some amount of electricity trade within the region. Lao PDR is the main exporter while Thailand is the main importer (with an import of 12.5 TWh in 2013). The trade takes place through bilateral exchanges at the national level and the ASEAN Power Grid is merely a collection of interconnected national grids but the region plans to have a functional regional grid. The Greater Mekong sub-region provides a successful example of power trading based mainly on hydropower generated in the region (IEA, 2015c). Large hydro resources in Lao PDR are being developed for export to neighbours as the local demand would not justify the development of such projects.

There is significant variation in terms of final energy consumption with the region. The residential sector plays an important role in Indonesia, whereas industry is a major player in Malaysia. In Indonesia, the residential sector accounted for 41% of electricity consumption in 2012. The commercial sector accounted for another 24% – thus, these two sectors were responsible for almost two-thirds of electricity consumption. Industry catered to the balance of the demand. On the other hand, industry accounted for 47% of electricity demand in Malaysia whereas residential and commercial sectors contributed 20% and 32% to the demand respectively. Chapters 22 and 23 provide further details on the power sector of South East Asian countries.

## Asian electricity outlook

In line with expected growth in economic activities of Asian countries, the electricity demand is likely to grow particularly in the developing Asia. The outlook normally considers the utility capacity and generation (as opposed to total capacity including the auto-production as indicated in Table 11.2). According to the International Energy Outlook 2016, the installed capacity in Asian countries will increase to 3281 GW in 2030 and 3786 GW in 2040 in the reference case (see Table 11.5). China and India will account for 73% of the capacity in 2030 and 74% in 2040. Capacity growth will be highest in India (3.1% per year on average), whereas China is likely to have a slightly slower growth rate in capacity (2.3% per year). The report suggests an annual growth rate of 2.1% in South Korea and other Asian regions. Japan is likely to have the lowest

*Table 11.5* Electricity outlook for Asia

| Region/country | Installed capacity (GW) | | Generation (TWh) | |
| --- | --- | --- | --- | --- |
| | *2030* | *2040* | *2030* | *2040* |
| China | 1937 | 2194 | 8145 | 9426 |
| India | 459 | 607 | 1966 | 2769 |
| Japan | 299 | 295 | 1124 | 1149 |
| Korea | 142 | 159 | 735 | 856 |
| Other Asia | 444 | 531 | 2086 | 2714 |
| Total | 3281 | 3786 | 14,056 | 16,914 |

*Data source*: EIA (2016).

*Table 11.6* Capacity outlook by fuel type in 2030 and 2040 (GW)

| Year | Fuel type | China | India | Japan | Korea | Other Asia |
|------|-----------|-------|-------|-------|-------|------------|
| 2030 | Coal | 886 | 197 | 46 | 38 | 112 |
|      | Oil | 7 | 7 | 44 | 4 | 36 |
|      | Natural gas | 124 | 31 | 97 | 39 | 160 |
|      | Nuclear | 95 | 34 | 30 | 36 | 13 |
|      | Others | 825 | 191 | 82 | 26 | 122 |
|      | Total | 1937 | 460 | 299 | 143 | 443 |
| 2040 | Coal | 845 | 255 | 43 | 41 | 136 |
|      | Oil | 6 | 6 | 40 | 4 | 33 |
|      | Natural gas | 186 | 45 | 104 | 49 | 195 |
|      | Nuclear | 152 | 41 | 25 | 36 | 14 |
|      | Others | 1004 | 261 | 84 | 30 | 154 |
|      | Total | 2193 | 608 | 296 | 160 | 532 |

*Data source*: EIA (2016).

growth in capacity addition over this period. Asia will account for 44 to 45% of global installed capacity in 2030 and 2040.

In terms of fuels, coal will still dominate in the reference case but the share is expected to fall to 39% by 2030 and 35% by 2040. On the other hand, hydropower and renewable energy capacity increases to 38% by 2030 and 40% by 2040. In fact, by 2040 the renewable energies are likely to become the dominant capacity for electricity generation in the region. Natural gas will account for 14–15% of the capacity, whereas nuclear power will contribute another 6–7% during this period (see Table 11.6). The rapid growth of renewable energy capacity is the main highlight of electricity outlook.

In terms of generation, China and India will account for 71–72% of the regional output by 2030 and 2040. Output is expected to grow at 3.5% per year in India, whereas output in China will grow at 2.5% per year during the above period. Japan will see a marginal growth in output during this period (see Table 11.5).

Clearly the region will experience a steady growth in capacity and generation. However, in the reference case the dependence on fossil fuel will remain reasonably high, which will have significant environmental implications. Accordingly, the efforts towards low-carbon transition and energy efficiency improvements will require further emphasis.

## Conclusions

The power sector in Asia has been growing rapidly over the past decades to meet the electricity needs of the growing economy. This trend is likely to continue in the future as countries in the region continue to develop. However, the sector is likely to grow at a moderate rate compared to the rapid growth in the past decade that was fuelled by China's power sector growth. As Chinese economy adjusts to its new normal growth pattern, India is likely to displace China as the major growth centre for electricity capacity.

The sector will gradually shift towards renewable energies, but the dependence on coal and fossil fuels is likely to continue over the next two decades or so. However, the environmental issues related to fossil fuel use in the sector will continue to exert pressure on the policymakers and the governments of the region as the growing middle-income class demand better local environmental policies. The regional support for climate change mitigation will also focus on more efficient production and distribution of electricity, efficiency improvements in end use and a transition to low-carbon fuels for power generation. The sector will have to navigate through these challenges in the future and this requires a strong power sector. A single solution may not fit all countries of the region and the search for an appropriate solution will have to continue.

## Notes

1  This chapter follows the same geographical coverage of Asia as indicated in Chapter 2.
2  www.renewable-ei.org/en/statistics/annual.php

## References

Boute, A., 2015. *Towards Secure and Sustainable Energy Supply in Central Asia: Electricity Market Reform and Investment Protection*, Brussels: Energy Charter Secretariat.

Cornot-Gandolphe, S., 2014. *China's Coal Market: Can Beijing tame 'King Coal'?*, Oxford: Oxford Institute of Energy Studies.

EIA, 2016. International Energy Outlook 2016, Energy Information Agency, Department of Energy, www.eia.gov/outlooks/ieo/pdf/0484(2016).pdf (accessed on 18 August 2017).

Henderson, C., 2015. *Coal-fired power plant efficiency improvement in India*, London: IEA Clean Coal Centre.

Hong, J., Kim, C. & Shin, H., 2014. *Power sector in developing Asia: Current statys and policy issues, Working Paper 405*. Manila: Asian Development Bank.

IEA, 2012. *Energy Policies of IEA Countries: The Republic of Korea*, Paris: International Energy Agency.

IEA, 2015a. *World Energy Outlook 2015*, Paris: International Energy Agency.

IEA, 2015b. *World Energy Balance (via UK Data Service)*, Paris: International Energy Agency.

IEA, 2015c. *Development Prospects of the ASEAN Power Sector: Towards an Integrated Electricity Market*, Paris: International Energy Agency.

IEA, 2016a. *Energy Policies in IEA Countries: Japan*, Paris: International Energy Agency.

IEA, 2016b. *Korea Energy System Overview*. [Online] Available at: www.iea.org/media/countries/Korea.pdf [Accessed 12 April 2017].

International Monetary Fund, 2015. *World Economic and Financial Surveys: Regional Economic Outlook: Sub-Saharan Africa Navigating Highwinds*, Washington, DC: International Monetary Fund.

Kanchan, S. & Kumarankandath, A., 2015. *The Indian Power Sector: the need of Sustainable Energy Access*, New Delhi: Centre for Science and Environment.

PFC, 2016. *The Performance of State Power Utilities for the years 2012–13 to 2014–15*, New Delhi: Power Finance Corporation.

UN, 2016. *2014 Electricity Profiles*, New York: United Nations.

# 12

# ON-GRID SOLAR ENERGY IN ASIA

## Status, policies, and future prospects

*Tania Urmee and S. Kumar*

### Introduction

The developing countries in Asia are currently poised for rapid economic growth and industrialisation. The energy demand of these countries has increased significantly in the past few decades, and is projected to almost double by 2030 (ADB, 2016). Most of the Asian countries have targeted at least 80% electrification access by 2030, and renewable energy sources and technologies are expected to play an important role in meeting this target, notably through solar, wind, biomass and micro hydro using grid, mini-grids and individual solar home systems (ADB, 2015). Large parts of Asia are endowed with high solar insolation levels, and because of this the electricity generation potential for both large-scale grid and off-grid applications are very high. With solar technology solutions diversifying rapidly in terms of application, efficiency and cost, solar energy could play a crucial role in the energy mix in the medium to long term.

Solar energy can refer to any phenomenon that traces its origin to energy from the sun and can be harnessed as useable energy, directly or indirectly. The sun is an average star of radius 0.7 million km and has a mass of about $2 \times 10^{30}$ kg. It radiates energy from an effective surface temperature of about 5760 K. From the fusion furnace of the sun, energy is transmitted radially, (i.e. outward) as electromagnetic radiation called 'solar energy'. The quantity of energy radiated by the sun can be estimated from knowledge of the sun's radius and its surface temperature (assuming it to be black body), which amounts to about $3.8 \times 10^{23}$ kW (Duffie and Beckman, 2013).

Harnessing the sun's energy on the surface of the earth includes a diverse set of technologies that range from simple sun drying of crops to direct generation of electricity using photovoltaic (PV) cells. Solar energy technologies can be divided into two broad categories based on their intended use/applications: solar thermal applications that convert solar radiation to thermal energy, which can then be directly used for heating or cooling (e.g. solar hot water systems, solar drying systems or solar absorption cooling) or conversion of thermal energy further into electricity (e.g. concentrating solar power (CSP)); and solar electricity applications using the photovoltaic effect that directly generates electricity from sunlight. Solar energy technologies have the advantages of being a renewable resource, local availability, the technologies can be modular and no or little impact to climate change, among others. Thus, solar technologies used for the generation of electricity ranges from W (Watt) to MW (Mega Watt) ranges, i.e. from street lights to off-grid and on-grid systems. The focus of this chapter is on-grid solar technologies that convert solar energy to electricity, and accordingly, this chapter presents an overview of the solar energy powered electricity generation in Asia and discusses its status, policies and market potential. This is discussed in detail country-wise to show the status of on-grid solar applications.

## Resource potential and technological options

Before the application of any solar technology, the solar resource potential needs to be assessed. This resource potential depends largely on the level of solar irradiation, the estimated land area suitable for solar technology installation and the efficiency of the solar energy systems. The solar energy potential can be assessed in terms of theoretical, technical and economic terms. The theoretical potential is based on land area available and current scientific knowledge, and considers only the geographic and climatic factors, while technical potential also takes into account the conversion technologies and its efficiency of conversion. The economic potential considers the cost of the competitive technologies. The solar radiation on a horizontal surface is composed of direct and diffuse components, and is usually available as monthly average values of hourly and/or daily radiation. In most of Asia, the average solar radiation is promising, as can be observed from the monthly average daily total solar insolation and yearly average for selected countries presented in Table 12.1. The average land use factor for a centralised PV system in South Asia is 1.92, East Asia is 2.14 and South East Asia is 0.51, and technical losses in conversion process are considered as 10% (ECOFYS, 2008).

Solar energy technologies can be categorised as: (1) passive and active; (2) thermal and photovoltaic; and (3) concentrating and non-concentrating. A number of text books discuss in detail the various solar energy technologies (Duffie and Beckman, 2013; Goswami et al., 2000; Kalogirou, 2009). Figure 12.1 shows one classification route of solar energy technologies.

The active solar energy technologies harness the energy from the sun that is either stored or converted to another application, which could be classified as photovoltaic or solar thermal. On the other hand, in passive form, energy collected from the sun is not converted or used in another form. It is essentially an approach to building design and features. Passive use of solar energy has been practiced for thousands of years and includes such considerations as site selection, placement of windows, dark walls and so forth, to maximise the collection of heat and light (Bradford, 2006; Chiras, 2002). Most of the solar energy for on-grid application is in active form, and is based on solar PVs or solar thermal-based electricity technologies.

### *Solar PV*

Though many technologies are currently under development, almost 85–90% of PV modules of the global annual market are made from wafer-based crystalline silicon (c-Si). The process of manufacturing c-Si modules involves growing ingots of silicon, slicing the ingots into wafers to make the solar cells, electrically interconnecting the cells and encapsulating the strings of cells to form a module. Modules currently use silicon in one of two main forms: single crystalline silicon (sc-Si) or multi crystalline silicon (mc-Si). Conversion efficiency of current commercial sc-Si modules is about 14–20% and is expected to increase to 23% by 2020 and to 25% in the longer term (IEA, 2014). Crystalline silicon PV modules are expected to remain a dominant PV technology until at least 2020, with a forecasted market share of about 50% by that time (IEA, 2008) due to their proven and reliable technology, and long lifetime.

Table 12.1 Average solar insolation for selected Asian cities and countries

| Country | City | Latitude | Longitude | Average insolation (10-year average) $(kWh/m^2)$ | | | | | | | | | | | | Yearly average |
|---|---|---|---|---|---|---|---|---|---|---|---|---|---|---|---|---|
| | | | | Jan | Feb | Mar | Apr | May | Jun | Jul | Aug | Sep | Oct | Nov | Dec | |
| Bangladesh | Dhaka | 23'42''N | 90'22''E | 4.44 | 5.08 | 5.87 | 6.06 | 5.5 | 4.41 | 4.09 | 4.37 | 4.17 | 4.5 | 4.37 | 4.13 | 4.75 |
| Cambodia | Phnom Penh | 11'33''N | 104'51''E | 5.27 | 5.78 | 6.02 | 5.76 | 5.09 | 4.3 | 4.55 | 4.07 | 4.34 | 4.41 | 4.88 | 5.03 | 4.85 |
| China | Beijing | 39'55''N | 116'25''E | 2.37 | 2.92 | 3.58 | 5.61 | 4.83 | 5.68 | 5.42 | 4.49 | 4.25 | 3.2 | 2.66 | 2.04 | 3.92 |
| | Nanjing | 32'03''N | 118'53''E | 2.04 | 2.22 | 2.65 | 4.5 | 3.84 | 4.47 | 4.93 | 4.5 | 3.67 | 3.02 | 2.88 | 2.09 | 3.40 |
| | Shanghai | 31'10''N | 121'28''E | 2.29 | 2.63 | 3.07 | 4.54 | 4.38 | 4.59 | 5.52 | 5.23 | 4.03 | 3.39 | 2.97 | 2.38 | 4.01 |
| | Hongkong | 22'18''N | 114'10''E | 2.59 | 2.56 | 3.06 | 3.93 | 4.13 | 4.74 | 5.81 | 4.95 | 7.68 | 4.05 | 3.56 | 2.93 | 4.18 |
| India | New Delhi | 28'N | 77'E | 3.68 | 4.47 | 5.5 | 6.6 | 7.08 | 6.55 | 5.01 | 4.62 | 5.11 | 4.99 | 4.15 | 3.42 | 5.10 |
| | Bombay | 18'33''N | 18'33''E | 5.22 | 6.03 | 6.66 | 7.05 | 6.77 | 4.59 | 3.54 | 3.4 | 4.72 | 5.39 | 5.15 | 4.8 | 5.28 |
| | Bangalore | 12'57''N | 77'37''E | 5 | 5.9 | 6.44 | 6.42 | 6.13 | 4.76 | 4.48 | 4.59 | 4.98 | 4.68 | 4.34 | 4.4 | 5.18 |
| Indonesia | Jakarta | 6'11''N | 106'50''E | 4.15 | 5.49 | 5 | 4.94 | 4.88 | 4.71 | 5.09 | 5.46 | 5.66 | 5.36 | 4.76 | 4.47 | 5.03 |
| Japan | Tokyo | 35'45''N | 139'38''E | 2.31 | 2.99 | 3.7 | 4.9 | 5.07 | 4.47 | 4.88 | 5.42 | 3.82 | 2.98 | 2.5 | 2.23 | 4.00 |
| Korea | Seoul | 37'31''N | 127'E | 2.62 | 3.4 | 4.29 | 5.24 | 5.83 | 5.15 | 4.26 | 4.55 | 3.99 | 3.64 | 2.6 | 2.24 | 4.16 |
| Laos | Vientiane | 18'07''N | 102'35''E | 4.3 | 4.94 | 5.52 | 5.74 | 5.11 | 4.24 | 5.22 | 4.19 | 4.61 | 4.26 | 4.21 | 4.24 | 4.63 |
| Malaysia | Kuala Lumpur | 3'07''N | 101'42''E | 4.54 | 5.27 | 5.14 | 5.05 | 4.8 | 4.98 | 4.91 | 4.78 | 4.54 | 4.51 | 4.23 | 7.07 | 4.70 |
| Mongolia | Ullanbaatar | 47'55''N | 106'54''E | 1.79 | 2.77 | 4.24 | 5.53 | 6.26 | 6.15 | 5.55 | 4.88 | 4.17 | 3 | 1.82 | 1.14 | 4.30 |
| Myanmar | Yangon | 16'47''N | 96'09''E | 5.4 | 6.06 | 6.65 | 6.69 | 5.14 | 3.24 | 3.3 | 2.99 | 4.12 | 4.51 | 4.82 | 5.05 | 4.65 |
| Philippines | Manila | 14'37''N | 120'58''E | 4.82 | 5.82 | 6.42 | 6.75 | 6.19 | 4.96 | 4.94 | 4.41 | 4.86 | 4.63 | 4.59 | 4.5 | 5.07 |
| Singapore | Singapore City | 1'N | 103'E | 4.43 | 5.52 | 5.05 | 5.05 | 4.62 | 4.66 | 4.51 | 4.61 | 4.49 | 4.5 | 3.98 | 3.93 | 4.61 |
| Thailand | Bangkok | 13'47''N | 100'30''E | 4.42 | 4.65 | 4.84 | 5.03 | 4.75 | 3.77 | 4.22 | 3.46 | 3.63 | 3.89 | 4.16 | 4.4 | 4.27 |
| | Chiang Mai | 18'N | 99'E | 4.79 | 5.51 | 6.11 | 6.29 | 5.53 | 4.44 | 4.16 | 4.18 | 4.5 | 4.34 | 4.28 | 4.48 | 4.88 |
| Vietnam | Hanoi | 18'N | 105'54''E | 2.52 | 2.94 | 3.81 | 4.34 | 4.66 | 4.51 | 4.62 | 4.62 | 4.57 | 3.64 | 3.29 | 3.17 | 3.89 |

Data source: NASA (2016).

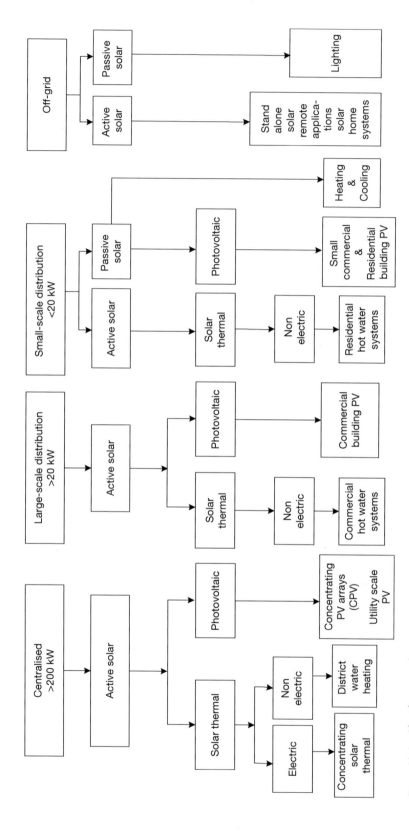

*Figure 12.1* Classifications of solar technologies

*Data source:* Adapted from Bradford (2006).

Thin-film technology is based on semiconductors that are extensively and cost-effectively applied to substrates such as glass, metal or plastic films. Thin films have relatively low consumption of raw materials and their production efficiency is higher. Their flexibility of integrating into building, good performance at high ambient temperature and reduced sensitivity to overheating makes them more attractive (Shukla and Khare, 2014). The major drawbacks are its lower efficiency and the industry's limited experience with lifetime performances. As this requires more area than crystalline silicon technologies in order to reach the same capacity, it faces challenges for large-scale generation. In recent years, thin-film production units have increased from pilot scale to 50 MW lines, with some manufacturing units in the gigawatt (GW) range. As a result, thin films technologies are expected to increase their market share significantly by 2020 (IEA, 2014).

### Solar thermal/concentrated solar power

The flat plate collector is the most common solar thermal technology, but it is only suitable for applications requiring temperatures of the order of up to about 80–85°C. When higher delivery temperatures are required (e.g. typically for electricity generation or for industrial process heat), as the input radiation cannot be increased, the reduction of heat losses could be possible by reducing the surface area of the absorber. In order to do that, an optical device is placed between the radiant source and the absorbing surface. Solar concentrators can be classified into three types: (1) planar and non-concentrating type; (2) line-focusing type that produces a high density of radiation on a line at the focus; (3) and point-focusing type that produces higher density of radiation in the vicinity of a point. Concentrating collectors need to follow the path of the sun during the day and according to the seasons continuously to focus the solar radiation on to the absorber. Likewise, trackers are used to follow the path of the sun in order to maximise the solar radiation incident on the photovoltaic surface. In the one-axis tracking, the array tracks the sun east to west, and so is used mostly with PV systems and with concentrator systems. On the other hand, in the two-axis tracking system, the panel or the concentrator points at the sun at all times.

The four types of CSP technologies are parabolic trough, Fresnel reflector, solar tower (heliostat) and solar dish. The first two systems are called line-focusing systems, while the latter are point-focusing systems. Currently, parabolic troughs are the most mature. They use synthetic oil, steam or molten salt to transfer the solar heat to a steam generator, and molten salt for thermal storage. Typical capacities of parabolic plants are in the range of 14–80 MWe, with efficiencies of the order of about 14–16%. The capacity factor depends on the location and is about 25–30%.

Though solar concentrators for electricity generation had been initiated in the 1980s with parabolic concentrators, a significant growth is observed only after 2008 (from less than 0.5 GW in 2008 to about 5.8 GW in 2015), and both parabolic and tower based technologies are used (REN21, 2016).

## Online solar PV market and current status in Asia

In recent years, renewable energy technologies have increased their contributions to the Asian energy supply portfolio. This contribution is expected to increase in the coming years and decades, due to the increasing number of countries that are creating favourable conditions for renewable energy applications by setting renewable energy targets, and the supporting policies being put in place to meet those targets. The specific details with regard to solar energy for a number of Asian countries are given in Box 12.1, which clearly indicates this favourable trend.

Among the active solar energy technologies, the first solar PV applications were initiated since the late 1950s when they were used on space satellites to generate electricity. Following the

## Box 12.1 Country-wise statistics of solar PV and CSP in Asia

**Bangladesh** *has more than 3 million solar home systems (SHS) operating at the end of 2014. The average size of the system is around 50–60 W for lighting, TV connections and mobile phone charging. Infrastructure Development Company Ltd. (IDCOL) has targeted 10,000 irrigation PV pumps (80 MW). The government started to introduce more PV power by setting a Solar Energy Program and is planning to introduce 500 MW of solar energy by 2017 (340 MW for commercial and 160 MW for grid connection). Bangladesh Power Development Board (BPDB) under the Ministry of Power, Energy and Mineral Resources signed a Power Purchasing Agreement (PPA) for a 60 MW PV power plant in July 2014 (BPDB, 2014).*

**China** *is expected to install approximately 19.5 GW in 2016 (Alex Nussbaum, 2015), a rise of 14.7% over 2015 and is expected to reach the target from 100 GW to 150 GW, which will bring about 21 GW of annual installation between 2016 through 2020 (Movellan, 2016). The target for the total installed solar PV capacity in 2050 is 1000 GW (Wang Sicheng, 2015). By 2018, a large number of concentrated solar power plants will be in place, and the 13th Five Year Electric Development Plan indicates that Concentrate Solar Power installation target is 5 GW.*

**India's** *PV market is driven by a mix of national targets and support schemes at various legislative levels. The Jawaharlal Nehru National Solar Mission aims to install 20 GW of grid-connected PV system by 2022 and an additional 2 GW of off-grid systems, including 20 million solar lights. Some states have announced policies targeting large shares of solar photovoltaic installations over the coming years; 2 GW of off-grid PV systems will be installed by 2017. However, in 2014 a new target of 10–60 GW of centralised PV and 40 GW of rooftop PV was announced (ADB, 2015). Total grid-connected solar power capacity in India is 8 GW at the end of the July 2016 (Mahapatra, 2016). The concentrated solar power installed by 2015 was 225 MW (REN21, 2016).*

**Japan** *had a total annual installed PV capacity of 9.7 GW (DC) in 2014, a 40% increase compared to 2013. The total cumulative installed capacity of PV systems in Japan reached 23.4 GW in 2014. About 60 GW of solar capacity has been approved but not been installed (ADB, 2015). The Japanese Ministry of Economy, Trade and Industry is discussing a revision of the current feed-in tariff (FiT) policy and the introduction of an auction process to promote lower cost operation (Movellan, 2016).*

**Korea** *enjoyed a record-breaking year in 2008 that saw 276 MW of PV installations. The PV market remained stagnant in the country during the following three years, mainly due to the limited FiT scheme. However, 230 MW in 2012, 530 MW in 2013 and finally 909 MW in 2014 were installed, reaching the highest level of installations so far. Thanks mainly to the newly introduced RPS scheme, at the end of 2014, the total installed capacity was about 2.4 GW, wherein the grid-connected centralised system accounted for around 87% of the total cumulative installed power. Korea installed 1 GW in 2015 (IEA PVPS, 2016b). The grid-connected distributed system amounted to around 13% of the total cumulative installed PV power. The share of off-grid non-domestic and domestic systems has continued to decrease and represents less than 1% of the total cumulative installed PV power.*

**Malaysia** *now has a total installed capacity of 168 MW. For the third year of its FiT system, the country installed 26.83 MW in 2015 (IEA PVPS, 2016b). The 2014 grid-connected distributed installations represented 86.7 MW compared to 48.2 MW in 2013. The residential segment remained stable in 2014 while the commercial segment doubled compared to 2013.*

**Myanmar** *plans to install a 220 MW solar PV plant, which is expected to be built in the Magway region.*
**Nepal's** *Electricity Agency planned to develop PV power plants totalling 325 MW by 2017.*

**The Philippines** *have installed 30 MW solar PV systems in 2014. The government approved 1.2 GW of utility-scale PV projects in 2014, and as in many countries, the tender was oversubscribed. Philippines PV market reached 110 MW in 2015 (IEA PVPS, 2016b).*

*Singapore* had a total PV installed capacity of 30 MW at the end of 2014. 15 MW of PV on rooftops have been installed in 2014, mostly in the commercial and industrial segments. The country has targeted 350 MW by 2020.

*Taiwan* installed about 227 MW, mostly as grid-connected roof top installations (IEA PVPS, 2016b). The total installed capacity at the end of 2015 is estimated at around 615 MW.

*Thailand's* cumulative grid-connected PV power reached to 1.3 GW at the end of 2014, with around 30 MW of off-grid applications. Thailand installed 121 MW in 2015 (IEA PVPS, 2016b). The concentrated solar power (CSP) installed by 2015 was 5 MW (REN21, 2016).

*Uzbekistan* has the intention to install 2 GW of PV plants and two utility-scale plants are being developed (100 MW and 130 MW).

*Vietnam's* solar PV capacity was approximately 4.5 MWp at the end of 2014, used typically for self-consumption purposes. However, its potential is estimated to be about 2–5 GW (residential and commercial rooftops), and 20 GW for ground-mounted PV power plants.

oil-shocks in 1970s, applications of PV technology expanded. For almost fifteen years, from 1983 to 1999, the PV industry maintained an upward, but not spectacular, growth trend of about 15% per year in the shipments of photovoltaics (Turkenburg, 2000). The grid-connected PV capacity dominated the market, by sustained dramatic growth rates in the early 2000s, and by 2008 this market sustained dramatic increases in cumulative installed capacity, growing from about 5.1 GW in 2006 to 7.8 GW in 2007, crossing 13 GW by the end of 2008 and reaching 123.2 GW at the end of 2013 (REN21, 2008, 2009, 2013; WSSD, 2002; IEA PVPS, 2014). As of 2015, the global installed capacity of solar photovoltaics reached 227 GW, an increase of about 25% (50 GW) from 2014 (REN21, 2016).

Figure 12.2 shows the total installed solar PV in Asia, which amounts to 83,860 MW (IRENA, 2016a).

Solar PV is by far the most popular solar technology for online grid-connected systems, through solar farms and roof top systems, and concentrated solar power systems are slowly making strong inroads. The trend is significant in South, East and South Asia. In Asia, a total of 87.75 GW PV systems, mainly grid-connected systems, have been installed by the end of 2015 (ADB, 2015, 2016; Alex Nussbaum, 2015; IEA PVPS, 2016a, 2016b; Mahapatra, 2016; REN21, 2016; IRENA, 2016b).

## Policies promoting solar electricity technologies in Asia

The price of solar PV modules has declined significantly over the past years (Figure 12.3), with a sharp drop from over US$3.25/Wp in 2006 to an average of US$0.72/Wp in 2014 – a drop of about 78%.[1] This was mainly driven by the oversupply in China, coupled with the declining demand in Europe, mainly in Germany. The shortage of polysilicon, which makes up a very significant part of the total module cost, increased the price to about US$400/kg. This led the producers of polysilicon to add additional capacity that dragged the price down to US$25/kg in 2010 (http://solarcellcentral.com/cost_page.html) and subsequently a drop in the cell and module price. This is expected to continue over the next few years until the excess supply of polysilicon is used up. The drop in cell and module price has also resulted in the decrease of installed cost of PV systems. For example, in the US, the cost of residential PV system has dropped from US$7.06/DC watt in 2009 to US$2.93/DC watt in 2016; the cost of commercial-scale PV dropped by about 60% between 2009 and 2016; and the cost of utility-scale PV (fixed tilt)

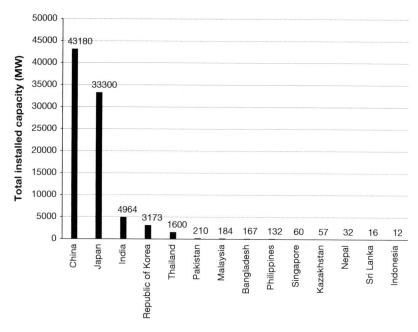

*Figure 12.2* Total installed solar PV capacity at the end of 2014

*Data source*: Adapted from IRENA (2016b).

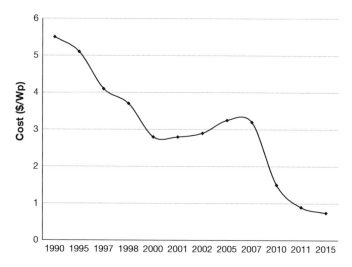

*Figure 12.3* Solar PV cost trend

*Data source*: Solar Cell Central (2015).

dropped by about 70% during the same period (Solar Cell Central, 2016). Along with the cost reduction and technology development of solar PV technologies, policies have also contributed to the increase in the installation of solar PV installations in Asia. A summary of country-wise status of policies promoting solar PV are given in Box 12.2.

## Box 12.2 Country-wise policy features for promoting solar PV

*Bangladesh is making a progress on grid-connected solar PV systems, mainly utility-scale, which has started to take-off recently. When the 200 MW grid-connected PV project is completed in 2018, the total grid-connected system would be about 300 MW (BPDB, 2014; Kenning, 2017; Photon.info, 2014). The government is attracting private sector investment through long-term power purchase agreement and aims to establish a FiT for the grid-connected system in the near future. The promotion of solar home systems in Bangladesh has been a great success with over 4.1 million SHS installed in the country (IDCOL, 2016). Through the Bangladesh Climate Change Strategy and Action Plan 2009 and with support from the World Bank and other donors, the government provides incentives schemes to encourage entrepreneurs who wish to start PV-based applications.*

*Brunei is considering a FiT policy.*

*China is using several schemes to incentivise the development of PV. They aim at developing utility-scale PV through rooftop PV in city areas and micro-grids and off-grid applications in un-electrified areas of the country. By 2014, China had a stable FiT scheme, which was financed by the surcharge paid by electricity consumers for utility-scale PV and rooftop PV. In that year, the National Energy Agency pushed for PV development on roofs, including PV power applications for large-scale industrial development districts and commercial enterprises with large roofs, high electrical load and high retail electricity prices, and for poverty reduction in Hebei, Shanxi, Anhui, Gansu, Qinghai and Ningxia provinces targeting 1.5 GW of installations.*

*India's solar energy projects, at the end 2016, had an aggregate capacity of over 8727.62 MW. The Ministry of New and Renewable Energy has initiated policy measures for achieving the target of renewable energy capacity to 175 GW by the year 2022, which includes: enforcement of Renewable Purchase Obligation and for providing Renewable Generation Obligation; setting up of exclusive solar parks; development of power transmission network through Green Energy Corridor project; identification of large government complexes/buildings for rooftop projects; provision of roof top solar and 10% renewable energy as mandatory under Mission Statement and Guidelines for development of smart cities; providing long tenor loans; and making roof top solar as a part of housing loan and others (MNRE, 2016).*

*Indonesia introduced a solar policy in 2014 which supports the purchase of solar photovoltaic power on the capacity quota offered through online public auction by the Directorate General of New Renewable Energy and Energy Conservation (MEMRI, 2014). The plant that wins the auction will sign a power purchase agreement with the National Electric Company at the price determined by the regulation. The maximum purchase price is 0.25 US$/kWh, which is increased to 0.30 US$/kWh in case of a local content requirement of 40%, and about 20 MW was installed in 2014 (MEMRI, 2014).*

*Japan established the FiT programme for solar PV in July 2012 – this has fostered the rapid growth of public, industrial application and utility-scale PV systems (IEA PVPS, 2016b). The breakdown of PV systems installed in 2014 is 1.4 MW for off-grid domestic application, and 9.7 GW for grid-connected distributed application. While the PV market in Japan developed in the traditional rooftop market which at the end of 2014 represented almost 5 GW of the cumulative capacity, 2013 and 2014 have seen the development of large-scale centralised PV systems, especially in 2014 with 3 GW up from 1.7 GW of centralised plants installed in 2013 (Hahn, 2014), and the market was balanced between residential (< 10 kW), commercial, industrial and large-scale centralised plants in 2014 (ADB, 2016).*

*Kazakhstan aims at installing 700 MW and has established a FiT programme in 2014.*

*Korea has developed various incentives to support PV development. In 2014, the "Fourth Basic Plan for the Promotion of Technological Development, Use, and Diffusion of New and Renewable Energy" based on the "Second National Energy Basic Plan" was issued. This plan includes many new subsidy measures including the development of "Eco-friendly Energy Towns," "Energy-independent Islands," and "PV Rental*

*Programs." The Renewable Portfolio Scheme launched in 2012 will be active until 2024 and is expected to be the major driving force for PV installations in Korea, with improved details such as boosting the small-scale installations (less than 100 kW size) by adjusting the Renewable Energy Certificate and multipliers, and unifying the PV and non-PV markets (IEA PVPS, 2016a, b).*

**Malaysia's** *long-term goals and commitment to renewable energy is explicated in The National Renewable Energy Policy and Action Plan. The Sustainable Energy Development Authority has the responsibility to implement and administer the FiT mechanism, which is financed by a Renewable Energy Fund. Building Integrated Photovoltaic installations were an additional premium on top of the FiT. The two main policy references were the Third Outline Perspective Plan and the Eighth Malaysia Plan (2001–2005) (Umar et al., 2014). Within the plan period, the Fifth-Fuel Diversification Policy 2001 has been established and the Small Renewable Energy Programme has been created (Ahmad et al., 2011). The Fifth-Fuel Diversification Policy 2001, is a major policy instrument that aims to increase the share of renewable energy in the power supply, that also includes solar among other six renewable energy resources (ADB, 2015).*

**Myanmar** *has seven line Ministries responsible for energy sector matters. The Ministry of Energy is a focal point for overall energy policy and planning coordination with the concerned Ministries; however, its major focus has been on planning and policies in the oil and gas sectors. National Energy Management Corporation published the National Energy Policy in 2014 where some key objectives were to develop energy resources that are accessible, while considering environmental and social impact, institute laws, rules and regulations to promote private sector participation, and implement sustainable energy development programmes that scale up use of renewable resources (MEMR, 2012; Suryadi, 2014; IEA, 2015). The National Electrification Plan started in June 2014 has a goal to electrify the whole country by 2030–2031 (World Bank, 2014).*

**Pakistan** *has approved 793 MW of solar plants to be commissioned in 2015. A FiT has been introduced for utility-scale PV in 2014. The initiatives aimed to boost the solar market include net metering programme and the introduction of a funding scheme allowing homeowners to borrow against their mortgage for solar installations (IFC, 2016; Government of Pakistan, 2006):*

*In* **Taiwan**, *the market is supported by a FiT scheme guaranteed for 20 years. This scheme is part of the Renewable Energy Development Act passed in 2009. The initial FiT was combined with capital subsidy. It has later been reduced and now applies with different tariffs to rooftops and ground-mounted systems. Larger systems and ground based systems have to be approved in a competitive bidding process based on the lowest FiT offered. Property owners can receive an additional capital subsidy. It is intended to favour small-scale rooftops at the expense of larger systems, in particular ground based installations. So far, agricultural facilities and commercial rooftops have led the market. The country targeted 842 MW of PV installations in 2015, 2.1 GW in 2020 and 6.2 GW in 2030 (3 GW on rooftops, 3.2 GW for utility-scale PV). In 2012, Taiwan launched the "Million Roof Solar Project" aiming at developing the PV market in the country, with the support of municipalities.*

**Thailand** *had introduced the feed-in premium or "adder" in 2007 aimed at promoting the development of grid-connected solar energy. This "adder" came in addition to the regular tariff of electricity, around 3 THB/kWh. It was phased out at the end of 2013 and has been replaced by a 25-year FiT scheme. In 2013, the solar power generation target was increased to 3 GW (and increased to 3.8 GW in 2015) together with the reopening of the solar PV rooftop Very Small Power Producer scheme with a new FiT. 1 GW has been granted for utility-scale ground-mounted PV systems. In addition, the Thai government also approved a generation scheme of 800 MW for agricultural cooperatives. With these schemes, Thailand aims at continuing the deployment of grid-connected PV in the rooftop segments, after a rapid start in the utility-scale segment (IEA PVPS, 2016a, b).*

*Table 12.2* Feed-in tariff in different countries in Asia

| Country | FiT rates (US$/kWh) | Comments |
|---|---|---|
| China | 0.14–0.16 | Different FiT for solar PV projects in different parts of China |
| India | 0.15 | The Indian government has set up power purchase tariffs for solar photovoltaic and solar thermal systems. The preferential tariffs are reviewed annually by the Central Electricity Regulatory Commission |
| Indonesia | 0.145–0.25 | Initiated in 2015, the tariff depends on project location. Java has been allocated the highest capacity of 150 MW but the lowest tariff, with individual project sizes capped at 20 MW |
| Japan | 0.369–0.371 (2013) 0.338–0.390 (2014) | Starting from 1 April 2014 to 31 March 2015, the FiT rates for solar energy have been slightly revised downwards (excluding tax) |
| Korea | 0.70 | The government set an upper limit of support for 20 MW for solar |
| Malaysia | 0.16–0.34 | The FiT is capped at a generation capacity of 30 MW |
| Mongolia | 0.15–0.18 | |
| Philippines | 0.22 | Government has guidelines for the Selection Process of Renewable Energy Projects under the Feed-in Tariff System and the Award of Certificate for Feed-in Tariff Eligibility. Projects are selected on the basis of this guideline |
| Thailand | 0.20 | It is called the "adder" scheme since the tariff is added to the base electricity price |

*Sources*: Pocci (2014), PVTech (2016), IEA (2014) and Mekhilef et al. (2014).

Table 12.2 summarises the FiT rates and related details in the selected Asian countries. Based on the observations on the promotion of solar-based electricity generation in Asia, it is clear that FiT has been the main instrument in Asia to have successfully increased the implementation of solar electricity generation technologies.

This has been contributed also by the falling costs of PV modules. The cost has seen a downward trend due to fall in polysilicon prices, technological innovations, scaling up and other factors, and have fallen by more than 80% since 2008 contributing to a reduction in the total system costs. In actual terms, the lowest tariff from a solar plant in Rajasthan (India) is lower than that of thermal (coal) power plants. These indicate the most favourable situations for the solar electricity growth in Asia in the coming decades.

A number of policies and measures have been employed by the countries (e.g. China, India, Japan, and Thailand) who have installed more than 1 GW installed capacity of online grid systems, for example: FiTs; competitive bidding; promotion of installation of solar PV in roof tops of various building types; and promotion of green cities and municipalities. Other countries with lower installed capacity (Malaysia, Singapore, Vietnam) as well as emerging economies (Bangladesh, Cambodia, Myanmar, etc.) are also introducing similar policies for solar PV promotion.

As of 2015, globally feed-in policies were the most widely adopted form of renewable power policy support. The market conditions in the country, type of technology (mature or less mature), and scale of implementation have contributed to the rates and the modalities of these policies. Moreover, market conditions are changing due to technological innovation, increasing deployment of renewable energy technologies, falling prices and shifting public opinion.

For example, with regard to mature technologies, countries have considered competitive bidding to support larger-scale projects (e.g. solar PV). The constant and continuing review and revision of the feed-in rate has occurred in China, Japan, the Philippines and Thailand, while new rates have been introduced in Malaysia and Pakistan.

The countries that adopted the global climate deal in Paris in December 2015 have identified renewable energy technologies deployment as the preferred option to reduce emissions. Of the 162 Intended Nationally Determined Contributions (INDCs) submitted, 106 countries have noted their intention to increase renewable energy, and 26 have set specific targets for electricity and other energy sources. Thus, some countries are using the INDC process to introduce more ambitious targets and strategies.

Roof tops are becoming increasingly attractive locations for placing solar PV systems/plants, especially in urban concentration countries (e.g. Singapore) or otherwise (e.g. China, India, Thailand). Different types of roofs are being considered for installation – from industry to government to private. This also resonates well with the "green cities, green town" initiatives of local and federal governments. In the recent auction of rooftop solar power projects in Rajasthan (India), the solar tariffs in 2016 reduced to Rs3 per unit, that is lower than that of thermal (coal) power plants. The rooftop projects will be installed on non-governmental organisation buildings, educational institutes, hospitals, trusts and not-for-profit companies. This trend, if found to be successful, is likely to find applicability in the other growing economies as well.

## Future potential for solar electricity technology promotion

The prospects for the deployment of solar energy-based electricity generation systems could be gauged considering the need for electricity in the region (given by the need to increase the energy access in terms of quantity and quality), availability of resources for the generation of electricity, technologies available and transfer of technologies, energy demand, country policies, costs of the technologies and financing models, greenhouse gas mitigation considerations, among others.

Energy access and provision of modern energy services, such as those provided by electricity, gasoline and liquefied petroleum gas is important for a country's economic development, as it can help provide lighting, communication, clean water and sanitation. However, at present, around 455 million people lack access to electricity in the Asia Pacific region, while in absolute terms, India, Bangladesh, Pakistan, Indonesia and Myanmar have the largest electricity access deficit (G20, 2016). Asia, in general, is well endowed with solar radiation availability, and access to data on the quantity of radiation for designing systems is also readily available. In the ASEAN region, for example, the solar radiation averages over 1500–2000 kWh per square meter annually, which allows for capacity factors of 20% and above.

The future electricity demand would be in rural and in peri-urban Asia that would require electricity generation and supply in various levels. Solar technologies, especially PV, has seen an important decreasing trend in cost in recent years, which along with the country-specific financial models through Energy Service Companies and Public Private Partnerships and government incentives (as indicated by the FiT presented in Table 12.2), is expected to contribute to the growth of the market for solar-based electricity systems at all levels.

It is thus clear that Asia is poised for an energy transition towards improving energy access and modern energy services. At the same time, INDCs have been submitted by the countries to reduce greenhouse gas emissions in 2015. The commitments made by Asian countries on their intended reductions of emissions is given in Table 12.3, and these emission mitigations would be mainly through the promotion of renewable energy and energy efficiency Moreover, Goal 7 of

*Table 12.3* A summary of the INDC targets (emission reduction) of selected Asian countries

| Country | INDC targets |
| --- | --- |
| Cambodia | Reduction of 3100 GgCO$_2$eq (27%) in the year 2030 compared to the baseline of 2010 |
| China | To lower CO$_2$ emissions per GDP by 60–65% from 2005 by 2030 |
| India | To reduce emissions intensity of GDP by 30–35% from 2005 levels in 2030 |
| Indonesia | Reduction of GHG emissions by 29% from the projected BAU level in 2030 |
| Malaysia | Reduce GHG emissions intensity of GDP by 35% (additional 10% with international aid) by 2030 from 2005 |
| Philippines | Reduction of GHG emissions by 70% by 2030 relative to its BAU scenario of 2000 |
| Singapore | Reduce its emission intensity by 36% in 2030 from the 2005 baseline |
| Thailand | Reduce GHG emissions by 20% (111 MtCO$_2$eq.) from the projected BAU level by 2030 |
| Vietnam | Reduce 8% (up to 25% with international aid) GHG emissions by 2030 from 2010 levels |

*Data source*: INDC Submissions to UNFCCC by the countries, compiled by the authors.

*Note*: BAU, business as usual; GDP, gross domestic product; GHG, greenhouse gases.

the 2030 Agenda for Sustainable Development aims to provide universal access to affordable, reliable, sustainable and modern energy services. For electricity, this would be through the measures of grid, off-grid and on-grid supply. Currently, solar energy-based electricity generation is observed in all these categories of measures.

According to ACE and IRENA (2016), "renewables are increasingly the least-cost option for electricity production, a trend that will accelerate over the coming decade". Comparing the levelised cost of electricity, it is estimated that by 2025 renewable power technologies (specifically, solar PV and wind) would have the fastest deployment in capacity growth, as for example, the levelised cost of electricity in 2025 could be US$40 per MWh for solar PV compared to about US$60 per MWh for coal.

The primary energy supply in the Asia is expected to increase to 8794 million tons of oil equivalent (Mtoe) in 2040 compared to 2012, while the demand for electricity is expected to almost double. Renewable energy consumption in Asia will also increase from 759 Mtoe in 2012 to 1174 Mtoe in 2040. The PV installed capacity could increase to about 17 TWh in 2040 (IEEJ, 2014).

Therefore, the Paris Agreement, and the implementation of the 2030 Agenda, along with the availability of solar radiation widely and the easy availability of PV panels, and cost factor is expected to further help deploy solar-based electricity generation in the region.

## Conclusions

The world now adds more renewable power capacity compared to all fossil fuels combined, and this amounted to about 60% in 2015. This significant growth of renewable energy technologies in the power sector is primarily due to cost competitiveness of solar technologies, favourable policies and attractive financing options, and concerns about greenhouse gas emissions. According to REN21 (2016), solar photovoltaics and concentrated solar power installations at the end of 2015 totalled 227 GW and 4.8 GW, respectively. In Asia, China, India and Japan together have installed more than 83 GW of PV, while India and Thailand have installed about 0.23 GW of concentrated solar power. The investment in solar energy technologies reached more than US$161 billion in 2015, compared to about US$286 billion for the entire renewable energy sector. The renewable energy sector's employment in 2015 was about 8.1 million jobs worldwide. The regional shifts in deployment to Asia resulted in this continent's share to the global employment to be about 60%, and, of this, more than 2.7 million were in the PV sector, an increase of 11% from the previous year (IRENA, 2016a).

Looking at the future, China is planning to install about 100 GW by 2020, while India plans to install a similar target by 2022 (60 GW of land mounted grid-connected solar power and 40 GW of rooftop grid interactive solar power). Japan and Thailand plan to install about 65.7 GW and 3 GW, respectively, by 2021 (ADB, 2012; Allendorf and Allendorf, 2013; Aung, 2014; Bodenbender et al., 2012; Cook, 2013; Fullbrook, 2013; Greacen, 2014; Greve, 1999; Kyaw et al., 2009; Martinot and Reiche, 2000; Practical Action, 2010, 2012, 2013a, 2013b). A major development in the industry has been the falling costs of PV modules due to fall in polysilicon prices, technological innovations, scaling up and other factors, and this has fallen by more than 80% since 2008 contributing to a reduction in the total system costs. In India's recent auction of rooftop solar power projects, the solar tariffs in 2016 reduced to Rs3 per unit, that is lower than that of thermal (coal) power plants.

During the last few years, advances in technology, reduction in costs, impetus from climate change, novel financing models and market transformations have contributed to the significant developments in the promotion of the solar and renewable energy-based electric power sector worldwide, and particularly in Asia. This is indicated by the fact that solar PV is cost competitive compared to fossil fuel-based electricity generation, and is dominated by large-scale generation, including the private sector. The government policies are also conducive and many countries have renewable energy targets. These factors, combined with the Sustainable Development Goal commitments and the Paris COP 21 agreements, indicate the significant role for solar and other renewable energy resource technologies in the coming decades.

## Note

1 http://solarcellcentral.com/cost_page.html

## References

ACE and IRENA 2016. *Renewable Energy Outlook for ASEAN: a REmap Analysis*, International Renewable Energy Agency (IRENA), Abu Dhabi and, ASEAN Center for Energy, Jakarta. Available: www.irena. org/DocumentDownloads/Publications/IRENA_REmap_ASEAN_2016_report.pdf.

ADB 2012. *Myanmar Energy Sector Initial Assessment*, Manila, Philippines: Asian Development Bank.

ADB 2015. *Renewable Energy Developments and Potential in the Greater Mekong Subregion*, Manila, Philippines: Asian Development Bank.

ADB 2016. *Energy Issues in Asia and the Pacific* [Online]. Asian Development Bank. Available: www.adb. org/sectors/energy/issues [Accessed 17 October 2016].

Ahmad, S., Kadir, M. Z. A. A. & Shafie, S. 2011. Current Perspective of the Renewable Energy Development in Malaysia. *Renewable and Sustainable Energy Reviews*, 15, 897–904.

Alex Nussbaum, B. 2015. *Chinese Solar to Jump Fourfold by 2020, Official Tells Xinhua* [Online]. Available: www.renewableenergyworld.com/articles/2015/10/chinese-solar-to-jump-fourfold-by-2020-official-tells-xinhua.html [Accessed 2 November 2016].

Allendorf, T. D. & Allendorf, K. 2013. Gender and Attitudes Toward Protected Areas in Myanmar. *Society and Natural Resources*, 26, 962–976.

Aung, H. N. 2014. Off Grid Solar PV Status & Its Potential in Myanmar. Off Grid Power Forum-Inter Solar Europe 2014, 4th June, Munich, Germany.

Bodenbender, M., Messinger, D. C. & Ritter, R. 2012. *Mission Report Energy Scoping Myanmar* [Online], Brussels: EUEI. Available: www.burmalibrary.org/docs14/EUEI_PDF_Myanmar_Energy_Scoping_Report_Final_June12.-red.pdf [Accessed 31 August 2017].

BPDB 2014. *Generation* [Online]. Dhaka: Bangladesh Power Development Board. Available: www.bpdb. gov.bd/generation.htm [Accessed 8 July 2016].

Bradford, T. 2006. *Solar Revolution. The Economic Transformation of the Global Energy Industry*, Cambridge, MA: The MIT Press.

Chiras, D. D. 2002. *The Solar House: Passive Heating and Cooling*. White River Junction, VT: Chelsea Green Publishing Company.

Cook, P. 2013. *Rural electrification and rural development* (pages 13–38) in *Rural Electrification Through Decentralised Off-grid Systems in Developing Countries*, Ed. S. C. Bhattacharyya, London: Springer.

Duffie, J. A. & Beckman, W. A. 2013. *Solar Engineering of Thermal Processes*, 4th edition, New Jersey, USA: Wiley-Interscience.

ECOFYS 2008. *Global Potential of Renewable Energy Sources: A Literature Assessment* [Online]. REN21— Renewable Energy Policy Network for the 21st Century. Available: www.ecofys.com/files/files/report_global_potential_of_renewable_energy_sources_a_literature_assessment.pdf [Accessed 10 October 2016].

Fullbrook, D. 2013. Power Shift: Emerging Prospects for Easing Electricity Poverty in Myanmar With Distributed Low-Carbon Generation. *Journal of Sustainable Development*, 6, 65.

G20. 2016. *Enhancing Energy Access in Asia and the Pacific: Key Challenges and G20 Voluntary Collaboration Action Plan* [Online]. Available: www.g20.utoronto.ca/2016/enhancing-energy-access-in-asia-and-pacific.pdf [12 December 2016].

Goswami, D. Y., Kreith, F. & Kreider, J. F. 2000. *Principles of Solar Engineering*, 2nd edition, Philadelphia, USA: Taylor and Francis.

Government of Pakistan 2006. *Policy for Development of Renewable Energy Generation*, Pakistan: Government of Pakistan.

Greacen, C. 2014. *SPP Regulatory Framework Options in Myanmar*, Washington, DC: International Finance Corporation (IFC).

Greve, H. R. 1999. The Effect of Core Change on Performance: Inertia and Regression Toward the Mean. *Administrative Science Quarterly*, 44, 590–614.

Hahn, E. 2014. *The Japanese Solar PV Market and Industry—Business Opportunities for European Companies* [Online]. EU-Japan Centre for Industrial Cooperation. Available: www.eu-japan.eu/sites/default/files/imce/minerva/pvinjapan_report_minerva_fellow.pdf [Accessed December 7 2016].

IDCOL 2016. *Solar* [Online]. Available: www.idcol.org/prjshsm2004.php [Accessed 4 November 2016].

IEA 2008. *Energy Technology Perspectives*, Paris: International Energy Agency.

IEA 2014. *Technology Roadmap. World Energy Outlook*, Paris: International Energy Agency.

IEA 2015. *World Energy Outlook Special Report*, Paris: International Energy Agency.

IEA PVPS 2014. *Snapshot of Global PV 1992–2013. World Energy Outlook*, Paris: International Energy Agency.

IEA PVPS 2016a. *National Survey Report of PV Power Applications in Korea 2015*, Paris: International Energy Agency.

IEA PVPS 2016b. *Trends 2016 in Photovoltaic Applications.*, International Energy Agency, Switzerland [Online]. Available: http://iea-pvps.org/fileadmin/dam/public/report/national/Trends_2016_-_mr.pdf [Accessed 20 February 2017].

IEEJ 2014. *Asia/World Energy Outlook*, Japan: The Institute of Energy Economics.

IFC 2016. *A Solar Developer's Guide to Pakistan*, Middle East and North Africa: International Finance Corporation.

IRENA 2016a. *Data and Statistics* [Online]. Available: http://resourceirena.irena.org/gateway/dashboard/ [Accessed 25 September 2016].

IRENA 2016b. *Renewable Capacity Statictics 2016* [Online]. Available: www.irena.org/DocumentDownloads/Publications/IRENA_RE_Capacity_Statistics_2016.pdf [Accessed 25 September 2016].

Kalogirou, S. A. 2009. *Solar Energy Engineering: Processes and Systems*, London, UK: Elsevier.

Kenning, T. 2017. *SunEdison Subsidiary Signs PPA for 200MW Solar Plant in Bangladesh* [Online]. PVTECH. Available: www.pv-tech.org/news/sunedison-subsidiary-signs-ppa-for-200mw-solar-plant-in-bangladesh [Accessed 20 February 2017].

Kyaw, U. H., Kyi, T., Thein, S., Hlaing, U. A. & Shwe, U. T. M. 2009. Myanmar: Country Assessment on Biofuels and Renewable Energy. Greater Mekong Subregion Economic Cooperation Program.

Mahapatra, S. 2016. *India's Grid-Connected Solar Power Capacity Tops 8 GW* [Online]. Available: https://cleantechnica.com/2016/08/18/indias-grid-connected-solar-power-capacity-tops-8-gw/ [Accessed 24 October 2016].

Martinot, E. & Reiche, K. 2000. *Regulatory Approach to Rural Electrifictaion and Renewable Energy: Case Studies from Six Developing Countries* (p. 16), Washington DC: World Bank.

Mekhilef, S., Aimani, M., Safari, A. & Salam, Z. 2014. Malaysia's Renewable Energy Policies and Programs with Green Aspects. *Renewable and Sustainable Energy Reviews*, 40, 497–504.

MEMR 2012. *Concerning on Electricity Tariff Provided by PT Perusahaan Listrik Negara*, Indonesia: Ministry of Energy and Mineral Resources.

MEMRI 2014. *Handbook of Energy and Economic Statistics of Indonesia* [Online]. Jakarta, Indonesia: Republic of Indonesia. Available: http://prokum.esdm.go.id/Publikasi/Handbook%20of%20Energy%20&%20Economic%20Statistics%20of%20Indonesia%20/HEESI%202014.pdf [Accessed 7 December 2015].

MNRE 2016. *Press Information Bureau*. Available: http://pib.nic.in/newsite/PrintRelease.aspx?relid=155612.

Movellan, J. 2016. *The 2016 Global PV Outlook: US, Asian Markets Strengthened by Policies to Reduce $CO_2$* [Online]. Renewable Energy World. Available: www.renewableenergyworld.com/articles/2016/01/the-2016-global-pv-outlook-u-s-and-asian-markets-strengthened-by-policies-to-reduce-co2.html [Accessed 2 November 2016].

NASA 2016. *Surface Meteorology and Solar Energy*, Washington, DC: NASA.

Photon.Info 2014. *Bangladesh Plans its Second Largest Project, Photon* [Online]. www.photon.info/en/news/bangladesh-plans-its-second-large-scale-pv-project [Accessed 21 February 2017].

Pocci, M. 2014. *Feed-In Tariff Handbook for Asian Renewable Energy Systems*, Hong Kong: Winston & Strawn.

Practical Action 2010. *Poor People's Energy Outlook 2010*, UK: Practical Action Publishing Ltd.

Practical Action 2012. *Poor People's Energy Outlook 2012: Energy for Earning a Living*, UK: Practical Action Publishing Ltd.

Practical Action 2013a. *Annual Report and Accounts 2012–13*, UK: Practical Action Publishing Ltd.

Practical Action 2013b. *Poor People's Energy Outlook 2013: Energy for Community Services*, UK: Practical Action Publishing Ltd.

PVTech 2016. *Indonesia Solar Fit Makes Java-Bali And Sumatra Attractive For Projects* [Online]. Available: www.pv-tech.org/news/indonesia-solar-fit-makes-java-bali-and-sumatra-attractive-for-projects-bne [Accessed 18 October 2016].

REN21 2008. *Renewables 2007 Global Status Report* [Online]. Paris: REN21 Secretariat and Washington, DC: REN21. Available: www.map.ren21.net/#fr-FR/search/by-technology/4,15,14,29 [Accessed 15 April 2016].

REN21 2009. *Renewables: Global Status Report 2009 Update*, Paris: Renewable Energy Policy Network for 21st Century (REN21) Secretariat and Washington, DC: REN21.

REN21 2013. *Renewables 2013 Global Status Report*, Paris, France.

REN21 2016. *GSR 2016 Key Findings* [Online]. REN21. Available: www.map.ren21.net/#fr-FR/search/by-technology/4,15,14,29 [Accessed 15 April 2016].

Shukla, P. N. & Khare, A. 2014. Solar Photovoltaic Energy: The State-of-Art. *International Journal of Electrical, Electronics and Computer Engineering*, 3, 91–100.

Sicheng, W. 2015. *PV Market in China and Rural Electrifications*. 2015 [Online]. Available: www.unece.org/fileadmin/DAM/energy/se/pp/eneff/6th_FESD_Yerevan_Oct.15/access/d3_s1/S1_6_Wang.Sicheng_CHI.pdf [Accessed 12 December 2016].

Solar Cell Central 2016. *Solar Electricity Costs* [Online]. Available: http://solarcellcentral.com/cost_page.html [Accessed 20 February 2017].

Suryadi, B. 2014. *ASEAN Electricity Tariff 2014* [Online]. Thailand. Available: http://asean.bicaraenergi.com/2014/05/asean-electricity-tariff-2014/ [Accessed 9 December 2015].

Turkenburg, W. C. 2000. *Renewable Energy Technologies*. World Energy Assessment: Energy And The Challenge Of Sustainability 2000 [Online]. Available: www.undp.org/content/dam/aplaws/publication/en/publications/environment-energy/www-ee-library/sustainable-energy/world-energy-assessment-energy-and-the-challenge-of-sustainability/World%20Energy%20Assessment-2000.pdf [Accessed 25 January 2016].

Umar, M. S., Jennings, P. A & Urmee, T. 2014. Generating Renewable Energy from Oil Palm Biomass in Malaysia: The Feed-in Tariff Policy Framework. *Biomass and Bioenergy*, 62, 37–46.

World Bank 2014. Myanmar—Development of a Myanmar National Electrification Plan Towards Universal Access 2015–2030 (English).

WSSD 2002. *Plan of Implementation of the World Summit on Sustainable Development* [Online]. Johannesburg. Available: www.un.org/esa/sustdev/documents/WSSD_POI_PD/English/WSSD_PlanImpl.pdf [Accessed 12 January 2017].

# 13

# WIND ENERGY DEVELOPMENT IN ASIA

*Christopher M. Dent*

## Introduction

Asia's industrialisation-driven economic development has been inherently energy-intensive, its manufacturing sectors requiring huge inputs of energy resources. The ensuing environmental problems caused by this process (most notably acute urban air pollution and climate change risks) has in turn created pressures upon Asian governments to promote renewable energy as an essential element of the transition to a low carbon economy. Wind energy is the world's largest non-hydro renewables sector, and has a key role to play in that transition over forthcoming years and decades. In recent years, the centre of gravity of the wind energy sector has moved increasingly to Asia both in terms of wind energy power generation and wind turbine production. The wind energy sector can be analysed from the following general perspectives:

- *Wind turbine manufacturing*: mainly whole-assembly turbine producers (e.g. Goldwind from China, Suzlon from India), component manufacturers (e.g. gearing systems, generators) and material suppliers, such as steel for turbine towers and rotors.
- *Installed wind energy capacity*: the construction of wind farms or installation of individual wind turbines by plant developers that actually produce electricity, otherwise known as power generation.
- *Wind energy consumption*: usually through a grid system, which can be national, regional or even a community-based micro-grid. Single installation turbines are often used by farms, small communities or individual households utilising very small wind turbines.
- *Wind energy stakeholders*: including wind turbine makers and their value-chain suppliers, wind farm developers, power generation companies that initially buy wind-generated electricity that sell to the grid companies who then sell on to end consumers, policy-makers, wind energy industry associations, and communities located in or near to wind energy installations.

Wind energy resources are distributed asymmetrically within Asian countries, and across the region as a whole. For example, much of Southeast Asia is located in the 'doldrums' zone, a low-pressure area spread over equatorial latitudes where relatively calm winds exist in normal weather conditions. Consequently, it has been difficult to develop commercially viable wind energy in Singapore, Brunei and large parts of Indonesia and Malaysia. However, Asia is generally rich in wind resources and produces more wind energy generated electricity than any other region. In this chapter, we explore how Asia's wind energy sector has developed over the years, what have been the key challenges facing this development, and the prospects for Asia's wind energy in the years to come.

## Wind energy: an overview

Humans have been harnessing the kinetic energy of wind in various ways across the world for centuries. Early windmill technologies in different civilisations were used to pump water and for basic agricultural processing. Vertical axis windmills are thought to have existed on the Persian–Afghan border around 200 BC, and horizontal axis windmills in the Netherlands and parts of the Mediterranean from around the 14$^{th}$ century onwards. Electricity-generating wind turbines were developed in Europe and the United States from the late 19$^{th}$ century onward, these having very-low-level kilowatt capacities. The first electricity-generating wind turbine was a 12 kilowatt (kW) device constructed at Cleveland, United States in 1888. In the early 20$^{th}$ century, a number of 25 kW devices were installed across Denmark. The Danish company Vestas was established in 1898 and for many years has been the world's largest wind turbine producer, although in 2015 it was displaced from the top spot for the first time by a Chinese company, Goldwind. Experimental, larger capacity wind turbines were developed around the mid-20$^{th}$ century, assisted by technological developments in aeroplane manufacturing (e.g. propellers, monoplane wings), with many turbines being grid-connected (Fleming and Probert 1984, Gipe 1991, Kaldellis and Zafirakis 2011).

Modern wind turbines convert the kinetic energy of moving air into mechanical and then electrical energy. The rotational energy of a moving turbine is converted into rated power by its generator, this then passing through a transformer before supplying power to an electricity grid or other system. The power output of wind turbines rises exponentially to wind speed, and areas that have constant and high wind 'densities' are naturally prime locations for developing wind farms (e.g. offshore and high altitude) to ensure the best potential and most predictable power generation. Around 3 to 4 metres per second is the threshold minimum 'cut-in' speed that modern large turbines require to produce electricity, and the maximum 'cut-out' speeds of most turbines are in the 20 to 25 metres per second range: higher speeds would cause structural and component damage. Due to wind-speed intermittencies and the non–dispatchability problem, wind energy is essentially a supplement to constant and predictable forms of energy generation.[1] However, its potential optimum share in the total energy mix will increase with grid expansion as more wind resources are harvested across larger inter-connected areas, whether within national or international spaces.

It was not until the late 1970s that an international wind energy industry began to emerge, as for other renewable energy technologies largely in response to the 1973/74 oil crisis. This development was initially concentrated in Europe and North America. However, investment and innovation in the sector remained relatively slow during the 1980s and much of the 1990s. This changed in the early 2000s due to a combination of two key factors: first, growing concerns about climate change as scientific consensus shifted increasingly towards belief of its substantive existence and significance; second, Asian governments and enterprises began to enter the market as serious industry players around this time. From the year 2000 to 2015, installed wind energy worldwide has experienced a 24-fold increase, from 18.0 GW[2] to 432.9 GW. An annual average growth rate of around 20 percent has been achieved since 2008, and the sector has added more new net installed capacity (312.0 GW) than any other energy sector during this time (Global Wind Energy Council/GWEC 2009, 2016). Wind energy generation capacity is around twice that of solar photovoltaic (PV), four times biomass generation and just under half that of hydropower globally (REN21 2016). It contributes to just over 3 percent of total worldwide power generation capacity. This is a small share but which has more than doubled in the last five years. A total of 107 countries now has installed wind energy (Windpower.net 2016).

The US initially led the way in global wind energy generation, by 1990 accounting for 79.9 percent of the global total, Europe 19.2 percent and the whole Asia and Oceania region a mere 0.9 percent. However, the positions of the US and Europe became reversed by the year 2000, the EU responsible for 73.3 percent, the US 14.2 percent while Asia now held a 9.8 percent share (Table 13.1). Thereafter, Asia began to make an increasing impact. By 2015, its share of global installed capacity had over quadrupled from its 2000 position, to 40.6 percent. In recent years, Asia has been responsible for half of global wind energy sector growth. The region's two main players have been China and India.

In 2000, China's installed capacity was just 352 MW, around the same as Britain's. From 2004 to 2009, China's capacity level doubled annually and by 2010 had overtaken the US as the world's leading installed capacity nation. By 2015 China's capacity level had reached 145.4 GW, 33.6 percent of the global total, remaining well ahead of the US (74.5 GW) and the leading European countries Germany (44.9 GW) and Spain (23.0 GW). Meanwhile, India's installed capacity level has increased from 1.3 GW in 2000 to 25.1 GW by 2015, and the Indian company Suzlon has emerged as a leading wind turbine producer. China now has five of the world's top ten turbine manufacturers – Goldwind, Guodian United Power, Mingyang, Envision and CSIC[3] (Table 13.2). Other Asian companies have become significant industry players, from Japan (Hitachi, Mitsubishi, Japan Steel Works and Komai Tekko), South Korea (Daewoo, Doosan, Hanjin, Hyundai, Samsung, STX and Unison), and Taiwan (TECO). Yet as Table 13.1 shows, Japan, South Korea and Taiwan have been relatively slow at actually installing wind turbines in their own territories. Moreover, wind energy development in Southeast Asia remains somewhat stunted: only Thailand and the Philippines have developed installed capacity of any significance although Vietnam is currently developing a number of large-scale wind farms (Windpower.net 2016). In other parts of Asia, Pakistan has plans to expand its installed capacity fivefold, from 256 MW in 2015 to 1.3 GW by 2018 based on the development of 21 projects, many of these using Chinese energy sector contractors and wind turbine equipment.[4] These will be mainly situated in the high wind potential province of Sindh, and this expansion in Pakistan's wind energy sector is due at least in part from China's broader investment in the bilateral economic and security relationship between the two nations. Pakistan is viewed as a key 'One Belt, One Road' partner to the Chinese government. Iran has meanwhile slowly built up its capacity to 118 MW, Sri Lanka to 63 MW and Mongolia to 51 MW (Table 13.1). Like in other Asian nations, a lack of capital investment and competition from other energy sectors has constrained wind energy development in these three countries.

Asia's wind energy sector is overall, then, dominated by China, its share of the region total rising from 19.8 percent in 2000, to 50.0 percent by 2008, and to 82.6 percent by 2015. Furthermore, China has driven the global growth of wind energy more than any other country. From 2008 to 2015 it added 133.2 GW of new installed capacity, equating to around 43 percent of the world total over the period. However, as we later discuss, China has faced a number of technical challenges and constraints regarding actual power generated from its burgeoning number of wind farms. Despite having installed more wind energy than any other country, and

Table 13.1 Wind energy development, Asia and global (2000–2015)

Installed capacity (MW)

| | 2000 | 2006 | 2007 | 2008 | 2009 | 2010 | 2011 | 2012 | 2013 | 2014 | 2015 |
|---|---|---|---|---|---|---|---|---|---|---|---|
| China | 352 | 2,599 | 5,912 | 12,210 | 25,810 | 44,733 | 62,364 | 75,324 | 91,412 | 114,609 | 145,362 |
| India | 1,267 | 6,270 | 7,850 | 9,587 | 11,807 | 13,066 | 16,084 | 18,421 | 20,150 | 22,465 | 25,088 |
| Japan | 142 | 1,309 | 1,528 | 1,880 | 2,083 | 2,304 | 2,501 | 2,614 | 2,661 | 2,794 | 3,038 |
| South Korea | 0 | 176 | 192 | 278 | 348 | 379 | 407 | 483 | 561 | 610 | 835 |
| Taiwan | 3 | 188 | 280 | 358 | 436 | 519 | 564 | 571 | 614 | 633 | 647 |
| Pakistan | 0 | 0 | 0 | 6 | 6 | 6 | 6 | 56 | 106 | 256 | 256 |
| Thailand | 0 | 0 | 0 | 0 | 0 | 0 | 8 | 112 | 223 | 233 | 233 |
| Philippines | 0 | 25 | 25 | 25 | 33 | 33 | 33 | 33 | 33 | 216 | 216 |
| Iran | 11 | 47 | 66 | 82 | 91 | 92 | 91 | 91 | 100 | 118 | 118 |
| Sri Lanka | 0 | 0 | 0 | 0 | 0 | 0 | 14 | 14 | 63 | 63 | 63 |
| Mongolia | 0 | 0 | 0 | 2 | 2 | 2 | 2 | 1 | 51 | 51 | 51 |
| Vietnam | 0 | 0 | 0 | 1 | 9 | 30 | 31 | 31 | 31 | 31 | 31 |
| **Asia total** | **1,775** | **10,614** | **15,853** | **24,429** | **40,625** | **61,164** | **82,105** | **97,751** | **116,005** | **142,079** | **175,938** |
| *% Share of world total* | *9.8* | *14.3* | *16.9* | *20.2* | *25.4* | *31.1* | *34.5* | *34.6* | *36.5* | *38.4* | *40.6* |
| Germany | 6,095 | 20,622 | 22,247 | 23,903 | 25,777 | 27,191 | 29,060 | 31,308 | 34,250 | 39,128 | 44,947 |
| Spain | 2,535 | 11,630 | 15,145 | 16,740 | 19,149 | 20,623 | 21,674 | 22,796 | 22,959 | 23,025 | 23,025 |

Table 13.1. (continued)

| | Installed capacity (MW) | | | | | | | | | | |
|---|---|---|---|---|---|---|---|---|---|---|---|
| | 2000 | 2006 | 2007 | 2008 | 2009 | 2010 | 2011 | 2012 | 2013 | 2014 | 2015 |
| Britain | 409 | 1,963 | 2,389 | 3,288 | 4,051 | 5,204 | 6,540 | 8,445 | 10,531 | 12,633 | 13,603 |
| France | 68 | 1,567 | 2,455 | 3,404 | 4,492 | 5,970 | 6,684 | 7,564 | 8,254 | 9,285 | 10,358 |
| Italy | 427 | 2,123 | 2,726 | 3,736 | 4,850 | 5,797 | 6,737 | 8,144 | 8,552 | 8,663 | 8,958 |
| Sweden | 241 | 571 | 831 | 1,067 | 1,560 | 2,163 | 2,907 | 3,745 | 4,470 | 5,425 | 6,025 |
| Portugal | 83 | 1,716 | 2,130 | 2,862 | 3,535 | 3,706 | 4,083 | 4,525 | 4,724 | 4,947 | 5,079 |
| Denmark | 2,417 | 3,136 | 3,125 | 3,160 | 3,465 | 3,749 | 3,871 | 4,162 | 4,772 | 4,881 | 5,063 |
| Other EU | 510 | 3,144 | 3,722 | 4,328 | 5,811 | 7,402 | 10,063 | 12,961 | 18,777 | 21,073 | 24,520 |
| **European Union** | **13,225** | **48,031** | **56,517** | **64,713** | **74,919** | **84,074** | **93,947** | **106,041** | **117,289** | **129,060** | **141,578** |
| *% share of world total* | *73.3* | *64.8* | *60.2* | *53.5* | *46.9* | *42.8* | *39.5* | *37.5* | *36.9* | *34.9* | *32.7* |
| United States | 2,564 | 11,575 | 16,823 | 25,237 | 35,159 | 40,180 | 46,919 | 60,007 | 61,091 | 65,877 | 74,471 |
| *% share of world total* | *14.2* | *15.6* | *17.9* | *20.9* | *22.0* | *20.4* | *19.7* | *21.2* | *19.2* | *17.8* | *17.2* |
| Rest of World | 490 | 3,948 | 4,799 | 6,613 | 9,161 | 11,310 | 14,817 | 19,067 | 23,989 | 32,689 | 40,896 |
| **World** | **18,040** | **74,122** | **93,927** | **120,903** | **159,766** | **196,630** | **237,669** | **282,587** | **318,105** | **369,705** | **432,883** |

*Data source:* Global Wind Energy Council/GWEC (2016, and other global annual reports), Windpower.net (2016).

*Table 13.2* World's top ten wind turbine producers

| 2005 | | | | 2015 | | | |
|---|---|---|---|---|---|---|---|
| Rank | Company | Country | Capacity (MW) | Rank | Company | Country | Capacity (MW) |
| 1 | Vestas | Denmark | 3200 | 1 | Goldwind | China | 7800 |
| 2 | Enercon | Germany | 2700 | 2 | Vestas | Denmark | 7300 |
| 3 | Gamesa | Spain | 1900 | 3 | GE Wind | United States | 5900 |
| 4 | GE Wind | United States | 1300 | 4 | Siemens | Germany | 3100 |
| 5 | Siemens | Germany | 1100 | 5 | Gamesa | Spain | 3100 |
| 6 | Suzlon Group | India | 900 | 6 | Enercon | Germany | 3000 |
| 7 | RE Power | Germany | 900 | 7 | Guodian United Power | China | 2800 |
| 8 | Goldwind | China | 700 | 8 | Mingyang | China | 2700 |
| 9 | Nordex | Germany | 500 | 9 | Envision | China | 2700 |
| 10 | Ecotecnica | Spain | 300 | 10 | CSIC | China | 2000 |

☐ Europe  ☐ China
☐ Other Asia  ☐ North America

*Data source*: Bloomberg New Energy Finance/BNEF (2011, 2016).

now also the EU as a whole, ongoing technical and policy problems have meant that a large proportion of its installed capacity has been idle. Consequently, while China has more installed wind energy capacity than the United States, the better quality of the US turbines located in more optimal wind flow areas, which are more effectively connected to the grid system, has meant that by 2015 it still produced higher levels of usable wind-generated electricity than China[5] (Global Wind Energy Council/GWEC 2016).

## Asia's wind energy sector in the global context

### *Technological and industrial perspectives*

Early grid-connected wind turbines had rated power capacities of around 20–30 kW but they now range from around 0.5 MW to 8 MW, the most common of which lie in the 1.5 MW to 3 MW range. Over time, the length of turbine blades have increased from an average of around 10 metres in the 1980s, to 30–40 metres by the late 1990s, to up to around 70–80 metres by the early 2010s. By 2016 there were three firms developing 10 MW turbines: Chinese firm Shandong Swiss Electric, Norwegian company Sway and US-owned Windtec, whose machine will have rotor blades over 90 metres long and a diameter of 190 metres (Windpower.net 2016). In the Shanghai metropolitan area, just one such 10 MW turbine could potentially meet the electricity demand of around 20,000 households and in West Europe between 5,000 to 7,000 households (World Wind Energy Association/WWEA 2012). There still exists some notable differences in turbine product quality, such as power conversion efficiency ratings, and Chinese company built turbines are still generally of a lower standard than their European, Japanese and US counterparts. Meanwhile, the very small wind turbine (i.e. normally sub 0.1 MW capacity rating) industry and market has steadily grown due to deepening worldwide demand among households, communities, organisations and construction firms. By the end of 2011, there was an estimated 730,000 such units installed, China accounting for close to 70 percent of these (World Wind Energy Association/WWEA 2013). Small wind turbines are providing electricity in many remote areas of developing countries, assisted by declining costs of grid-connected inverter technology.

The average size of wind farms has also consistently increased over time, and the largest have increased from around 190 MW capacity during the early 2000s period to around 1 GW by the early 2010s, and are expected to rise to 1.5 GW in the 2016–2020 period (REN21 2014, Bloomberg New Energy Finance/BNEF 2011). In addition, there is a growing trend of community wind farm projects in Europe, North America and parts of Asia, most notably in Japan. Improvements in material technologies and component supply chain systems are extending wind turbine life-spans, which is currently around 20 to 25 years. As wind turbine technology has improved and reduced both production and operations and maintenance (O&M) costs, so wind farms in lower-resource areas have become more cost competitive, thus extending the geographic scope for wind energy development. Technological leader firms are developing cheaper and lighter composite materials, sensors related to extreme elements, and advanced blade coatings: these all mainly relate to developing more robust offshore turbines (Lee *et al.* 2009, Roland Berger 2010). Overall, wind energy is subject to high rates of techno-innovation where efficiency rating levels are improving on a constant basis (Global Wind Energy Council/GWEC 2016).

The wind energy industry has technology cluster linkages to various other high-tech sectors, such as nanotechnology, aerospace, energy storage electronics, meteorology software and marine engineering. Developed country firms still retain prominent technology advantages in wind energy, though this is likely to be challenged over time by companies from China and other emerging Asian nations (Lee *et al.* 2009). Leadership in wind energy manufacturing has shifted increasingly from Europe and North America to Asia. In 2005, only two Asian firms (Goldwind and Suzlon) were ranked in the world top ten wind turbine producers, but as Table 13.2 shows by 2015 there were five, all from China (Bloomberg New Energy Finance/BNEF 2016). By the early 2010s, the country was not only producing around half the world's wind turbines, but had also become major exporters and were improving techno-innovation capabilities (China National Renewable Energy Centre/CNREC 2013, Global Wind Energy Council/GWEC 2014, Zhou *et al.* 2012). Indian, Japanese, Taiwanese and Korean firms were also expanding their wind turbine production operations as the industry internationalised. For example, Korean conglomerate Samsung has set up operations in Britain to develop new offshore wind technology testing facilities in Europe's North Sea.[6]

The wind energy sector has matured into a mass production industry. There are now over 200 wind turbine manufacturers and over 16,000 wind farms either operational or under development worldwide. Asia has the greatest number of these in all fields. Techno-innovatory advances and intensifying competition has helped drive production and technical efficiencies.[7] In many countries, wind energy can now compete quite effectively on price against fossil fuels, and at its most efficient is on cost parity with grid-connected hydropower (Dent 2014). The wind sector may also be considered a mainstream advanced technology industry attracting fast growing research and development (R&D) funding and venture capital investment since the early 2000s. In 2004, a global total of US$14.5 billion was invested in wind energy but this had risen to an annual US$109.6 billion in 2015. Although less than solar PV this still represents higher levels of investment directed towards net additional power capacity in any fossil fuel sector or nuclear.

Offshore wind farms were first developed by Denmark in the early 1990s, and accounted for just 2.8 percent of global installed wind capacity by 2015 (Global Wind Energy Council/GWEC 2016). Although much more expensive to develop than onshore, offshore wind offers higher capacity loads due to stronger and more constant sea winds. The largest wind turbines in development are for offshore farms. They are less prone to siting conflicts (visual and noise related) and land-based transportation constraints in the wind farm construction process as offshore wind turbines are invariably manufactured in nearby seaports. Although O&M costs are normally higher than for onshore, offshore farms offer much greater scope for capturing scale economies. Offshore wind farms are situated in relatively shallow water areas (normally up to 20 metres deep) and typically up to 20 km from shore. Both distance parameters have increased over time and will continue to do so with improvements in engineering technology, thus extending offshore wind into higher energy yield zones. Many of the world's largest proposed new wind farms are offshore, and the growing involvement of oil, gas and large civil engineering firms in the offshore sub-sector has intensified competition. For Northeast Asia's densely populated economies of Japan, South Korea and Taiwan in particular, offshore wind is likely to prove a strategically important renewables option. Over 90 percent of global offshore wind energy is installed around the waters of Europe, especially in the North Sea. China's global share had steadily risen, to 8.4 percent by 2015 and is now the world's fourth largest market for offshore wind with 1 GW operational capacity (Global Wind Energy Council/GWEC 2016). However, the Chinese government failed to reach their 5 GW installed offshore wind target by the end of the 12th Five-Year Plan (FYP) period.

Scaling up the wind energy industry in Asia and globally will face certain resource challenges. For instance, motors in advanced wind turbines use a highly magnetic rare earth element, neodymium, which is almost exclusively produced at present in China, and in relatively small quantities. The larger wind turbines become, the higher the demand generated for steel, aluminium, copper and other metals and materials like carbon fibre and cement (Wiser and Bolinger 2010). The rising cost of steel was a notable cause of rising prices for wind energy between 2005 and 2009 (REN21 2012). Furthermore, continued disputes over intellectual property rights infringements between developed country companies and Chinese firms in particular may hamper future efforts at fostering techno-innovatory collaboration on wind energy development. In addition, bilateral disputes over wind turbine trade at the World Trade Organization (WTO) have soured inter-firm relations. China has again been the main target of US and EU complaints in cases taken to the WTO Dispute Settlement Mechanism, Chinese firms being accused of trade dumping practices.

## Government policy and strategy perspectives

Governments at all levels have played an essential role at promoting the development of wind energy worldwide. Policies were initially aimed at financing new R&D in fledgling wind energy technologies as well as helping fund small demonstration projects. As industrial and technology capacity strengthened in the 2000s, state support progressed increasingly towards enabling utility-scale wind energy development, providing direct state support, regulatory environments and a growing range of market-based incentives for investment in new wind farms. International institutions (e.g. the International Energy Agency) and emerging renewable energy organisations (e.g. REN21) helped promote best policy practice on wind sector development where mutual learning of successful policy practice grew. In East Asia in particular, the promotion of wind energy and other renewables occurred in the broader political economic context of 'new developmentalism' (Dent 2014). This concerned the conflation of East Asia's state capacity tradition of active government involvement in guiding the economic development process with the growing influence of ecological modernisation ideas on East Asian governments. Ecological modernisation essentially concerns reconciling capitalism with sustainable development, and thus incentivising companies to invest in green industries while engaged in their usual pursuit of profit and market expansion. In a region that has long demonstrated a predilection for industrial policies, the appeal to East Asian states was that they could promote wind energy and other low carbon technologies as emerging strategic industries in addition to environmental reasons for doing so. By way of illustration, let us compare the evolution of wind energy policies in China and Japan, comparing and contrasting successes, failures and challenges faced in each country.

The Chinese government's prioritisation of wind energy over most other renewable energy sectors (the key exception being hydropower) is due in part to existing competitive advantages the country possessed in relevant engineering industries. In Japan, comparative strengths in electronics and other related industries led to the prioritisation of solar energy over wind. However, it was the Japanese government that was the first in Asia to introduce wind energy policies, deriving out of the multi-sector 1974 Sunshine Project. This, though, was more or less limited on the wind energy side to funding a handful of demonstration installations (Harborne and Hendry 2009, Ushiyama 1999). After the Sunshine Project programme review of 1990, the government sought to expand their wind energy partnerships with Japan's ten regional power companies (*Denjiren*) and emerging turbine manufacturers. It was not until the mid-1990s that the government introduced exclusive wind energy policy measures. The 1996 New Renewable Energy Target initiative set an installed capacity objective of reaching 3 GW by 2010 that subsequently the nation failed to attain (Table 13.1). In 1998 a new subsidy scheme for R&D and wind farm development was introduced, helping initially spur national installed capacity (Maruyama *et al.* 2007). By the mid-2000s, state support for wind energy R&D was terminated and funds redirected to improve grid performance and power quality (Harborne and Hendry 2009).

It took until the 1990s for China to launch its own wind energy policy, beginning with various state support measures for promoting indigenous turbine production and technology development, these though relying on joint venture arrangements with European firms such as Vestas and Gamesa. China's wind turbine manufacturers developed out of this process. In 1995, the government introduced its first power purchasing arrangement for wind energy as part of the China Electric Power Act, obliging the national grid operator to procure wind-generated electricity at prices and profit levels aimed at sustaining the growth of wind farm developers and turbine producers (Global Wind Energy Council/GWEC 2007). This was later supported in 1997 by the 'Ride the Wind' programme and parallel policy mechanisms that provided loan finance to foster indigenous industry development, install larger wind turbines in the initial four designated wind resource provinces (Inner Mongolia, Xinjiang, Zhejiang and Hebei) over the 9th and 10th FYP periods, and a local content requirement on foreign investor producers to source a 20 percent minimum of components and materials from domestic suppliers (Li 2010, Xia and Song 2009). The programme also encouraged further foreign technology transfers, German company Nordex being the first to participate in the scheme with local firm Xian Aero Engine Company (International Renewable Energy Agency/IRENA 2012). Goldwind soon followed by signing technology licensing agreements with German firms Jacobs (600 kW turbines) and REpower (750 kW) in the late 1990s, and Sinovel a 1.5 MW technology license with Furlander in 2004 (Ru *et al.* 2012).

Four key policy initiatives of the early to mid-2000s helped lay a firmer foundation for wind energy development in China. First, the 2000 National Debt Wind Power Programme offered financial support for wind farms constructed using locally made turbines and other equipment.

Second, in 2002 the government restructured the monopoly State Power Corporation into five separate power generation companies and two grid companies. Consequently, the new power generation companies became among the world's largest wind farm developers, with Guodian also establishing a wholly-owned subsidiary wind turbine producer company – Guodian United Power – currently ranked seventh in the world. However, there was now a multiplication of state actors to co-ordinate across the wind energy sector, which became exacerbated by the proliferation of private and state-owned wind turbine industry producers in subsequent years. We later discuss the implications of this and the problems arising for China's national wind energy development. Third, the 2003 Wind Power Concession Programme introduced China's first feed-in tariff (FiT) scheme[8] and targets to create up to 20 wind farms with between 100 MW and 200 MW capacity through competitive bidding, as well as a national 20 GW target by 2020. A 50 percent local content rule on turbine production was applied to the concession projects, revised upward in 2004 to 70 percent but later abolished in 2009 after China's trade partners initiated a WTO dispute case against this policy action. Fourth, the 2006 Renewable Energy Law legally obliged the two newly formed state grid companies to purchase at least 5 percent of their electricity from wind energy produced from the five new power generation companies (Global Wind Energy Council/GWEC 2007).

Whereas China's five power generation companies responded on the whole positively to the government's wind energy policies, this was not the case for their counterparts in Japan, whose ten regional *Denjiren* power companies have long proved somewhat resistant to developing this particular renewable. Up until the early 2000s, they had exercised their own 'introduction limitation quotas' on wind energy to limit their exposure to financial risk (then still relatively high generation costs) and technical risks relating mainly to its intermittent power supply. Any wind energy generated above these quota levels was competitively tendered. Collectively, *Denjiren* quotas for wind by 2003 only amounted to 330 MW, whereas wind farm developers submitted tenders totalling 2,400 MW, indicating a significant mismatch between what developers wished to supply to the market and what the power companies actually wanted to supply to the electricity grid. However, new renewable portfolio standard (RPS) legislation introduced in 2003 changed this, obligating the *Denjiren* to generate minimum quantities of electricity from renewables set by national government via the Agency for Natural Resources and Energy (Inoue and Miyazaki 2008, Maruyama *et al.* 2007). Yet the RPS targets set by the Japanese were relatively unambitious and a period of policy inertia on wind energy policy followed. In the aftermath of the global financial crisis, Japan ended its subsidies for wind energy development in 2009, and a year later the government failed to realise its 3 GW installed capacity by 2010 target for the sector.

In contrast, the Chinese government had to constantly revise its wind energy targets upward as the national market and industry continued to expand rapidly. The 12th FYP (2011–2015) set a 100 GW target for the end of the plan period, this including 70 GW from the Wind Base Programme launched in 2008 comprising mega-scale projects in Inner Mongolia, Xinjiang, Gansu, Hebei, Jilin and Jiangsu – the largest Wind Base project being at Jiuquan in Gansu province, capacity rated at 6.8 GW and over twice that of Japan's entire wind energy capacity (Global Wind Energy Council/GWEC 2014, National Development and Reform Commission/ NDRC 2011). A longer term target of 150 GW grid-connected capacity by 2020 was initially set by the 12th FYP, the great majority of this (138 MW) to come from the Wind Base provinces (Kang *et al.* 2012). The 2020 target of the 12th FYP was later revised upward to 200 GW, and a special technology-oriented development plan for wind energy was introduced in 2012. This was essentially a strategic industry policy aimed at fostering capacity in high-tech large turbines. China's provincial and city governments have also played a particularly important role in developing the country's wind energy sector, exercising significant decision-making autonomy within the national policy framework (e.g. approving projects below 50 MW capacity) as well as operating their own local policies. Inner Mongolia was China's first province to devise its own distinct provincial-level policy, in 2006 issuing rules on wind farm planning, facilitating new meteorological surveys on wind resources and feasibility studies on possible new projects, as well as specifying budget resources, administrative processes and timetables for project development. Local institutionalisation of wind energy policy has been replicated in other parts of China, albeit often with a different emphasis. For example, whereas Inner Mongolia has concentrated on optimising installed capacity, Jiangsu Province has focused on promoting domestic industry production (Liu and Kokko 2010).

Most recently, the Chinese government announced it was going to gradually phase out its current FiT scheme for wind – introduced in 2009 – as the sector was becoming more commercially self-supporting and China's leadership was looking to tighten the country's fiscal position. This led to a new spurt of wind farm development applications: in 2015 China added 30.8 GW of new wind energy capacity, around half the global total. The problem is that the slowdown in the Chinese economy, and consequently in energy demand, led to much of China's expanded new wind capacity being under-utilised. Another reason for this has been the 'curtailment' problem, where wind power is available but grid operators refuse to take the electricity, instead having a preference to accept power from other sources, such as fossil fuel plants. Curtailment and other grid connectivity issues has meant that around a sixth of China's wind energy capacity has been idle in recent years. This peaked at 17 percent in 2012, falling to 8 percent in 2014 but rising again to 15 percent by 2015 (Global Wind Energy Council/GWEC 2016). The 13th FYP (2016–2020) meanwhile further revised China's 2020 wind installed target from 200 GW set under the previous plan to 250 GW, and included programmes to promote high-tech development in the sector. Despite the above noted problems, this demonstrated the Chinese government's continued commitment to expanding wind energy, which is seen as a core element of its 'energy transition' strategy.

In the early 2010s, the Japanese government again returned to promote the wind energy sector. In the wake of the Fukushima nuclear disaster, in July 2012 it extended its FiT scheme to include wind and other renewables (not just solar). While new wind FiT rates were considered generous by international comparison[9] and relative to other renewable energy sectors in the same scheme, a loophole reportedly existed that allowed the *Denjiren* to refuse wind farms connection to the grid (Ushiyama 2012). There additionally remain strict regulations on using land in forest reserves, farmland and nature reserves, and environment laws protecting against bird strikes. Previously, a 2007 building code that classified wind turbines 60 metres high or taller as buildings had effectively paralysed the Japanese wind market for around a year due to compliance to the very complicated and time-consuming planning procedures involved. Although this process was later streamlined, in October 2012 a new law required more stringent environmental impact assessments of wind farms over 10 MW capacity, in effect applying to more or less every proposed new project. These assessments were expected to take around three to five years to complete and add an anticipated extra JP¥100 million (US$1.3 million) cost to each project investment plan.[10] In July 2015, the Japanese Ministry of Economy, Trade and Industry (METI) launched its 2030 Energy Mix Plan but only set a 10 GW (1.7 percent of total power generation) future target for wind energy (Global Wind Energy Council/GWEC 2016).

In sum, the Japanese government's predominant technology-oriented approach to renewable energy generally has meant it has provided more industrial policy support to the country's wind turbine production and export rather than energy policy support to support the rapid deployment of wind energy installations within its own territory. By way of comparison, Britain – another developed, densely populated but much smaller island nation – has now over four times the wind energy installed capacity than Japan after being at quite similar capacity levels just a decade ago (Table 13.2). Offshore installation could be – like Britain – the best future option for wind energy development for Japan as well as other densely populated, advanced Asian economies like South Korea and Taiwan.

## Infrastructural challenges

Almost all wind energy installations require connection with electricity grid infrastructure. This can often be a technical, financial and policy challenge. For example, most of China's best wind resources and large-scale wind farms (i.e. Wind Base plants) lie in remote areas far from the nation's main 'load centre' cities in the coastal and southern provinces – over half of the national wind capacity is located in Inner Mongolia, Gansu and northern Hebei provinces (Global Wind Energy Council/GWEC 2014). Thus, grid connectivity distances are great and the required investments in new grid infrastructure have been considerable. Zhang and Li (2012) summarise the infrastructural challenges of integrating China's now almost a thousand wind farms[11] across the country as follows: the uncoordinated development of capacity and power grids; lack of appropriate technical standards for integration; insufficient clarity over corporate responsibility for grid connection; and poor economic incentive structures for grid companies to use wind energy. Consequently, a significant level of China's installed wind energy is not actually delivering electricity to grid. Part of the problem is that investment in China's grid infrastructure has not kept pace with rapidly expanding power generation capacity generally, resulting in many power stations operating at low generation load levels. While in contrast Japan and most other Organisation for Economic Co-operation and Development nations have invested comparatively more in grid infrastructure than power generation since the late 1970s, the converse is true for China (Li *et al.* 2012). It is a predicament that applies equally to the grid's geographic coverage and transmission capacity: for example, the Northeast China grid network is based on a 500 kV system which has proved rather limited in the wind base zones (Zhao *et al.* 2012).

According to Luo *et al.* (2012), the main root of China's problem stems from the afore-mentioned restructuring of the energy sector in 2002 and its consequent subdivision of interests among power plant operators, electricity generation firms, grid companies, national government and local government. This created co-ordination difficulties on grid planning and construction, as well as various actors blaming others for not sufficiently investing in new grid infrastructure. Luo *et al.* (2012) also argue that a key reason for the almost uncontrollable expansion of wind farm development in China is due to the aforementioned 50 MW capacity approval rule. To evade coming under relatively stricter NDRC approval at this threshold capacity level, a large number of wind farm projects rated at 49.5 MW have emerged that only required comparatively looser local government approval, further weakening national government grip on the sector's development overall. The approval process for wind farms (especially smaller ones) has also hitherto been much quicker than for constructing new transmission lines, and wind farm developers willing to wait some months or even years for grid connectivity that has contributed to the capacity–generation gap.

A key part of the problem concerns policy incentives and strategic planning geared towards installation rather than power generation output, leading to 'excess' installed capacity problems. Zhang *et al.* (2013) contend that this approach was primarily driven by strategic industry policy motives in that setting ambitious targets on installed wind energy was conceived as an important driver for expanding China's turbine manufacturing base. Poor co-ordination between government agencies (urban, environment, industrial policy) on wind farm development was also clearly evident (Luo *et al.* 2012). Given that wind energy is a fast emerging and potentially very profitable industry in China, wind farm developers have been able to attract considerable investment. The lack of appropriate grid connectivity technical standards or codes specifically for wind farms, and poor compliance by operators where they do exist, has too meant an underuse of wind energy generated and bad management of the grid transmission system during high wind-speed periods, leading to power generation cut-outs in parts of the grid (Kang *et al.* 2012).

There have too been weak incentives for new grid infrastructure investment. In China's electricity generation industry, a planned market applies for base power needs and an open trade market for incremental power needs between producers and consumers. This affects how electricity is traded on an inter-provincial basis, and thus how wind energy generated in remote areas is sold to large load centres some distances away, as only incrementally produced wind energy is effectively tradable across the regions. Moreover, as there has been no national standard price formula for inter-provincial power transmissions, prices are determined by bilateral negotiations between provincial authorities, which have created significant transaction costs. By the early 2010s there was also still no mechanism to compensate grid companies for the inevitable power losses arising from long-distance transmissions, thus offering the grid companies little incentive for facilitating inter-provincial wind energy trade (Zhao *et al.* 2012).

The 12th FYP sought to address problems regarding China's grid infrastructure and regulatory gaps on wind energy. In 2012 the National Energy Association introduced the Wind Farm Development and Management Interim Rules and Regulation that stipulates wind farms must acquire complete approval from the authorities before commencing operation, and other measures that aim to strengthen quality control over installations. A new Safety Management of Wind Farms procedure has been brought into force, and the government is now encouraging development of wind farms closer to main load centres albeit in low wind-speed areas. Furthermore, a new grid code and 17 other technical standards have been introduced, and operators' use of low voltage ride through technology (Global Wind Energy Council/GWEC 2012). The central government also took more direct control of wind farm development within 12th FYP strategic plans, wrestling power away from local government. Investment totalling RMB3.8 trillion (US$580 billion) for 'strong and smart grid' development up to 2020 has additionally been implemented that should in time alleviate many of these grid connectivity problems (Global Wind Energy Council/GWEC 2013). However, for smart grids to perform optimally they should be allowed to operate on certain decentralisation principles and being reactive to consumer demand, which will require the Chinese power generation sector to become more open and competitive rather than staying monopolised by a few state-owned enterprises (Solidiance 2013).

Japan, South Korea and Taiwan face similar challenges to China regarding the exploitation of the nation's best wind resources, which are located in remote, sparsely populated areas far from main load centres. Around 70 percent of all three territories are mountainous, presenting certain technical and logistical challenges when developing wind farms in non-urbanised areas. A result of many wind energy developers having to build on hilly terrain has meant that installation costs are around twice the levels as the United States.[12] After the 2011 Fukushima disaster there has been growing political support for rationalising and expanding Japan's grid infrastructure to improve electricity trading among regions (Global Wind Energy Council/GWEC 2012). However, even with political backing it could take up to a decade before the nation's grid infrastructure was sufficiently extended into remote northern (Hokkaido and Tohoku) and southern (Kyushu) areas, where Japan's best wind resources are located. The government's aforementioned 2013 JP¥25.0 billion plan for new grid extension investment is perhaps the first important step in this direction.

## Institutional and socio-technical perspectives

Institutional support for wind energy development at the national level across Asia has generally strengthened over time as government policy towards the sector has strengthened. A growing number of national business associations, environmental organisations, research institutes and other relevant bodies have helped promote wind energy as a part of their agendas for a low carbon energy transition. International energy and climate-related institutions such as the International Energy Agency, International Renewable Energy Agency, Inter-governmental Panel on Climate Change and United Nations Framework Convention on Climate Change (UNFCCC) have together with the Global Wind Energy Council and miscellaneous business alliances (e.g. the Breakthrough Energy Coalition) are also strongly supporting wind energy development as part of their push on renewables generally. For example, the Clean Development Mechanism programme introduced in 2005 under the auspices of the UNFCCC's Kyoto Protocol helped fund 1,517 wind energy projects in China, and 805 projects in India, and around 40 others elsewhere in Asia by September 2013 (Dent 2014).

As well as having a conducive institutional environment for promoting wind energy development, there are also important socio-technical factors to consider. Generally speaking, most wind energy installations are wind farms, these being 'utility-scale' plants for electricity generation fed into a grid system. This is in contrast to the more 'distributed' nature of solar PV where a myriad of much smaller scale roof-top installations form the sector's aggregated whole. Moreover, solar PV panels have a much lower spatial profile and visible imprint than wind turbines, which as discussed earlier have gradually become larger and taller. The siting of wind farms near inhabited areas have become a socio-politically sensitive issue in Asia and globally. In the densely populated and liberal democratic states of Japan, South Korea and Taiwan there has been strong well-organised local protests against wind farm development. Furthermore, their governments have tough planning permission regulations to help protect the already scarce publicly accessible countryside. In South Korea, it has been reported that only a quarter of wind farm proposals are approved for this reason (Global Wind Energy Council/GWEC 2012). There has, though, been evidence of local opposition to wind farm projects waning in the aftermath of the 2011 Fukushima nuclear disaster (Mizuno 2014). Future advances in small wind turbine technology aimed at integrating devices into buildings and other constructions would pose a different set of socio-technical challenges yet also make the public more direct stakeholders in the sector akin to the solar PV 'prosumer' (i.e. simultaneously producer and consumer) development over the last few decades. The small-scale application possibilities offered by renewable energy technologies generation pose the prospect of a 'new energy societies' being created in the low carbon revolution (Dent 2014).

## Conclusion

Asia is the most carbon-intensive part of the planet. This is not surprising given it continues to be by far the world's most populous region or continent, and its burgeoning twin processes of industrialisation and urbanisation. Over recent years and decades, Asian societies have consumed increasing levels of energy and this trend shows little sign of abating. The challenge is, then, to promote renewable and other low carbon energy development that will not just keep pace with Asia's energy consumption growth but at an even faster pace than this, if the region is to make an effective and successful transition to decarbonised energy and economic systems. Asian countries have a huge impact on global carbon emissions and thereby climate change, and most are also highly susceptible to climate change risks. Wind energy is now a mainstream energy sector that has attracted rising levels of public and private investment, and has grown at around 20 percent annually during the last decade. It has a key role to play in Asia and the world's low carbon future.

This chapter has examined the development of Asia's wind energy sector against the backdrop of its global development. It was shown how Asia only became a major player in the industry in the early 2000s but that it now enjoys an ever dominant position. China in particular has driven Asia's development, now accounting for over 80 percent of the region's installed capacity, and a third of the world total. The country has in addition become a major manufacturing hub for wind turbine assembly and component production, with five of the world's top ten turbine companies. Although India has too made significant investments in the sector, and now has the world's fourth highest installed capacity level, other large Asian economies – most notably Japan, South Korea, Taiwan and Indonesia – have fallen somewhat behind in installed power generation capacity from both a global and regional perspective. However, it should be noted that certain Japanese and Korean companies have become significant wind turbines producers and technology developers. Nonetheless, the pattern of Asia's wind energy development is highly asymmetric and ever more skewed towards China.

The factors behind Asia's successes, asymmetries and barriers to wind energy development were explored under four different thematic headings: technological and industrial; government policy and strategy; infrastructural challenges; institutional and socio-technical perspectives. To summarise, high rates of techno-innovation and market optimism concerning wind energy have made it a highly dynamic sector. However, in China the rapid expansion of wind energy installations has caused various problems and challenges. The country is developing wind farms at a very fast rate but large proportions of their capacity are not being absorbed by grid operators. As discussed, this has been as much a policy challenge as a technical infrastructural one, and where tensions between different energy sector stakeholders persist. Other forms of resistance to wind energy development in Asia is evident. In Japan, which in contrast to China and India has a relatively weak industry lobby for wind energy, the country's ten regional power companies (*Denjiren*) have reluctantly engaged with 'intermittent' wind power, even after the 2011 Fukushima nuclear disaster. In South Korea and other Asian nations, local opposition to proposed onshore wind farms and tough planning permission regulations have proved a notable obstacle to the sector's development. Yet, public support for cleaner energy in Asia is growing, especially in cities with acute air pollution problems. This is increasingly where Asia's population is concentrated. Furthermore, most Asian countries have signed the December 2015 COP21 Paris Agreement on climate change as well as implemented green energy and industrial policies, committing them to decarbonise their economic development as the 21st century progresses. Wind energy is currently one of the most important options for Asian and other countries to fulfil these obligations, and help create a cleaner and more sustainable energy future for all.

## Notes

1  Dispatchable sources of electricity generation are those able to immediately respond to fluctuations in energy demand from power grids. This is not possible for renewables such as wind, solar and ocean energy given their generation intermittencies stemming from their dependence on variable meteorological phenomena such as wind flows and sunshine levels. However, hydropower, geothermal and bioenergy are not faced with this problem due to them offering constant energy streams like fossil fuels and nuclear.

2  Gigawatt. 1,000 kilowatts (kW) equal 1 megawatt (MW), and 1,000 MW equals 1 GW.

3  China Shipbuilding Industry Corporation, an example of how companies with relevant engineering competences have diversified into the wind energy sector.

4  *Clean Techia News*, 1 September 2016: https://cleantechnica.com/2016/09/01/pakistan-will-add-1-gw-wind-energy-capacity-2018/

5 These figures were 185.6 terrawatt-hours for the US and 185.1 terrawatt-hours for China.

6 *Business Green news*, 16 December 2015: www.businessgreen.com/bg/news/2439362/samsung-7mw-turbine-set-to-become-offshore-training-hub

7 In 1980, the typical cost range of wind energy generated electricity was US$0.60 to US$0.70 p/kWh but by 2011 had fallen to as low as US$0.05 p/kWh (IEA 2004; REN21 2012). Taking an illustrative example from China, operating generation tariffs in large-scale wind farms in Hebei and Xinjiang provinces were 3,850 yuan per MW in 2011, which compared to 6,200 yuan per MW capacity just a few years earlier in 2008.

8 A feed-in tariff (FiT) scheme normally involves government tariffs paid to producers both the generation or consuming of renewable energy electricity and/or supplying excess amounts back to the grid.

9 *Sun Wind Energy News*, 21st June 2012 [www.sunwindenergy.com/news/japan-feed-tariff-scheme-confirmed, accessed on 8 September 2016].

10 *Bloomberg News*, 30th March 2012 [www.bloomberg.com/news/2012-03-29/floating-windmills-in-japan-help-wind-down-nuclear-power-energy.html, accessed on 8 September 2016].

11 Windpower.net (2016) reported that by September 2016 China had 988 operational or under development wind farms, the average size of these farms being many times larger than in Germany, the US and other large producer countries.

12 *Recharge News*, 21st March 2012 [www.rechargenews.com/regions/asia_pacific/article307337.ece, accessed 7 September 2016].

# References

Bloomberg New Energy Finance/BNEF (2011) *The Geopolitics of Clean Energy*, West Hartford: BNEF.

Bloomberg New Energy Finance/BNEF (2016) *2015 Global Wind Turbine Market Shares Report*, West Hartford: BNEF.

China National Renewable Energy Centre/CNREC (2013) *China Renewable Energy*, Vol 2(1), Beijing: CNREC.

Dent, C.M. (2014) *Renewable Energy in East Asia: Towards a New Developmentalism*, London: Routledge.

Fleming, P.D. and Probert, S.D. (1984) 'The Evolution of Wind Turbines: An Historic Review', *Applied Energy*, Vol 18(3), pp 163–177.

Gipe, P. (1991) 'Wind Energy Comes of Age: California and Denmark', *Energy Policy*, Vol 19(8), pp 756–767.

Global Wind Energy Council/GWEC (2007) *Global Wind Report 2006*, Brussels: GWEC Secretariat.

Global Wind Energy Council/GWEC (2009) *Global Wind Report 2008*, Brussels: GWEC Secretariat.

Global Wind Energy Council/GWEC (2012) *Global Wind Report: Annual Market Update 2011*, Brussels: GWEC Secretariat.

Global Wind Energy Council/GWEC (2013) *Global Wind Report: Annual Market Update 2012*, Brussels: GWEC Secretariat.

Global Wind Energy Council/GWEC (2014) *China Wind Power Report and Outlook*, Brussels: GWEC Secretariat.

Global Wind Energy Council/GWEC (2016) *Global Wind Report 2015*, Brussels: GWEC Secretariat.

Harborne, P. and Hendry, C. (2009) 'Pathways to Commercial Wind Power in the US, Europe and Japan: The Role of Demonstration Projects and Field Trials in the Innovation Process', *Energy Policy*, Vol 37, pp 3580–3595.

Inoue, Y. and Miyazaki, K. (2008) 'Technological Innovation and Diffusion of Wind Power in Japan', *Technological Forecasting and Social Change*, Vol 75, pp 1303–1323.

International Energy Agency/IEA (2004) *Renewable Energy: Market and Policy Trends in IEA Countries*, Paris: IEA.

International Renewable Energy Agency/IRENA (2012) *30 Years of Policies for Wind Energy*, Masdar City: IRENA.

Kaldellis, J.K. and Zafirakis, D. (2011) 'The Wind (R)evolution: A Short Review of a Long History', *Energy Policy*, Vol 36, pp 1887–1901.

Kang, J., Yuana, J., Hud, Z., and Xu, Y. (2012) 'Review on Wind Power Development and Relevant Policies in China During the 11th Five-Year-Plan Period', *Renewable and Sustainable Energy Reviews*, Vol 16, pp 1907–1915.

Lee, B., Iliev, I., and Preston, F. (2009) *Who Owns Our Low Carbon Future?: Intellectual Property and Energy Technologies*, London: Chatham House.

Li, J. (2010) 'Decarbonising Power Generation in China: Is the Answer Blowing in the Wind?' *Renewable and Sustainable Energy Reviews*, Vol 14(4), pp 1154–1171.

Li, X., Hubacek, K., and Siu, Y.L. (2012) 'Wind Power in China: Dream or Reality?', *Energy*, Vol 37, pp 51–60.

Liu, Y. and Kokko, A. (2010) 'Wind Power in China: Policy and Development Challenges', *Energy Policy*, Vol 38, pp 5520–5529.

Luo, G.L., Zhi, F., and Zhang, X. (2012) 'Inconsistencies Between China's Wind Power Development and Grid Planning: An Institutional Perspective', *Renewable Energy*, Vol 48, pp 52–56.

Maruyama, Y., Nishikido, M., and Iida, T. (2007) 'The Rise of Community Wind Power in Japan: Enhanced Acceptance Through Social Innovation', *Energy Policy*, Vol 35, pp 2761–2769.

Mizuno, E. (2014) 'Overview of Wind Energy Policy and Development in Japan', *Renewable and Sustainable Energy Review*, Vol 40, pp 999–1018.

National Development and Reform Commission/NDRC (2011) *China's 12th Five-Year Plan for Economic and Social Development*, Beijing: NDRC.

REN21 (2012) *Renewables 2012 Global Status Report*, Paris: REN21 Secretariat.

REN21 (2014) *Renewables 2014 Global Status Report*, Paris: REN21 Secretariat.

REN21 (2016) *Renewables 2013 Global Status Report*, Paris: REN21 Secretariat.

Roland Berger (2010) *From Pioneer to Mainstream: Evolution of Wind Energy Markets and Implications for Manufacturers and Suppliers*, Berlin: Roland Berger Consultants.

Ru, P., Zhi, Q., Zhang, F., Zhong, X., Li, J., and Su, J. (2012) 'Behind the Development of Technology: The Transition of Innovation Modes in China's Wind Turbine Manufacturing Industry', *Energy Policy*, Vol 43, pp 58–69.

Solidiance (2013) *China's Renewable Energy Sector: An Overview of Key Growth Sectors*, Shanghai: Solidiance.

Ushiyama, I. (1999) 'Wind Energy Activities in Japan', *Renewable Energy*, Vol 16, pp 811–816.

Ushiyama, I. (2012) 'Wind Power Development in Japan', presentation at the Japan Wind Energy Association, 9th March.

Windpower.net (2016) *Wind Energy Database*, Buc: Windpower.net (www.thewindpower.net)

Wiser, R. and Bolinger, M. (2010) *2009 Wind Technologies Market Report*, Washington DC: US Department of Energy.

World Wind Energy Association/WWEA (2012) *Quarterly Bulletin*, Vol 2(June), Bonn: WWEA Secretariat.

World Wind Energy Association/WWEA (2013) *Quarterly Bulletin*, Vol. 2(June), Bonn: WWEA Secretariat.

Xia, C. and Song, Z. (2009) 'Wind Energy in China: Current Scenario and Future Perspectives', *Renewable and Sustainable Energy Reviews*, Vol 13, pp 1966–1974.

Zhang, S. and Li, X. (2012) 'Large Scale Wind Power Integration in China: Analysis from a Policy Perspective', *Renewable and Sustainable Energy Reviews*, Vol 16, pp 1110–1115.

Zhang, S., Andrews-Speed, P., and Zhao, X. (2013) 'Political and Institutional Analysis of the Successes and Failures of China's Wind Power Policy', *Energy Policy*, Vol 56, pp 331–340.

Zhao, X., Zhang, S., Yang, R., and Wanga, M. (2012) 'Constraints on the Effective Utilization of Wind Power in China: An Illustration from the Northeast China Grid', *Renewable and Sustainable Energy Reviews*, Vol 16, pp 4508–4514.

Zhou, Y., Zhang, B., Zou, J., Bi, J., and Wang, K. (2012) 'Joint R&D in Low-Carbon Technology Development in China: A Case Study of the Wind Turbine Manufacturing Industry', *Energy Policy*, Vol 46, pp 100–108.

# 14

# HYDROPOWER IN ASIA

*Arthur A. Williams*

## History and growth

Asia is a continent with diverse physical geography and climate. The key resources for hydropower (i.e. mountains and precipitation) are available in wide areas of the continent. Asia is almost certainly the continent from which hydropower technology originated in the form of horizontal axis *noria* employed in Persia (Iran) and recorded by Lucretius in first century BC (Reynolds, 1983), and vertical axis paddle wheels in China which were widespread by $2^{nd}$ century AD (Siebert, 2012). A recent historian, Needham, suggested India as the common source for each of these inventions, but the evidence is unclear, especially given that the devices are completely different and therefore likely to be from two separate sources.

In its modern form, as a technology for generating electricity, hydropower is well established as a renewable energy resource which contributes around 16% of all electricity production across the continent of Asia. The chart in Figure 14.1 shows how hydropower production in Asia has increased over the 30 years up to 2015. This is based on data from BP (2016), cross-checked against those from International Energy Agency (2015). However, in either case, the data for Asia as a whole has been resolved based on additional information, for example taking into account that 70% of hydropower capacity in the Russian Federation is in Asia, and only 30% in Europe.

A clear observation from the chart is the dominance of China as the main location for new hydropower plants. During the first 15 years of the new millennium, hydropower production in the rest of Asia grew by 50%, which is an increase of nearly 3% per year. Nevertheless, in China, the growth in the same period was by a factor of 5, equivalent to 11% per year. For many of these years, China has added more hydropower capacity than the rest of the world combined (See Table 14.1). Most of this increase has been through the completion of large hydropower projects, including the largest power plant in the world at Three Gorges (22.5 GW capacity).

Other Asian countries with more than 4 GW of hydropower capacity are Taiwan, Thailand, Laos, North Korea and Myanmar. The capacity factor gives the ratio between average power production and full capacity. This varies according to the design of the scheme and the variation of water flow through the seasons. It tends to be higher in countries with all-year rainfall or those with snow-fed river systems. The capacity factor is also affected by the timing of the data given, particularly for

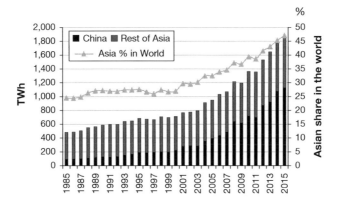

*Figure 14.1* Hydropower production in Asia (TWh)

*Data source*: BP (2016).

*Table 14.1* Hydropower data for key Asian countries in 2015

| Country | Hydropower production (TWh/yr) | Hydropower capacity (GW) | Hydropower as % of total electricity supply | Hydropower capacity factor (%) | Growth in production 2000–2015 (p.a.) (%) |
|---|---|---|---|---|---|
| China | 1126 | 319 | 19 | 40 | 16 |
| Russia | 167 | 51 | 16 | 38 | 0.2 |
| India | 124 | 52 | 10 | 28 | 4 |
| Japan | 97 | 50 | 9 | 22 | 0.1 |
| Turkey | 67 | 26 | 26 | 29 | 6 |
| Viet Nam | 64 | 15 | 39 | 48 | 17 |
| Pakistan | 35 | 7 | 31 | 54 | 4 |
| Iran | 18 | 11 | 6 | 19 | 6 |
| Indonesia | 16 | 5.3 | 7 | 35 | 7 |
| Malaysia | 15 | 5.5 | 10 | 31 | 8 |
| Philippines | 10 | 4.2 | 12 | 27 | 4 |
| South Korea | 3 | 6.4 | 1 | 5 | –2 |

*Data sources*: Author based on International Hydropower Association (2016); BP (2016); International Energy Agency (2015).

countries with high growth rates. An Indonesian government agency, for example, stated an expected hydro capacity of 8.7 GW for the end of 2015, a much higher value than that shown above given by International Hydropower Association (IHA). The potential for the whole country is stated as 75 GW and there are plans by the state electricity company PLN for massive growth (AAAT (Agency for the Assessment and Application of Technology), 2015). Indonesia has widespread hydropower potential, and some parts of the country (e.g. West Papua) with less than 30% electrification rates. Currently the main focus is on meeting the expanding demand in more developed parts. For example, there is 500 MW currently under construction in Aceh and North Sumatra. The largest of these plants is Asahan III, with a capacity of 174 MW, due for completion in 2019.

Economically exploitable potentials for key Asian countries are shown in Table 14.2. Only two countries, Japan and South Korea, have already developed most of their potential

*Table 14.2* Hydropower potential and scope for development in key Asian countries in 2015

| Country | Exploitable potential (TWh/yr) | % Developed |
| --- | --- | --- |
| China | up to 2400 | 47 |
| Russian Federation | 852–1700 | 10–20 |
| India | 442 | 29 |
| Indonesia | 260 | 6 |
| Tajikistan | 264 | 4 |
| Turkey | 170 | 39 |
| Nepal | 140 | 3 |
| Japan | 135 | 72 |
| Bhutan | 95 | 8 |

*Data sources*: Author based on International Hydropower Association (2016); BP (2016).

hydropower resources. Other countries still have considerable scope for expansion, for example India has developed less than one-third of its "economically feasible potential". Many other countries in South and South-east Asia have high proportions of untapped hydropower potential. Nevertheless, there is considerable debate about whether this is a good strategy for sustainable development, taking into account the impacts, as discussed in the "Impacts of Hydropower" section. Nepal is currently looking for new investment, having recently suffered from disruption to its power development due to an earthquake (Awale, 2016). Meanwhile, Bhutan has implemented a longer term plan to develop its hydro resources in order to export power (Tshering & Tamang, 2004). Further information on current developments can be found in the Renewable Energy Policy Network for the 21st Century (REN21) (2016) Global Status Report.

The variation of hydropower output due to changing weather conditions is one of its limitations as a reliable source of electricity. It is noticeable in Table 14.1 how power production can vary widely from one year to another, even in large countries with diverse geography. From 2010 to 2011, the production in China dropped from 722 TWh to 699 TWh despite additional capacity being added to the grid. In Japan, which has had very little change in capacity over the last 15 years, the production of electricity from hydropower varied from 72.5 to 96 TWh/yr. The percentage variation in South Korea, which has the highest concentration of large dams in Asia, has been even greater. In 1987 hydropower production was greatest at 5.35 TWh, whereas in 1994 and again in 2001 it was less than 2.35 TWh. Although hydro capacity has not decreased, Table 14.1 shows a reduction in output due to the limitations of water available to run the plants when comparing 2000–2015 to the previous 15 years.

A similar pattern can be seen for Russia, but in this case there is an additional reason for the limited output from hydropower plants from 2010 to 2012. In August, 2009 RusHydro's largest hydroelectric power plant – Sayano-Shushenskaya in Khakassia (South-West Siberia) suffered a catastrophic turbine failure which resulted in partial destruction of the power house, with the loss of 75 lives. The accident was due to a combination of poor maintenance and additional stress put on the turbine. At the time of the accident, the turbine set that failed was being controlled to adjust output in response to grid demand, using a control system that had not been fully tested. The whole plant, consisting of ten 640 MW units, was out of action for several months and took several years to fully repair. In 2015, an additional transmission line was completed to the site to enable it to deliver up to 5100 MW.

The events at Sayano-Shushenskaya highlight concerns about the potential impact of large hydropower plants when the plant is not adequately managed throughout its lifetime. The first units at the plant were installed in 1978, so were towards the end of their expected 30-year lifetime. Many of the bolts holding down the turbine casing should have been replaced due to fatigue cracking, but inspection, monitoring and maintenance regimes proved to be inadequate. The long lifetime of hydropower plants is one of their advantages over other sources of power, but alongside this is a requirement for adequate inspection and maintenance over several generations.

## Impacts of hydropower

Although providing much needed electricity with negligible carbon emissions, there have been significant impacts from the construction of large hydropower projects, which have been clearly elaborated in the report of the World Commission on Dams (2000). Impacts can be broadly divided into three types, according to the source of the impact. There are those due to the dam itself, those due to the land flooded by the reservoir and those due to the process of construction. A dam creates a barrier across the natural river flow, blocking and controlling the natural flow and thus changing the river ecosystem. The reservoir inundates land that may include people's homes, farms, forest and other natural ecosystems or existing infrastructure. The construction process has impacts due to the dust and transport of materials, but also due to the setting up of new infra-structure and the influx of a workforce that may outnumber the local population. As with all large energy projects, developers of new schemes are required to produce a detailed environmental impact assessment, which will include the various aspects outlined above. However, environ-mental impact assessments are governed mostly by national legislation, which means that the scope of the assessment will rarely consider impacts beyond the border of the particular country.

Some of the impacts of large dams may be predicted and to a certain extent mitigated against, but the sheer size of such schemes makes it difficult to predict all of the impacts and to provide adequate mitigation or compensation for displaced populations. For the Three Gorges project, over a million people were displaced from their homes, and whole cities relocated. A review of hydropower in China at the time that Three Gorges had been newly commissioned, emphasises the positive contribution of hydropower as a low-cost and low-carbon resource (Chang et al., 2010). While it mentioned some of the negative impacts of such large hydropower projects, and acknowledged the debate about these projects, it concluded that hydropower is a sustainable resource which should be further promoted and developed. In contrast, many other evaluations of the impacts of the Three Gorges after completion of the plant demonstrate that valid criticisms have been raised during the development stage, especially with regard to changes in water quality (Magee, 2014). Furthermore, one impact had not been identified before construction (Stone, 2011) perhaps because the potential for increased water-borne diseases from large reservoirs was thought to be a problem confined to tropical regions. More recently, a planned 1 GW hydro-power plant at Xiaonanhai on a tributary of the Yangtze was refused planning permission on the basis of environmental concerns (Harris, 2015b).

There are genuine concerns about planned further expansion of hydropower in Asia, especially within environmentally sensitive regions such as the Mekong River Basin. This is one of the largest rivers in South-east Asia, which rises in China, but then flows south, impacting on populations in Myanmar, Laos, Cambodia, Thailand and Viet Nam. The recommendations of a strategic environmental assessment compiled by International Centre for Environmental Management (2010) were to defer implementation of dams that would cut across the whole river for ten years and in the meantime to investigate alternatives, taking into account the needs for electricity production alongside the needs for sustainable river ecology and impacts on large rural populations dependent on the river for their food and livelihoods.

For many years, papers have warned about the difficulty of mitigating the impacts of dams across the Mekong, due to the vast quantities of fish involved in migrations, which in turn affect the livelihoods of millions of people (Jensen, 2001). Also, there are other significant ecological impacts of dams that are not so easily understood or avoided. These include the long-term changes due to alteration of water flows and temperatures, together with reduced sediment in flows downstream of a dam. In March 2016, China allowed additional water to flow into the drought-affected lower Mekong in advance of a meeting of the Mekong River Commission. Although an important action at the time, this also demonstrates the control that can be wielded by the owners of hydropower dams in the upper reaches of such an international river.

A key environmental impact that may not be included in standard environmental assessments is the greenhouse gas emissions from hydropower plants, which are often assumed to be negligible. The main source of emissions is the decomposition of trees that were flooded in the reservoir bed. Although the exact flows of carbon in river and forest systems are difficult to measure, it is clear that where reservoirs flood tropical forest, there are significant levels of methane which build up in the lower levels of the reservoir and are then dissipated from spillways and powerhouse, particularly in the first few years after flooding the area (Barros et al., 2011). Methane has around forty times the global warming potential of carbon dioxide, so relatively small carbon flows can have a big effect. Although much of the research in this context has been done in South America, it is equally relevant to South-east Asia where similar climate and vegetation exist. If taken properly into account, this will require reconsideration of the role to be played by hydropower in providing low-carbon electricity and meeting carbon reduction targets.

Another potential impact of dams that has been the subject of considerable debate is the topic of induced seismicity, where earthquakes have occurred close to recently filled large reservoirs. Two sites in Asia are of particular interest. One is at Koyna in India, a site in a low-risk area for seismic events. Three years after the completion of the dam, which started filling the reservoir of 2780 million m$^3$, an earthquake of magnitude 6.5 was recorded in 1967. In more recent times, the 2008 earthquake in Sichuan, China, was also linked by some with the construction of the Zipingpu Reservoir, which was filled between 2004 and 2006. This dam is situated 500m from a fault line which moved during the earthquake, though the epicentre was 10 km from the dam and 5.5 km below the surface. A paper by Li et al. (2011) discusses the evidence in the light of geological characteristics and analysis of aftershocks. They established a potential link and concluded that further research was needed to understand more.

Finally, the process for development of large hydropower projects has been reviewed and the outcomes compared with predictions and promises made by the project planners. Sovacool has investigated a number of projects, among them a scheme in east Malaysia. This study highlights how such projects can be implemented at the expense of environmental degradation and disruption of the livelihood and social structure for the local population, whose needs are overridden in the interests of "development" (Sovacool & Bulan, 2011).

With such large projects, decisions are often made on the basis of the overall cost-benefit ratio, but the projected costs and benefits may not be matched by real costs and benefits once the project has been implemented. In a recent study of such projects worldwide, Ansar et al. (2014) investigated factors affecting time and cost over-runs. Of the 247 projects in their study, which had a mean output of 500 MW, 40% were in Asia. Three-quarters of these plants had cost and time over-runs (with a significant correlation between the two). Projects with the worst record tended to be in countries with low gross domestic product per person – i.e. there is a high probability of a mismatch between initial estimates and actual delivery where the country has limited resources and infrastructure. For such countries, implementation of smaller scale schemes tends to be more appropriate. Some examples of how small and micro-hydropower has been used to benefit rural populations are described later in this chapter.

## Hydropower and pumped storage – supporting other renewable energy on the grid

Large hydropower plants with reservoir storage have important features in supporting the overall electrical grid, because they are able to start and change power output rapidly in order to follow changes in overall demand. They are also capable of initiating generation without an external power supply, for example in the case of a power failure, which is known as "black-start" capability. Pumped hydro, which is mainly a means of energy storage rather than energy generation, is particularly flexible in this respect. When sudden increases of demand occur, pumped hydro plants are capable of starting up and feeding the network within about 30 seconds. They can start up equally rapidly in pumping mode, absorbing excess power from the grid. Historically, the main purpose of pumped hydro, was to carry out load levelling, to balance supply and demand on the power network. Demand from industrial activities and consumers is in continuous fluctuation, so that there is a need to balance supply with demand within periods of minutes, as well as hour-by hour through the day. Pumped hydro is the main form of energy storage on the grid, which can reduce the peaks of demand by storing the energy excess when demand is low (usually at night), and feed the grid when there is insufficient supply from conventional generation (Figure 14.2).

Japan has had for a long-time, a very significant level of pumped hydro storage. It is currently the country with the largest pumped storage capacity in the world, but may be soon overtaken by China. However, in relation to the size of the grid, Japan has much greater storage capacity, representing 24% of average grid load. This played an important role in relation to the large generation from nuclear plants, which are best run day and night. Following the reduction in nuclear power production after the Fukushima accident, pumped storage plants play an important role in balancing increasing renewable energy, especially solar power. In a market-based electricity supply system, pumped storage has the potential to play a significant role within the balancing operations, as explained by Barbour et al. (2016). However, unless the market is set up

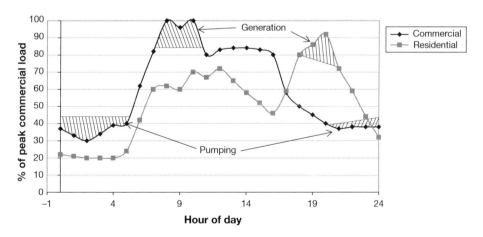

*Figure 14.2* Load levelling by pumped hydro, based on a typical daily load profile in China

*Data source*: Author, using daily load data for China from Yang and Fan (2014).

to reward the delivery of such "grid services" it is difficult for privately owned pumped storage plants to generate revenue from them. These services provide an opportunity to avoid costs elsewhere in the grid system, but those avoided costs will only benefit the grid operator unless there is a specific market mechanism to provide an incentive for other plant operators to invest. In Asian markets, there is generally a strong role played by the grid operator and national government in setting incentives. The development of new pumped storage capacity is partly a reflection of the various mechanisms that have been established to reward pumped hydro operators. Since pumped hydro is a very long-term investment, long-term stability in revenues is essential to justify expansion of capacity.

Six Asian countries are in the top twenty countries worldwide for pumped storage capacity; they are listed in order of decreasing capacity in Table 14.3. Although Russia has five pumped storage plants, these are all in Europe, where also the main electricity consumption is located.

Over the last decade, with the development of other renewable energy generation, pumped hydro has an even greater role to play. Wider use of intermittent energy sources, such as solar and wind, causes large swings in generation which are not easy to predict. Pumped hydro can allow greater integration of these sources, without the need for curtailment when production exceeds grid demand. The US Department of Energy holds a database of energy storage projects worldwide, which includes interesting information about the purpose of the energy storage project. In the case of pumped hydro projects, the database in 2016 listed fourteen projects that had previously been shown as "under construction" in Asia, adding another 15 GW (30%) to the previous total. Other countries are also planning development of pumped storage capability, including Nepal where a 140 MW plant is planned at Tanahu, 100 km from the capital Kathmandu (Harris, 2015a).

Apart from helping to integrate the renewable sources into the grid, pumped storage has other roles in maintaining the grid stability and power quality. Many of the more recently commissioned schemes are listed as contributing to real-time load levelling, frequency regulation and voltage control. This shows how the role of pumped storage is being expanded alongside the changes to the grid due to integration of other renewable energy sources. One technology that is particularly helpful in this respect is the use of variable-speed motor-generators. This was an innovation from Japan that was tested at a conventional hydro plant at Narude in 1981, but has since been implemented at three further Japanese plants, at least two plants in Europe, and will be a feature of the 1 GW plant currently being constructed at Tehri in India.

*Table 14.3* Pumped storage hydropower by country

| Country | No. of pumped storage plants | Rated power of pumped storage (GW) | Ratio of storage power to average supply (%) | Mean capacity of pumped storage plants (MW) | Largest pumped storage plant capacity (MW) |
|---|---|---|---|---|---|
| China | 34 | 32.0 | 5 | 941 | 2448 |
| Japan | 44 | 28.7 | 24 | 652 | 1932 |
| India | 8 | 5.1 | 4 | 644 | 1670 |
| S. Korea | 7 | 4.7 | 8 | 671 | 1000 |
| Taiwan | 2 | 2.6 | 9 | 1300 | 1600 |
| Thailand | 3 | 1.4 | 7 | 467 | 720 |
| Iran | 1 | 1.0 | 3 | 1040 | 1040 |
| Philippines | 1 | 0.7 | 7 | 709 | 709 |

*Data source*: Data from US DoE (2016) and BP (2016).

For frequency control, which depends on the real-time matching of total supply and demand on the grid, speed governors are used for each generator set. These regulators operate differently depending on whether it is a fixed-speed pumped hydro or a variable speed one. In conventional pumped hydro in which the speed does not vary, the frequency control can be performed by varying the flow through the guide vanes (or wicket gates) to control the fluid power. This operation could last about 10 s and is normally only possible in generating mode (Lung et al., 2007). In contrast, a plant with variable-speed operation can respond to frequency changes in a much shorter period of time (150 ms), taking advantage of the rotational inertia of the pump-turbine. Variable-speed generator technology is commonly used in large wind turbines, with various options for the generator and converter. One of the most popular configurations is the Doubly-fed Induction Generator, which has the advantage that the frequency converter only switches rotor currents, not the full power of the generator, so is less costly. However, the standard configuration used in wind turbines of up to 5 MW is not suitable for pumped storage plants with powers of 100–300 MW per unit. Pumped storage plants use vector control of the rotor field, which limits the speed variation to $\pm 10\%$ or less. As with wind turbines, there is now a move towards full-scale converters with synchronous generators, which completely isolate the generator and grid frequencies, enabling a much larger speed range. One plant of this type has been operating at Grimsel in Switzerland (Schlunegger & Thöni, 2013) and there is now research into this technology in Asia, too (Chang et al., 2015). Such plants are able to produce controllable reactive power and therefore contribute to grid voltage regulation more effectively, which is one of the reasons that they are commonly used in the largest (5–8 MW) wind turbines. However, unlike a wind turbine, the input power to a hydro plant is predictable and controllable, so variable-speed pumped storage units can play a vital balancing role where large numbers of wind farms are supplying the grid.

A further advantage of using variable speed for pumped hydro plants is to increase the range of head for efficient operation. With fixed speed, the efficiency drops off either side of optimum head, which means that there is a limited change of depth in the reservoir. With variable speed, a small but deep reservoir can be used (for which sites are more available in mountainous areas) and efficiency maintained at a high level. In all the plants that use variable speed, there are a number of parallel units, and some of them operate at fixed speed. This reduces costs but enables the plant to retain its black-start capability.

Another innovation for pumped storage in Japan was the implementation of a coastal plant at Yanburu, Okinawa using seawater and a cliff-top reservoir constructed for the scheme. This demonstrated the feasibility of this technology at medium scale (30 MW). Although no further plants have been built in Asia yet, this is a technology that could have wider use because it does not depend on availability of specific site conditions in a mountainous area.

## Small hydropower in Asia

The terms "small", "mini", "micro" and "pico" hydropower have been used to describe projects of descending size from the largest ones. However, the definitions of these terms is relatively loose, even within the context of the Asian continent, as explained by Tong et al. (1996). There are two main reasons behind the delineation of these smaller projects from larger ones, despite the grey area in the definitions. One is that small hydropower, and especially micro and pico hydropower, have been used to describe plants set up with the aim of boosting energy resources in rural areas, particularly in more remote parts, beyond the reach of national electricity grids. Asia has been the main centre for the successful development of such projects, which have brought electrical power to remote towns and villages through the use of local grids, as in China and

Nepal, or through dissemination of household-size units, as in Viet Nam. A second reason is that in several respects, small hydropower avoids some of the problems of environmental and social costs associated with large hydropower projects. Firstly, as noted above, the benefit of the plant is for the local population. Secondly, and particularly for the smallest scale schemes (micro, usually defined as <100 kW), the plants are of the "run-of-river" type, i.e. with limited storage of water, resulting in lower impacts on downstream flows and overall river ecology.

Although Europe and North America have a long history in the development of smaller hydropower projects, Asia is now the continent which makes most use of this technology. By 2012, 61% of all hydropower plants below 10 MW were in Asia (a total of 46 GW) with an average annual growth in capacity since 2003 of nearly 4% (Liu et al., 2013). As with large hydro, a significant part of this is in China, where the government developed a specific policy to use hydropower for rural electrification during the 1980s. According to Tong et al. (1996), more than one-third of 2300 "counties" in China could be electrified mainly on the basis of small hydropower. Liu et al. (2013) state the potential for economic development of small hydropower in each region of the world. It is still very large in Asia as they estimate only 40% has been exploited so far, which contrasts with over 85% in North America and Western Europe. The countries with largest untapped potential include China, Turkey, India, Nepal and Viet Nam.

In several Asian countries, as the grid infrastructure has developed, isolated hydropower plants have later been connected into the grid. Depending on the size of the plant, the local loads that it meets, and the reliability of the grid, different options are possible for this connection and for the control of the plant. Legal barriers have sometimes prevented community or privately owned mini hydropower plants from selling surplus power into the grid, or even to connect into the grid at all. These barriers have generally been reduced over the last 20 years, with the acceptance in many Asian countries of the role of independent power providers and the development of standard grid codes alongside smarter control equipment. An interesting example of this is in Sri Lanka, where hydropower was traditionally used to supply remote tea estates, but has also been developed for village electrification. Around 5% of all electrical power in Sri Lanka is supplied from small hydro power plants. Atputharajah (2011) describes five different options for grid connection of such schemes ranging from occasional use for back-up through to permanent supply, depending on the level of the local loads, the requirements of the grid and the contract between the independent supplier and the grid operator.

For isolated schemes, cost reduction has been possible through standardisation of equipment and of design procedures (see Williams & Simpson, 2009). A good example of this is the use of direct-drive Pelton turbines, with fixed nozzles and made to a specific range of runner diameters. These are sold as a complete unit with an induction motor used as a generator and an electronic "induction generator controller" based on IGBT technology (Smith, 1996). In Nepal these are commonly sold as "Peltric sets", which have wide uptake due to their price, as low as $600/kW (Bhattarai & Joshi, 2011). Others have tried to improve on this combination of turbine and generator using more complex schemes, but these are often not cost-effective, because they lose the savings due to use of standard equipment. Atputharajah suggests that C-2C connection requires 50% de-rating of the generator, which might be correct if all 3-phase currents in the generator need to be exactly balanced. However, since a small imbalance is normal for most three-phase generators, the actual limit is the current loading in the windings, which enables a de-rating of only 20% to be employed, making the C-2C connection effective and economic. Nevertheless, developments to the electronic load controllers that are used in micro-hydro plants have potential for improving reliability, reducing costs and (important for Nepal, where equipment has to be brought to many sites by human porters) reducing weight.

When discussing the environmental and social impacts of hydropower, for rural electrification projects it is important to take into account that the main benefit is to provide a good quality energy source to the local population. In many countries with relatively undeveloped rural areas, such as Nepal, the electricity from micro-hydro schemes is mainly used for lighting, as a substitute for expensive and dangerous kerosene fuels, or lights based on burning pine resin, as described by Zahnd & McKay (2008). A comparison by the author of a pressurized kerosene lamp shows that the light output is less than a modern 3 W white LED, whereas the input power, calculated from the specific energy of kerosene multiplied by the rate of consumption, is around 800 W. The benefits of such low-power electric lighting for rural households have been monitored and evaluated in several different Asian countries. They show a positive impact by improvements in education opportunities, indoor air quality, income generating activities (especially for women) and where communal lighting is installed, for safety. Where larger hydro resources are available, as in many remote parts of China, electricity can also be used for cooking, which reduces dependence on local wood resources and potentially limits deforestation. Nevertheless, the sustainability of off-grid rural electrification projects depends on local ownership and control (Williams & Simpson, 2009). Published case studies from across Asia describe various models for successful implementation of village hydropower projects. Common themes include the importance of villager contribution, usually in terms of construction labour, but also in the planning and long-term management. Clear agreements need to be made between villagers and between the village management and external stakeholders, such as development organizations, rural finance institutions and equipment installers, in order to ensure that tariffs are collected, maintenance carried out and equipment repaired if faults do occur.

Where appropriate resources are available, micro-hydropower has been identified as the lowest cost option for rural electrification, as long as operational costs are taken into account. Unfortunately, diesel generators are usually the option with the lowest initial cost, but fuel and maintenance soon cause a heavy burden of expenditure. Hydro plants have high initial costs, but low running costs. They have a disadvantage that they are more or less site specific, which means that a higher level of engineering knowledge is required to implement a successful project, relative to diesel or even solar photovoltaic (PV). There have been several studies in recent years comparing the relative costs of various options for rural electrification. Different researchers have derived different costs, partly because they have looked at different countries (even within Asia) and partly because costs have changed and are dependent on the size of plants. A study by the World Bank (2006) predicted that by 2010, costs of pico-hydro and diesel generators of 300–1000 W would be in the range 10–18 c/kWh and 42–72 c/kWh, respectively. A more recent study of costs for village grid systems in Indonesia (Blum et al., 2013) found a higher range of 18–23 c/kWh for micro-hydro, but comparable range for solar PV (39–56 c/kWh). These are probably realistic values, whereas a study comparing renewable energy with coal in India shows micro-hydro costs at only 5 c/kWh and solar PV 24 c/kWh. Investigating the sources, the costs for micro-hydro are actually based on MW sized "mini-hydro" data collected by IRENA, but these are optimistic for village-scale projects. Of more relevance is a prediction based on analysis of learning curves applied to costs in Thailand (Huenteler et al., 2016), which suggests that by 2021, costs for solar PV could be lower than micro-hydro at less than 15 US c/kWh. Village hydro plants have limited scope for further cost reduction, but the reduction in cost of solar PV and batteries could be significant, as shown by the arrows in Figure 14.3. A significant cost reduction of these technologies will make a hybrid system more attractive for an isolated mini-grid than a single energy source. A hydro-solar-battery system can provide reliable power all year round, meeting the demand during long periods of dry (and sunny) weather by solar power.

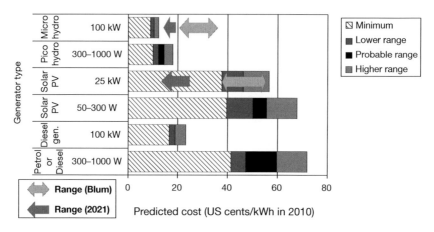

*Figure 14.3*  Comparison of hydro, photovoltaic and diesel electricity costs

*Data source*: World Bank (2006), Blum et al. (2013) and Huenteler et al. (2016).

Alternatively, a number of hydro and other renewable technologies could power mini-grids, as proposed for Nepal by Maskey et al. (2012).

## Conclusion

The future of hydropower in Asia is likely to be one of continued growth, given the large resources and growing economies. This is true both for small-scale power plants as well as grid-connected larger hydropower plants. For small-scale plants, hydropower can make a significant contribution in bringing power to areas that currently lack electricity. These areas tend to be the mountainous parts of the continent with good water resources, where remote communities are lacking electrical power. For larger plants, the contribution of hydropower lies in its compatibility with other renewable energy sources, providing a controllable energy supply. The greatest benefit for water power in future will be as part of a system in which other renewable energy sources, such as wind and solar, which are often much more intermittent but less site specific, are integrated to provide low-carbon electricity, both on and off-grid.

## Acknowledgement

The section on pumped hydro storage has been developed from the work of Masters student Miriam Diez Dago, from University of Oviedo, who carried out her project under the author's supervision while in Nottingham in Spring Semester, 2016, funded through the Erasmus programme.

## References

AAAT (Agency for the Assessment and Application of Technology), *Indonesia Energy Outlook 2015*, Center for Energy Resources Development Technology, Jakarta, 2015.
Ansar, A, Flyvbjerg, B, Budzier, A, Lunn, D, "Should we build more large dams? The actual costs of hydropower megaproject development", *Energy Policy* 69 (2014) 43–56.
Atputharajah, A, "Operational Challenges of Low Power Hydro Plants", Chapter 19 in *Handbook of Renewable Energy Technology*, by AF Zobaa and RC Bansal (Eds) World Scientific, 2011, pp 469–483.

Awale, S, "Nepal Seeks Investors for 10 GW of Hydropower Projects by 2026", *Renewable Energy World*, 2016, available at: www.renewableenergyworld.com/articles/2016/06/nepal-seeks-investors-for-10-gw-of-electricity-by-2026.html

Barbour, E, Wilson, IAG, Radcliffe, J, Ding, Y, Li, Y, "A review of pumped hydro energy storage development in significant international electricity markets", *Renewable and Sustainable Energy Reviews* 61 (2016) 421–432.

Barros, N, Cole, JJ, Tranvik, LJ, Prairie, YT, Bastviken, D, Huszar, VLM, del Giorgio, P, Roland, F, "Carbon emission from hydroelectric reservoirs linked to reservoir age and latitude", *Nature Geoscience* 4 (2011) 593–596.

Bhattarai, BP, Joshi, CB, "Economic Sustainability of Peltric System; A case Study in Nepal", Conference, Stanford University, California, USA, 2011.

Blum, NU, Wakeling, RS, Schmidt, TS, "Rural electrification through village grids—Assessing the cost competitiveness of isolated renewable energy technologies in Indonesia", *Renewable and Sustainable Energy Reviews* 22 (2013) 482–496.

BP, *Statistical Review of World Energy 2016*, London, 2016.

Chang, X, Han, M, Zheng, CH, "Power Control Analysis for Variable Speed Pumped Storage with Full-Size Converter", IECON2015-Yokohama, 2015, Paper 001327 (IEEE).

Chang, X, Liu, X, Zhou, W, "Hydropower in China at present and its further development", *Energy* 35 (2010) 4400–4406.

Harris, M, "Article for Renewable Energy World, 30 July 2015", 2015a, available at: www.renewableenergyworld.com/articles/hydro/2015/07/nepal-awards-contract-for-first-large-scale-pumped-storage-hydropower-project.html

Harris, M, "Chinese Ministry Stops 1,000-MW Xiaonanhai Hydroelectric Project", 2015b, article for HydroWorld, 4 Sept, 2015, available at: www.hydroworld.com/articles/2015/04/chinese-ministry-stops-1-000-mw-xiaonanhai-hydroelectric-project.html.

Huenteler, J, Niebuhr, C, Schmidt, TS. "The effect of local and global learning on the cost of renewable energy in developing countries", *Journal of Cleaner Production* 128 (2016) 6–21.

International Centre for Environmental Management, "Strategic Environmental Assessment of Hydropower on the Mekong Mainstream: Final Report" prepared by ICEM for the Mekong River Commission, 2010.

International Energy Agency, *Key World Energy Statistics 2015*, IEA, Paris, 2015.

International Hydropower Association, *Briefing: 2016 Key Trends in Hydropower*, IHA, London, 2016.

Jensen, JG, "Fish passes in Australia—A National success: can it be copied in the Mekong?" Catch & culture 6 (2001) 3.

Li, H, Xu, X, Ma, W, Xie, R, Yuan, J, Xu, CH, "Seismic structure of local crustal earthquakes beneath the Zipingpu reservoir of Longmenshan fault zone", *International Journal of Geophysics*, Hindawi Publishing Corporation, Article ID 407673 (2011) 6.

Liu, H, Masera, D, Esser, L, Eds, "World Small Hydropower Development Report 2013 Executive Summary", United Nations Industrial Development Organization; International Center on Small Hydro Power, 2013, available at: www.smallhydroworld.org.

Lung, J-K, Lu, Y, Hung, W-L, Kao, W-S, "Modeling and dynamic simulations of doubly fed adjustable-speed pumped storage units", *IEEE Trans. Energy Conversion* 22 (2) (2007) 250–258.

Magee, D, "Dams in East Asia: Controlling Water but Creating Problems", *Routledge Handbook of East Asia and the Environment*, by P Harris and L Graeme (Eds) Routledge, 2014.

Maskey, RK, Bhandari, V, Adhikary, B, Dahal, R, Shrestha, N, "Prospects for Small Hydro Power Plants Based Mini-Grid Power Systems in Nepal Developments", Renewable Energy Technology (ICDRET), 2012 (2nd International Conference) IEEE, 2012.

Renewable Energy Policy Network for the 21st Century (REN21), "Renewables 2016: Global Status Report", Paris: REN21 Secretariat, 2016, available at: www.ren21.net/status-of-renewables/global-status-report/

Reynolds, TS, *Stonger than a Hundred Men—A History of the Vertical Water Wheel*, Johns Hopkins University Press, 1983, p 17.

Schlunegger, H and Thöni, A, "100 MW full-size converter in the Grimsel 2 pumped-storage plant", Hydro 2013 Conference, Innsbruck, Austria, Session 14a, 2013.

Siebert, M, "Making Technology History", *Sinica Leidensia: Cultures of Knowledge: Technology in Chinese History*, by D Schäfer (Ed) Brill, 2012, p 273.

Smith, NPA, "Induction generator for stand-alone micro-hydro systems", Proceedings of IEEE International Conference on Power Electronics, Drives Energy System for Industrial Growth 2 (1996) 669–712.

Sovacool, BK, Bulan, LC, "Behind an ambitious megaproject in Asia: the history and implications of the Bakun hydroelectric dam in Borneo", *Energy Policy* 39 (2011) 4842–4859.

Stone, R, "The legacy of the three Gorges dam", *Science* 333 (6044) (2011) 817.

Tong, J, Zheng, N, Wang, X, Hai, J, Ding, H, *Mini Hydropower*, UNESCO Energy Engineering Series, John Wiley & Sons, 1996.

Tshering, S, Tamang, BH, "Hydropower: Key to Sustainable Socio-Economic Development of Bhutan", United Nations Symposium on Hydropower and Sustainable Development, Beijing, China, 2004, available at: www.un.org/esa/sustdev/sdissues/energy/op/hydro_tsheringbhutan.pdf

US DoE (Department of Energy), "Global Energy Storage Database—Office of Electricity Delivery and Energy Reliability" [Online], available at: www.energystorageexchange.org; accessed: July 2016.

Williams, AA, Simpson, RG, "Pico hydro—Reducing technical risks for rural electrification", *Renewable Energy*, 34 (8) (2009) 1986–1991.

World Bank, "Technical and economic assessment of off-grid, mini-grid and grid electrification technologies – summary report", World Bank Energy Unit, Washington, DC, September 2006.

World Commission on Dams (WCD), *Dams and Development: A New Framework for Decision-making*, Earthscan Publications, 2000.

Yang, F, Fan, M-T, "Network responds to distributed resources", *Transmission & Distribution World Magazine*, 2014.

Zahnd, A, Mckay, K "A mountain to climb? How pico-hydro helps rural development in the Himalayas", *Renewable Energy World* 11 (2) (2008) 118–123.

# PART 3

# Energy policy issues in Asia

# 15

# RETHINKING ELECTRICITY SECTOR REFORM IN SOUTH ASIA

## Balancing economic and environmental objectives

*Anupama Sen, Rabindra Nepal, and Tooraj Jamasb*

### Introduction

Many of the electricity reform programmes currently being undertaken across developing economies in non-OECD (Organisation for Economic Co-operation and Development) countries in Asia originated in the experiences of a group of OECD countries in the 1980s and 1990s.[1] These were, primarily, the United Kingdom (England and Wales), Norway, the USA, and Chile – often highlighted as forerunners (Pollitt, 2004). Reform entailed restructuring the sector from a scenario characterised by state-owned, vertically integrated entities, into one where smaller, and in most cases privately owned, firms competed for the provision of electricity supplies (Sen et al., 2016). From these experiences emerged a raft of basic reform measures or 'blueprint', which included:

- opening the electricity sector to independent power producers (IPPs),[2]
- corporatisation[3] of vertically integrated state-owned utilities and the commercialisation of their functions,
- enactment of electricity legislation,
- establishment of an independent electricity regulator,
- unbundling of vertically integrated utilities into competitive (generation and supply) and regulated (transmission and distribution) segments, and
- divestiture or privatisation of the unbundled utilities.

Collectively, these measures came to be known as the 'standard model' of electricity reform.[4] Wholesale markets featured prominently in the details of the standard model, as did retail competition and consumer choice. The standard model was based on the implicit assumptions of well-functioning markets, developed institutions, and stable political frameworks found in developed countries.

Although the arguments in favour of electricity market reform via the OECD model primarily focused on its economic benefits,[5] the literature presents mixed evidence on whether the predominant drivers behind initial reform, particularly in the UK, were purely economic or ideological (Newbery, 2013; Rutledge, 2010; Helm, 2010; Keay, 2009, Rutledge and Wright, 2010).[6] It has been suggested that the economic rationale for market liberalisation evolved *ex post* and was secondary to ideological and political considerations (Rutledge, 2010).[7]

Evidence on the success of reforms in developed economies is mixed; the literature suggests that it is difficult to attribute the improvements in operational efficiency to reform *per se* when there has been a combination of external factors.[8] There is a vast literature documenting the early experience of reforms in developed countries (Sen, 2014).[9] However, growing environmental pressures and international consensus towards climate change mitigation since 2009[10] have prompted fresh debates over the effectiveness and the mission of electricity reforms, as decarbonisation has been incorporated into reforms as an important environmental objective. This is because the electricity sector presents the biggest opportunity to bring about the single largest emissions reduction in many countries (IEA, 2015).

Consequently, in OECD countries, reforms have been under review to incorporate emissions reduction goals. In developing non-OECD Asia, however, this poses an additional challenge as governments already struggle to successfully square efficiency objectives of electricity reform with access and distributional objectives.

South Asia alone accounts for 25% of global population yet just 5% of global electricity consumption. Just under a third of India's population of 1.2 billion still subsists on non-commercial energy sources. Moreover, electricity demand in non-OECD Asia as a whole is predicted to double over the next two decades: the International Energy Agency (IEA) predicts that it will rise from 6,317 terawatt hours (TWh) in 2012 to 13,982 TWh by 2035 (IEA, 2014). Total installed capacity (gigawatts) in non-OECD Asian economies comprises 30% of global installed capacity, and is predicted to grow to 44% by 2040, but will be insufficient to meet rising demand (IEA, 2014). Emissions from non-OECD Asia, however, already comprise around 38% of global emissions, and given rising demand, there will be increasing pressure from tightening post-COP21 global climate architecture for these countries to tackle environmental objectives. While meeting this demand will be a challenge, doing so in a sustainable way will be even more difficult.

In this chapter we review the progress in adoption of the OECD model of electricity reform in non-OECD Asian developing countries, highlighting the fact that most countries are midway towards the implementation of liberalised electricity markets. However, we also argue that pursuing the OECD model to completion can lead to adverse environmental outcomes in these countries. We conclude that a new paradigm of electricity reform is needed in Asia, which balances economic and environmental objectives.

The remaining sections are organised as follows. The second section reviews the implementation of the OECD reform model, the third section analyses the environmental dimension and its implications for electricity reform in developing Asia, the fourth section discusses the experiences of three Asian countries – India, Nepal and Bhutan, highlighting not just the contradictions between economic and environmental objectives, but their complexities with cross-border electricity trade. The final section concludes.

## The standard reform model in non-OECD Asia

From the 1990s, there was a gradual and widespread adoption of the standard model of electricity reform (or its variants) across developing non-OECD Asia. The drivers behind this have been categorised into 'pull factors' and 'push factors' (Nepal and Jamasb 2011). The 'pull' factors included a demonstration effect following experiences in the OECD (Zhang et al., 2008). The 'push factors' were twofold; the first related to the adoption of structural adjustment programmes, e.g. in India as a condition of multilateral financial assistance following balance of payments crises.[11]

The second 'push factor' was related to endemic problems within the electricity sectors of developing countries, and a genuine need for reform. The sector in most developing countries was publicly owned, through vertically integrated entities which carried out generation, transmission, and distribution, as well as infrastructure creation. Vertically integrated monopolies, based on the economies of scale and scope[12] argument, were deemed to be the best way of extending electricity to the majority of the populations that lacked access to it. Similarly, public ownership was justified on the basis that the state was the custodian of public interest, the enabler of necessary coordination among different segments (generation, transmission, and distribution), and necessary to the strategic nature of the sector, given its role in development (Gratwick and Eberhard, 2008).

The concentration of all functions in singular state-owned entities, with no effective self-imposed or independent oversight, led to technical and financial problems (Victor and Heller, 2007). Transmission and distribution losses in developing economies averaged 20% prior to reform, in comparison with a world average of 9% (Gratwick and Eberhard, 2008). The number of employees per million units of electricity sold was 5 for India, compared with 0.1 for Norway[13] (Sen, 2014). There is a large literature attributing the underlying reasons for these inefficiencies to the politicisation of the sector and its capture by politicians to seek votes through the promise of low-priced electricity, partly financed through government subsidies, and partly by the state-owned enterprises themselves. Such actions forced state-owned enterprises to maintain the prices below costs of supply, constraining the amount by which prices could be adjusted upwards against increases in the cost of supply. This in turn limited the amount of capital available for reinvestment in infrastructure and access – thus defeating the original purpose of state-controlled electricity provision and engendering a culture of wastefulness and political opportunism (Tongia, 2003; Dubash and Singh, 2005). In India, this circular problem was demonstrated by the fact that, despite successive increases in expenditure allocated to electricity within the Five-Year Plans, capacity addition targets were rarely achieved and state-owned utilities continue to accumulate losses.

Therefore, in developed countries, the underlying logic and objectives for reform were higher efficiency, lower prices, consumer choice, and national competitiveness (Williams and Ghanadan, 2006). In developing countries, the objectives were more to do with the declining state of utilities' finances, propagation of private investment to enable infrastructure, technology upgrades, and removal of the electricity supply constraint on growth. (Williams and Ghanadan, 2006). Developing economies also faced a capacity shortage, which meant that the risks in reform were higher at the outset in comparison with developed economies.

At the time of initial reform, the OECD or standard model was presented as a solution for technical and financial inefficiencies and at infusing transparency into the operations of state-owned enterprises in developing economies. Technological advancements had also rendered the economies of scale argument redundant. Thus, the standard reform model effectively represented a 'common path' but to 'different goals' for the OECD and non-OECD economies. Further, although there was a general sequence that had been intended to shape the basic reform measures,[14] non-OECD Asian economies ended up adopting variations of it, as reform tended to occur in fits and starts.

Table 15.1 depicts the progress of reforms set against the 'standard' reform model used to assess its milestones for 17 prominent non-OECD Asian countries. The entry of IPPs (Independent Power Producers) into generation is the most widely adopted measure (Figure 15.1), but only four countries have progressed to distribution privatisation. Further, open (or third party) access,

*Table 15.1* Electricity reforms in non–OECD Asia, 2013

| | Independent power producers | Regulator | Unbund-ling | Corporatisation | Open/third party access[a] | Distribution privatisation |
|---|---|---|---|---|---|---|
| Bangladesh | × | × | × | × | | |
| Bhutan | × | × | × | × | | |
| Brunei | | × | | | | × |
| China | × | × | × | × | | |
| India | × | × | × | × | × | × |
| Indonesia | × | | × | × | × | |
| Laos | × | | | | | |
| Malaysia | × | × | × | × | | |
| Maldives | × | × | | × | | |
| Myanmar | × | × | | | | |
| Nepal | × | × | × | × | | |
| Pakistan | × | × | × | × | | |
| Philippines | × | × | × | × | × | × |
| Singapore | × | × | × | × | × | × |
| Sri Lanka | × | × | | | | |
| Thailand | × | × | × | × | × | |
| Vietnam | × | × | × | × | | |

*Data source*: Sen et al. (2016).

*Note*: [a]Open access has been implemented to varying degrees; in the majority it has been confined to large consumers.

a fundamental enabler of retail competition, has been adopted in just five countries, and restricted to large consumers. Experiences have been mixed across countries.

One of the main motivations for the introduction of IPPs, the most popular measure was they there were a quick way to introduce competition without significant restructuring.[15] They transferred investment risks to utilities and in some cases ultimately to consumers (through higher tariffs) through the 'take or pay' clauses prevalent in many contracts. While some countries (e.g. Malaysia and Singapore) coped by evolving their sectors to adapt to this risk, many (e.g. India and Pakistan) struggled to harness IPPs to fit with their fiscal and institutional contexts leading to a spate of renegotiations and cancellations (Sen et al., 2016).

However, political factors have markedly impeded IPPs' success. For instance, the Philippines in the 1990s successfully contracted IPPs for 40% of generation capacity, as did Indonesia – however, following the Asian financial crisis a spate of renegotiations uncovered allegations of patronage in the awarding of IPP contracts in both countries (Henisz and Zelner, 2002; Wu and Sulistiyanto, 2013).

Among smaller countries, Laos, Bhutan, and Nepal have significant hydropower potential, some of which has been developed through IPPs. However, concerns over property rights and sovereignty have prevented their progress. In China, the lack of grid integration meant that despite the early introduction of IPPs, capacity surpluses could not be spread to deficit regions, making IPP investments susceptible to regional supply and demand fluctuations (Wu, 2005). The reorientation of multilateral financing towards clean energy has stalled IPPs in newly hydro-carbon-rich countries such as Bangladesh and Myanmar.

Although the majority of non-OECD Asian countries have established some type of electricity regulator, in most cases these are not independent from government. In countries where

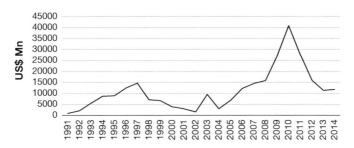

*Figure 15.1*    IPP investments in Asia, 1990–2014 (graphed according to the year of financial closure)

*Data source*: World Bank PPI Database.

electricity reform is lagging, the main issue faced by regulators relates to reforming tariffs to reflect costs (e.g. India, Pakistan). Where markets have largely developed, issues include the mitigation of market power, which in many cases is exercised by state-owned companies (e.g. Thailand, Philippines). Some countries such as China have aimed to consolidate electricity regulation with other energy-related sectors. In smaller oil import-dependent countries such as Maldives, the regulator plays a critical role in the country's trade balance.[16] Although the majority of countries have implemented unbundling and corporatisation, public sector provision largely dominates and in many cases the finances of distribution companies have not improved. Indeed, Nepal and Jamasb (2012) argue that in smaller systems (e.g. Nepal), the creation of an independent regulatory authority may be more important than unbundling, particularly for politically unstable countries and especially where hydropower is predominant.

Open access has been implemented in just five countries, with some obstacles. In India, for instance, the main impediment has been the imposition of 'surcharges' by public utility companies on large industrial consumers to compensate for the loss in revenue.[17] In Indonesia, the state company PLN continues to be the sole owner of transmission and distribution assets as it is given priority rights under the law to conduct its business. In Thailand, despite open access, public sector companies operate geographically segregated oligopolies and have majority shares in private generation companies (Wisuttisak, 2012).

Four Asian countries (Brunei, India, Philippines, and Singapore) have implemented distribution privatisation. Singapore is arguably the most advanced, with seven electricity retailers and the Market Support Services Licensee competing for (contestable) retail consumers. Privatisation in the Philippines' electricity distribution sector has on the other hand resembled the switch from a public to a private monopoly. India has had a mixed experience with distribution privatisation – Odisha in 1996, carried out without restructuring, and Delhi in 2002, where bids were awarded on the basis of the largest promised reductions in average commercial and technical losses, with the gains shared with consumers. In May 2015 Odisha's state electricity regulator revoked the licenses of the three private distribution utilities, citing 'gross failure in raising their performance and financial health, reducing distribution loss, preventing theft of energy and running the organisation in a financially viable manner' (Mohanty, 2015). Distribution assets were consequently taken over by the state-owned grid management company GRIDCO.

It is therefore evident that developing Asian economies (and developing countries in general – see Figure 15.2) have implemented hybrid versions of the standard OECD reform model, as their reform processes have often been influenced by country-specific heterogeneity.

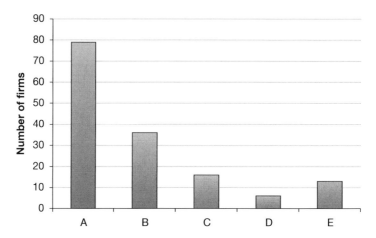

*Figure 15.2* Structure of the electricity sector in 150 developing countries

*Data source*: Gratwick and Eberhard (2008); Besant-Jones (2006).

*Note*: GenCo, TransCo, DistCo refer to generation, transmission and distribution companies.

A – vertically integrated monopolies; B – vertically integrated monopolies with IPPs; C – single buyer as national GenCo, TransCo or DisCo OR a combined national GenCo–TransCo or TransCo–DisCo + IPPs; D – multiple DisCos and GenCos, including IPPs, TransCo as single buyer with third party access; E – power market of GenCos, DisCos and large users, TransCo and independent system operator.

## Electricity reforms and environmental objectives

As Table 15.2 shows, electricity (and heat) production account for the largest proportion of emissions from fuel combustion across most economies, implying that the electricity sector presents a significant opportunity for potential emissions reduction.[18] This is why in OECD countries (particularly the EU) environmental objectives have mainly been pursued through decarbonisation of the electricity sector.

Under the OECD model, electricity prices in liberalised wholesale markets are set according to system marginal cost – the short-term marginal cost of the last (and, following the merit order, typically most expensive) plant that is required to be brought onto the system in order to meet demand.[19] Generation companies are incentivised to compete on costs, as those with lower short-term marginal costs than the system marginal cost will gain from this. These markets are also referred to as 'energy-only markets' in that the incentives to encourage more investment in generation are considered as being built into the price signals, rather than through externally imposed generation adequacy standards.[20] The market effectively determines how much generation capacity is required (UKERC, 2010). Energy-only markets are largely designed to match the characteristics of conventional (fossil) fuel generation and investment (Keay et al., 2013a, 2013b; Keay, 2010).

Renewable energy has high capital costs but a zero marginal cost of operation. Further, renewable energy is intermittent[21] and while during periods of abundant availability supply could be sufficient to match demand, the intermittency requires that 'backup' generation be available to ensure continuity of supply. It is important that this backup is flexible and able to adjust quickly to demand – implying the use of fossil fuels, hydro or nuclear (Sen, 2014).

Two problems emerge here: first, in order to encourage the capital-intensive investments in renewables, governments have to offer support schemes such as Feed-in Tariffs (FiTs) or other subsidies that undermine the role of the liberalised market in setting prices and in motivating investment. Second, the unpredictability of renewables (e.g. wind) implies that market prices will

*Table 15.2* $CO_2$ emissions by sector, 2013 (million tonnes of $CO_2$)

| | Electricity and heat production | Other energy industry own use | Manufacturing and construction | Transport | Residential | Other | Total $CO_2$ emissions from fuel combustion |
|---|---|---|---|---|---|---|---|
| OECD | 4,866 | 722 | 1,378 | 3,384 | 932 | 755 | 12,038 |
| Asia | 6,069 | 504 | 3,665 | 1,428 | 483 | 481 | 12,630 |
| *India* | 4,386 | 367 | 2,806 | 754 | 330 | 334 | 8,977 |
| *China* | 945 | 43 | 493 | 222 | 87 | 79 | 1,869 |
| Africa | 435 | 88 | 137 | 289 | 69 | 58 | 1,075 |
| Middle East | 639 | 121 | 344 | 386 | 124 | 35 | 1,648 |

*Data source*: IEA (2015).

either be set equal to zero marginal costs (e.g. during periods of abundant wind) or set at very high levels at times when renewables are low or unavailable (in order to allow backup generators to recover their costs and justify capital investments in backup generation). This could imply shorter periods of zero or low prices and longer periods of very high prices (Keay et al., 2013a, 2013b); these high prices would be difficult to justify to a public which already questions high energy prices, as market liberalisation was largely propagated on the basis of 'competitive' prices (Sen, 2014).

Further, if renewables are integrated into wholesale markets in their current form, they may not even recover their fixed costs (without extra-market payments) because when they run, prices would be low or zero (Robinson, 2013). It has also been argued that support schemes such as FiTs discourage innovation through 'cherry picking technologies' thereby impeding reductions in the capital costs of renewable energy (Keay et al., 2013a; Keay, 2013b). Arguably, even if the technical costs of renewable energy were to decline over time, system or resource costs could continue rising, as the resource may tend to get costlier the more it is exploited (Keay, 2013a).[22]

These problems reflect some of the inherent contradictions between the OECD model and newer environmental objectives of reforms in which the sector needs to rely increasingly on renewable energy sources. The solutions that have been advocated to this problem in OECD countries can be categorised into two: the greater use of markets to ensure investments in both renewables and backup generation (that is, the setting up of separate capacity and balancing markets in addition to energy-only markets), or conversely, the establishment of a single-buyer agency to coordinate the integration of renewables into the electricity sector (Keay et al., 2013a, 2013b; Newbery, 2013).

The shift to a sustainable, low-carbon electricity system does not preclude the original hurdle of cost-reflective pricing. Asian countries continue to grapple with important distributional issues, and in the past the direct use of the pricing system to address these issues has led to regressive outcomes, which have ended up benefitting the rich. Systems of direct cash transfers to eligible consumers have been adopted in some countries (e.g. India, Indonesia) as a method of removing pricing distortions and targeting poorer consumers for subsidies.

### *Economic versus environmental objectives in Asian developing countries*

One way of viewing the contradictions between economic and environmental objectives in the OECD reform model is that markets and institutions created through reforms may gradually evolve to resolve these conflicts. However, three factors suggest that a paradigm shift in the reforms process is critical in Asian developing economies. The first and most obvious one is rising electricity demand. The second is that for developing countries, the amount of effort required to bring about a policy outcome can be disproportionate to the value of that outcome (Sen and Jamasb, 2013),[23] necessitating proactive policy choices in the short term.

The third is that these countries stand to lose the most from climate change in terms of its human costs (IPCC, 2007), and it can be assumed that their citizens will hold governments accountable for inaction – in other words, environmental policy is becoming as much a political issue as cost-reflective pricing. This has indeed been seen recently in countries such as China and India. For instance, in late 2015 citizens in Delhi launched litigation against municipal authorities on the grounds of their inaction to combat rising urban air pollution, proven to be detrimental to human health.[24]

As Table 15.1 showed, any serious effort at decarbonisation would need to have the electricity sector at its core, as electricity comprises the largest source of emissions across the board, and provides the most direct way to reduce emissions (Keay, 2009). In this regard, non-OECD Asian countries face a series of challenges in balancing economic and environmental objectives in the process of electricity sector reforms. We summarise the main ones below.

## Investment in generation

While energy-only markets focus on short-term marginal costs, they do not separately take into account long-run marginal costs, which are directly related to investments in generation. The problem of adequate investment in generation under the OECD reform model stretches across developed and developing countries; however, the nature of the investment problem is different for both. In developed countries, the concern is whether markets will deliver investments in renewables and backup generation, or whether a single buyer agency should take on the task. In Asian economies, the single-buyer (central planning) model has arguably failed to deliver adequate investment in both conventional and renewable generation.

The investment problem faced by Asian countries is complex: it requires a solution that (a) ensures universal access to electricity; (b) is based on cost-reflective pricing; (c) integrates renewables onto the system; and (d) is free from political appropriation. The design of investment mechanisms for developing economies necessitates a closer look at the hybrid market structures that have emerged in these economies – in terms of identifying all these structures and working out how they could function in conjunction with each other towards a defined set of goals. Essentially, this strategy could comprise a looser, less sophisticated approximation of the 'multiple markets' solution being considered in the UK.

## Access and the proliferation of coal

Most Asian developing countries continue to struggle with extending access to electricity. If developing countries were to fully implement the OECD model (full electricity market liberalisation, assuming that the problem of cost-reflective pricing and finances of utilities were addressed along the way), then the system could arguably be successful in balancing electricity generated from renewables with electricity from conventional (fossil) fuels. The irony here is that backup generation is likely to be coal – which is cheap and plentiful but environmentally far more damaging than say gas, which is its closest substitute – constituting a significant policy contradiction.

Despite this limitation of markets, it is by no means certain that a central planning agency would advocate the exclusion of coal from the energy mix, particularly as energy has been viewed as a balance of payments problem in net-importing Asian countries (Sen, 2014). Countries therefore need to implement some form of a (binding) carbon price to square the balance between economic and environment reform objectives.

An added dimension is the role of multilateral financial institutions, which provided the original catalyst for electricity reforms. Although many have discontinued lending to coal projects, International Financial Institutions (IFIs) accounted for roughly US$51 billion in the financing of coal-related projects from 2007 to 2013.[25] Some multilateral finance institutions have announced that they will cease funding inefficient coal projects. By some accounts, bilateral finance to coal-fired power from Exim banks and export credit agencies, has formed the larger proportion of IFI financing over the last five years.

### Cross-border electricity trade

One approach being pursued in OECD countries with a view to the promotion of clean energy, the reduction of costs in its provision, and the availability of electricity to countries that face a deficit, is regional electricity market integration (Buchan, 2013). 'Large regional markets', achieved through cross-border integration, have been proposed as a way of ensuring greater flexibility in the response of generation to demand, as a larger pool of plants is available to call upon for dispatch (Riesz et al., 2013).

For instance, the EU's climate and energy package sets binding legislation for its '20–20–20' targets – namely, a 20% reduction in EU greenhouse gas emissions from 1990 levels, raising the share of EU energy consumption from renewables by 20%, and a 20% improvement in the EU's energy efficiency – to be achieved by 2020, with a longer-term goal of an 80–95% reduction in emissions by 2050. This is to be achieved through a set of policy tools based on greater market integration[26] combined with nationally-driven targets on emissions reduction. Several regional electricity market integration initiatives exist in Asia, supported by multilateral finance (such as the ADB); for example, electricity trading in the Greater Mekong sub-region (GMS) – comprising Cambodia, Laos, Myanmar, Thailand, and the Yunnan Province and Guangxi Zhuang Autonomous Region of China – primarily takes place through bilateral exports. In South Asia, interconnections exist between India, Bangladesh, and Bhutan.

The OECD reform model has, however, led to conflicts between national and regional targets in the EU. For EU countries such as France, where 75% of electricity is generated from nuclear energy and is low-priced by European standards, greater market integration is likely to lead to higher prices (at least initially), which is unlikely to be accepted by French consumers who initially had to pay higher tariffs to finance the early capital investments in nuclear energy (Percebois, 2013). Also, there is some ambiguity over EU regulations against national governments providing 'state support' to emissions reduction schemes that are unsanctioned at EU level, as this could be seen as undermining the principle of competitive markets.

## Balancing economic and environmental objectives – the case of three South Asian countries

The current debates over the suitability of the OECD model in delivering on both economic and environmental objectives presents an opportunity for Asian developing countries, most of which have progressed little, or midway, towards implementing the model, to identify solutions that can address both objectives *without* mimicking the experience of OECD countries. Essentially, from the above discussion it can be argued that this would entail delivering electricity through a system which encompasses: (a) low emissions; (b) a suitably flexible baseload to intermittent renewables; (c) a competitive price, particularly with regard to low-income consumers; and (d) adequate market incentives (or pricing signals) for investors.

*Table 15.3* Key electricity indicators: India, Bhutan, and Nepal

| | Installed capacity (MW) | Peak demand (MW) | Electricity deficit (% peak demand) | IPPs as capacity (%) | Access rate (%) | Technical losses (% consumption) | Per capita consumption (kWh) | Hydro potential (GW) | Coal reserves (Bt) |
|---|---|---|---|---|---|---|---|---|---|
| India | 2,430,218 | 129,815 | | 34.0 | 75 | 23.65 | 917 | 148 | 126 |
| Bhutan | 1,615 | 336 | | | 95.5 | 9.3 | 2,600 | 33 | – |
| Nepal | 787 | 1,200 | 40 | 33.3 | 76 | 25.03 | 106 | 83 | – |

*Data source*: Yangki and Tashi (2016); World Development Indicators; USAID (2014); Authors.

A prominent means of balancing these objectives has been attempted through inter-regional electricity market integration,[27] such as in the cases of the EU (described above) and in other parts of the world.[28] In this section, we explore this further through a comparative case study using three countries based on: (a) geography; (b) shared experiences of electricity reform; and (c) ongoing efforts towards greater regional market integration. The three South Asian countries analysed here are: India (which accounts for the largest regional share of emissions and abundant low-cost coal reserves), Bhutan, and Nepal (which both account for the highest reserves/potential of hydropower in the region, mostly underdeveloped). In such a market, given that hydro reserves are subject to sovereign property rights, the role of private investment and markets (relatively advanced in India) would largely be concentrated on *infrastructure* and other renewables capacity (e.g. solar).

In this section, we first review country experience on electricity reforms against the context of economic and environmental objectives for India, Bhutan, and Nepal (see Table 15.3). We then analyse the argument for greater market integration as a potential solution to achieving the 'balance', and barriers to the same.

## *India*

Power sector reforms in India have been undertaken in three phases (Victor and Heller, 2007). The first phase, in 1991, focused on the introduction of IPPs into generation, with little success. Enron's attempt to set up an IPP is widely cited as the failure of India's 1991 effort to open up the power generation sector (Mukherjee, 2014).[29] The second phase, in the mid-1990s, constituted state-level efforts at restructuring State Electricity Boards into unbundled companies and setting up independent electricity regulatory commissions.[30] The third phase was marked by the legislation of the Electricity Act in 2003, which consolidated and replaced all previous federal laws governing the electricity sector, and was a fairly momentous step forward in India's hitherto unsteady reform progress. The provisions of the Act aimed to transform the electricity market from a non-competitive, single-buyer model to a multiple-buyer model with several competing participants in the generation, transmission and distribution segments. Sen and Jamasb (2013) provide a comprehensive analysis of power sector reforms in India.

Consequently, India has a hybrid version of the OECD/standard model; state-owned utilities exist alongside private companies in generation and distribution, and power is mostly traded through Power Purchase Agreements while a relatively small but growing proportion of electricity is traded on short-term markets through power exchanges and bilateral contracts.[31] In Figure 15.3, the top row represents generation – comprising a mix of federal and state government utilities, private and privatised companies, IPPs, and captive generators. Two power exchanges and the state-owned Power Trading Corporation of India (second row) facilitate a

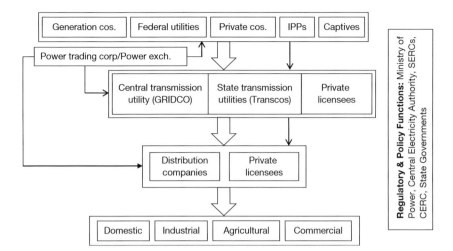

*Figure 15.3*   Power sector restructuring in India

*Data source*: Singh (2010).

small percentage of market-based trading, whereas the majority of trade is carried out through longer-term Power Purchase Agreements. The third and fourth rows depict transmission and distribution companies (retail supply and distribution are integrated) which include government (federal and state)-owned companies as well as private licensees, and the last row shows categories of consumers.

There are three distinctive characteristics of India's electricity reform experience. First, despite the early failures with IPPs, the latter have provided a functional alternative to the lack of public sector capacity additions. The government's planned programme of adding supercritical coal-fired generation plant capacity through 'ultra-mega power plant' projects[32] have faced similar issues with tariff levels and the economic viability of distribution utilities to purchase power from IPPs. These issues have been partially resolved through the increase of low-cost domestic coal production and the auctioning of domestic coal supply linkages to IPPs.[33] Notably, briefly India achieved a 'surplus' of power for the first time in decades in 2016–2017.

A second characteristic is that India is beginning to demonstrate problems similar to those seen currently in the OECD, where a system based on marginal cost pricing cannot cope with the intermittency of renewables, while linking in with long-term incentives for investment in backup generation. Attempts at incorporating renewables have run into problems. For instance, a 'renewables purchase obligation' (RPO) requires distribution utilities to purchase a certain proportion (set by state regulators) of electricity from renewables, or alternatively an equivalent amount in 'renewable electricity certificates' (each equating to 1 MWh) on India's two main power exchanges. However, the exchanges have reported a growing inventory of untraded certificates since the programme's inception. For instance, the Indian Electricity Exchange, which accounted for over 90% of the domestic electricity trading market in 2015, reported that of 9.6 million certificates available in fiscal 2014–2015, only 3.1 million had been traded.

A third characteristic is the continued failure of state-owned distribution utilities to enforce cost-reflective pricing, implying that they are often unable to fulfil their RPOs. A debt restructuring programme announced in November 2015, asked state governments to appropriate 75% of distribution utilities' debts over a period of 2 years. Following this the debts were to be included as part of state fiscal deficits, thereby necessitating that states take parallel measures to increase power sector tariffs and eliminate subsidies. The incentive to states which implement this reform was the according of priority in other federal funding. However, two previous attempts at debt restructuring have failed, as it is incumbent upon state governments (often reluctant to increase tariffs for their constituents) to implement reform.

Despite the penetration of competitive markets into India's electricity sector, problems continue to exist with regional imbalances and constraints to the evacuation of power, the most prominent illustration of which was the simultaneous failure of three of India's five regional transmission grids in July 2012. In February 2016, the percentage composition of India's 289 GW of installed capacity was: coal (61%), oil (0.34%), gas (8%), nuclear (2%), hydro (15%) and renewables (13%). Within renewables, solar and wind were 13% and 65%, respectively. According to the IEA, $CO_2$ emissions from electricity comprise the single largest proportion (44%) of total $CO_2$ emissions from fuel combustion in India. Given that roughly a third of its population lack access to modern commercial energy, India's government in 2014 set a target of providing '24×7 electricity to all' by 2019. Despite the environmental implications, India's government plans to triple coal production to 1.5 billion tonnes by 2019 to meet this target. While India plans to add 100 GW of solar, 60 GW of wind and 15 GW of other renewables to installed capacity by 2022, these require an exponential growth in the renewable energy market that is unlikely to be achieved.[34] Further, it faces issues over land acquisition and resettlement pertaining to new (as well as existing) hydro power projects. The viability of this target is therefore contingent not just upon the continued expansion of installed capacity in a sustainable manner, but also on further technical and financial reforms to the electricity sector. Given its dual emissions and development implications, the power sector therefore represents the biggest hope for a successful confluence of India's electricity and climate policy, but at the same time, its biggest hurdle.

One way of achieving this is through regional electricity market integration with neighbouring South Asian countries which have considerable hydro and gas resources. India already has an interconnection with Bangladesh which supports 500 MW and is being upgraded to 1000 MW. There have been discussions since 1970 on a submarine India–Sri Lanka HVDC link, and India has interconnection and trading agreements in place with Nepal and Bhutan (discussed below). More importantly, with its status as South Asia's largest (and growing) electricity market and quasi-competitive structure (particularly in generation), India will play a pivotal role in facilitating South Asia's regional electricity market integration.

## Bhutan

Bhutan joined the global reform bandwagon in the mid-1990s. Key drivers of reforms in the Bhutanese power sector were: (a) the institutional weakness of the sector, (b) inability to assume the role of key growth driver in promoting the country's economic development through power exports, (c) providing access to electricity for rural consumers; and (d) poor cost recovery and dependence on donor financing for new investments (ADB, 2010). Table 15.4 summarises the major reform timelines and the current status of the Bhutanese power sector. The wholesale market is based on a single-buyer model (SBM) consisting of multiple buyers and single seller. The electricity supply industry is vertically integrated with only functional separation of generation, transmission and distribution. Private participation in the generation segment was initiated in 2008 while independent regulation was instituted in 2010.

Regulatory and institutional reforms in the energy sector have successfully contributed to the development of hydropower in Bhutan. The economy that was once solely reliant on foreign aid now is self-reliant due to hydropower revenue which has been ploughed back into social and

*Table 15.4* Reform timeline in Bhutan

| | |
|---|---|
| 2001 | Electricity Act passed |
| 2002 | Functional unbundling of the power sector |
| | Bhutan Power Corporation made responsible for transmission and distribution |
| 2008 | Druk Green Power Corporation Limited was established in January 2008 and made responsible for developing, maintaining, and operating hydropower plants owned by the Royal Government of Bhutan (RGoB) |
| 2010 | Bhutan Electricity Authority made responsible for regulation and became fully autonomous in January, 2010 |

*Data source*: Yangki and Tashi (2016).

industrial sectors. Domestic tariff has been kept low to increase energy affordability and stimulate economic growth. Rural electrification is high priority in Bhutan implying universal urban electrification and 99% rural electrification. Notably, a continuous supply of electricity is available to industrial consumers.

India and Bhutan signed a bilateral agreement in 1996 under the Indo-Bhutan Power exchange. The agreement provided that all surplus power would be sold to the Government of India (GOI), and in turn GOI is committed to purchase all the surplus power (Saran, 2014). An umbrella agreement was signed in 2006 under which India provides project investigation, design and engineering services, constructional supervision and highly concessional finance for upcoming hydro projects which allows India to supply power to Bhutan under a swap arrangement. Electricity sales (domestic and export) amounts to over 45% of the total internal government revenues, and 20% of GDP. Around 75% of electricity was exported to Bhutan's largest trading partner India, in 2014 (Yangki and Tashi, 2016).

Hydroelectric energy is considered to be a clean form of renewable energy in Bhutan with a potential of 83 GW. Hydroelectricity exports have allowed Bhutan to offset 4.4 million tonnes of $CO_2$ annually with the potential to offset up to 22.4 million tonnes of $CO_2$ per year by 2025 (Arora and Dema, 2016). Most projects are run of the river and therefore environmentally benign, due to the rivers flowing through deep valleys while rich forests cover 70% of the land. Bhutan's constitution also mandates that its territory be covered by at least 60% by forests (Arora and Dema, 2016). This target may need revisiting considering Bhutan's growing population and concurrent energy and development requirements.

## *Nepal*

Power sector reform in Nepal has been unsuccessfully pursued since the early 1990s, and marked by political instability resulting in discontinued policies and an environment that has disincentivised investment. The original drivers of reforms included the inability of the sector to extend electricity access to all consumers, inability to attract investments in electricity generation, high transmission and distribution losses, and electricity theft. Nepal and Jamasb (2012) and Singh et al. (2015) summarise the historical contexts, drivers and various attempts towards reforming the Nepalese power sector (see Table 15.5 for the timeline). After more than two decades of the initial reform attempt, the state-owned Nepal Electricity Authority (NEA) still remains a vertically integrated entity with minimal functional separation among generation, transmission, and distribution, operating under a single-buyer model. A legally independent regulator was introduced 1993, but subsequently revoked, and then reinstated in 2011. IPPs in electricity generation were originally introduced in 1992.

*Table 15.5* Reform timeline in Nepal

| | |
|---|---|
| 1984 | Nepal Electricity Authority Act, 1984 enacted to set up Nepal Electricity Authority (NEA) |
| 1992 | Hydropower Development Policy 2049 (1992) issued |
| 1992 | The NEA Act amended to 'enable the NEA to function autonomously' NEA transformed from being as a sole player to a licensee to buy electricity generated by private IPPs |
| 1993 | Electricity Regulations, 1993, introduced to operationalise the Electricity Act, 1992 to enable entry of IPPs |
| 2001 | Hydro-Power Development Policy (2058) 2001 issued to develop country's hydro resources including those for export purposes |
| 2006 | Rural Energy Policy 2006 issued to address, among others, the energy needs of the rural population, creation of a rural energy subsidy scheme |
| 2013 | Subsidy Policy for Renewable Energy (2069) issued to increase access to renewable energy to low-income households through subsidy and access to credit, to support rural electrification and to attract private investors |

*Data source*: Nepal and Jamasb (2012).

The performance of the Nepalese power sector after more than two decades of reforms is not satisfactory when assessed against the anticipated outcomes. Transmission and distribution electricity losses are relatively high at 25%. Around 50% of the population is connected to the grid, while around 25% have off-grid access. However, the supply and demand gap was about 410 MW resulting in load shedding up to 14 hours a day in 2013 (World Bank, 2015). As such, the lack of access to reliable grid electricity is one of the key obstacles in overcoming poverty reduction and a major constraint to an export-led economic growth. Electricity tariffs continue to remain below cost-recovery levels. However, notable progress has been made with IPPs, which account for 33% of total installed capacity.

A Bilateral Power Trading Agreement between Nepal and India commenced in 1966 with an initial exchange of 5 MW in 1971. Three river treaties (Koshi, Gandak, and Mahakali) were also signed between these countries to support power generation and trade (Saran, 2014). Power Trading Corporation representing the GOI has signed a power purchase agreement with Nepal for supply of 150 MW for 25 years through the upgrade of existing 132 kV transmission links. In 2014, a Power Trade Agreement (PTA) was signed between Nepal and India although its implementation continues to be pending (Bhat, 2016). Hence, the PTA has not yet delivered the anticipated benefits arising from market accessibility due to slow progress in implementation although the hopes are high.

Like Bhutan, Nepal also relies on hydropower which is considered to be clean and renewable, originating mostly from run of river projects. Nepal has the potential to generate 83 GW (with around 40 GW commercially feasible) of hydroelectricity. Hence, the environmental benefits in terms of carbon offsets by generating hydroelectricity and trading with India (and even South Asia) are large (Timilsina et al., 2015). Nonetheless, unlike Bhutan, Nepal does not have any mandate to ensure any minimum forest coverage of its territory.

It is evident that regional electricity market integration through hydropower in the three South Asian countries presents one solution that meets the four objectives laid out at the beginning of Section 3. The electricity sector/market structures of the three countries arguably facilitate this – for instance, Bhutan and Nepal represent a sector structure similar to Models B and C, whereas India represents a market similar to D and E. Market integration particularly

*Table 15.6* Gains from increased South Asia electricity integration

| | |
|---|---|
| Changes in countries' total installed generation by 2040 (GW) | Afghanistan (+4), Bangladesh (− 11), Bhutan (+9), India (− 35), Nepal (+52), Pakistan (− 13), Sri Lanka (− 1) |
| Changes in regional installed generation capacities by 2040 by technologies (GW) | Hydro (+72), Coal (− 54), gas (− 6), Wind (− 7) |
| Changes in cross-border & inter-grid transmission capacities (GW) | Net increase in cross-border transmission capacity (+95); Inter-grid capacity in India (− 37) |
| Reduction of regional power sector $CO_2$ emissions | 8% |

*Data source*: Toman and Timilsina (2016).

presents significant gains for emissions reduction. It is estimated that every GWh of coal-fired generation in India releases close to 1000 tonnes of $CO_2$ (see Figure 15.2) (Wijayatunga et al., 2015). Indeed, some studies attempting to quantify the benefits of such as solution arrive at positive results. Toman and Timilsina (2016), for instance, set out the gains in terms of a more balanced distribution of regional installed capacity, composition of installed capacity – reductions in thermal (coal) and increases in renewable (primarily hydro) capacity, increases in transmission capacity, and resultant reductions in emissions, relative to a baseline scenario (see Table 15.6).

Wijayatunga et al. (2015) examines the economic benefits of six interconnections within South Asia (Bhutan–India power transmission reinforcement, India Nepal 400 kV; India–Sri Lanka Submarine Link; Bangladesh–India HVDC Link; India–Pakistan 220/400 kV; CASA 1000 Link) by comparing their investment costs and their gains – importantly, the latter includes the benefits of unserved energy reductions (i.e. alleviating power shortages and facilitating greater access). It finds that the six interconnections could lead to a total benefit of over US$4 billion against a cost of one-tenth of that amount. When unserved energy reductions are excluded, the benefits are largely confined to fuel (or dispatch) related cost savings; the break-even utilisation level for a high-cost link to recover a decent return on investment would be very high (Wijayatunga et al., 2015). Unserved energy reductions are therefore an important distinction in developing countries as opposed to developed countries, and imply that regional electricity market integration in developing Asian is not subject to the same caveats as in OECD markets.

## Conclusions

This chapter has described and analysed the seemingly contradictory objectives of electricity sector reform: between economic efficiency and environmental sustainability via decarbonisation, with specific regard to developing Asia. While electricity sector reforms have aimed at economic efficiency, there has also been a recent move towards incorporating decarbonisation as an objective of reforms, as an effective way of reducing emissions.

Our review of electricity sector reform with reference to economic efficiency concluded that despite the initial promise of electricity reforms, the 'one-size-fits-all' approach has not worked in Asian developing economies. One reason for this is that, while standard reform models implicitly assume stable institutional, regulatory, and political frameworks, many developing countries lack strong institutions required to underpin successful reform (Gratwick and Eberhard, 2008). In addition, factors such as initial resource endowments and the size of electricity systems have also been identified as having the potential to affect outcomes (Jamasb et al. 2015; Nepal and Jamasb, 2011). Other differences in starting conditions are also important (e.g. reforming in a situation of excess capacity where average costs are above marginal costs, as opposed to deficits, where they are below marginal costs).

Two factors have stood out on why developing Asian economies have been unsuccessful in implementing reforms to achieve economic objectives. The first is the continual inability of governments to implement cost-reflective pricing. This has given rise to a circular but crucial problem. The absence of cost-reflective pricing to begin with implied that reform was likely to lead to rising prices; thus public opposition to these price increases typically created an *ex ante* impediment (Sen and Jamasb, 2013; Littlechild, 2000; Newbery, 2000).

A second factor is a shift, sometime in the 2000s, in the focus of reforms from operating efficiency of utilities, to capacity addition, given growing concerns over 'energy security' and shortages. For example, in India captive generation was encouraged as a way of circumventing the lack of grid capacity addition;[35] it has been argued that this 'dichotomy' in electricity sector structure – the state sector coexisting alongside the private sector – had an underlying political basis (Joseph, 2010). It allowed a separation of the problem of supply and shortages from the problem of deteriorating finances in state-owned utilities, thus reinforcing the 'electricity–politics nexus' and circumventing the issue of cost-reflective pricing (Joseph, 2010).[36]

As a result, despite over two decades of reform involving the standard model and its variations, few developing countries, let alone developing Asian countries, have successfully progressed to the extent of full liberalisation. These intrinsic obstacles have in the past been regarded as an evolving part of the reform process, overcome through the gradual strengthening of institutions and competition. However, a recently added dimension to the discourse relates to suitability of the OECD model to achieve not just economically efficient but also environmentally sustainable electricity systems, with renewables playing a greater role.

Using three South Asian country case studies, we have argued that the two goals can be potentially balanced by utilising the synergies brought about via greater regional electricity market integration. There are, however, hurdles to this, such as infrastructure and efforts required to translate bilateral trading agreements into competitive market arrangements. Arguably, one of the determinants of whether market integration will be realised is the cost of off-grid electrification solutions versus the cost of enabling infrastructure and market arrangements required towards greater regional market integration. This arguably presents a whole new set of challenges for Asian countries, and also presents an avenue for further research.

## Notes

1 Some of the arguments in this chapter were first developed in a working paper: Sen (2014). 'Divergent Paths to a Common Goal: An Overview of Challenges to Electricity Sector Reform in Developing versus Developed Countries', EL10, Oxford Institute for Energy Studies.
2 Privately owned electricity generation companies which produce electricity for sale to utilities.
3 The creation of separate legal entities.
4 Also, the 'textbook' or 'prescriptive' approach (Gratwick and Eberhard, 2008; Joskow, 2008; Victor and Heller, 2007).
5 Specifically pertaining to improvements in the efficiency of operation of utilities brought about by greater competition, which could then be passed on to consumers through competitive (potentially lower) prices and better quality of service.
6 For instance, the political factors underpinning electricity reform in the UK included weakening the power of the coal unions and the CEGB. But there was also the expectation that competition would lead to improvements in the efficiency of operations. Further, privatisation and restructuring in a system with excess supply involved very little risk.
7 It should also be noted that these political considerations may have varied, although reform converged around the same set of principles. Thus for instance, while in Norway reforms were aimed at a pragmatic restructuring of the electricity sector and were initiated by a Labour government, privatisation was off the

table from the very beginning. In the EU, reforms were a part of the wider integration process, although member-states had the freedom to decide upon their policy paths. And in the UK, there are still debates around whether it was politics or economics that influenced initial reforms (Rutledge, 2010).

8 For example, falling prices of fuel inputs could have coincided with reform; similarly, improved labour productivity could simply be attributed to cuts in the labour force following privatisation (Rutledge, 2010). The existence of excess capacity at the outset in most of these early reformer countries was also indicative of lower risks (Sen, 2014). Further, technological changes within the industry which were to some extent endogenous to the reform process itself, also played an important role.

9 Newbery (2013); Keay et al. (2013a, 2013b); Rutledge (2010); Bye and Hope (2005); Magnus (1997); Newbery and Pollitt (1997).

10 This was a watershed year as the Copenhagen climate change summit saw major developing countries like India adopt 'national missions' on climate change mitigation, laying the ground for COP21 in December 2015.

11 The literature indicates that much of this was tied to the introduction of private investment through IPPs (Williams and Ghanadan, 2006).

12 That is, limiting the risks associated with large-scale investment. Governments can also use a public or privately-owned monopoly to finance public objectives that the company may otherwise not undertake, such as rural electrification.

13 Norway carried out reforms entirely within the public sector.

14 A 'scorecard' for electricity reforms, citing the different steps and the sequence, was developed to broadly measure progress (Bacon, 1999). Also see Jamasb et al. (2004) for evidence on the determinants and performance of reforms.

15 IPPs were also a way of adding capacity.

16 Maldives aims to achieve 'carbon neutrality' in energy by 2020.

17 Industrial consumers cross subsidises agricultural consumers; hence in the absence of tariff reform, open access has serious financial consequences for public utilities.

18 Followed by transport – however, many countries are aiming for the electrification of transport to mitigate emissions.

19 This section draws from Keay et al. (2013a, 2013b); Buchan and Keay (2014); Keay (2009; 2013a, 2013b), Robinson (2013), Rhys (2013) and Sen (2014).

20 Such as by a government or regulator.

21 E.g. wind, and solar in the absence of large-scale commercial storage technologies.

22 In other words, the easiest and best sites for the development of renewable energy are likely to be used up early on (Keay, 2013a).

23 Referring, for instance, to the difficulties governments have had with implementing cost-reflective pricing, despite the last two decades of electricity reform.

24 See 'Notice to Centre, Delhi government, after plea on pollution by Toddlers', Indian Express, 8 October 2015. www.ndtv.com/delhi-news/notice-to-centre-delhi-government-after-plea-on-pollution-by-toddlers-1229910.

25 See www.huffingtonpost.com/jake-schmidt/too-much-public-funding-i_b_4314333.html.

26 See http://ec.europa.eu/clima/policies/package/index_en.htm for details. The EU Emissions Trading System has been widely criticised for its ineffectiveness.

27 See Singh et al. (2015) for a discussion of how political economy factors can sometimes constrain regional electricity market integration, using a case study of South Asia.

28 Latin America, and the Greater Mekong region in East Asia.

29 Enron's early attempt to set up an IPP in India ran into significant problems, with allegations over excessively high tariffs in the Power Purchase Agreement with the State Electricity Board (which the Board allegedly could not afford to pay) leading to the state government of Maharashtra reneging on its agreement with Enron. After several years mired in expensive litigation, Enron exited the project, shortly before its own financial collapse. The project has since been taken over by a public consortium, which is still struggling to revive it to its originally envisaged full capacity. See Mukherjee (2014) for a discussion of private participation in generation in India and the Enron experience and Sant and Dixit (1995) for arguments in favour of the cancellation of the project at the time.

30 Orissa in 1996, Haryana in 1997 and Andhra Pradesh in 1998.

31 Approximated at 11% of the total electricity market. The short-term power market in India covers contracts of less than a year's duration transacted through trading licensees, power exchanges, direct trading between distribution companies, and through 'unscheduled interchanges'. (CERC, 2013).

32 The details of this programme can be found at http://powermin.nic.in/upload/pdf/ultra_mega_ project.pdf.
33 The Indian government in 2014/15 carried out a series of 'reverse auctions' for coal to IPPs. This has been criticised as an unsustainable solution from the point of view of climate/environmental goals, as it encourages the use of coal.
34 For instance, in order to meet the solar target (solar comprised 5GW of installed capacity in 2016) the solar energy market has to add 12 GW/year, from current levels of 1 GW/year. (See Sen, 2016).
35 In India, the Value of Lost Load (VOLL) was estimated at Rs. 34–122 per kWh, compared with the weighted average of short-term prices on power exchanges at Rs. 4.96 (CERC, 2013). Captive capacity is believed to account for a third of India's total installed capacity.
36 Industries can generate independently, whereas state utilities which serve residential and agricultural consumers struggle with cost-reflective pricing.

# References

ADB (2010). *Sector Assessment (Summary): Energy, Country Partnership Strategy: Bhutan, 2014–2018*, Manila, Philippines: Asian Development Bank.
Arora, V. and Dema, C. (2016). Bhutan Should Come Clean on Hydropower Megaplan, *The Diplomat*, Available at: http://thediplomat.com/2016/02/bhutan-should-come-clean-on-hydropower-megaplan/
Bacon, R. (1999). 'A Scorecard for Energy Reform in Developing Countries', *Viewpoint*, note no. 175, World Bank, Washington, D.C.
Besant-Jones, J.E. (2006). 'Reforming Power Markets in Developing Countries: What Have We Learned?' Paper No. 19, Energy and Mining Sector Board Discussion paper, The World Bank Group.
Bhat, S. (2016). Post-PTA Indo-Nepal Power Trade, SAARC Dissemination Workshop on Study for Development of Potential Regional Hydro Plant in South Asia, Kathmandu, 09–10 May 2016.
Buchan, D. (2013). 'Limiting State Intervention in Europe's Electricity Markets', Oxford Energy Comment, Oxford Institute for Energy Studies, November.
Buchan, D. and Keay, M. (2014). 'The EU's New Energy and Climate Goals for 2030: Under-Ambitious and Over-Bearing?' Oxford Energy Comment, Oxford Institute for Energy Studies, January.
Bye, T. and Hope, E. (2005). 'Deregulation of Electricity Markets—The Norwegian Experience', Discussion Paper No 433, Research Department, Statistics Norway.
CERC (2013). *Report on Short Term Power Market in India, 2012–13*, Economics Division, Central.
Dubash, N.K. and Singh, D. (2005). 'Of Rocks and Hard Places: A Critical Overview of Recent Global Experience with Electricity Restructuring', *Economic and Political Weekly*, 10–16 December.
Gratwick, K.N. and Eberhard, A. (2008). 'Demise of the Standard Model for Power Sector Reform and the Emergence of Hybrid Power Markets', *Energy Policy*, 36(10), 3948–60.
Helm, D. (2010). 'Credibility, Commitment, and Regulation: Ex Ante Price Caps and Ex Post Interventions', in Hogan, W. and Sturzenegger, F. (eds.), *The Natural Resources Trap Private Investment without Public Commitment*, Cambridge, MA: MIT Press.
Henisz, W.J. and Zelner, B.A. (2002). 'Political Risk Management: A Strategic Perspective', WP 2002-06, Reginald H. Jones Centre, The Wharton School, University of Pennsylvania. Available at: http:// citeseerx.ist.psu.edu/viewdoc/download?doi=10.1.1.513.4456&rep=rep1&type=pdf
IEA (2014). *World Energy Outlook*, Paris: International Energy Agency.
IEA (2015). *CO₂ Emissions from Fuel Combustion*, Paris: International Energy Agency.
IPCC (2007). *Climate Change 2007*, in Parry, M.L., Canziani, O.F., Palutikof, J.P., van der Linden, P.J. and Hanson, C.E. (eds.), *Contribution of Working Group II to the Fourth Assessment Report of the Intergovernmental Panel on Climate Change*, Cambridge, UK and New York: Cambridge University Press.
Jamasb, T., Mota, R., Newbery, D., and Pollitt, M. (2004). 'Electricity Sector Reform in Developing Countries: A Survey of Empirical Evidence on Determinants and Performance', Cambridge Working Papers in Economics 0439, Faculty of Economics, University of Cambridge.
Jamasb, T., Nepal, R., and Timilsina, G.R. (2015). 'A Quarter Century Effort Yet to Come of Age: A Survey of Power Sector Reform in Developing Countries', Policy Research Working Paper 7330, June, Development Research Group, The World Bank Group, Washington, D.C.
Joseph, K.L. (2010). 'The Politics of Power: Electricity Reform in India', *Energy Policy*, 38(1), 503–11.
Joskow, P. (2008). 'Lessons Learned from Electricity Market Liberalization', *The Energy Journal*, 29(2), 9–42.

Keay, M. (2009). 'Electricity Market Liberalisation in the UK: the End is Nigh', OIES presentation, to the Electricity Policy Research Group in Cambridge on 26 January, published 1 February.

Keay, M. (2010). 'Can the Market Deliver Security and Environmental Protection in Electricity Generation?', in Rutledge, I. and Wright, P. (eds.), *UK Energy Policy and the End of Market Fundamentalism*, Oxford: OIES/OUP.

Keay, M. (2013a). 'Renewable Energy Targets: the Importance of System and Resource Costs', Oxford Energy Comment, Oxford Institute for Energy Studies, February, Oxford, UK. Available at: www.oxfordenergy.org/wpcms/wp-content/uploads/2013/02/Renewable-energy-targets-the-importance-of-system-and-resource-costs.pdf

Keay, M. (2013b). 'UK Electricity Market Reform and the EU', Oxford Energy Comment, Oxford Institute for Energy Studies, April.

Keay, M., Rhys, J., and Robinson, D. (2013a). 'Electricity Market Reform in Britain: Central Planning Versus Free Markets', in Sioshansi, F.P (ed.), *Evolution of Global Electricity Markets: New Paradigms, New Challenges, New Approaches*, London: Academic Press, Elsevier.

Keay, M., Rhys, J., and Robinson, D. (2013b). 'Decarbonization of the Electricity Industry—Is There Still a Place for Markets?', OIES Working Paper EL9, Oxford Institute for Energy Studies, November.

Littlechild, S. (2000). 'Privatization, Competition and Regulation in the British Electricity Industry, with Implications for Developing Countries', Report 226/00 Energy Sector Management Assistance Programme (ESMAP)/World Bank, Washington, D.C., February.

Magnus, E. (1997). 'Competition without Privatisation—Norway's Reforms in the Power Sector', *Energy for Sustainable Development*, 3(6), 55–61.

Mohanty, D. (2015). 'Orissa Govt Cancels Licence of 3 Reliance Infra Power Discoms', *The Indian Express*, 5 May.

Mukherjee, M. (2014). 'Private Participation in the Indian Power Sector: Lessons from Two Decades of Experience', *Directions in Development: Energy and Mining*, World Bank Group.

Nepal, R. and Jamasb, T. (2011). 'Reforming the Power Sector in Transition: Do Institutions Matter?', Cambridge Working Papers in Economics (CWPE) 1125 & Electricity Policy Research Group Working Paper (EPRG) 1109, Electricity Policy Research Group, Faculty of Economics, University of Cambridge, February.

Nepal, R. and Jamasb, T. (2012). 'Reforming Small Electricity Systems under Political Instability: The Case of Nepal', *Energy Policy*, 40, 242–51.

Newbery, D. (2000). *Privatization, Restructuring and Regulation of Network Utilities*, Cambridge, MA: MIT Press.

Newbery, D.M. (2013). 'Evolution of the British Electricity Market and the Role of Policy for the Low-Carbon Future', in Sioshansi, F.P (ed.), *Evolution of Global Electricity Markets: New Paradigms, New Challenges, New Approaches*, London: Academic Press, Elsevier.

Newbery, D.M. and Pollitt, M.G. (1997). 'The Restructuring and Privatization of the U.K. Electricity Supply—Was it Worth it?' *Public Policy for the Private Sector*, Note No. 124, The World Bank, September.

Percebois, J. (2013). 'The French Paradox: Competition, Nuclear Rent, and Price Regulation', in Sioshansi, F.P (ed.), *Evolution of Global Electricity Markets: New Paradigms, New Challenges, New Approaches*, London: Academic Press, Elsevier.

Pollitt, M.G. (2004). 'Electricity Reform in Chile: Lessons for Developing Countries', CMI Working Paper 51 & Cambridge Working Papers in Economics (CWPE) 0448, Electricity Policy Research Group, Faculty of Economics, University of Cambridge.

Rhys, J. (2013). 'Current German Energy Policy—The "Energiewende": A UK and Climate Change Perspective', Oxford Energy Comment, Oxford Institute for Energy Studies, April.

Riesz, J., Gilmore, J., and Hindsberger, M. (2013). 'Market Design for the Integration of Variable Generation', in Sioshansi, F.P (ed.), *Evolution of Global Electricity Markets: New Paradigms, New Challenges, New Approaches*, London: Academic Press, Elsevier.

Robinson, D. (2013). 'Living with Intermittent Renewable Power: Challenges for Spain and the EU', Oxford Energy Comment, Oxford Institute for Energy Studies, June.

Rutledge, I. (2010). 'UK Energy Policy and Market Fundamentalism: A Historical Overview', in Rutledge, I. and Wright, P. (eds.), *UK Energy Policy and the End of Market Fundamentalism*, Oxford: OIES/OUP.

Rutledge, I. and Wright, P. (eds.) (2010). *UK Energy Policy and the End of Market Fundamentalism*, Oxford: OIES/OUP.

Sant, G. and Dixit, S. (1995). 'The Enron Controversy: Techno-economic Analysis and Policy Implications', *Prayas*, September 1995.

Saran, H. (2014). 'Role of Trading Companies in Cross-Border Power Trade, South Asia Regional Workshop on Competitive Electricity Markets—Design, Implementation and Benefits', 18–20 March, Colombo, Sri Lanka.

Sen, A. (2014). 'Divergent Paths to a Common Goal? An Overview of Challenges to Electricity Sector Reform in Developing versus Developed Countries', EL10, Oxford Institute for Energy Studies.

Sen, A. (2016). 'India's Climate and Energy Goals: Contradictory or Complementary?', *Oxford Energy Forum*, May 2016.

Sen, A. and Jamasb, T. (2013). 'Not Seeing the Wood for the Trees? Electricity Market Reform in India', in Sioshansi, F.P (ed.), *Evolution of Global Electricity Markets: New Paradigms, New Challenges, New Approaches*, London: Academic Press, Elsevier.

Sen, A., Nepal, R., and Jamasb, T. (2016). 'Reforming Electricity Reforms? Empirical Evidence from Non-OECD Asia', Oxford Institute for Energy Studies, February.

Singh, A. (2010). 'Towards a Competitive Market for Electricity and Consumer Choice in the Indian Power Sector', *Energy Policy*, 38(8), 4196–208.

Singh, A., Jamasb, T., Nepal, R., and Toman, M. (2015). 'Cross-Border Electricity Cooperation in South Asia', World Bank Policy Research Working Paper, WPS 7328. World Bank, Washington, D.C.

Timilsina, G.R., Toman, M., Karacsonyi, J.G., and de Tena Diego, L. (2015). 'How Much Could South Benefit from Regional Electricity Cooperation and Trade?' Policy Research Working Papers WPS7341, The World Bank, Washington, D.C.

Toman, M. and Timilsina, G. (2016). 'The Benefits of Expanding Cross-Border Electricity Cooperation and Trade in South Asia', IAEE Energy Forum, Second Quarter 2016.

Tongia, R. (2003). 'The Political Economy of Indian Power Sector Reforms', PESD Working Paper #4, Stanford University.

UKERC (2010). 'Electricity Market Design for a Low Carbon Future', UK Energy Research Centre Working Paper, UKERC/WP/ESM/2005/004, October.

USAID (2014). 'Regional Energy Security For South Asia, Energy For South Asia, Regional Report', South Asia Regional Initiative/Energy. Available at: http://pdf.usaid.gov/pdf_docs/pnads866.pdf

Victor, D. and Heller, T.C. (eds.) (2007). *The Political Economy of Power Sector Reform: The Experiences of Five Major Developing Countries*, Cambridge: CUP.

Wijayatunga, P., Chattopadhyay, D., and Fernando, P. (2015). *Cross Border Power Trading in South Asia: A Techno-Economic Rationale*, Asian Development Bank.

Williams, J.H. and Ghanadan, R. (2006). 'Electricity Reform in Developing and Transition Countries: A Reappraisal', *Energy*, 31, 815–44.

Wisuttisak, P. (2012). 'Regulation and Competition Issues in Thai Electricity Sector', *Energy Policy*, 44, 185–98.

World Bank (2015). *Nepal—Power Sector Reform and Sustainable Hydropower Development Project*, Project Appraisal Document, Washington, D.C.

Wu (2005). 'China's Electric Power Market: The Rise and Fall of IPPs' Working Paper 45, Stanford Programme on Energy and Sustainable Development.

Wu, X. and Sulistiyanto, P. (2013). 'Independent Power Producer (IPP) Debacle in Indonesia and the Philippines: Path Dependence and Spillover Effects', National University Singapore. Available at: http://lkyspp.nus.edu.sg/wp-content/uploads/2013/03/IPP_debacle1.pdf [Accessed 8 November 2015].

Yangki, T. and Tashi, T. (2016). Bhutan Country Presentation, SAARC Dissemination Workshop on Study for Development of Potential Regional Hydro Plant in South Asia, Kathmandu, 9–10 May.

Zhang, Y., Kirkpatrick, C., and Parker, D. (2008). 'Electricity Sector Reform in Developing Countries: An Econometric Assessment of the Effects of Privatisation, Competition, and Regulation', CRC Working Paper 31/2002, Institute for Development Policy and Management, University of Manchester.

# 16

# DEREGULATION, COMPETITION, AND MARKET INTEGRATION IN CHINA'S ELECTRICITY SECTOR

*Yanrui Wu, Xiumei Guo, and Dora Marinova*

## Introduction

China has been the world's largest electricity user as well as producer since 2011. The country has also been engaged in cross-border trading in electricity with several neighbours, namely, Laos, Myanmar, and Viet Nam. China is a member of the East Asia Summit group (EAS) which aims to interconnect their power grids and hence develop an integrated regional electricity market in East Asia (Wu et al., 2014). Internally, in order to develop an integrated domestic market and improve efficiency, China's electricity sector has undergone dramatic changes and reforms. This process is not complete yet and further restructuring is anticipated in the near future. Thus, a study of China's electricity sector may help elicit important insights into issues such as deregulation, competition, and market integration. The findings may also have implications for other developing economies that are undertaking a similar trajectory of reforms.

Several existing studies have focused on China's electricity sector. For example, the role of the private sector in China's power generation was the theme of a World Bank (2000) conference. Also, an Asian Development Bank (ADB) report examined electricity demand and investment requirements (Lin, 2003). Several years later, a study by the International Energy Agency or IEA (2006) discussed further reforms after the 2002 restructuring and provided policy recommendations for the Chinese government, while Yang (2006) presented a brief review of China's electricity sector.

More recently, a short report by ADB (2011) provided observations and suggestions about China's electricity sector; an IEA (2012) project explored the policy options for low-carbon power generation in China; and an ERIA discussion paper (Sun et al., 2012) examined barriers to private and foreign investment in China's power sector. However, these existing research works are either outdated or concerned with a specific issue. Thus, this study aims to present an updated examination of various issues in China's power sector, especially on reforms and market integration. It begins with a review of China's electricity industry in the second section. This is followed by a discussion of major reforms in the sector in the third section. More recent reform initiatives in several regions are highlighted in the fourth section. The challenges of further reforms are then explored in the fifth section. The chapter concludes with discussion of policy implications in the final section.

## China's electricity sector

Demand for electricity has shown robust growth for decades in China (Figure 16.1). In particular, it doubled between the years 1990 and 2000, and trebled between 2000 and 2010. Growth has, however, slowed down as the Chinese economy is restructuring in response to population ageing, falling export demand, and rising domestic wages. In 2015 electricity consumption recorded the lowest annual growth of 0.5 percent in recent decades. In 2011, China overtook the United States to become the world's largest power consumer. According to the latest statistics, China's consumption share of the world's electricity was 22.9 percent in 2013, while the US share continuously declined to 18.5 percent (Figure 16.2). Power demand in China is now more than the combined total consumption in Japan, Russia, India, Germany, Canada, Brazil, and France. However, on a per capita basis, China's power consumption is only a fraction of that in major economies such as the United States and Japan (Figure 16.3).

As the Chinese economy flourishes, there remains considerable room for further growth in both per capita and total electricity consumption. For example, electricity demand in China will reach 8767 terawatt hours (TWh) in 2035, according to the ADB (2013). That level would double China's total consumption in 2010. In terms of per capita consumption, China would only proximate the current level of demand in Russia or Japan. According to Wu (2013a), China's per capita consumption of electricity in 2050 will reach 9300 kilowatt hours (kWh), which is close to the current consumption level in high-income OECD economies in 2011 (WDI, 2013).

At the sector level, manufacturing still accounts for the lion's share of China's total electricity consumption due to the ongoing rapid industrialisation (Figure 16.4). In 2013, the manufacturing sector used 73.5 percent of China's total electricity consumption, which is slightly smaller than its 79.3 percent share in 1990. Therefore, while manufacturing's share of China's electricity consumption is still high, it is declining. In comparison, the Japanese manufacturing sector's share dropped from 70.2 percent in 1973 to 29.7 percent by 2011.

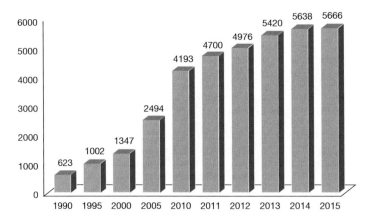

*Figure 16.1* Electricity consumption in China, 1990–2015

*Data source*: NBS (various issues) and NEA (2014).

*Note*: The unit on the *y*-axis is terawatt hours (TWh).

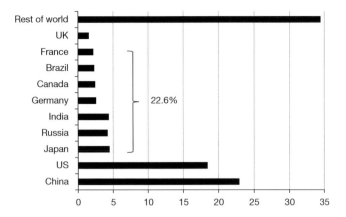

*Figure 16.2*   Electricity consumption shares (%) in major economies in 2013

*Data source*: The numbers are calculated using data from WDI (2016).

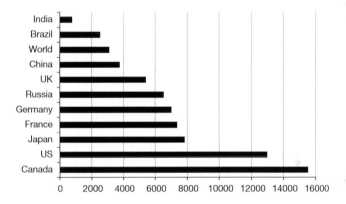

*Figure 16.3*   Electricity consumption per capita in major economies in 2013

*Data source*: WDI (2016).

*Note*: The unit of measurement is kilowatt hours (kWh/person).

Likewise, that of South Korea slid from 69.0 percent in 1973 to 52.3 percent by 2011 (OECD, 2014). If these are any indications of China's own trajectory, then the country's share of manufacturing sector in electricity consumption is also expected to continue to fall in the coming decade.

However, power consumption in the service and household sectors is growing faster than that in the primary and manufacturing sectors. For example, the average percentage growth rates during 2005–2013 are 3.4 percent for the primary; 9.6 percent, industrial; 12.1 percent, service; and 11.3 percent, residential sector. As a result, consumption shares of households and services increased from 7.7 percent and 6.2 percent in 1990, to 12.8 percent and 11.8 percent in 2013, respectively. During the period 1973–2011, these shares respectively rose from 19.1 percent and 10.5 percent, to 30.9 percent and 38.8 percent in Japan; and from 12.1 percent and 18.3 percent, to 13.1 percent and 32.3 percent in South Korea (OECD, 2014). There is, hence, considerable room for growth in the electricity consumption of China's own household and service sectors.

One of the features in China's electricity sector is the uneven distribution of resources across its regions. In particular, the coastal regions tend to be net importers of electricity while the western regions are net exporters. Thus, cross-regional electricity trade in China is inevitable. This requires efficient transmission lines and an integrated market. For example, Xinjiang's power grid was connected with the northwest power grid in 2010 and has since exported electricity to the rest of the country, including Jiangsu and Zhejiang (CP, 2013). In 2013, the total power exported from Xinjiang amounted to 6 TWh, according to Xinhua News Agency (2014a).

There is also some cross-border power trading between China's Yunan province and Laos, Myanmar, and Viet Nam. The first cross-border transmission between China and Laos took place in 2001, and that between China and Viet Nam in 2004. China reportedly exported 3.2 gigawatt hours (GWh) to Viet Nam and 0.2 GWh to Laos in 2013. In the same year, Yunan also imported about 1.9 GWh from Myanmar (MOC, 2014). So far, the total power exchanges are valued at about US$1.5 billion. Heilongjiang in Northeast China has also been importing electricity from Russia amounting to about 13 GWh since 1992.[1] Imported Russian electricity is anticipated to reach 3.6 GWh in 2014.

By 2013, China's total installed generation capacity amounted to 1247 gigawatts (GW), of which 862 GW are sourced from thermal, 280 GW from hydro, 75 GW from wind, and 15 GW from nuclear power plants (NEA, 2014). Clearly, thermal power facility takes the dominant share (see Figure 16.6). According to a Bloomberg (2013) report, China's generation capacity will be more than double in 2030, with large expansions in wind and solar energy-powered generations. This changing trend is already taking place. Of the newly installed generation capacity in 2013, more than a half is based on non-thermal sources (Figure 16.5).

The structure of production output is generally consistent with the pattern of generation capacity. Coal-fired generators still dominate thermal production and account for the largest share, followed by hydropower (Table 16.1). The market is divided between fossil fuel generation (coal, oil, and gas) with a share of 80.9 percent, and non-fossil fuel production with a share of 19.1 percent in 2011. Over time the trend is a modest decline in fossil fuel generation and an increase in non-fossil fuel electricity as shown in Table 16.1.

In the near future, coal will remain a main fuel in China. Coal-fire power is projected to still secure about 43 percent of the market share in China by 2050 (Wu 2013a). This has serious environmental consequences. It also leaves China far behind its neighbours in terms of

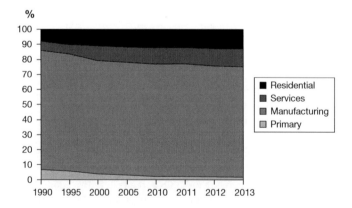

*Figure 16.4*   China's electricity consumption shares by sector, 1990–2013

*Data source*: Authors' own estimates using data from the NBS (various issues) and NEA (2014).

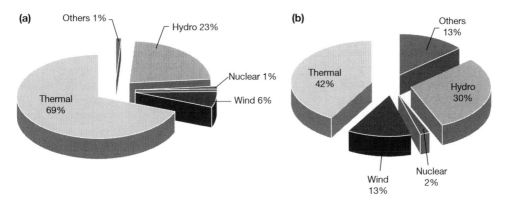

*Figure 16.5*   Structure of China's generation capacity, 2013. (a) Total installed capacity. (b) Newly installed capacity

*Data sources*: NEA (2014).

*Table 16.1* China's electricity output shares (%) in 2011 and 2013

| Fossil fuels | 2011 | | 2013 | Non-fossil fuels | 2011 | | 2013 |
|---|---|---|---|---|---|---|---|
| Coal | 78.9 | } | | Nuclear | 1.8 | | 2.1 |
| Gas | 1.8 | | 78.2 | Hydro | 14.8 | | 16.9 |
| Oil | 0.2 | | | Wind | 1.5 | | 2.6 |
| | | | | Solar | 0.1 | | |
| | | | | Biofuels | 0.7 | } | |
| | | | | Waste | 0.2 | | 0.2 |
| | | | | Others | 0.0 | | |
| Sub-total | 80.9 | | 78.2 | Sub-total | 19.1 | | 21.8 |

*Data source*: Data for 2011 are drawn from IEA (2013) and those for 2013 from NBS (various issues).

international environmental perspectives. For example, Germany will reportedly reduce its use of coal in electricity generation and increase the share of renewables from the current 25 percent to 80 percent in 2050 (The Economist, 2014). Meanwhile, in that same year, China's electricity production is projected to still be divided equally between fossil fuels and non-fossil fuels (Wu, 2013a).

## Evolution of reforms in the power sector

China's electricity sector began with a single vertically integrated utility, which the government through then its Ministry of Power Industry owns and operates. Following the global trend of deregulation, a series of reform initiatives were implemented. The first reform initiative in China's power sector was the introduction of independent power producers (IPPs) into the generation sector in the 1980s (IEA, 2006). At one point, IPPs in China cornered a 14.5 percent market share (Sun et al., 2012). By the late 1990s, all non-state generators provided more than half of the country's total electricity supplies (Wu, 2013b; Du et al., 2009). The participation of IPPs and other non-state generators were argued to play a critical role in the growth of China's power generation. While fuel and equipment prices increased dramatically, competition helped reduce the cost of generation and boosted output growth to overcome investment inadequacy and power shortage in the country in the 1990s.

The second major change was the corporatisation of the electricity businesses, thus establishing the State Power Corporation (SPC) in 1997 (Sun et al., 2012). This represents the first move to separate businesses from regulatory activities. The SPC was state-owned and a typical vertically integrated power supplier. It later became the main focus of electricity sector reforms in China.

The third wave of reforms was initiated in 2002. China's ambitious program involved the unbundling of power distribution, grid management, and generation. The goal was to introduce competition into the electricity industry. Due to this round of reforms, the SPC was divided into two grid companies, five generation companies, and two auxiliary companies (i.e. the Power Construction Corporation of China and China Energy Engineering Group Co Ltd). The two grid companies are the State Grid Corporation (SGC), which owns five regional grids; and South China Grid Corporation (SCGC), which operates the grid that interconnects five southern regions. The five power generation companies are China Huaneng Group, China Huadian Group, China Datang Co., China Guodian Co., and China Power Investment Co. (Shi, 2012). These five power providers together captured a market share of about 40 percent in 2006 (Zhang, 2008).

In the area of institutional development, the promulgation of the Electricity Act in 1995 was a hallmark. The Act laid the foundation for reforms in 1997 and 2002. To strengthen regulatory functions, the State Electricity Regulatory Commission (SERC) was formed in 2003. Its role is to promote reforms and create a market-based power industry with competing players and to set prices according to supply and demand situations in the market. Following the formation of SERC, a series of regulatory rules were released in 2005, including the first major revision of the 1995 Electricity Act (Table 16.2). Those rules and the Act have since guided the supply and demand of electricity, grid access, infrastructure development, and energy preservation in China.

However, it is argued that after a decade, SERC as an independent regulatory body still falls behind its stated goals (Shi, 2012). For example, open bidding for grid access was pilot-tested in two regional markets (Northeast and East China) but was later suspended. Government also still

*Table 16.2* China's electricity sector reform initiatives

| Year | Reform initiatives |
| --- | --- |
| 1979 | Establishment of the Ministry of Power Industry |
| 1980s | Introduction of IPPs |
| 1995 | Release of the Electricity Act |
| 1997 | Establishment of SPC |
| 2002 | Split of SPC into SG and SCG |
| 2003 | Formation of SERC |
| 2005 | Revision of the Electricity Act |
| 2008 | Formation of NEA |
| 2010 | Establishment of NEC |
| 2013 | Merger of SERC and NEA |
| 2014 | Pilot reforms in Yunnan and Inner Mongolia |
| 2015 | Release of "Document No.9" on Electricity Reform |
| 2016 | Draft of "power generation and distribution reform plan" |

*Data source*: Authors' own work.

plays the key role in price setting. In 2013, SERC and National Energy Administration (NEA) merged to form the current NEA.

In March 2014, right after the National People's Congress (NPC) and Political Consultative Conference, reforms in the electricity sector gained new momentum. During the two political gatherings, a consensus was reached to deepen economic reforms, including those in the power sector. On 18 April 2014, the National Energy Commission (NEC) held the second meeting of its kind after the first gathering in 2010. The NEC, which is led by China's prime minister, is the most powerful energy institution. Its board consists of officials from the central bank; other government bodies responsible for the environment, finance, and energy; state-owned enterprises (SOEs); and so on. This latest meeting stressed the need to construct ultra-high-voltage (UHV) electricity transmission lines as well as China's commitment to the use of nuclear energy. In addition, the NEC reaffirmed the reform of the electricity sector, particularly by introducing the direct purchase and sale of electricity between generators and large consumers. Yunnan province was designated to pilot test the scheme immediately which is to discuss in the next section.

A year later, in March 2015, China's central government released a document on "Further Strengthening the Institutional Reform of the Electric Power Industry" (or Document No.9) (Liu and Kong, 2016). On 26 November 2015, six supporting documents were published. These provide reform guidelines specifically for six key areas, namely transmission and distribution tariffs; electricity market development; electricity trading; electricity dispatch plan; retail markets; and regulation of small coal-fired power plants.[2] The latest development is to focus on reforms of electricity trading, dispatch plan and transmission and distribution tariffs. Two national trading centres in Beijing and Guangzhou are established with 13 provinces having set up local trading centres (Yan, 2016). In 2016, due to slow economic growth, Chinese power market is essentially a buyer's market which could offer the best opportunity for the implementation of further reforms. It is reported that price reforms have been "experimented" in 12 provinces and are to be extended to all regions in 2017 (Yan, 2016). More comprehensive reforms are expected in Yunnan and Guizhou provinces (Qi and Wang, 2016).

## Reform initiatives in Yunnan, Inner Mongolia, and Shenzhen

In 2014, China's policymakers gave Yunnan, Inner Mongolia and Shenzhen the go-signal to implement then the latest reform initiatives. These initiatives include the direct purchase and sale of electricity between large consumers and generators and the development of smart grids. There are two main reasons for the selection of these regions for pilot reforms. That is, (a) the markets in Yunnan and Inner Mongolia were characterised with the presence of an oversupply of power; and (b) Shenzhen is China's largest special economic zone and the country's traditional test ground for economic reforms. Yunnan's power supply is dominated by hydroelectricity, which accounted for over 80 percent of the total production in the area and maintained a growing trend though production output was almost flattened in 2015 (Figure 16.6). In 2014, total production and consumption of electricity in Yunnan reached about 235 TWh and 153 TWh, respectively. Thus one-third of Yunnan's electricity was transmitted to other regions, including exports to neighbouring countries. In 2015, electricity export earnings in Yunnan amounted to US$130 million (YNPSB, 2016).

Oversupply coupled with inadequate transmission facility means that some hydro power plants could not operate at full capacity. As the reforms allow the users and suppliers to negotiate electricity sale prices directly, such negotiation is expected to lower the price of electricity so that the region's consumers (both residential and commercial) may benefit from the low cost of electricity. Meanwhile, transmission prices are currently set according to past practices. The sum

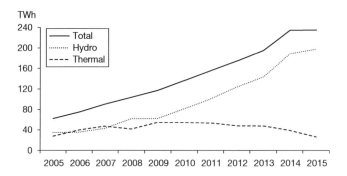

*Figure 16.6*  Electricity production in Yunnan

*Data source*: YNPSB (2016) and YNPSY (2015).

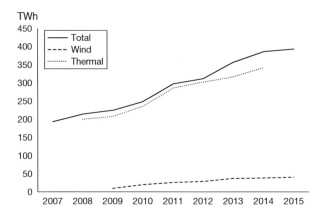

*Figure 16.7*  Electricity production in Inner Mongolia

*Data source*: IMSY (various issues) and IMSB (2016).

of the two (sale and transmission prices), plus some considerations to account for transmission power losses, would be the final electricity price. It is speculated that the prices of electricity transmission will be set through a public consultation process. After the release of Document No. 9 in March 2015, Yunnan province was approved to implement further reforms in the electricity sector. Its provincial power trading centre has been expanded. It is estimated that up to 35 percent of the region's electricity demand in 2016 will be traded in the market (Fan 2016). However, there are still complaints about government intervention in the market particularly electricity price setting (Fan 2016).

Inner Mongolia also experienced a rapid growth in electricity supply, although slower than that in Yunnan (Figure 16.7). The region is dominated by thermal power with a share close to 90 percent. Wind power accounted for about 10 percent of electricity output in recent years. The Inner Mongolia region has also faced the problem of excessive supply. In 2014 one-third of the supply had to be transmitted to other regions. Reforms in this region will focus on developing smart grids to accommodate the growth of renewable energies (REs) as well as promoting power exports to other regions. However, the progress in reforms has been slow. One known area needing immediate action is the excess supply of wind power in Inner Mongolia. This needs to be resolved so that wind farms will not have to shut down, as what had occurred in recent years. Thus, the connectivity between REs and inter-regional transmission are the priorities in this region.

In 2014, pricing reform was introduced in Inner Mongolia to allow for direct negotiation between generators and large power consumers. Under the new round of reform initiatives in 2016, direct sales and purchases have been expanded. In particular, transmission companies used to be the sole seller. Under the new initiatives more power marketing companies have been set up to distribute electricity and transmission companies are gradually withdrawing from marketing business. Thus competition becomes possible at both ends, namely generation and distribution. Prices for transmission are heavily regulated and reforms will focus on improving the transparency in price setting. While these reforms are one step forward towards a free market system, the force of market is still very limited due to the existence of electricity subsidies for residential and farming sectors in Inner Mongolia. Further reforms are needed in the future.

While Yunnan and Inner Mongolia represent two regional economies with excess electricity supply, Shenzhen is an example of a large city and net importer of electricity. Shenzhen is also a special economic zone without an agricultural sector and hence does not have to deal with the problem of agricultural power subsidy. Shenzhen was selected to implement the pilot reforms at the beginning of 2015. In the old system, the grid companies are also distributors of electricity. Thus, transmission prices are often set as the difference between the average retail prices and average generation costs. The monopolistic role of grid companies effectively guarantees high profits for these companies and the financial burden is often passed on to consumers. The reform aims to separate transmission from other electricity businesses. Essentially, in the new system, the grid companies become electricity "highway management corps" which only collect regulated toll fees from the "highway" users (Policy Watch, 2015). The new reform initiatives also allow the participation of the private sector in electricity distribution and marketing. It is estimated that the revenue of China's electricity market in 2014 was about 2720 billion RMB or US$450 billion (Shenzhen Electric Power Information, 2015). Thus, there is plenty of opportunity for multiple players in the market. However, how effective the current reforms are has to be assessed in the coming years.

## Challenges ahead

Further reforms in China's electricity sector have been well articulated by policymakers as well as scholars. But actions were stalled for years in the aftermath of the power blackout in California and supply interruption at home during severe winter weather in 2008. The current energy policy priorities include the commitment to invest in nuclear power plants along the coastal area and the construction of UHV power lines for long distance power transmission. As mentioned earlier, several regions, including Yunnan and Inner Mongolia, were selected to adopt pilot reform initiatives for direct power sales and purchases in 2014. Similar reforms were extended to more regions in 2015 and 2016. It is speculated that these reforms may be adopted nationwide in 2017. However, this new round of reforms still faces several challenges.

While the Electricity Act was promulgated in 1995 and revised in 2005, the Chinese power regulatory body (SERC) is vested with lesser authority compared to its supposed counterparts such as the Federal Electricity Regulatory Commission in the United States. The SERC has to work with two other powerful institutions; namely, the National Development and Reform Commission (NDRC) and State-owned Asset Supervision and Administration Commission (SASAC). Through its offices, the NDRC is essentially responsible for energy pricing, strategic planning, project approval, and energy efficiency. Meanwhile, SASAC is the shareholder of the power sector's state-owned enterprises (SOEs), including the SGC and SCGC. Thus, the first challenge posed is how to strengthen the autonomy and authority of the regulatory body, the SERC, so as to truly separate regulation from business activities. In 2013, the State Council merged SERC with the NEA. This consolidation demonstrates the government's intent to have a single independent regulatory body for the electricity sector. Nonetheless, the NEA still has to continue to work with NDRC and SASAC in one way or another. The recent actions indicate policymakers' resolve to carry out reforms in the power sector. As for its effectiveness, one just has to wait and see.

The second challenge is to unbundle truly power generation and transmission. In the 1990s, IPPs and other non-state invested power plants owned a large market share in power generation. This was due to incentives such as guaranteed returns, and prices and purchases offered to the private sector in the 1980s, when the Chinese economy was experiencing severe power shortage. Since the late 1990s, China's electricity market has become a buyers' market. When China became a World Trade Organization (WTO) member in 2001, the business environment for the private sector completely changed. Foreign investors were hit hard and started withdrawing from the Chinese market. Between 1998 and 2002, foreign investment share in the electricity sector fell from 14.3 percent to 7.5 percent (Chen, 2012). By the late 2000s, this share dropped to almost zero.

In the recently introduced scheme in Yunnan, the electricity price for a large power user is composed of two parts. One part is the agreed price directly negotiated between a generator and the consumer. The second part is the transmission cost determined currently by using historical information and eventually through public consultation. This mechanism has been effective to boost demand and reduce electricity prices. However, little has been discussed about the practice and conduct of public consultation. The commonly accepted practice is to regulate power transmission and deregulate generation and distribution as it is currently trialled in Shenzhen.

Third, pricing reform has been debated for years, but no action was ever taken. Several pilot tests for grid access bidding had been abolished. Since electricity generation is dominated by coal-fired technology, the price of coal matters in the determination of electricity prices. The coal market is now deregulated; hence, coal price is very much set by market conditions. However, the electricity price is still regulated. Thus, the upstream and downstream prices in the electricity sector are delinked. This delink has caused a lot of problems.

Urgent pricing reforms are therefore needed. As a first step, large electricity users, initially in seven provinces, have been allowed since 2004 to directly purchase electricity from the generators. By 2013, this reform was expanded to more than ten provinces (Smartgrids, 2014). However, the direct purchase arrangement did not catch on, and in fact was stopped in most regions by 2014. The main problem stemmed from the lack of coordinated reforms in other aspects of the electricity business (such as unbundling). In early 2014, Chinese policymakers and their advisors initiated the same reform measure anew in Yunnan. They remain convinced that large electricity users should be allowed to directly purchase power from generators and that this practice could lead to further deregulation. In 2015 and 2016, more regions have in turn been permitted to trial the new reform initiatives. It seems there may be enough momentum for extending the trial to all regions in China in 2017.

Finally, while electricity market integration is the key for effective reforms, China's power market remains fragmented due to several factors:

1    Cross-regional trade in electricity is still limited, and institutional facilities for cross-regional trade are underdeveloped. Currently cross-regional trade occurs mainly through the offi-cially designated "west-east power exchange" channel. Trading outside this system hardly exists.

2    The price of electricity has been controlled by the government for a long time. The invisible hand of the market forces plays no role in price setting nor in affecting supply and demand.

3    Although the country's grid networks are interconnected, the capacity and efficiency of long distance transmission of electricity is still constrained. Hydropower stations in Yunnan cannot operate at full capacity as surplus output cannot be sent out of the province. This is the same constraint seen in the wind and solar power production in Inner Mongolia, where the lack of smart grids hindered the utilisation of the existing facilities recently.

## Conclusions and policy implications

China has made substantial progress in the electricity sector's deregulation, competition, and market integration. Major changes took place particularly in the late 1990s and early 2000s. These changes helped China overcome power shortage, complete the construction of a national grid, and introduce multiple players in the electricity sector in a short period of time. However, the reforms seem to be stalled in the 2000s and 2010s. China still has a long way to catch up with developed economies such as the United States, the United Kingdom, and Australia in market and institutional development in the electricity sector. Although the national grids are physically interconnected, the country's electricity market remains fragmented. Thus, the electricity sector has not realised the maximum benefits of an integrated market.

Because of the dominance of state-owned enterprises in the market, governments at various levels can always find ways to intervene in businesses. As a result, electricity pricing and business activities are still tightly controlled and the role of the markets' invisible hands is limited, not to mention complete unbundling of generation, transmission and distribution of electricity. The findings in this study can be used to draw several policy implications. These cover pricing reform, institution-building, market integration, private participation and foreign investment, and renewable power sources.

China has made major efforts to improve the pricing mechanism of main fuels such as coal and oil. The domestic prices of fuels now move closely with international prices. However, electricity price in China is still tightly controlled, and hence cannot respond in a timely manner to the changing conditions in the fuel markets. This situation can affect the generation sector gravely when the fuel prices are very volatile. It is important to further reform electricity pricing so as to get the electricity price right. A gradual approach could be adopted. The first step may be to allow direct negotiations between generators and large power users as currently being trialled in some regions. The second step could involve the separation of the transmission business from the distribution side. The third step may be to expand the direct negotiation of sales to medium-size power users and allow for bidding for transmission. The policy makers' endorsement of the pilot schemes in Yunnan, Shenzhen, and Inner Mongolia is encouraging and a step towards the right direction.

Successful implementation of electricity sector deregulation in major economies such as the United States and the United Kingdom started with the establishment of an independent regulatory body. In China, the electricity sector is now composed of multiple players. China has been successful in the corporatisation of the electricity businesses initially. In terms of regulatory responsibility, multiple parties (NEA, NDRC, and SASAC) are also involved. None of those institutions can function independently of each other. This has come about partly due to the historical role of NDRC in central planning. Formerly called the National Planning Commission (NPC), the NDRC was responsible for the country's economic plans and strategies. Under the current regime, the NDRC maintains some of the functions of the old NPC. Therefore, vested interests make it impossible for either of the trio to have the ultimate authority in electricity regulation. Here is where there is a need to consolidate the regulatory tasks for execution by a single, independent body. China's telecommunication sector has been relatively successful in deregulation and may be able to offer lessons for the electricity sector.

While the main power grids in China are physically interconnected, the Chinese market is still fragmented. This is largely due to the monopoly of the grid companies and the highly regulated nature of the entire sector. An integrated electricity market would help smooth demand and use regional resources more effectively. Also, given China's vast land area, infrastructure development becomes vital for the efficient transmission of power over long distances. The country's current plan to build several ultra-high voltage transmission lines across the nation seems to be the right move. A more integrated market can help maintain stable supply and price of electricity, which is often a prerequisite for the introduction of drastic reforms. Thus, market integration and reforms mutually re-enforce each other.

In the 1990s, the private sector (particularly foreign IPPs) played an important role in helping overcome supply shortage and capital inadequacy in the Chinese market. However, ever since China became a WTO member in 2001, preferential policies towards private investment have been removed, leading almost all the private players to move out of the country's electricity sector. State-owned enterprises have now become the main players, mainly because their government connection helped them cope with large losses during bad times. This outcome is against the aim of reform efforts in the electricity sector. Thus, government policies are urgently needed to remove barriers to private participation and to invite non-SOEs back to the power sector.

China still overwhelmingly relies on fossil fuels for electricity generation. To control environmental pollution and meet the country's international climate change commitments, renewable energy should play an important role. In particular, China is currently enjoying the growth of hydropower, which is the main non-fossil source of power. When hydropower resources are exhausted, renewables will be the only source of growth in non-fossil energy. Renewable resources are, however, only available in certain conditions and their exploration only becomes economically feasible if technology is available or if supported by specific government policies. In the case of Inner Mongolia; for example, wind farms are not fully utilised because of infrastructure deficiency or lack of government support. Thus, it is important to support renewable energy growth, in particular infrastructure development such as smart grids.

## Notes

1 These import statistics were reported by Xinhua News Agency (2014b).
2 The small coal-fired power plants are often constructed by firms to provide electricity for internal use.

## References

ADB (2011), 'People's Republic of China Electricity Sector Challenges and Future Policy Directions', *Observations and Suggestions No. 2011-4*, Asian Development Bank, Manila.
ADB (2013), *Energy Outlook for Asia and the Pacific*, Asian Development Bank, Manila.
Bloomberg (2013), 'The Future of China's Power Sector: From Centralised and Coal Powered to Distributed and Renewable', *New Energy Finance*, 27 August.
Chen, Nan (2012), 'Foreign Investors Withdrawn from China's Electricity Sector', *Nanfang Zhoumo* newspaper (*South China Weekend*), 5 May.

CP (2013), 'Xinjiang Electricity to the East China Power Grid for the First Time', China Power Network Online News. Retrieved from chinapower.com.cn, 16 February.

Du, Limin, Jie Mao and Jinchuan Shi (2009), 'Assessing the Impact of Regulatory Reforms on China's Electricity Generation Industry', *Energy Policy 37(2)*, pp. 712–720.

Fan, Ruohong (2016), 'Disappointing Electricity Reform in Yunnan', *Caixin Weekly 704*, May 16.

IEA (2006), *China's Power Sector Reform: Where to Next?* Paris: International Energy Agency.

IEA (2012), *Policy Options for Low Carbon Power Generation in China*, Insights Series 2012 by International Energy Agency (Richard Barron, Andre Aasrud, Jonathan Sinton and Nina Campbell), Paris, and Energy Research Institute (Kejun Jiang and Xing Zhuang), Beijing.

IEA (2013), *China: Electricity and Heat 2011*, online statistics retrieved from www.iea.org, Paris: International Energy Agency.

IMSB (2016), *2015 Statistical Communique of Social and Economic Development in Inner Mongolia*, Inner Mongolia Statistics Bureau, March 3, www.nmgtj.gov.cn/nmgttj/tjgb/jjshfztjgb/webinfo/2016/03/1455760440407284.htm (accessed 4 August 2016).

IMSY (various issues), *Inner Mongolia Statistical Yearbook*, Beijing: China Statistics Press.

Lin, Bo Q. (2003), 'Electricity Demand in the People's Republic of China: Investment Requirement and Environmental Impact', *ERD Working Paper Series No. 37*, Economics and Research Department, Asian Development Bank, Manila.

Liu, Xiying and Lingcheng Kong (2016), 'A New Chapter in China's Electricity Market Reform', *Policy Brief 13*, Energy Studies Institute, National University of Singapore.

MOC (2014), 'Grid Connection Between the Laos, Myanmar, Vietnam and Yunnan', www.mofcom.gov.cn, Ministry of Commerce, People's Republic of China.

NBS (various issues), *China Statistical Yearbook compiled by the National Bureau of Statistics*, Beijing: China Statistics Press.

NEA (2014), '2013 Electricity Situations in China', National Energy Administration, 14 January. Retrieved from neb.gov.cn on 22 January 2013.

OECD (2014), *Electricity Information 2013*. Retrieved from www.oecd.org, Paris: OECD.

Policy Watch (2015), 'Regulate the Grids and Deregulate Generation and Distribution', *Information of Shenzhen Electric Power* 10(4), p. 7.

Qi, Hui and Yichen Wang (2016), 'A New Stage of Electricity Reforms', *Economic Daily*, May 30.

Shi, Yaodong (2012), 'China's Power Sector Reform: Efforts, Dilemmas and Prospects', unpublished presentation, Development Research Centre and Harvard Electricity Power Group, California.

Shenzhen Electric Power Information (2015), 'Private Capital Can Enter the Electricity Retailing Sector', *Information of Shenzhen Electric Power* 10(4), pp. 11–12.

Smartgrids (2014), 'China Electricity Reform and Its Impact on Vested Interest Groups', online document. Retrieved from www.smartgrids.ofweek.com, 13 January.

Sun, Xuegong, Liyan Guo and Zheng Zeng (2012), 'Market Entry Barriers for FDI and Private Investors: Lessons from China's Electricity Market', *ERIA Discussion Paper Series ERIA-DP-2012-17*, Economic Research Institute for ASEAN and East Asia, Jakarta.

The Economist (2014), 'Germany's Energy Transition: Sunny, Windy, Costly and Dirty', *The Economist*, 18 January. Retrieved from www.economist.com/news/europe/21594336-germanys-new-super-minister-energy-and-economy-has-his-work-cut-out-sunny-windy-costly

WDI (2013), *World Development Indicator* Database, October version, Washington, DC: The World Bank.

WDI (2016), *World Development Indicator 2016* (Database), Washington, DC: The World Bank.

World Bank (2000), *The Private Sector and Power Generation in China*, Washington, DC: The World Bank.

Wu, Jingyu (2013a), 'Outlook for the Electricity Sector During 2012-2050', online report from chinapower.org.cn, 25 February, China Electricity Industry Network.

Wu, Yanrui (2013b), 'Electricity Market Integration: Global Trends and Implications for the EAS Region', *Energy Strategy Reviews 2(2)*, pp. 138–145.

Wu, Yanrui, Fukunari Kimura and Xunpeng Shi (eds.) (2014), *Energy Market Integration in East Asia: Deepening Understanding and Moving Forward*, London: Routledge.

Xinhua News Agency (2014a), 'Cross-regional Transmission of Xinjiang Electricity Exceeded 6 TWh in 2013'. Retrieved from online news www.gov.cn/jrzg/2014-01/22/content_2572988.htm, 22 January.

Xinhua News Agency (2014b), 'Heilongjiang Electricity: Imports of Russian Power Reached 1.675GWh in the First Half of the Year'. Retrieved from online news http://news.xinhuanet.com/power/dw/2014-07/10/c_1111545769.htm, 10 July.

Yan, Yaowu (2016), 'Great Leap Forward in Electricity Sector Reform', *China Marketing Newspaper*, July 19.

Yang, Hongliang (2006), 'Overview of the Chinese Electricity Industry and Its Current Issues', CWPE 0617 and EPRG 0517, Electricity Policy Research Group, University of Cambridge.

YNPSB (2016), *2015 Statistical Communique of Social and Economic Development in Yunnan*, www.stats.yn. gov.cn/TJJMH_Model/newsview.aspx?id=4125442, Yunnan Provincial Statistics Bureau, April 18 (accessed 4 August 2016).

YNPSY (2015), *Yunnan Provincial Statistical Yearbook 2015*, Beijing: China Statistics Press.

Zhang, Qing (2008), 'Regulatory Framework for the Electricity Industry in China', unpublished presentation slides, China University of Politics and Law.

# 17

# ENERGY SECTOR REFORM IN CHINA SINCE 2000 FOR A LOW-CARBON ENERGY PATHWAY

*Songli Zhu, Ming Su, and Xiang Gao*

## Overall progress of energy reform since 2000

Energy reform in China, starting from 1980s when the world entered into the period of energy privatization and liberalization (Pollitt, 2012), is an essential part of China's overall reform and opening up process. This section reviews the reform process from three perspectives: institutional arrangements, energy marketization and energy pricing.

### Reforms of institutional arrangements

China's reform of institutional arrangements for energy administration system is basically about adjusting three fundamental relationships, namely the relationship between the energy industry and other macro-economic regulatory departments, the relationship between central and local governments, and the relationship between governments and enterprises.

#### Energy administration

China has adjusted the energy administration institutions for a number of times since the end of 1980s with the general purpose to transfer to a market-oriented economy (see Table 17.1 for the timeline). The pathway was not smooth, indicated by the cycle of "establishment–elimination–re-establishment" of several administrations. Overall, in the first round of institutional reform initiated in early 1990s, government administrative functions were largely separated from enterprise business operation. By the new millennium, the government has shifted to strengthen macro-management and supervision of energy. In 2002, State Electricity Regulatory Commission (SERC), as an independent regulator, was set up to perform its functions and duties of supervising the national electric power market. In 2003, National Development and Reform Commission (NDRC) was established, under which the Energy Bureau was set up to comprehensively administer national energy affairs. In 2008, National Energy Administration (NEA) was established to take charge of energy development. In 2014, the SERC was folded into NEA to achieve the combination of administration and supervision. In addition, National Energy Leading Group and National Energy Commission were set up in 2005 and 2010, respectively, with the Premier of State Council serving as the group leader and commission director, indicating China's high value of energy management. All these restructuring actions attempt to streamline the

*Table 17.1* The brief history of energy administration in China

| Year | Energy administration and its change | Driver for the change |
|---|---|---|
| Prior to 1988 | MoC, MoP, MoEP and MoN were online | Energy supply security is the primary concern |
| 1988 | The four ministries dismantled and MoE established<br>Meanwhile NPCs and NNC set up | To separate administration and business; enhancing the role of the companies in market |
| 1993 | MoE abolished; MoC and MoEP re-established. NPCs and NNC took over the administrative responsibility of oil and nuclear industry | MoE failed to achieve a good coordination among coal, oil, power and nuclear industry |
| 1998 | MoC re-dismantled; SACI established and affiliated to SETC<br>MoEP re-dismantled either; SPC established to take over the business operation of the sector and administrative functions transferred to SDPC (now-called NDRC since 2003) | To further separate the administrative functions from the business functions of the enterprises |
| 2002 | SPC divided into five generating companies and two grid companies; SERC set up | First round of power industry reform to unbundle generation and transmission |
| 2003 | SETC abolished; EB set up under NDRC to manage energy affairs | Streamline governmental institutions |
| 2005 | NELG established and led by State Council | Energy shortage calling for a higher administrative management on energy affairs |
| 2008 | NEA set up with vice-ministerial level to integrate the administration of energy industries | Promote the position of energy administration |
| 2010 | National Energy Commission set up and led by State Council | For a better long-term strategic research on energy development |
| 2014 | SERC folded into NEA and second round of power industry reform launched afterwards to unbundle transmission and distribution | Combining administration and supervision of power industry |

*Data source*: Authors.

*Note*: MoC: Ministry of Coal; MoP: Ministry of Petroleum industry; MoEP: ministry of Electric Power; MoN: Ministry of Nuclear; MoE: Ministry of Energy; NPCs: National Petroleum Corporations; NNC: National Nuclear Corporation; SACI: State Administration for the Coal Industry; SETC: State Economic and Trade Commission; SPC: State Power Corporation; SDPC: State Development and Planning Commission; SERC: State Electricity Regulatory Commission; EB: Energy Bureau; NELG: National Energy Leading Group; NEA: National Energy Administration.

administrative and regulatory authorities among various agencies, making energy department more resourceful to be in a better position to influence sector reforms and policies.

## Central–local relationship

After decades and several rounds of institutional reform, local governments are now gradually having greater autonomy. At the end of the last century, the central government transferred the management of coal enterprises to local governments. Since 2014, the central government has further promoted the streamlining and decentralisation of the administration by delegating approval authority to local governments, including most projects related to coal, thermal power, hydro power, wind power and solar photovoltaic (PV) power and part of projects related to grid and oil-gas pipelines. Meanwhile, the central government also pays attention to overall planning and regional coordination to prevent disorder and mass over capacity.

However, there are still certain deficiencies in China's energy system management at both central and local levels, such as paying more attention to project construction rather than anti-monopoly, information transparency/availability to public and oversight on environmental externality, lack of standards, and insufficient institutionalisation. Meanwhile, energy regulation currently focuses on power and nuclear safety, whereas regulations of oil and gas have just started, and regulations of heat supply network, coal and renewable energy need to be supplemented (Jiao and Wang, 2012; Fisher-Vanden and Ho, 2007; Cherni and Kentish, 2007; Fan et al., 2014).

## Government and enterprises

The energy industry is extremely crucial to national security and the lifeline of the national economy, and state-owned enterprises (SOEs) have always held the dominant position. Since 2000, SOE reforms focusing on "improving quality and efficiency" have been unswervingly pushed forward. Through a series of reforms, including introduction of corporate and share-holding systems, equity division, separation between primary and auxiliary business and freeing enterprises from performing social functions, state-owned energy enterprises are becoming more efficient. Currently, there are more than 100 state-owned energy enterprises that have gone public and built modern corporate systems.

It is still common to see regular businesses carrying the burden of social responsibilities, high debt ratios and redundant employees in state-owned energy enterprises, so the burdens on them have not been fundamentally addressed. Meanwhile, with regard to ensuring national security of oil, gas and power supply, setting national standards for oil products and regulating industry order, several large state-owned energy enterprises are still performing some government functions. Division of functions between government and enterprises should be further clarified.

## Reforms in energy marketization

With the advance of reform and opening up of economy to market forces, China's energy sector is also trying to cultivate a modern market structure with sufficient competition.

### Coal

The transformation began in coal industry in late 1970s and enhanced in 1992 (Zhu and Cherni, 2009). Further market-oriented reforms are going on in the new century. In 2001, China reformed its coal investment system by abolishing direct investment from government in coal mine construction and allowing all investors to enter the coal sector. After that, the coal industry saw a competitive landscape where multiple ownership and companies of all sizes were actively involved. By the end of 2005, the implementation of policy (No.[2005]18) freed enterprises completely from performing social functions, separated main and auxiliary business and reformed the auxiliary business; meanwhile, a modern market system for coal trading with the National Coal Transaction Center as the main body and regional coal markets as supplement was set up. This policy document further liberated coal enterprises and almost declared the completion of coal reform, which was totally marketized by the end of 2012 marked by the NDRC's withdrawal from the annual Coal Ordering Meeting (Mou, 2014).

However, this liberalization process introduced a highly fragmented coal industry featuring heavy involvement of town and village coal enterprises that tended to ignore the environmental impact of coal mining (Zhu and Cherni, 2009). Since late 2008, a consolidation process was launched leading to a marked increase in the proportion of the total output produced by large and medium-size coal mines. Liu et al. (2016) commented that the consolidation process had a negative impact on coal productivity, but significantly improved the environmental performance of the coal industry.

## Electricity

The electricity reform dates to the mid–1980s when power shortages seriously troubled industrial growth and residential livelihoods. Hence, the aim of this early reform was to expand supply, not so much the administrative management. In 2002, when economic growth was tentatively stable, the State Council passed a comprehensive set of electricity sector reforms under its Electricity Reform Plan (No.[2002]5), initiating the real market reform with the aim to break the monopoly and introduce competition, more specifically, to separate "generators" from "grid" to facilitate non-discriminatory access to the monopoly network. As part of a reform package common in electricity sector restructuring in the world, the major part of the reform is to reshape the market structure by vertically unbundling the dominant integrated firms. As a result, the State Power Corporation (SPC) was "unbundled" into five generating companies and two grid companies. This was supported by creation of market-based pricing mechanisms and regional wholesale power trade.

After ten years of reform, though the progress was impeded almost immediately by severe electricity shortages that began in late 2002 and continued through 2006(Williams and Kahrl, 2008), a competitive electricity market with multi players was generally established, witnessed by continuously growing investment, noticeable efficiency gain (Gao and Biesebroeck, 2014; Du et al., 2009) and improving service. However, new challenges are continuously emerging. First, the unreasonable income induced by high on-grid price led to the irrational investment on the generation side and over capacity (Zeng et al., 2016). Second, insufficient support was given to renewable industry and energy-saving generators, so that the high-carbon electricity sources dominate the industry with rapidly growing emission of pollutants and greenhouse gases (Zhang and Gao, 2016). Third, the responsibility of grid enterprises is ambiguous. Unlike the unbundling of generation enterprises in 2002 reform, vertically bundled systems are still found in transmission, distribution and retail components, which are largely controlled by grid companies (see Figure 17.1). The income of these natural monopoly companies comes from the price gap between on-grid and retail price, the latter of which is decided by government. Therefore, grid companies become the major reason for the curtailment of clean powers in recent years as they tend to let the generators with low on-grid price to generate more power in order to achieve enlarged profits.

Noticing all these deficiencies, China's government has carried out a new round of power industry reform in 2015 by issuing *Relative Policies on Deepening the Reform of Power Industry* (No.[2015]9), aiming to further reform pricing and planning mechanism, promote renewable generation utilization and reposition of grid enterprises (see more details in Zeng et al. (2016)). The launch of a new round of reforms in power system means China is forging ahead towards a modern

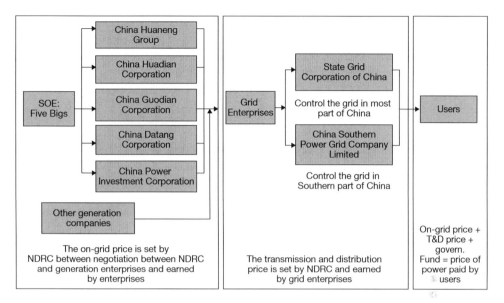

*Figure 17.1*   The structure of China's power industry after the reform of 2002

*Data source*: Authors.

and deregulated power market. Good signal has been observed in the pilot area of Yunnan Province where the on-grid price has dropped by 0.0165–0.164 RMB/kWh (Zeng et al., 2015).

## Oil and gas

The market for oil and natural gas industry is gradually opening up, but monopoly is still obvious in exploration, development, storage and transportation. Since 2004, retail and wholesale of crude and refined oil has been successively opened up to foreign and non-state traders together with a concession system. Afterwards, private refinery enterprises have been continuously growing, forming a tripod situation involving China National Petroleum Corporation (CNPC), China Petroleum & Chemical Corporation (Sinopec) and private refinery enterprises. In 2007, China National Offshore Oil Corporation (CNOOC) built its first gas station in mainland China, indicating the increasingly fierce competition in retail market of refined oil. The number of gas supply companies is increasing in all cities. All these facts show that China's oil and gas industry is going towards diversification, reflecting the vitality of the market. However, the dominant position of CNPC, Sinopec and CNOOC in the whole oil and gas system has not been shaken. The three giants almost monopolize the exploration, development and supply of oil and gas in China, and consolidate its dominance by the highly integrated operation that integrates exploitation, transportation, refining and sale, both upstream and downstream. Private and foreign invested enterprises can only operate in middle and downstream areas of the industry, often in a disadvantaged position when competing with the three giants. Currently, China is making efforts to promote a new round of reforms in oil and gas industry, which is expected to speed up the marketization of China's oil and gas industry.

## Reforms in energy pricing systems

While prices of most fossil fuels in China are still subject to regulations more or less by national, regional and local authorities, in most recent years, the extent of fuel price liberalization and consistency with international market have been further improved through the introduction of significant changes in its pricing regulation both nationally or in pilot regions, with the goal of moving towards a market-based system more closely tied to international commodity prices (OECD, 2014).

## Coal

The price of coal is largely unregulated. Since 2000, the price of coal has been basically determined by the market, except thermal coal. Since 2008, thermal coal production has only been subject to intermittent price control. In 2012, the State Council issued *Guiding Opinions on Deepening Market-oriented Reforms for Thermal Coal* (No.[2012]57), which removed key contracts and the dual-pricing system of thermal coal, marking the complete dominance of market force in coal pricing system.

## Electricity

In order to promote the reform initiated in 2002, the *Scheme for Reforming Electricity Pricing System* (No.[2003]62) was issued in 2003. According to the Scheme, the *long-term goals* of power pricing reform were: power pricing is divided into on-grid pricing, transmission /distribution pricing and retail pricing; the price for generation and retail is determined by the market; and price for transmission and distribution is determined by government. Later on, implementation plans and measures supporting the power pricing reform were issued; policies for on-grid benchmark pricing and mechanism of coal–electricity price linkage were introduced; standards for pricing of power transmission and distribution were published; pilot projects were conducted, for example, bidding for access to grid, energy-saving generation dispatching and direct power-purchase for large consumers; policies for pricing of desulfurization and de-nitration of power and renewable energy were formulated; policies for differential power pricing and demand side management power pricing were implemented and improved; and policies for ladder-type pricing for residential power consumption and direct power supply for large consumers were also introduced. Despite all these trials, some were soon suspended because of a nationwide shortage of power, as mentioned above, and electricity price is still largely controlled by government in reality, and both wholesale and retail markets are far from mature. Zeng et al. (2016) even commented that the electricity pricing mechanism became seriously distorted after the reform in 2002. In the new round of reform launched in 2015, pricing mechanism is still key to further streamlining the whole system.

## Oil

In December 2008, the State Council issued *Notice on Implementing the Price and Tax Reform of Refined Oil* (No.[2008]37), which highlighted the principle of linking price of domestic crude oil and refined oil with international oil price. In the light of the Notice, the price of domestic refined oil should be indirectly linked with crude oil price in international market with adjustment by governmental pricing competent authority according to fluctuations of international oil price. Price adjusting period has been shortened over time to reflect the frequent fluctuation of international oil price.

Nevertheless, when international crude oil price dropped significantly in 2015, the policy of "floor price" was introduced by issuing of *Price Management Approach for petroleum products*

(No.[2016]64) at the beginning of 2016. When the international crude oil price is under 40 dollars per barrel, there will be no adjustment to domestic refined oil price.

### Gas

In late 2011, the pilot of "netback market value" pricing for natural gas was demonstrated in Guangdong and Guangxi province, which altered the past "cost–plus" pricing system (see Figure 17.2), in which prices were mainly set by NDRC and local governments and only took account of the natural gas providers' costs and profits, neglected the consumers' specific needs and affordability, and also failed to guide the rational use of gas (Zhu et al. 2016). The new approach selects benchmark price and alternative energy to establish a mechanism linking natural gas price with alternative energy. Table 17.2 makes a comparison between the applications of these two methods in China. In July 2013, the NDRC extended the reform to the rest of the country and raised the natural gas price by 15% for all consumers except the residential sector (Lin et al. 2015).

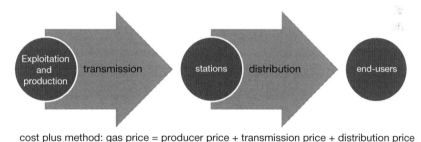

cost plus method: gas price = producer price + transmission price + distribution price

*Figure 17.2*   Old natural gas pricing mechanism based on "cost-plus" system in China

*Table 17.2* Comparison of cost-plus and netback market value method in China

| Pricing method | Cost-plus | Netback market value |
|---|---|---|
| Pricing | Producer and transmission price set by gas sources and lines | 'Station price' set by province. Benchmark price identified in Shanghai whose LPG and fuel oil taken as alternative energies to link with natural gas |
| Pricing agency | Producer and transmission price set by NDRC; distribution price set by local government | Station price cap set by government, and trade price negotiated between supplier and consumers within the cap |
| Adjustment mechanism | Adjusted by NDRC (three adjustments undertaken since 2005) | Dynamic adjustment mechanism designed in order to adjust the price at least once a year |
| Unconventional gas | No specific regulations | For shale gas, coal-bed gas and town-gas, producer price set by negotiation between producer and consumer |

*Data source*: Authors.

### *Consideration of environment externality by taxing and removing subsidies*

China is formulating corresponding tax policies specifically for the externality in exploration and utilization of energy. There are different tax rates for the exploration of coal, oil and natural gas, which have been gradually increased. Since 2014, China has been constantly raising its fuel taxes. The tax for gasoline has reached 1.52 RMB yuan per litre, and tax for diesel is 1.20 RMB yuan per litre; whereas the gasoline and diesel price is about 5.93 RMB yuan per litre and 5.55 RMB yuan per litre, respectively, in September 2016. Based on the assessment of International Energy Agency (IEA) (2007), by 2006, subsidies had been dramatically reduced and energy prices in China increasingly reflected actual cost.

However, the exploitation and utilization of coal that has the most significant impact on environment is not well integrated into the coal pricing. According to relevant calculations, the externality of the exploitation and utilization of coal is over 200 RMB per ton (Mao et al., 2008), but there is no tax policy available to control it.

Overall, deregulation of energy pricing has seen progress, but still has not yet fully reflected the scarcity of resources, let alone externalities caused by damages to ecological environment and human health.

### *Conclusion*

Although much progress has been achieved, the market-oriented reform of the energy sector generally lags behind the overall marketization process of China. China still has a long way to go in building a modern energy market system. As President Xi Jinping stated at the meeting of the Central Leading Group on Financial and Economic Affairs in June of 2014, in the future, China needs to focus on "promoting energy system revolution, unswervingly pushing forward reform, restoring commodity attribute of energy, building market structure and system featuring effective competition, forming the mechanism of market determining energy price, transforming government supervision on energy and establishing and improving energy law systems".

## Energy reforms for a low-carbon energy system

One of the most important aims is to integrate the environmental consideration into the system since the reform and liberalization re-initiated after 2000. The drivers also include, but are not limited to, energy security, climate change and being a technological leader in global energy transition to a low-carbon system (Boyd, 2012). Similar to many other countries, policies and measures (PAMs) are concentrating on renewable energy and energy efficiency.

### *Introduction to implemented PAMs*

The initiation of renewable energy (RE) in China could be traced back to 1950s when household biogas was developed in rural regions (Ling, 2009), though on a small and trial scale. During 1990s, the industrialization of RE became practical and development outlines were tentatively initiated, highlighted by the program of "Tibet Sun" (Shi, 2009). Similarly, energy efficiency policy was formally started in 1980s and achieved significant progress by central government working closely with provincial and municipal authorities during 1980–2000. Entering into the new century, a set of holistic PAMs have been adopted and implemented to reverse the trend of rapidly increasing energy demand, inferior energy structure and consequent environmental degradation.

## *Overarching PAMs*

### LEGISLATION AND PLANNING

The Renewable Energy Act (REA) was approved in February 2005 and taken into effect in January 2006, followed by the amending in December 2009 (Central government of China (CGC), 2005; 2009). The Act provided a favourable situation for RE industry by giving the priority in energy development, mandating the setting of the overall quantitative target and relevant PAMs (Schman and Lin, 2012). Meanwhile, the Energy Conservation Act (ECA), originally enacted in 1998, was amended in 2007. The newly revised ECA lays the legal basis for the measures identified in the 11th five-year plan (FYP) and further in 12th and 13th FYP.

Leading by these laws and supporting regulations, the low-carbon energy development plan and specific targets were formulated in order to guide the updating of energy system from the top-down perspective.

### QUANTITY-BASED POLICY (TARGET)

China announced the overall targets for RE development up to 2020 and 2030, before the Copenhagen Conference in 2009 (COP15) and the Paris Conference in 2015 (COP21), promising the non-fossil fuel should produce 15% and 20%, respectively, in total energy consumption (the share is 12% in 2015). In addition, a set of progressive and sub-sectorial targets are adopted. Particularly, in the Medium to Long Term Renewable Energy Development Plan (MLTREDP) issued in 2007, the principles, key areas and PAMs are particularly raised to instruct the project construction from 2007 to 2020 (NDRC, 2007). The sub-sectorial targets listed in the plan are provided in Table 17.3.

As introduced in Chapter 3, beginning at the late 2005, the Chinese government has set mandatory energy-saving targets that aim for a 20% reduction in the average of national energy intensity by 2010, using 2005 data as the benchmark, and an additional 16% reduction by 2015, compared to 2010. Owning to this ten-year effort, the energy intensity in 2015 has dropped

*Table 17.3* Targets for RE development for the period 2007–2020

|  | *Indicator* | *Target* |
|---|---|---|
| General target | Share of RE in TPEC | 10% (2010); 15% (2020) |
| Hydro power | Installed capacity | 190 GW (2010); 300 GW (2020) |
| Biomass energy | Installed power capacity | 5.5 GW (2010); 30 GW (2020) |
|  | Biomass briquette | 1 Mt (2010); 50 Mt (2020) |
|  | Biogas | 19 bn m³/year (2010); 44 bn m³/year (2020) |
|  | Bio ethanol | 2 Mt (2010); 10 Mt (2020) |
|  | Biodiesel | 0.2 Mt (2010); 2 Mt (2020) |
| Wind power | Installed capacity | 5 GW (2010); 30 GW (2020) |
| Solar energy | Installed power capacity | 300 MW (2010); 1800 MW (2020) |
|  | Solar heating | 30 Mtce (2010); 60 Mtce (2020) |
| Geothermal | Geothermal heating | 4 Mtce (2010); 12 Mtce (2020) |
| Tidal power | Installed capacity | 100 MW (2020) |

*Data source*: Authors; data from NDRC (2007).

*Note*: TPEC = total primary energy consumption

33.8% from 2005 to 2015 (at an annual rate of 3.7%), whereas energy intensity in the US decreased by only 11.7% (i.e. at an annual rate of 1.6%) during 2005–2013 (Zhao and Wu, 2016).

In the latest FYP (13th FYP) approved in March 2016, an overall cap on total energy consumption in 2020 with legal force was raised for the first time, as 5000 Mtce, implying an average of 3.1% growth rate per year as it was 4300 Mtce in 2015. However, compared with the recent low energy growth, this is a rather moderate target.

## *Pricing policies*

Favourable pricing for RE has shifted from a case-by-case basis to a regular feed-in-tariff (FIT) basis. In 2003, NDRC promulgated the wind power concession policy, under which the government signed the contract with the bid winner and the grid company purchased the wind power based on the contract that can guarantee a return on investment (Zhao et al., 2016). Recognizing the harmful impact of this concession policy to the sustainability of the whole wind power industry, NDRC set baseline for wind tariffs in 2009, ranging from 0.51 RMB yuan/kWh to 0.61 RMB yuan/kWh, depending on the location of the wind farm (NDRC, 2009; Wang et al., 2012). National solar PV FIT was regulated in 2011(NDRC, 2011a).

In 2015, NDRC issued the "Notification on further improving FIT of on-shore wind and solar PV" (NDRC, 2015a), in which revised and step-down FITs for wind and increasing FITs for PV are set for two periods, namely 2016-2018 and the period after 2018, as shown in Figure 17.3.

In order to promote energy conservation among electricity-intensive industries, "differentiated power price" has been used among selected industries from 2004, in which energy-intensive enterprises with outdated production technologies and/or a poor environmental record were required to pay electricity surcharges between 0.1 and 0.3 RMB yuan/kWh. However, the real effectiveness of this policy is complex. On the one hand, the effects of the policy varied across different industries, with the largest impact on the non-ferrous metal smelting and rolling industry, followed by the chemical industry (Lo, 2014); on the other hand, the implementation of policy in provincial level is rather difficult since local interest is hurt to some extent and may manage to deviate from central orders (Chen, 2011).

Regarding fossil fuels, based on the assessment of International Energy Agency (IEA) (2007), by 2006, subsidies had been dramatically reduced and energy prices in China are increasingly

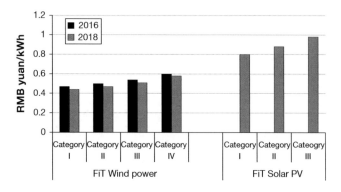

*Figure 17.3*   Feed-in tariff of wind power and solar PV power in China

*Data source*: Authors; data from NDRC (2015a).

*Note*: The classification of category is based on the assessment on resource abundance; see more details in Zhao et al. (2016).

reflecting the actual cost. In most recent years, the extent of fuel price liberalization and consistency with the international market has been further improved by significant changes in its pricing regulation both nationally or in pilot regions, with the goal of moving towards a market-based system more closely tied to international commodity prices (OECD, 2014), such as shortening of price adjusting period of petroleum products, abolishing of "dual pricing" for natural gas system (NDRC document of No.[2015]351) and shifting cost-plus approach to market netback value approach, extending the scope of coal markets, and eliminating pricing subsidies in energy-intensive industries. Taking the retail price of gasoline and diesel in China in October 2014 as an example, it was 30% and 20%, respectively, higher than that in US.

## Fiscal and tax incentives

Stable capital source for RE development is established in China through a special fund from central government budget and "renewable electricity surcharge" initiated in 2006. The surcharge was increased from 0.001 RMB yuan/kWh in 2006 to 0.004 RMB yuan /kWh in 2009, and again to 0.008 RMB yuan /kWh in 2011, and further to 0.015 RMB yuan /kWh in 2013, though the level remains low compared with international standards, particularly for solar PV (Lo, 2014).

Tax incentives are also provided to RE. For example, the Notice on the taxation of the comprehensive utilization of resources and other value-addition (No. [2008]156) stipulates that 50% of the value-added tax imposed on enterprises utilizing wind power electricity should be immediately refunded (Zhao et al., 2016). Direct subsidies are also used, particularly for but not limited to solar PV. Two programs were rolled out in 2009, the Solar Roofs program and Golden Sun Demonstration project, both supported by substantial subsidy from Ministry of Finance (MoF). For example, in the later program, subsidies up to 50% of the total cost for on-grid systems and 70% for off-grid system in rural areas were provided (Ministry of Finance, 2009).

Three tax-related measures, namely corporate income tax deductions, vehicle and fuel taxes collection, and reduction of export tax rebates, have been used to promote energy efficiency in China since 2006 (Zhou et al., 2010). Overall, the subsidies provided to energy efficiency improvement and pollution abatement increased rapidly over last several years, from 23.5 billion RMB yuan in 2007 to 47.8 billion RMB yuan in 2015,[1] growing by an average rate of nearly 10% per year. Most of this funding goes to the launching of the Ten Key Projects, elimination of inefficient facilities, and installation of environmental protection measures. More effective management of the funding is expected after the issuing of "notice on interim methods for energy-saving and pollution abating fund" in 2015 by MOF (No. [2015]161).

## Renewable and energy efficiency obligation

Renewable obligation (also known as renewable portfolio standard in China, RPS) was introduced in China by the MLTREDP in 2007, assigning the RE target for grid companies as 1% non-hydro renewable power by 2010 and 3% by 2020; for generators, the target in terms of installed capacity is 3% non-hydro renewable power by 2010 and 8% by 2020 (NDRC, 2007). However, many aspects of the policy were far from success and it was reported in 2010 that none of the six largest generators met the 3% RE target (Lo, 2014); by the end of 2014, the collective installed capacity of wind power in the five groups has amounted 6.75% of total installed capacities, which is on track for meeting the 2020 target, however, in terms of electricity generation, the share is less than 3% of overall power generation in these five companies. After long debate and preparation, the NEA re-initiated the program by issuing "Opinions on establishment of RPS to fulfil the target of RE development" (No. [2016]54) (NEA (National Energy

*Table 17.4* Provincial targets for non-hydro renewable power consumption in
2020 (share in the total power consumption)

| Area | Target (%) | Area | Target (%) |
| --- | --- | --- | --- |
| **Nationwide** | **9** | Henan | 7 |
| Beijing | 10 | Hubei | 7 |
| Tianjin | 10 | Hunan | 7 |
| Heibei Province | 10 | Guangdong | 7 |
| Shanxi | 10 | Guangxi AG | 5 |
| Inner Mongolia | 13 | Hainan | 10 |
| Liaoning | 13 | Chongqing | 5 |
| Jilin | 13 | Sichuan | 5 |
| Heilongjiang | 13 | Guizhou | 5 |
| Shanghai | 5 | Yunnan | 10 |
| Jiangsu | 7 | Tibet | 13 |
| Zhejiang | 7 | Shaanxi | 10 |
| Anhui | 7 | Gansu | 13 |
| Fujian | 7 | Qinghai | 10 |
| Jiangxi | 5 | Ningxia | 13 |
| Shandong | 10 | Xinjiang | 13 |

*Data source*: Authors; data from NEA (National Energy Administration) (2016).

Administration), 2016). This policy document sets the general and specific target by provinces on the share of non-hydro power electricity consumption in total power demand up to 2020 (see details in Table 17.4). More importantly, the green certification trade mechanism has been established, allowing generators to fulfil their obligation by green trading. Drawing lessons from previous trials, a monitoring and assessment system is also built to make the progress available to the public. However, punishment for non-compliance is still ambiguous.

Energy efficiency obligation (EEO) was first used in China in 2006 by assigning energy-saving target for enterprises whose energy consumption is over 180,000 tce per year. The program, called the Top-1000 Energy-consuming Enterprise Program (Top-1000 program), targeted the 1008 largest enterprises, accounting for 33% of China's total energy use in 2005 and approximately 50% of the total industrial sector energy consumption. In 2011, the program was significantly expanded to the Top-10000 program, reflecting the increased number of enterprises brought under the regulatory net. By the end of the 11th FYP (2010), 866 out of the 881 firms assessed (98.3%) met their energy-saving obligation; overall the program saved 165 Mtce of energy, which was the 165% of the original target (100 Mtce) (NDRC, 2011b). In the first four years of the 12th FYP (2011–2014), the extended program saved 309 Mtce of energy, which was 121% of the target (255 Mtce), by all 13,328 firms (NDRC, 2015b). Nevertheless, how these firms estimated the energy saved and how to ensure the comparability is still in question and it is proposed to cap the overall energy demand of these firms, rather than setting energy-saving targets (Zhao and Wu, 2016).

## Command and control: forced closure

It is one of major tasks to phase-out the small and inefficient production facilities with the purpose to promote energy saving in industrial sectors. From 2006, phasing-out plans were set year by year for fossil-fired power generation, and iron steel, coking, cement, electrolytic aluminium, ferroalloy plants, etc. To alleviate the negative economic and social impacts of the forced closures, funding was made available to local governments to support affected firms and workers. For example, 21.91 billion RMB was raised to support the closure in 11th FYP. In the period of 2006–2014, 100.48 GW of thermal power generator, 199.72 million ton of iron smelting capacity, 149.24 million of steel production capacity and 1015 million ton of cement production capacity were retired in total, with details shown in Figure 17.4.

The program has helped to significantly increase the energy efficiency in energy-intensive industries. Taking the iron and steel sector as an example, the average volume of blast furnaces and basic oxygen furnaces increased by 81% and 84%, respectively, during the period of 2005–2014. As a result, the comprehensive energy consumption per ton of steel produced dropped by 16% at the same time (Zhao et al., 2015).

### Specific good practices: target responsibility system

The energy-saving target responsibility system (TRS) assigns the overall targets to lower level of government and key energy-consuming industry pursuant to an allocation system and has proven effective in reducing national energy intensity (Zhao et al., 2014). More specifically, the 20% and 16% of national energy intensity reduction target in 11th FYP and 12th FYP, respectively, was met by using "responsibility contracts" which hold government officials and enterprise leaders accountable for target performance through an evaluation system, linking the local implementation of central policies to financial bonuses and career advancement of local cadre. The GHG intensity reduction target formulated for 12th FYP is implemented by the same approach.

Linking closely with TRS is the energy-saving target allocation system which disaggregates the overall target to provincial (Table 17.5) and sector level. After ten years, the allocation system has become more and more comprehensive, attempting to reflect correctly the responsibility, capacity and potential of various provinces and sectors (Zhang et al., 2015). Although the environmental governance in China follows a paradigm of authoritarian environmentalism dictated by mono-centric, non-participatory policy process, from the evolution of energy-saving allocation system, it could be concluded that Chinese policymakers simultaneously learned from previous experiences, local practice and expert knowledge to improve the system step by step, starting being collaborative and participatory (Zhao and Wu, 2016).

### Impacts of these PAMs and further improvement expected

The massive and rapid growth of RE and improvement of energy efficiency has been highlighted in Chapter 3. In terms of carbon emission, there was an estimated change of

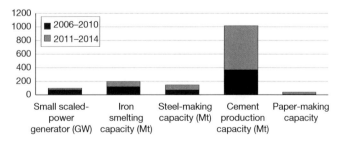

*Figure 17.4*   Closure of inefficient capacity in 11th FYP and the first four years of 12th FYP

*Data source*: Authors.

*Table 17.5* Disaggregated energy-saving targets by provinces in 11th FYP and 12th FYP

| Province | 11th FYP (%) | 12th FYP (%) | Province | 11th FYP (%) | 12th FYP (%) | Province | 11th FYP (%) | 12th FYP (%) |
|---|---|---|---|---|---|---|---|---|
| Beijing | 20 | 17 | Anhui | 20 | 16 | Sichuan | 20 | 16 |
| Tianjin | 20 | 18 | Fujian | 16 | 16 | Guizhou | 20 | 15 |
| Hebei | 20 | 17 | Jiangxi | 20 | 16 | Yunnan | 17 | 15 |
| Shanxi | 22 | 16 | Shandong | 22 | 17 | Tibet | 12 | 10 |
| Inner Mongolia | 22 | 16 | Henan | 20 | 16 | Shaanxi | 20 | 16 |
| Liaoning | 20 | 17 | Hubei | 20 | 16 | Gansu | 20 | 15 |
| Jilin | 22 | 16 | Hunan | 20 | 16 | Qinghai | 17 | 10 |
| Heilongjiang | 20 | 16 | Guangdong | 16 | 18 | Ningxia | 20 | 15 |
| Shanghai | 20 | 18 | Guangxi | 15 | 16 | Xinjiang | 20 | 10 |
| Jiangsu | 20 | 18 | Hainan | 12 | 10 | | | |
| Zhejiang | 20 | 18 | Chongqing | 20 | 16 | | | |

*Data source*: Authors.

approximately −3.9%(−4.6% to −1.1%) for China's emissions (including $CO_2$ from fossil fuel combustion and cement production) in 2015, amounting to an absolute decrease of 0.4 $GtCO_2$ and contributing significantly to −0.6% growth of world emission (Jackson et al., 2016), though there are still uncertainties around reduction in coal use and related $CO_2$ emissions (Korsbakken et al., 2016).

Behind the achievement lies a huge investment program. It is estimated that the average of initial cost of RE electricity capacity is about 12,214 RBM yuan/kW, 8619 RBM yuan/kW, 10,639 RMB yuan/kW, 10,140 RMB yuan/kW and 17,763 RBM yuan/kW for solar PV stations, wind farms, biogas generators, biomass-fired power plants and garbage incinerators, respectively (CNREC (China National Renewable Energy Center), 2015). Based on these data, at least 1.25 trillion RMB has been invested to RE industry in China, around 1.85% of the gross domestic product (GDP) in 2015. Forced closure of inefficient facilities far earlier than their lifetime made the previous investment redundant, and social cost is also significantly influenced by the now unemployed workers that were once working at these firms.

The effective integration of RE electricity into existing grids is one of the problems. Curtailment of wind power, solar power and hydro power was observed frequently with increasing trend in recent years, particularly in North China where wind resources are most abundant. In 2014, over 14.8 billion kWh of wind electricity was curtailed, accounting for 8% of total wind power generation, with 20% in North-East region (CNREC (China National Renewable Energy Center), 2015); in 2015, the curtailment increased to 33.9 billion kWh. No compensation was received by these RE enterprises whose power generation has to be disposed. The failure of implementing "fully purchase" rules regulated by REA imposes challenges on the further development of RE.

The reasons are complex. The excess subsidy is criticized as one of the main reasons for the redundant construction and overcapacity of wind and solar enterprises; therefore a moderate subsidy scale was proposed by Zhang et al. (2016). Zhao et al. (2013) pointed out that the lack of incentive policies on stakeholders of wind power development led to the high curtailment ratios of wind power in China. Probably the most important reason is the absence of a liberalized market since both FIT and working hours are regulated by the government, so lack of flexibility in the power system hinders the maximizing of integrating potential for intermittent RE power; on the other hand, there is no incentive for regular generators to be the peak load capacity for

RE power, making the latter hard to really integrate into the grid. It is obvious that there are still mechanisms that are protecting fossil-fired capacities from the perspectives of local interest or sector interest. The dilemma calls for the accelerating market reform and well-coordinated power construction and generation planning for a low-carbon system.

# Conclusion

Low carbon has increasingly been one of the 'key words' of energy policies and market liberalization since 2005. Consequently, a set of pricing and non-pricing policies have been adopted and implemented to improve RE industry and energy efficiency. Slowing down of energy growth and $CO_2$ emission shows the effectiveness of these policies, together with problems and areas requiring further refining. Competitive and efficient wholesale market should be established, accelerating to promote the flexibility of electricity system in order to integrate more green power. In the area of energy efficiency, an absolute cap on big energy consumers and the whole system should be implemented rather than an intensity concept, which has been highlighted for 10 years; there also needs to be a shift towards a more transparent accounting system and accelerated decoupling of GDP from energy demand. Nevertheless, we emphasize the conclusion from Pollitt (2012): "it is not liberalization per se that will determine the movement towards a low-carbon energy transition, but the willingness of societies to bear the cost, which will be significant no matter what the extent of liberalization".

# Note

1 Interpreting 2015 national budge (解读 2015 年国家账本：节能减排补助资金增 40.4%.). 2015. www.ceweekly.cn/2015/0313/105539.shtml

# References

Boyd OT. 2012. China's energy reform and climate policy: the ideas motivating change. CCEP (Centre for Climate Economics and Policy) working paper 1205.

Central government of China (CGC). 2005. Renewable Energy Act (可再生能源法). www.gov.cn/ziliao/flfg/2005-06/21/content_8275.htm (in Chinese).

Central government of China (CGC). 2009. Renewable Energy Act (可再生能源法-2009年修订). www.gov.cn/fwxx/bw/gjdljgwyh/content_2263069.htm (in Chinese).

Chen J. 2011. China's experiment on the differential electricity pricing policy and the struggle for energy conservation. *Energy Policy*. 39:5076–5085.

Cherni JA and Kentish J. 2007. Renewable energy policy and electricity market reforms in China [J]. *Energy Policy*. 35(7):3616–3629.

CNREC (China National Renewable Energy Center). 2015. Renewable Energy Data Handbook (可再生能源数据手册 2015). Available at: www.cnrec.org.cn/cbw/zh/2015-10-23-488.html (2016-06-17) (in Chinese).

Du L, Mao J and Shi J. 2009. Assessing the impact of regulatory reforms on China's electricity generation industry. *Energy Policy*. 37(2):712–720.

Fan J, Zhao D, Wu Y, et al. 2014. Carbon pricing and electricity market reforms in China [J]. *Clean Technologies & Environmental Policy*. 16(5):921–933.

Fisher-Vanden K and Ho MS. 2007. How do market reforms affect China's responsiveness to environmental policy? [J]. *Journal of Development Economics*. 82(1):200–233.

Gao H and Biesebroeck JV. 2014. Effects of deregulation and vertical unbundling of the performance of China's electricity generation sector. *The Journal of Industrial Economics*. LXII(1):41–75.

International Energy Agency (IEA). 2007. *Energy Prices and Taxes*. Paris: IEA.

Jackson RB, Canadell JG, Quere CL, Andrew RM, Korsbakken JI, Peters GP and Nakicenovic N. 2016. Reaching peaking emissions. *Nature Climate Change*. 6(1):7–10.

Jiao J and Wang XJ. 2012. Electricity market reforms and sustainable development in China [J]. *Advanced Materials Research*. 433–440:1710–1714.

Korsbakken JL, Peter GP and Andrew RM. 2016. Uncertainties around reductions in China's coal use and $CO_2$ emissions. *Nature Climate Change*. 6:687–691.

Lin B, Liu C and Lin L. 2015. The effect of China's natural gas pricing reform. *Emerging Markets Finance and Trade*. 51(4):812–825.

Ling Y. 2009. Milestones of the renewable energy development in China (中国可再生能源发展历程大事记). *Sino-Global Energy*. 3:76 (in Chinese).

Liu J, Liu H, Yao X and Liu Y. 2016. Evaluating the sustainability impact of consolidation policy in China's coal mining industry: a data envelopment analysis. *Journal of Cleaner Production*. 112:2969–2976.

Lo K. 2014. A critical review of China's rapidly developing renewable energy and energy efficiency policies. *Renewable and Sustainable Energy Reviews*. 29:508–516.

Mao Y, Sheng H and Yang F. 2008. *The True Cost of Coal*. Beijing: China Coal Industry Publishing House.

Ministry of Finance. 2009. Notification on initiating golden sun demonstration project (No.[2009]397). (关于实施金太阳示范工程的通知) Available at: www.mof.gov.cn/zhengwuxinxi/caizhengwengao/2009niancaizhengbuwengao/caizhengwengao200907/200911/t20091118_233416.html (2016-04-23).

Mou D. 2014. Understanding China's electricity market reform from the perspective of the coal-red power disparity. *Energy Policy*. 74:224–234.

NDRC. 2007. Mid- and long-term development plan for renewable energy (可再生能源中长期发展规划). Available at: www.sdpc.gov.cn/zcfb/zcfbghwb/200709/W020140220601800225116.pdf (2016-05-15).

NDRC. 2009. The notification to improve the feed-in-tariff of wind power (No.[2009]1906) (国家发展改革委关于完善风力发电上网电价政策的通知). Available at: http://zfxxgk.ndrc.gov.cn/PublicItemView.aspx?ItemID={bef52635-547c-490f-8334-7fdbdb5d057a} (2016-06-01).

NDRC. 2011a. The notification to improve the feed-in-tariff of solar PV (No.[2011]1594) (国家发展改革委关于完善太阳能光伏发电上网电价政策的通知). Available at: www.nea.gov.cn/2011-08/01/c_131097437.htm (2016-06-01).

NDRC. 2011b. Announcement of NDRC No.31 of 2011: implementation of top-1000 program in 11[th] FYP(国家发展和改革委员会关于千家企业``十一五''节能目标完成情况的公告). Available at: www.gov.cn/zwgk/2011-12/27/content_2030819.htm (2016-07-07).

NDRC. 2015a. The notification to further improve the feed-in-tariff of in-land wind and solar PV power (No.[2015]3044) (国家发展改革委关于完善陆上风电光伏发电上网标杆电价政策的通知). Available at: www.sdpc.gov.cn/gzdt/201512/t20151224_768582.html (2016-06-01).

NDRC. 2015b. Announcement of NDRC No.34 of 2015: progress of top-10000 program in 2014 (国家发展和改革委员会关于2014年万家企业节能目标责任考核结果的公告). www.sdpc.gov.cn/gzdt/201601/t20160107_770747.html (2016-07-7).

NEA (National Energy Administration). 2016. Direction to the establishment of provincial renewable energy development targets (No.[2016]54). (国家能源局关于建立可再生能源开发利用目标引导制度的指导意见) http://zfxxgk.nea.gov.cn/auto87/201603/t20160303_2205.htm.

OECD. 2014. Inventory of estimated budgetary support and tax expenditures for fossil fuels—a report on selected non-member countries.

Pollitt MG. 2012. The role of policy in energy transitions: lessons from the energy liberalization era. *Energy Policy*. 50:128–137.

Schman A and Lin A. 2012. China's renewable energy law and its impact on renewable power in China: progress, challenges and recommendations for improving implementation. *Energy Policy*. 51:89–109.

Shi D. 2009. Thirtieth anniversary of the Chinese Renewable Energy Society (中国可再生能源学会三十年). *Solar Energy*. 12:21–24 (in Chinese).

Wang H, Burene J and Kurdgelashvil L. 2012. Review of wind power tariff policies in China. *Energy Policy*. 53:41–50.

Williams JH and Kahrl F. 2008. Electricity reform and sustainable development in China. *Environmental Research Letter*. 4:1–14.

Zeng M, Yang Y, Fan Q, et al. 2015. Coordination between clean energy generation and thermal power generation and thermal power generation under the policy of "direct power-purchase for large users" in China. *Utilities Policy*. 33:10–22.

Zeng M, Yang Y, Wang L, Sun J. 2016. The power industry reform in China 2015: policies, evaluations and solutions. *Renewable and Sustainable Energy Reviews*. 57:94–110.

Zhang LX, Feng YY and Zhao BH. 2015. Disaggregation of energy-saving targets for China's provinces: modeling results and real choices. *Journal of Cleaner Production*. 103:837–846.

Zhang Y and Gao P. 2016. Integrating environmental considerations into economic regulation of China's electricity sector. *Utilities Policy*. 38:62–71.

Zhang H, Zheng Y, Ozturk UA and Li S. 2016. The impact of subsidies on overcapacity: a comparison of wind and solar energy companies in China. *Energy*. 94:821–827.

Zhao H, Zhang B and Hao JW. 2015. Closing outdated and redundant capacity in iron and steel industry in 13th Five-year-plan ("十三五"钢铁行业淘汰落后和过剩产能的政策建议). *Metallurgical Economy and Management*. 6:51–53 (in Chinese).

Zhao X and Wu L. 2016. Interpreting the evolution of the energy-saving target allocation system in China (2006–2013): a view of policy learning. *World Development*. 82:83–94.

Zhao XL, Zhang SF, Zou YS and Yao J. 2013. To what extent does wind power deployment affect vested interests? A case study of the Northeast China Grid. *Energy Policy*. 63:814–822.

Zhao X, Li H, Wu L and Qi Y. 2014. Implementation of energy-saving policies in China: how local governments assisted industrial enterprises in achieving energy-saving targets. *Energy Policy*. 66:170–184.

Zhao X, Li S, Zhang S, Yang R and Liu S. 2016. The effectiveness of China's wind power policy: an empirical analysis. *Energy Policy*. 95:269–279.

Zhou N, Levine MD and Price L. 2010. Overview of current energy efficiency policies in China. *Energy Policy*. 38(11):1–37.

Zhu S and Cherni JA. 2009. Coal mining in China: policy and environment under market reform. *International Journal of Energy Sector Management*. 3(1):9–28.

Zhu Z, Zhang H and Tao G. 2016. Effects of gas pricing reform on China's price level and total output. *Natural Hazards*. 84:S167–S178.

# 18

# ENERGY SECURITY ISSUES IN ASIA

*Vlado Vivoda*

## Introduction: the concept of energy security

As an analytical concept, energy security has been established since the 1970s, when two oil crises shocked the major industrialised nations. The heavy external energy dependence and supply risks threatened their economic interest and national security, and shaped energy security thinking and behaviour. A traditional energy security approach characterised by a supply-sided, primarily oil-oriented strategic thinking, as well as a foreign policy response to safeguard energy security, has remained the mainstream until the end of the century. Most early studies conceptualised energy security in terms of security of oil supplies (Fried and Trezise 1993; Stringer 2008).

Over the past two decades, energy security has become an emerging area of policy focus, with high energy prices, the increased demand and competition for geographically concentrated resources, the fear of resource scarcity and concerns with the economic, environmental, social and geopolitical effects of climate change. According to Victor and Yueh (2010), the extra-ordinary shift in energy security concerns has challenged hitherto dominant policy orthodoxies. Accordingly, with a broader set of issues, the concept has been expanded to include many new factors and challenges (Yergin 2006). With increasingly global, diverse energy markets and increasingly global problems resulting from energy transformation and use, issues such as climate change have become increasingly salient. In the academic literature, the substance of emerging challenges has been incorporated into a new concept of energy security, commonly defined as the availability of energy at all times in various forms, in sufficient quantities and at affordable prices, without unacceptable or irreversible impacts on the economy and the environment (Vivoda 2010).

The chapter proceeds as follows. The first two sections outline key concepts, drivers and trends in energy policy approaches, providing a conceptual and analytical framework for analysis of major energy security issues in Asia. The discussion in the third section centres on five major energy security issues in the region: perceptions and competitive energy security approaches, regional geography and market fragmentation, bilateralism and strategic compe-tition, fossil fuel subsidies and the demand for coal and environmental sustainability. The concluding section proposes avenues for greater regional energy cooperation in light of common energy and climate challenges.

## Energy policy objectives and approaches

National energy policy frameworks summarise existing policies and formulate strategies to support delivery of the core objectives. Traditionally, the main objective was to secure access to reliable energy supplies at affordable or competitive prices. Increasingly, in addition to availability and affordability, policies are aimed at improving environmental sustainability of energy choices. Looking for simultaneous progress towards the goals of energy availability, affordability and sustainability, policy makers face complex and sometimes contradictory choices (Maurin and Vivoda 2016). The challenge of concurrently meeting these three objectives is referred to as the 'energy trilemma', a term coined by the World Energy Council.

The three objectives are regularly in tension and difficult trade-offs are often required. For example, improved sustainability requires a trade-off with affordability when significant capital expenditure is directed at reducing greenhouse gas emissions. Alternatively, improved supply security requires a trade-off with affordability when an importing state decides to diversify supply sources and transportation routes. Ideally, policies should be framed so that the three objectives are tackled concurrently with none (explicitly) given precedence over the others. However, political reality dictates otherwise and one objective is often privileged over the other two. Most often policy outcomes involve compromises among interested parties.

The relative significance of the three objectives is influenced by the national and international policy setting in which states and markets interact. This also includes the national views towards the teleology of the markets and the degree of optimal state intervention. In state capitalist systems, the extent of government intervention in energy markets is greater than in countries that have a market capitalist tradition (Vivoda 2015a). National governments also vary greatly in the importance they attach to environmental sustainability of energy choices relative to supply security and affordability.

Two ideal-type policy approaches are based on a diametrically opposing view towards the role of the state in energy markets: strategic and market-based. In reality, there is always a degree of intervention even by the most market-oriented governments (Sidortsov and Sovacool 2015). The strategic approach posits that leaving energy to market forces does not provide optimal outcomes. Government intervention in energy-related activities is necessary in order to steer the markets towards the state's best interests. The thrust is that energy is too important strategically to be left to market forces alone. Governments use a range of strategies and regulatory instruments to steer the market towards desired objectives. For example, subsidies and taxes can promote or curb the use of a specific energy source; state ownership of energy companies and infrastructure may lead to greater control across the value chain; diplomatic activity and provision of foreign assistance to resource-rich governments can improve access to energy resources (Vivoda and Manicom 2011; Stoddard 2013; Hancock and Vivoda 2014).

In contrast, according to the market-based approach, energy markets should be exposed to the same conditions as other commodity markets. The belief is that open and competitive markets deliver energy at best prices and ensure adequate and reliable supplies. Government interference is only needed in times of market failure (Vivoda and Manicom 2011). The market approach is characterised by agnosticism regarding the source of energy imports; eschewal of policies that seek to promote the interests of national over foreign firms; liberalisation of domestic resource sectors and integration with international markets through open trade and investment policies; and foreign policy cooperation with other states to improve the functioning of international markets on a multilateral basis (Hancock and Vivoda 2014; Wilson 2014).

## Energy policy trends

During the 1970s and the early 1980s, the energy sector was heavily politicised and security of supply was on top of the political agenda. The involvement of major consuming governments in the energy markets reflected the prevailing interventionist approach to economic management, as well as concerns over security of supply. In industrialised economies, state-owned companies held statutory monopolies as importers and wholesale traders. Long-term contracts between well-established parties, with secure prices, were essential to protect both the supplier and buyer. The suppliers regarded the powerful position of state-owned monopolies as a guarantee that the purchase obligations under long-term contracts would be fulfilled (Radetzki 1999).

The 1986 oil price collapse contributed to a changed government attitude towards energy. This began already in the early 1980s, as a consequence of Ronald Reagan's and Margaret Thatcher's crusades in favour of politically unhampered market solutions and competition. As the decade evolved, there was increasing disillusion with the far-reaching energy policies implemented in the preceding years. The oil price collapse was seen as a confirmation that energy supplies were ample and that public interventions to assure supply security, e.g. in the form of national mon-opolies, were costly and unnecessary. Prevailing attitudes, values and beliefs during the 1990s were based on the idea that the markets were more efficient than governments (Helm 2004).

Since the turn of the century, intensified competition, high and volatile prices and ambitious renewable energy targets have motivated governments to adopt a proactive approach to energy issues. This is reflected in the policy trend to consider energy increasingly as a strategic issue in both exporting and import-dependent countries. As a consequence, energy supply and demand is increasingly shaped by geopolitical developments and government intervention.

This policy trend is particularly pronounced in major Asian economies where energy security has emerged as a critical issue.[1] There are three major challenges that affect all Asian states to a varying degree: (1) continued, or in some cases increasing, dependence on non-renewable energy sources to meet demand; (2) the environmental impact of energy use, as seen by the environ-mental repercussions from the heavy coal use; and (3) underdeveloped energy infrastructure and transportation networks both within and between countries. With the epicentre of global energy demand moving from industrialised Western economies to Asia, regional energy security atti-tudes and approaches carry major consequences for global energy geopolitics, markets and efforts to tackle climate change.

The spectacular economic growth in Asia has spurred a vast expansion in the need for energy services, and an expansion in the demand for the fuels that help to supply these services. Asian energy demand has doubled since 2001, with the region's share in global energy demand increasing from 29% to 42% (BP 2016). Most projections suggest that Asia's voracious thirst for energy will further expand in the coming decades. Moreover, regional states are becoming increasingly reliant on imported fossil fuels. For example, the region hosts four of the world's five largest oil importers and top three liquefied natural gas (LNG) and coal importers, respectively (BP 2016). Finally, the increased fossil fuel demand has had a significant carbon footprint, with the region's $CO_2$ emissions doubling since 2001 mainly caused by China's rapidly growing coal demand (BP 2016).

## Issues

### *Regional perceptions and competitive energy security approaches*

Independent self-interested behaviour can result in undesirable or suboptimal outcomes, refer-ring to these situations as dilemmas of common interests and dilemmas of common aversions

(Stein 1982). According to this 'security dilemma', many of the means by which a state tries to increase its security, decrease the security of others (Herz 1951; Jervis 1978; Booth and Wheeler 2007). When one state pursues a market interventionist strategy to secure sufficient energy supplies this may result in enhanced national energy security, but may simultaneously deteriorate energy security of its regional neighbours. In Asia, energy is 'securitised' as governments perceive possible conflicts over availability and pricing of energy as threats to national security (Prantl 2011). The regional trend towards energy securitisation has aggravated tensions among major powers (Lai 2009; Phillips 2013). Energy security in Asia continues to be dominated by strategic approaches due to continued prevalence of nationalism, political tensions and distrust among major economic powers. Regional energy dynamics can be characterised as a 'security dilemma' in which each country acts in its own interests and thereby inflicts greater insecurity on the region. This security dilemma is illustrated by examining China's and Japan's approaches to international energy markets, which are dominated by competitive and often costly pursuits of bilateral energy agreements with key suppliers. Given that these two countries are the region's largest economies and energy importers, evaluating their energy security approaches is indicative of broader regional trends.

Since the mid-1990s, China's government has been particularly alert to the country's growing energy deficit and demand for imported supplies. As a rapidly growing economy that has been a net energy importer since only 1996, examining the nexus of energy and national security has become an urgent task for scholars and policy makers alike. Mainstream thinking on energy security in China shares characteristics of the traditional energy security approach (Downs 2004). Chinese scholars and policy makers largely equate energy security to the security of external energy supply, often referring to oil imports transported through key sea lines of communication, such as the Malacca Strait. Government-supported research has concluded that energy security more accurately is oil security. Therefore, a common perception in China is that the country's energy security problem is its growing oil deficit (Vivoda 2015b). The possibility of a deliberate interruption to oil supplies by potential adversary poses one of the biggest threats to China's energy security (Kong 2011). In response to threat perceptions, China has adopted an energy security approach emphasising the security of imported supplies, particularly oil. As early as 1993, the then-Premier Li Peng set the stage for this approach by defining the objectives of the country's energy policy as to secure the long-term and stable supply of oil to China (Chang 2001). This fundamental objective continues to guide China's energy strategy (Vivoda 2015b).

In Japan, the state is an active participant in the energy markets. The government regularly sets the overarching energy strategy, including specific targets, with the aim of ensuring access to reliable and affordable energy supplies for Japanese consumers. The oil crises of the 1970s marked a major shift towards government intervention in the energy markets. Reducing Japan's external dependency played an important role in the move to more active government management of Japan's energy supply and demand mix. The surge in energy prices since the turn of the century associated with the perceived competition with China for energy supplies has given further impetus to government intervention. A general view among Japanese energy policy makers is that a 'leave it to the market' approach is not a solution since this could render oil and gas security more vulnerable (Nakatani 2004; Vivoda 2014a). Instead, energy in Japan remains conceptualised as a national security issue (Phillips 2013). The Fukushima disaster and the ensuing energy crisis have cemented this view (Vivoda 2012).

The rapid economic growth in the region has intensified competition for increasingly scarce energy resources, elevating energy security to the top of the regional agenda. China and Japan used to have strong energy relations before China transformed into a net oil importer in the early 1990s. The past two decades have witnessed an increasingly intensive competition between the two countries over oil and gas supplies. When China became a net oil importer, Sino-Japanese energy cooperation weakened dramatically (Dorian 1995). China's transformation to a net oil importer and its growing economic potential turned the two countries into competitors over oil and gas supplies and also rendered cooperation less likely because of their increasing concerns over relative gains. The end of the Cold War not only weakened their shared strategic and political interests, but also posed China as one of the major threats in Japan's security agenda. China and Japan have no concrete form of policy adjustment towards energy suppliers at the bilateral level (Itoh 2009).

## Regional geography and energy market fragmentation

Asian oil and gas markets are fragmented. Unlike their European and North American counterparts, Asian markets are not well connected by pipelines, and are dominated by seaborne trade. For example, in 2015, Asian states imported 70.5% of internationally traded LNG, but only 8.7% of piped gas (BP 2016). Regional pipeline infrastructure includes China's import pipelines from Myanmar and the former Soviet Union and several international gas pipelines in Southeast Asia (Vivoda 2014b). China is the only major regional oil and gas importer that has significant domestic pipeline connections to major exporters. China's pipeline gas imports from Turkmenistan and Myanmar commenced in 2010 and 2013, respectively (Vivoda 2014c). In addition, in May 2014, China and Russia signed a US$400 billion natural gas deal, with Russia to supply 38 billion cubic metres of natural gas per annum via pipeline for 30 years (Russell 2014). Chinese oil imports via the Kazakhstan-China oil pipeline, from Russia, via the Eastern Siberia-Pacific Ocean (ESPO) oil pipeline, and in transit via the Sino-Myanmar crude oil pipeline commenced in 2006, 2011 and 2014, respectively.

The Chinese government's concern about the security of seaborne energy imports and a desire to diversify oil supplies away from the Middle East, as well as diversify the energy transport routes away from the Malacca Strait have triggered long-term supply agreements with Russia and Central Asian countries in parallel with Chinese-financed construction of long-distance oil and gas pipelines from the former Soviet Union and Myanmar to China. Most Chinese analysts regard participation in the development of Russian and Central Asian energy resources as an important source of energy security, as oil and gas imported by pipeline is perceived as less vulnerable to disruption than seaborne imports (Vivoda 2015b).

With the exception of Chinese pipelines, due to geographic separation and long distances between major producers and between major producers and consumers, there are limited prospects for establishing regional connectivity through cross-border oil and gas pipeline infrastructure. Where geography does not present an insurmountable obstacle, politics have prevented pipeline projects from materialising. The most notable examples include three proposed gas pipelines to India – from Myanmar (via Bangladesh), Turkmenistan (via Afghanistan and Pakistan) and Iran (via Pakistan) – the Trans-Korean gas pipeline and the Sakhalin-Hokkaido gas pipeline (Vivoda 2014b).

## Bilateralism, strategic competition and energy affordability

Bilateralism is a historically developed practice of realising the security of supplies (Aalto and Korkmaz Temel 2014). Over the past decade, China, Japan and South Korea have used strategic partnerships and bilateral free trade agreements with energy exporters in order to lock-in future supplies and thus improve their energy security (Wilson 2012). There is ample evidence that Asian states routinely pay higher prices for energy because of the hoarding practices of their regional neighbours (Vivoda and Manicom 2011). Most notably, China and Japan have implemented strategies that aim to enhance energy security by promoting bilateral cooperation with major suppliers (Sovacool and Vivoda 2012). Both China and Japan engage in the practice of promoting overseas investment in order to secure equity, or self-developed, oil and gas stakes in the belief that this enhances their energy security and provides protection against supply disruptions (Zhang 2012; Vivoda 2014a). With a distrust of the international energy market, both Beijing and Tokyo believe that special relationships with energy producers will guarantee reliable access to energy imports (Vivoda 2015b).

In Japan's case, the first national target of securing self-developed oil supplies was set during the 1970s and downwardly revised several times; however, none of the targets were achieved. Historically, self-developed overseas ventures never generated more than 15% of Japan's oil imports (Drifte 2005; Koike et al. 2008; Vivoda 2014a). Supported by the government, Chinese national oil companies (NOCs) commenced their quest to acquire overseas oil and gas assets in the mid-1990s. While China's aggressive pursuit of self-developed oil has been much more successful than Japan's, oil produced by China's NOCs is neither cheaper nor more available to the Chinese customers in a supply crisis (Kennedy 2010; Zhang 2012).

At minimum, the Sino-Japanese competition has raised the costs of oil imports for both countries by delaying the development of domestic and regional energy infrastructure. For example, Tokyo's original impetus for participation in the Azedegan project in Iran was to regain a foothold in the Middle East following initial Chinese gains; the loss of the stake was seen in zero-sum terms (Shaoul 2005). Similarly, the competition over the preferential route of Russia's ESPO pipeline developed a soap opera-like quality (Vivoda and Manicom 2011). Japan's involvement in influencing Russia to preference the longer pipeline route to the Pacific Ocean strengthened Beijing's perception that supply security of energy supply is an important aspect of strategic competition. The ever-shifting pipeline route delayed the final investment decision and raised import costs through expensive stop-gap measures, such as China's oil imports by rail and shipment by rail to the Pacific. The dispute over gas and oil exploration rights in the East China Sea has delayed full production at Chunxiao field (Vivoda and Manicom 2011; Manicom 2014). Japan contested China's right to produce resources in the area and on several occasions China has halted development of the fields for diplomatic purposes, which has delayed full production.

The tussle between China and Japan over oil and gas reserves in the East China Sea demonstrates that rigorous competition over energy sources can result in diplomatic disputes and deepen mutual mistrust between countries. Despite the compelling common interest in developing energy resources for mutual benefit, energy concerns have emerged as a significant source of friction in the Sino-Japanese relationship. Both parties have conceptualised energy as a security issue and have approached the issue against the backdrop of an already tense and deteriorating bilateral relationship. While Japan's tendency to conceptualise energy as a national security issue is far from new, its increasingly assertive regional energy diplomacy does constitute a significant departure from Cold War precedents. China's newly found dependence on energy imports has seen it follow Japan's longstanding example in securitising energy as a policy issue (Phillips 2013).

The recent escalation of the South China Sea dispute indicates that resource nationalism extends beyond the two major regional powers. Maritime disputes elsewhere in Asia have historically stemmed from unsettled territorial and maritime claims. However, in recent years, energy security concerns have increasingly inflamed these disputes. Rising energy prices, fears of supply scarcity and rapid increases in oil-import dependency in China and other regional powers, such as India and Indonesia, have perpetuated resource nationalism (Collins and Erickson 2011). The South China Sea dispute illustrates that resource nationalism increasingly drives the involvement of the more "institutionalised" ASEAN member states. Growing energy security concerns, historical and political tensions that continue to fuel nationalism and distrust, and the diversity in political, economic and social conditions among ASEAN member states continue to hinder progress towards mutually beneficial outcomes.

## Fossil fuel subsidies in developing Asia

Developing Asia is home to a majority of the world's energy poor, with about 615 million people having no electricity and 1.8 billion burning firewood, charcoal and crop waste to meet their basic daily needs (Asian Development Bank 2016). Energy is essential for economic growth and human progress and the region's energy demand will continue to rise as economies grow and living conditions improve. The continued growth in demand for affordable energy underlies the widespread use of subsidies on fossil fuels (oil, gas and coal) and electricity across Asia. Ten of the top 25 countries in the world that subsidise fossil fuel are located in Asia. Developing Asian countries also account for close to a third of global subsidies on fossil fuel consumption, equivalent to about 2.5% of gross domestic product (GDP). Fossil fuel subsidies in India, Indonesia and Thailand stand at 2.7%, 4.1%, and 1.9% of GDP, respectively (Asian Development Bank 2016). Low-priced petroleum products account for over half of these subsidies in each country.

In 2009, the Group of Twenty (G20) and Asia-Pacific Economic Cooperation (APEC) committed to rationalising and phasing out inefficient fossil fuel subsidies. Unfortunately, there has been little progress on subsidy reforms. As people get used to low prices and powerful beneficiaries oppose reforms, governments are unable to push their reform agendas fearing social unrest and political upheaval when prices rise. In some Asian countries, government expenditure on fossil fuel subsidies, which covers the gap between global and domestic prices, exceeds public spending on education or health. High fossil fuel subsidies also reduce incentives for investment in renewable energy and energy efficiency. Moreover, fossil fuels (coal, oil, and gas) are major carbon emitters, and burning coal, the most carbon-intensive energy source, has serious climate change implications (Asian Development Bank 2016). Consequently, energy subsidy reform remains as one of the most important policy challenges for developing Asian economies.

## Coal use and environmental sustainability

Although the global energy system showed a decarbonising trend before the turn of this century, in recent years, this trend has been reversed mainly because of growing coal use in Asia (Intergovernmental Panel on Climate Change 2014). The carbonisation of the global energy system poses a severe challenge for efforts to reduce carbon emissions. The increase in the carbon intensity of global energy use is caused by the increased use of coal, mainly in China, but also other rapidly growing Asian economies. The relatively low price and widespread resource abundance are key drivers of increased coal use in Asia. This underlines the importance of cheaply available energy for economic growth and suggests that viable alternatives to cheap coal are required to ensure the participation of developing countries in global climate change mitigation (Steckel et al. 2015).

The low price of coal relative to gas and oil has played an important role in accelerating coal consumption since the turn of the century. The Intergovernmental Panel on Climate Change (IPCC) Fifth Assessment Report identified the replacement of coal-fired power plants by less carbon-intensive energy technologies as one of the most cost-efficient options to reduce global greenhouse gas emissions (Intergovernmental Panel on Climate Change 2014). Given that emerging Asian economies mainly meet their growing appetite for energy with coal raises serious doubts about whether current development trajectories are compatible with climate change mitigation. If future economic growth in developing Asia is fuelled mainly by coal, ambitious mitigation targets will become unfeasible. Building new coal-fired power plants will lead to lock-in effects for the next few decades. If that lock-in is to be avoided, international climate policy must find ways to offer viable alternatives to coal for developing countries (Steckel et al. 2015). Without government intervention, it is highly unlikely that coal use will decline drastically in the near future.

Focusing on synergies between climate policy and other policies, McCollum et al. (2013) point out that ambitious climate measures would reduce the costs of clean air policies and energy security measures by US$100–600 billion (0.1–0.7% of GDP) annually by 2030. However, most major Asian governments consider energy in isolation from climate policy. According to Jakob et al. (2014), the best incentive for switching to alternative sources of energy lies in policy objectives other than climate policy, such as those addressing energy security, localised air pollution and energy access (Jakob et al. 2014). Measures that would discourage coal use and encourage the use of low-carbon technologies as a co-benefit of other policies require identifying country-specific policy goals and opportunities. A salient example of a policy that serves objectives that are not national security or climate related but that nevertheless could reduce coal use is China's implementation of the "Action Plan for Air Pollution Prevention and Control". Even though it is aimed at improving ambient air quality, this policy could lead to declining $CO_2$ emissions from 2020 onwards (Sheehan et al. 2014). Other examples include Vietnam's recent Green Growth policies that include a reform of implicit fossil fuel subsidies in the power sector (Zimmer et al. 2015) and India's climate discourse, which largely revolves around energy security (Dubash 2013).

## Concluding remarks

Christensen (1999) argued that Asia is characterised by skewed distributions of economic and political power within and between countries, political and cultural heterogeneity, anaemic security institutionalisation, and widespread territorial disputes that combine natural resource issues with nationalism. He suggested that if security dilemma theory is applied to the region, the chance for spirals of tension in the area seems great (Christensen 1999). Similarly, Calder (1997) warned that energy competition in Asia will lead to strategic rivalry and represents a recipe for conflict.

Reducing competition for fossil fuels and mitigating the climate change risks posed by increased greenhouse gas emissions are desirable regional (and global) outcomes. Security of energy supply and climate change mitigation are inherently cross-border concepts – trade behaviour and emissions of or in one state affects others – and many important influences (e.g. transnational business cooperation and broader relations between states) are situated on the transnational and international level.

While major Asian economies are highly dependent on uninterrupted supply of energy, and share common interests as major energy consumers, regional cooperative mechanisms remain underdeveloped. Indeed, since the turn of the century, regional security dynamics have deteriorated both in terms of prospects for energy and environmental security and broader regional cooperation. While the importance of Asian energy security cooperation has been consistently emphasised as a means to cope with uncertainties in the global energy market, and to avert potential conflicts regarding energy supply among major consumers, the lack of progress is reflective of the competitive nature of national energy security strategies.

Regional cooperation is limited to information sharing, confidence building measures and the setting of aspirational targets (Ravenhill 2013). Major regional powers are reluctant to commit to mutually binding multilateral rules and principles that effectively constrain state sovereignty. Nationalism and political tensions remain the major obstacles for deeper regional cooperation. States perceive that their interests in the energy sector are best served by national rather than collective action and that cooperative arrangements fail to provide sufficient incentives to prevent states from succumbing to opportunistic behaviour in the event of a short-term clash of interests (Ravenhill 2013). The absence of deep trust is an additional constraint to deeper forms of regional cooperation (Foot 2012). As illustrated in the case of China and Japan, this absence of deep trust reinforces a more deeply rooted and continuing preference for state autonomy.

Competitive energy security approaches adopted by major Asian economies are not conducive to addressing the urgent and complex energy challenges and promoting transformation to alternative and sustainable energy systems. Energy cooperation in Asia is essential for the prevention of potential conflicts stemming from competitive energy procurement (Lee 2010). Given that major Asian economies share some concerns with regards to energy security, it is in their interests that they pool their resources and jointly strive for collective energy security. For that reason, multilateral initiatives are preferable to bilateral efforts.

How can regional states improve mutual cooperation in light of common energy and climate challenges? Potential areas for cooperation include energy efficiency and renewable energy technology, energy poverty reduction strategies, regional natural gas market integration, reform of domestic markets (including fossil fuel subsidies), nuclear energy cooperation and the management of nuclear waste.

For example, industrialised countries, such as Japan and South Korea, should assist developing economies in improving energy efficiency, conservation and fuel-switching. Particularly appealing in this context is Japan's experience in energy conservation, efficiency and fuel substitution following the oil crises and, more recently, the Fukushima disaster (Calder 2012; Vivoda 2014a).

According to Prantl (2011), one of the most promising issue areas for collective action is research & development investment in clean energy technologies. Japan's advanced coal-fired power plant efficiency technology and China's low-cost solar and wind energy technology are particularly appealing. In this context, it is crucial that any newly constructed coal-fired power plants are "capture-ready", i.e. that they can be retrofitted with carbon capture and storage (CCS) technology in order to avoid emissions. In the longer run, such a scheme could be complemented with subsidies for CCS (Kalkuhl et al. 2015).

Importantly, it is crucial that regional governments broaden their perception of energy security risks, moving away from traditional approaches, which focus on reducing risks associated with oil supply security. In addition to diverting more attention to demand management, governments should focus their attention on environmental and social sustainability of energy use when devising energy policies and energy security strategies.

## Acknowledgements

Research for this paper was supported by the National Research Foundation of Korea Grant, funded by the Korean Government (NRF-2015S1A3A2046684).

## Note

1 For analytical purposes of this chapter, Asia encompasses Bangladesh, Bhutan, China (including Hong Kong and Macau), East Timor, India, Japan, Maldives, Mongolia, Nepal, North Korea, Pakistan, South Korea, Sri Lanka, Taiwan and ten ASEAN member states, namely Brunei, Cambodia, Indonesia, Laos, Malaysia, Myanmar, Philippines, Singapore, Thailand and Vietnam.

# References

Aalto, P., and D. Korkmaz Temel. 2014. "European Energy Security: Natural Gas and the Integration Process." *Journal of Common Market Studies* 52 (4): 758–774. doi:10.1111/jcms.12108.

Asian Development Bank. 2016. *Fossil Fuel Subsidies in Asia: Trends, Impacts, and Reforms—Integrative Report.* Mandaluyong City, Philippines: Asian Development Bank. www.adb.org/sites/default/files/publication/182255/fossil-fuel-subsidies-asia.pdf.

Booth, K., and N.J. Wheeler. 2007. *The Security Dilemma: Fear, Cooperation and Trust in World Politics.* Basingstoke: Palgrave Macmillan. https://he.palgrave.com/page/detail/?sf1=barcode&st1=9780333587447.

BP. 2016. *Statistical Review of World Energy 2016.* London: BP. www.bp.com/content/dam/bp/pdf/energy-economics/statistical-review-2016/bp-statistical-review-of-world-energy-2016-full-report.pdf.

Calder, K.E. 1997. *Asia's Deadly Triangle: How Arms, Energy and Growth Threaten to Destabilize Asia–Pacific.* London: Nicholas Brealey Publishing. www.allenandunwin.com/browse/books/general-books/military/Asias-Deadly-Triangle-Kent-E-Calder-9781857881615.

Calder, K.E. 2012. *The New Continentalism: Energy and Twenty–First–Century Eurasian Geopolitics.* New Haven: Yale University Press. http://yalebooks.com/book/9780300171020/new-continentalism.

Chang, F.K. 2001. "Chinese Energy and Asian Security." *Orbis* 45 (2): 211–240. doi:10.1016/S0030-4387(01)00069-2.

Christensen, T.J. 1999. "China, the U.S.-Japan Alliance, and the Security Dilemma in East Asia." *International Security* 23 (4): 49–80. doi:10.1162/isec.23.4.49.

Collins, G., and A.S. Erickson. 2011. "Energy Nationalism Goes to Sea in Asia." The National Bureau of Asian Research, NBR Special Report #31, September. www.andrewerickson.com/wp-content/uploads/2011/09/Energy-Nationalism-Goes-to-Sea-in-Asia_NBR_201109.pdf.

Dorian, J.P. 1995. *Energy in China: Foreign Investment Opportunities, Trends and Legislation.* London: Financial Times Energy Publishing.

Downs, E.S. 2004. "The Chinese Energy Security Debate." *The China Quarterly* 177: 21–41. doi: 10.1017/S0305741004000037.

Drifte, R. 2005. "Japan's Energy Policy in Asia: Cooperation, Competition, Territorial Disputes." *Oil, Gas & Energy Law* 4. www.ogel.org/article.asp?key=2035.

Dubash, N.K. 2013. "The Politics of Climate Change in India: Narratives of Equity and Co-benefits." *Wiley Interdisciplinary Review of Climate Change* 4 (3): 191–201. doi:10.1002/wcc.210.

Foot, R. 2012. "Asia's Cooperation and Governance: The Role of East Asian Regional Organizations in Regional Governance: Constraints and Contributions." *Japanese Journal of Political Science* 13 (1): 133–142. doi:10.1017/S1468109911000326.

Fried, E.R., and P.H. Trezise. 1993. *Oil Security: Retrospect and Prospect.* Washington, DC: The Brookings Institution. www.brookings.edu/book/oil-security/.

Hancock, K.J., and V. Vivoda. 2014. "International Political Economy: A Field Born of the OPEC Crisis Returns to its Energy Roots." *Energy Research & Social Science* 1: 206–216. doi:10.1016/j.erss.2014.03.017.

Helm, D. 2004. *Energy, the State, and the Market: British Energy Policy since 1979.* Oxford: Oxford University Press.

Herz, J.H. 1951. *Political Realism and Political Idealism: A Study in Theories and Realities.* Chicago: University of Chicago Press.

Intergovernmental Panel on Climate Change. 2014. "Climate Change 2014: Mitigation of Climate Change. Contribution of Working Group III to the Fifth Assessment Report of the Intergovernmental Panel on Climate Change." Edited by O. Edenhofer, R. Pichs-Madruga, Y. Sokona, E. Farahani, S. Kadner, K. Seyboth, A. Adler, I. Baum, S. Brunner, P. Eickemeier, B. Kriemann, J. Savolainen, S. Schlömer, C. von Stechow, T. Zwickel, and J.C. Minx. Cambridge, UK: Cambridge University Press. www.ipcc.ch/pdf/assessment-report/ar5/wg3/ipcc_wg3_ar5_full.pdf.

Itoh, S. 2009. "Russia's Energy Policy towards Asia: Opportunities and Uncertainties." In *Energy and Security Cooperation in Asia: Challenges and Prospects,* edited by C. Len, and E. Chew, 143–166. Stockholm: Institute for Security and Development Policy. http://isdp.eu/content/uploads/images/stories/isdp-main-pdf/2009_len-chew_energy-and-security-cooperation-in-asia.pdf.

Jakob, M., Steckel, J.C., Klasen, S., Lay, J., Grunewald, N., Martínez-Zarzoso, I., Renner, S., and O. Edenhofer. 2014. "Feasible Mitigation Actions in Developing Countries." *Nature Climate Change* 4 (11): 961–968. doi:10.1038/nclimate2370.

Jervis, R. 1978. "Cooperation under the Security Dilemma." *World Politics* 30 (2): 167–174. www.jstor.org/stable/2009958.

Kalkuhl, M., Edenhofer, O., and K. Lessmann. 2015. "The Role of Carbon Capture and Sequestration Policies for Climate Change Mitigation." *Environmental and Resource Economics* 60 (1): 55–80. doi:10.1007/s10640-013-9757-5.

Kennedy, A.B. 2010. "China's New Energy Security Debate." *Survival* 52 (3): 137–158. doi:10.1080/00396338.2010.494881.

Koike, M., Mogi, G., and W.H. Albedaiwi. 2008. "Overseas Oil–development Policy of Resource–poor Countries: A Case Study from Japan." *Energy Policy* 36 (5): 1764–1775. doi:10.1016/j.enpol.2008.01.037.

Kong, B. 2011. "Governing China's Energy in the Context of Global Governance." *Global Policy* 2 (1): 51–65. doi:10.1111/j.1758-5899.2011.00124.x.

Lai, H., ed. 2009. *Asian Energy Security: The Maritime Dimension.* New York: Palgrave Macmillan. www.palgrave.com/us/book/9780230606425.

Lee, J.-S. 2010. "Energy Security and Cooperation in Northeast Asia." *Korean Journal of Defense Analysis* 22 (2): 217–233. doi:10.1080/10163271003744462.

Manicom, J. 2014. *Bridging Troubled Waters: China, Japan, and Maritime Order in the East China Sea.* Washington, DC: Georgetown University Press. http://press.georgetown.edu/book/georgetown/bridging-troubled-waters.

Maurin, C., and V. Vivoda. 2016. "Shale Gas and the Energy Policy 'Trilemma'." In *Handbook of Shale Gas Law and Policy*, edited by T. Hunter, 369–382. Cambridge, UK: Intersentia. http://intersentia.com/en/handbook-of-shale-gas-law-and-policy.html.

McCollum, D.L., Krey, V., Riahi, K., Kolp, P., Grübler, A., Makowski, M., and N. Nakićenović. 2013. "Climate Policies can Help Resolve Energy Security and Air Pollution Challenges." *Climatic Change* 119 (2): 479–494. doi:10.1007/s10584-013-0710-y.

Nakatani, K. 2004. "Energy Security and Japan: The Role of International Law, Domestic Law, and Diplomacy." In *Energy Security: Managing Risk in a Dynamic Legal and Regulatory Environment*, edited by B. Barton, C. Redgwell, A. Rønne, and D.N. Zillman, 413–427. Oxford: Oxford University Press.

Phillips, A. 2013. "A Dangerous Synergy: Energy Securitization, Great Power Rivalry and Strategic Stability in the Asian Century." *The Pacific Review* 26 (1): 17–38. doi:10.1080/09512748.2013.755362.

Prantl, J. 2011. "Crafting Energy Security Cooperation in East Asia." S. Rajaratnam School of International Studies, Centre for Non-traditional Security Studies, Policy Brief, No. 9, April. www.ciaonet.org/catalog/21765.

Radetzki, M. 1999. "European Natural Gas: Market Forces Will Bring about Competition in Any Case." *Energy Policy* 27 (1): 17–24. doi:10.1016/S0301-4215(98)00040-8.

Ravenhill, J. 2013. "Resource Insecurity and International Institutions in the Asia-Pacific Region." *The Pacific Review* 26 (1): 39–64. doi:10.1080/09512748.2013.755364.

Russell, C. 2014. "Cheaper Asian LNG Depends on Coal, Japan Nuclear." *Reuters*, 25 March.

Shaoul, R. 2005. "An Evaluation of Japan's Current Energy Policy in the Context of the Azadegan Oil Field Agreement Signed in 2004." *Japanese Journal of Political Science* 6 (3): 411–437. doi:10.1017/S1468109905001970.

Sheehan, P., Cheng, E., English, A., and F. Sun. 2014. "China's Response to the Air Pollution Shock." *Nature Climate Change* 4 (5): 306–309. doi:10.1038/nclimate2197.

Sidortsov, R. and B.K. Sovacool. 2015. "State–market Interrelations in the US Onshore and Offshore Oil and Gas Sectors." In *States and Markets in Hydrocarbon Sectors*, edited by A.V. Belyi and K. Talus, 171–197. Basingstoke: Palgrave Macmillan. doi:10.1057/9781137434074_9.

Sovacool, B.K., and V. Vivoda. 2012. "A Comparison of Chinese, Indian, and Japanese Perceptions of Energy Security." *Asian Survey* 52 (5): 949–969. doi:10.1525/as.2012.52.5.949.

Steckel, J.C., Edenhofer, O., and M. Jakob. 2015. "Drivers for the Renaissance of Coal." *Proceedings of the National Academy of Sciences of the United States of America* 112 (29): E3775–E3781. doi:10.1073/pnas.1422722112.

Stein, A.A. 1982. "Coordination and Collaboration: Regimes in an Anarchic World." *International Organization* 36 (2): 299–324. www.jstor.org/stable/2706524.

Stoddard, E. 2013. "Reconsidering the Ontological Foundations of International Energy Affairs: Realist Geopolitics, Market Liberalism and a Politico-Economic Alternative." *European Security* 22 (4): 437–463. doi:10.1080/09662839.2013.775122.

Stringer, K.D. 2008. "Energy Security: Applying a Portfolio Approach." *Baltic Security & Defence Review* 10: 121–142. doi:10.1.1.529.3357.

Victor, D.G., and L. Yueh. 2010. "The New Energy Order: Managing Insecurities in the Twenty-First Century." *Foreign Affairs* 89 (1): 61–73. www.jstor.org/stable/20699783.

Vivoda, V. 2010. "Evaluating Energy Security in the Asia-Pacific Region: A Novel Methodological Approach." *Energy Policy* 38 (9): 5258–5263. doi:10.1016/j.enpol.2010.05.028.

Vivoda, V. 2012. "Japan's Energy Security Predicament Post-Fukushima." *Energy Policy* 46: 135–143. doi: 10.1016/j.enpol.2012.03.044

Vivoda, V. 2014a. *Energy Security in Japan: Challenges after Fukushima.* Abingdon: Routledge. www.routledge.com/Energy-Security-in-Japan-Challenges-After-Fukushima/Vivoda/p/book/9781409455301.

Vivoda, V. 2014b. "Natural Gas in Asia: Trade, Markets and Regional Institutions." *Energy Policy* 74: 80–90. doi:10.1016/j.enpol.2014.08.004.

Vivoda, V. 2014c. "LNG Import Diversification in Asia." *Energy Strategy Reviews* 2 (3/4): 289–297. doi:10.1016/j.esr.2013.11.002.

Vivoda, V. 2015a. "State-Market Interaction in Hydrocarbon Sector: The Cases of Australia and Japan." In *States and Markets in Hydrocarbon Sectors*, edited by A.V. Belyi and K. Talus, 240–265. Basingstoke: Palgrave Macmillan.

Vivoda, V. 2015b. "Energy Security in East Asia." In *Non-Traditional Security in East Asia: A Regime Approach*, edited by J. Reeves and R. Pacheco Pardo, 143–165. London: Imperial College Press. www.worldscientific.com/worldscibooks/10.1142/p1008.

Vivoda, V., and J. Manicom. 2011. "Oil Import Diversification in Northeast Asia: A Comparison between China and Japan." *Journal of East Asian Studies* 11 (2): 223–254. doi:10.1017/S1598240800007177.

Wilson, J.D. 2012. "Resource Security: A New Motivation for Free Trade Agreements in the Asia-Pacific Region." *The Pacific Review* 25 (4): 429–453. doi:10.1080/09512748.2012.685098.

Wilson, J.D. 2014. "Northeast Asian Resource Security Strategies and International Resource Politics in Asia." *Asian Studies Review* 38 (1): 15–32. doi:10.1080/10357823.2013.853027.

Yergin, D. 2006. "Ensuring Energy Security." *Foreign Affairs* 85 (2), 69–82. www.jstor.org/stable/20031912.

Zhang, Z. 2012. "The Overseas Acquisitions and Equity Oil Shares of Chinese National Oil Companies: A Threat to the West but a Boost to China's Energy Security?" *Energy Policy* 48: 698–701. doi:10.1016/j.enpol.2012.05.077.

Zimmer, A., Jakob, M., and J.C. Steckel. 2015. "What Motivates Vietnam to Strive for a Low-carbon Economy? — On the Drivers of Climate Policy in a Developing Country." *Energy for Sustainable Development* 24: 19–32. doi:10.1016/j.esd.2014.10.003.

# 19

# SUSTAINABLE ENERGY INFRASTRUCTURE FOR ASIA

## Policy framework for responsible financing and investment

*Artie W. Ng and Jatin Nathwani*

## Introduction

Asia has been undergoing a long period of economic growth and development supported by an energy infrastructure fueled by a variety of traditional energy resources. For rapidly developing economies like China and India, fossil fuels in particular have played a crucial role in provision of stable and reliable energy supplies consistent with primary economic development objectives. Continued reliance on fossil fuels is problematic in light of greenhouse gas emissions and consequential impacts on climate change. The global consensus emerging from Paris COP21 now calls for greener, low carbon development in the energy sector among the sizable advanced and emerging economies. In the meantime, countries in Asia are urged to seek a balanced portfolio of electric power generating facilities for their sustainable development in lieu of merely considering nuclear power.[1]

In fact, the Paris Agreement executed on Earth Day, April 22, 2016, advocated substantial further investments in renewable energy technologies for timely deployment on a global basis (UNEP, 2016). A dilemma for reducing the use of fossil fuels is that there are still many developing countries struggling with underdeveloped infrastructure for the generation of electricity. Despite the recent decades of global efforts, it is worthwhile to note that many people in developing nations still live without electricity while many others struggle with unreliable access. It is predicted by the International Energy Agency (IEA) that the number of people without electricity in the year 2030 will exceed 1 billion (IEA, 2014). Adverse impact on the environment resulting from greenhouse gas emissions by both the developed and developing nations remains a threat despite the efforts to reverse the ongoing climate change inertia.

To deal with these environmental and social economic challenges, Asian countries need to strategically reformulate their energy policies to cope with the global political initiatives on combating climate change while building up the capacities of sustainable energy infrastructure to support their continuing economic developments. However, increasingly there are time constraints over such redevelopment of energy infrastructure as there are impending environmental risks and disparities in social economic developments associated with climate change. A dilemma remains as developing nations require steady energy supply, traditionally of fossil sources such as coal, to fuel their economic growth and development.

As the lifespan of the existing energy infrastructure expires and new capacity is required to meet expanding demand, Asian countries need to swiftly formulate a range of interrelated economic and finance policies to incentivize significant amounts of capital for allocation of resources into the development of sustainable energy infrastructure. Such policies should enable collaboration of both public and private capital with consideration of responsible financing as the principle. This requires a comprehensive package of economic incentives and regulatory measures as well as an integrated risk-based decision making that takes into account both internal and external costs associated with un-sustainability over time. For instance, continuing adverse impact on the environment and human health should be factored in by the governments in their policy making. This chapter proposes a policy framework for financing sustainable energy infrastructure in Asia under such an imminent, global concern.

## Sustainable energy infrastructure for sustainable development

Development of a sustainable energy infrastructure has been identified as a critical pursuit within the international community. The United Nations Office for Project Services (UNOPS) advocates the design and development of sustainable infrastructure projects as a way to mitigate the problems faced by "vulnerable individuals and communities" while providing opportunities for the enhancement of human rights and fundamental freedoms as well as protection of the environment (UNOPS, 2012). Energy infrastructure as a key component of the overall societal infrastructure can be broadly viewed as a system of facilities that enable power generation from various sources with zero or low greenhouse gas emission as well as the related distribution network for delivering reliable power to the end-users. Typically it has a long period of useful life (20–30 years) before it becomes obsolete and requires replacement. To facilitate development of sustainable energy, such new infrastructure needs to enable a variety of sustainable energy sources, namely solar, wind, geothermal and other renewable sources that allow low to zero emission of greenhouse gases. In addition, smart electricity grids can be incorporated into the system to enable end-users to select their choice of sustainable energy from independent sustainable energy producers.

A reliable supply of a sufficient amount of energy will continue to sustain a developing nation's economic growth. However, developing countries are naturally less advanced in renewable energy technologies and know-how than those developed economies which have been investing in pertinent research and development (R&D) projects. There are, however, opportunities for technology transfer to highly dense urban regions of developing nations in Asia for large-scale deployment of advanced renewable energy technologies. To enable development of these large-scale infrastructure projects, viable financing and investment schemes are necessary given the amount of financial resources required.

As explained by Bhattacharya et al. (2015), "Over the coming 15 years, the world will need to invest around (US)\$90 trillion in sustainable infrastructure assets, more than twice the current stock of global public capital . . . the bulk of these investment needs will be in the developing world." Such investments in sustainable energy infrastructure would enable the world to develop with a low-carbon trajectory leading to economic prosperity rather than a pathway to high-carbon and a less sustainable environment (Bhattacharya et al., 2015).

In order to develop a sustainable energy infrastructure, it is important to perform a comprehensive lifecycle production analysis in its project development. Under the framework developed by UNOPS (2012), it is necessary to perform assessments on environmental and social impacts as part of a systematic infrastructure project lifecycle. For energy infrastructure projects, it is even more critical to perform lifecycle analysis to look into all the external costs, such as adverse impacts on the environment and human health, associated with their development, financing and implementation (Fthenakis and Kim, 2007; Sovacool, 2008). Regulatory bodies instrumental to development of sustainable energy policy and accountable to their stakeholders need to take into account careful analysis of the lifecycle beyond the economic benefits in the near term.

## Existing regulatory models for energy infrastructure in Asia

There are different regulatory models for energy among the Asian economies as they have evolved along their particular social and economic trajectories as influenced by their own histories. As a result, there are various extents of regulatory controls over the Asian power markets as imposed by their respective regulatory policies. As analyzed by Baker & McKenzie (2014), it is observed that there are largely four main types of power markets in the region with respect to the extent of their regulations namely (i) Most Regulated Model, (ii) Less Regulated Model, (iii) Further Deregulated Model and (iv) Deregulated Model. Their key characteristics are summarized in Table 19.1 (Baker & McKenzie, 2014).

The four models of regulated markets in Asia suggest that these power markets are to a large degree still highly regulated by their governments. There is limited access to these markets by overseas and regional investors/developers unless they collaborate with the key local players.

*Table 19.1* Regulatory models in the electric power market of Asia

| Types | Most Regulated Model | Less Regulated Model | Further Deregulated Model | Deregulated Model |
|---|---|---|---|---|
| Characteristics | "Generation, transmission, distribution and sales of electricity to end-users are controlled through a government-owned monopoly or a designated regional public utility." | "Private entities are authorized to generate and sell electricity to a central government-owned utility through a power purchase agreement. The government utility in turn transmits to regional distributors, either government or publicly-owned distributors which sell electricity to the end-users." | "Private generators and IPPs can sell electricity to a central government owned transmitter or a regional government or publicly-owned distributor to sell electricity to the end-users." | "Generation, transmission, distribution and retail are conducted by separate entities, which are typically commercially owned, which trade electricity through a fairly open, competitive market." |
| Examples | Hong Kong, Japan, Philippines | China,[a] Indonesia, Laos, Malaysia, Taiwan, Vietnam | India | Australia |

*Data source*: Baker & McKenzie (2014).

*Note*: [a]While China's electricity market is largely controlled by its state-owned enterprises in the supply chain, it allows participation of private operators through a highly regulated regime.

Development of sustainable energy infrastructure in these countries would very much rely on top-down policy formulation by their governments to initiate changes in regulations so as to attract investments in the renewal of their energy infrastructure. Technology transfer of advanced renewable energy technologies could be very much confined through collaboration with designated government-owned enterprises in these regulated markets. Alternatively, the deregulated markets would facilitate more timely adoption of advanced technologies under an open electricity market platform that allows overseas investments in sustainable energy infrastructure embedded with such new technologies for more efficient and effective energy solutions. Such openness of the market will enable development of new knowledge about sustainable energy infrastructure and thereby improved absorptive capacity for technological innovation of renewable energy applications over time. However, due to the complex process of deregulation involving multiple political and economic stakeholders, complete reliance on market deregulation among the Asian jurisdictions is unrealistic.

## Socially responsible financing and investment for sustainable energy infrastructure and clean technology

Despite the constraints under such sectoral regulations, finance and investment communities around the world have become interested in sustainable investing over recent years. In fact, socially responsible investment (SRI) has been acknowledged by these communities in light of concerns about risks and opportunities associated with ongoing climate change and the challenges associated with globalization. Corporate managers are increasingly expected to take measures to assess the external risks associated with climate change and to respond strategically as part of their fiduciary duties to the shareholders (Viederman, 2008).

In particular, Hopwood et al. (2010) point out the growing trend of integrating environmental, social and governance (ESG) issues which are likely to have an impact on the access to both equity and debt in the financial markets. It is found that institutional fund managers have developed investment strategies to integrate ESG issues in order to enhance their investment decision-making process (Robins, 2008). The concerned communities have expanded to include international credit-rating agencies and other stakeholders in financial services, including professional accountants, as a new set of values and expectations are developed seeking long-term, appropriately risk-adjusted investment returns. This increasingly perceived relationship between ESG and investment decision making would affect future access to debt financing, perceived risk and therefore overall cost of capital; as a consequence, returns to shareholders would be affected as well (Hopwood et al., 2010). This growing international trend for responsible investing is supported by multiple stakeholders of profit-making enterprises, namely institutional investors, credit-rating agencies, lenders, individual investors, employees and customers, among others, as suggested in prior studies (Bhattacharya et al., 2008; Renneboog et al., 2008; Attig et al., 2013; Ioannou and Serafeim, 2015).

These primary stakeholders, as resource providers through international financial markets, have been advocating the trend for pertinent corporate social responsibility (CSR) and sustainability disclosures as a means to enhance decision making. Increasingly, CSR has been revealed both as an outcome and as part of reputation risk management processes (Bebbington et al., 2008). Corporations are encouraged to adopt an integrated approach to report on information about sustainability to the stakeholders as there are implications for business performance (Perrini and Tencati, 2006; Schaltegger and Wagner, 2006).

Increasing interest in responsible and sustainable investing means that more investors could be keen to pursue opportunities with sustainable energy infrastructure. In fact, there has long been significant interest in financing energy projects in emerging economies given the expected favorable economic and development opportunities, despite the risks involved (Razavi, 1996). Augmenting legitimacy in tackling greenhouse gas emission, there are diversified interests in pursuing investment projects that are complementary to the development of sustainable energy infrastructure and complementary clean tech, namely a smart grid system that enhances electricity distribution, solar power technology, wind power technology, fuel cell technology and electric vehicles, etc. These investment opportunities in clean tech are pursued by a wide range of institutional investors, from venture capitalists that pursue hi-tech ventures with equity funds to commercial banks that issue debts for financing infrastructure projects.

For example, clean coal technologies have been utilized in developing countries as an environmentally friendly approach to improve the energy efficiency of using coal as a fuel. Coal was adopted in China's energy system but caused tremendous amounts of greenhouse gas emissions. In particular, it was revealed that coal accounted for about 90% of the $SO_2$ emissions, 70% of the dust emissions, 67% of the NOx emissions, and 70% of the $CO_2$ emissions in China (Chen and Xu, 2010). Since coal is still one of the most abundant energy resources in developing countries like China and Indonesia, utilization of clean coal technologies is envisaged to enable sustainable development in these countries in an economical manner. Such clean coal technologies, such as IGCC (integrated gasification combined cycle) and carbon capture and storage (CCS), are further enhanced by local R&D initiatives and related financial support initiatives by the government (Chen and Xu, 2010). Clean coal technologies that contribute to energy saving and emission reduction in a country can be further leveraged by effective implementation of domestic public policy and technology transfer through similar applications in other developing countries (Yue, 2012). To promote R&D initiatives facilitating technological innovation and applications, developing countries are encouraged to formulate domestic policy measures, including a national science and technology strategy with funding programs for clean tech R&D and a supportive policy environment to incentivize investments in clean tech (Tan, 2010).

## Relevance of United Nations' responsible investment principles as a global reinforcement

Investing in sustainable energy infrastructure has been reinforced by the global initiatives for responsible investing as endorsed by the United Nations (UN). In response to the growing attention to an enhanced role of the global capital market in relation to sustainability concerns, the UN has developed its Principles for Responsible Investments (UNPRI) in an attempt to encourage more responsible investments around the world. According to UNPRI (2016),

> Responsible investment is an approach to investment that explicitly acknowledges the relevance to the investor of environmental, social and governance factors, and of the long-term health and stability of the market as a whole; it recognises that the generation of long-term sustainable returns is dependent on stable, well-functioning and well governed social, environmental and economic systems.

Such a philosophy of investment synchronizes with the long-term benefits generated from investing in sustainable energy infrastructure. The six initiatives advocated by UNPRI (2016) are:

(i) To incorporate ESG issues into investment analysis and decision-making processes.
(ii) To be active owners and incorporate ESG issues into ownership policies and practices.
(iii) To seek appropriate disclosure on ESG issues by the entities in which we invest.

(iv) To promote acceptance and implementation of the Principles within the investment industry.

(v) To work together to enhance our effectiveness in implementing the Principles.

(vi) To report on our activities and progress towards implementing the Principles.

In fact, there is empirical evidence suggesting that the costs for equity and debt capital could be associated with ESG, in that businesses involved in "good" CSR activities and projects even exhibit lower financing costs on average (Attig et al., 2013; Schroder, 2014). Other studies reveal such linkage with long-term performance and sustainability of business organizations. For instance, a study of listed companies in China demonstrates such a relationship with business performance over time (Wut and Ng, 2015). Another study by Plumlee et al. (2015) supports a positive relationship between quality of environmental disclosure and value of a firm. There is a growing interest in the ability of SRI funds to attain both their financial performance and social goals in the longer term (van Dijk-de Groota and Nijhof, 2015). Given the positive implications for the environment, the returns in renewable energy financing and investments should be assessed with consideration of upsides as the risks of renewable energy projects could be over-estimated during their initial development (Donovan, 2015). Such overestimation of risk is largely due to unfamiliarity with the relatively new technologies and operations involved in renewable energy and the lack of a comprehensive lifecycle analysis.

## A multitude of funding sources and the roles of financial institutions

Development of the next generation of sustainable energy infrastructure requires a range of financial capital given the variety of investment risks involved. Such sustainable energy infrastructure needs to be composed of breakthrough, disruptive renewable energy technologies for generation of emission-free power as well as smart grids that enable us to distribute energy intelligently with reduction of wasted energy. There are higher risks associated with adopting new technologies as it takes time for them to become financially viable or fully commercialized. More traditional infrastructure, such as power generating facilities, with power purchase agreements in place, is considered less risky for the stable cash flows generated from operations.

To finance such a wide range of infrastructural developments, various financial instruments or funding sources are required for the variety of risks involved. For instance, high-tech ventures are typically evaluated by venture capitalists for their in-depth experience with investments in emerging technologies and managing the risks involved for attractive financial returns (Neshiem, 2000). In recent years, private equity has played an increasingly important role in clean energy projects as favorable policy mechanisms enable attractive investment returns from deployment of renewable capacity (Burer and Wustenhagen, 2009; Potskowski and Hunt, 2015).[2] Different investors and financial institutions would have their specific preferences in energy projects with variations in risks and returns. As explained by Potskowski and Hunt (2015), private equity funds have extended appetites of risks and are willing to invest in a spectrum of projects, ranging from low-risk investments in infrastructural projects, such as power networks with about 5–10% target returns, to growth or expansion type of power generating facilities with about 10–25% target returns. Alternatively, crowdfunding has been considered as a viable method to finance local projects by raising capital swiftly from various local sources to support initial growth and development. This method could be useful for renewable energy projects of pilot scales among smaller and developing nations in Asia. It also eliminates the foreign exchange risk exposures in cases of foreign direct investments and financing with a foreign currency.

*Table 19.2* Funding sources for development of sustainable energy infrastructure

| Funding sources | Sustainable energy projects |
| --- | --- |
| Venture capital | Investing in startups of disruptive technologies for renewable energy, namely solar, wind and fuel cell, that enhances future developments of sustainable energy infrastructure with improved efficiency and effectiveness |
| Private equity | Investing in various stages of the development and expansion of sustainable energy infrastructure; mergers and acquisitions; restructuring of assets, including smart grids and networks |
| Corporate equity funding of listed companies | Equity investments in a variety of energy projects; R&D investments for commercialization of new products and services |
| Green bonds (international commercial banks) | Project finance and debt financing in large-scale green projects based on preset criteria for climate-related projects |
| Multilateral development bank (ADB, AIIB, World Bank) | Co-financing arrangements in project financing and debt financing in large-scale projects based on preset criteria |
| Sovereign wealth funds | Cross-border equity investment, project financing and debt financing in large-scale projects based on preset criteria |

*Data source*: Authors' compilation.

In addition, there are other institutions and related financial instruments that are willing to allocate their investments into renewable energy infrastructure. One innovative initiative is the development of Green Bonds that are issued to finance infrastructure projects of renewable energy installations, energy efficiency, new technologies in waste management and agriculture that reduce greenhouse gas emissions, as well as forest and watershed management (World Bank, 2016). Accordingly, the World Bank has so far initiated 60 green bond transactions in sustainable energy infrastructure worth more than US$4 billion. Table 19.2 summarizes a multitude of funding sources for development of sustainable energy infrastructure.

With respect to multilateral development banks, the Asian Development Bank (ADB) and Asian Infrastructure Investment Bank (AIIB) play a critical role in financing sustainable energy infrastructure in developing Asian countries. Based in Manila, Philippines ADB was established as a key regional development bank in the 1960s with a mission to support social and economic development in Asia. The bank has 67 country members largely associated with the United Nations Economic and Social Commission for Asia and the Pacific (UNESCAP, formerly the Economic Commission for Asia and the Far East or ECAFE), but some of them are non-regional developed countries. With a similar mission to the World Bank but with a regional focus, ADB has a weighted voting system in proportion with members' capital subscriptions. Among the country members, Japan has appeared to dominate in the management leadership position in recent years.

According to ADB (2016), 80% of its lending activities are related to five core operational areas summarized as follows:

(i) Infrastructure

- Water Operational Plan
- Energy Policy
- Sustainable Transport Initiative Operational Plan
- Urban Operational Plan
- Toward E-Development in Asia and the Pacific: A Strategic Approach for Information and Communication Technology

(ii) Education by 2020: A Sector Operations Plan
(iii) Environment

- Environment Operational Directions 2013–2020
- Addressing Climate Change in Asia and the Pacific: Priorities for Action
- Operational Plan for Integrated Disaster Risk Management 2014–2020

(iv) Regional Cooperation and Integration Strategy
(v) Financial Sector Operations

ADB developed its Energy Policy in 2009 in alignment with ADB's operations with an aim to meet energy security needs in the region while enabling a transition to a low-carbon economy. Such an initiative also upholds the vision of a region free of poverty. As explained by ADB (2016), such policy is to "prioritize energy-related objectives and identify the institutional capabilities needed for the future within a changing regional, global, and technological context." Moreover, it has an objective to help ADB's regional members provide reliable, adequate and affordable energy for inclusive growth in a socially, economically and environmentally sustainable way. The three main pillars of ADB's Energy Policy are: (a) promoting energy efficiency and renewable energy; (b) maximizing access to energy for all; and (c) promoting energy sector reform, capacity building and governance (ADB, 2016).

Alternatively, AIIB is a multilateral development bank established through a series of participatory processes among founding members based on a set of core philosophies, principles, policies, value systems and operating platforms (AIIB, 2016). Initiated by China, AIIB focuses on the development of infrastructure and other pertinent utility sectors in Asia, including energy and power, transportation and telecommunications, rural infrastructure and agriculture development, water supply and sanitation, environmental protection, urban development and logistics. AIIB emphasizes the principles of being lean (with a small efficient management team and highly skilled staff), clean (critically being an ethical organization with zero tolerance for corruption) and green (being an institution built on respect for the environment). The AIIB aims to build strong policies on governance, accountability, environmental and social frameworks. It admits regional and non-regional countries, and developing and developed countries that seek to contribute to Asian infrastructure development and regional connectivity through mutual agreements (AIIB, 2016).[3]

AIIB is expected to carry out its strategy in collaboration with the other existing multilateral development banks to jointly address the increasing infrastructure needs in Asia. In fact, AIIB has executed a memorandum of understanding (MOU) with ADB for collaboration in joint financing projects, including projects for co-financing in the road and water sectors. The two regional multilateral development banks agree to strengthen cooperation at strategic and technical levels on the basis of complementarity in terms of institutional strengths and comparative advantages in addressing the massive infrastructure financing needs in Asia (AIIB, 2016). Moreover, the scope of such collaboration will include co-financing of infrastructure projects relating to energy, transportation, telecommunications, rural and agriculture development, water, urban development and environmental protection. Both institutions aspire to promote the implementation of the Sustainable Development Goals and the COP21 climate agreement.

While ADB and AIIB are two strategic regional financial institutions for the development of sustainable energy infrastructure in Asia, it is worthwhile to note that the private sector still plays a critical role in investing in and operating clean tech projects through commercial arrangements with regional enterprises that embrace the economic risks and opportunities involved. For instance, China Light and Power (CLP), a listed company based in Hong Kong, has been diversified into a regional enterprise in Asia that are all-around in owning and operating major sustainable energy projects. It also acts as an operator of sustainable energy infrastructure that actively manages a wide range of risks involved. Further, it evaluates technologies in terms of financial and technical feasibilities. Another noticeable effort is the MTR Corporation of Hong Kong (MTR), which provides mass transit through clean public transportation powered by electricity. As a listed company partially owned by the government, it has successfully issued its first green bond to finance its new railway developments.

As highlighted by Hong Kong's Under-Secretary for the Environment, "The Hong Kong SAR Government sees MTR Corp as an essential stakeholder and facilitator in Hong Kong's low-carbon transition, and the good response to this green bond issuance shows the financial community's growing interest in investing in responsible and sustainable companies" (Railway Gazette, 2016).

The complementary role of public–private collaboration in financing such projects can hardly be underestimated.

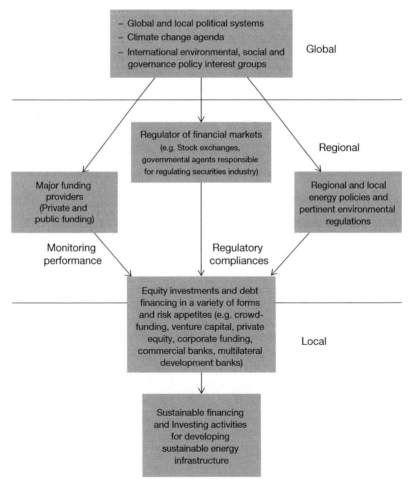

*Figure 19.1* Holistic framework for financing sustainable energy infrastructure

*Data source*: Authors' illustration.

## Proposed holistic framework

Observing the current global trend of investing in sustainable energy, a framework is proposed to integrate various driving forces for the development of sustainable energy infrastructure. In particular, policy intervention would play a critical role to incentivize renewable energy directly or "disadvantage fossil fuels in the form of carbon tax, emission trading scheme or direct legislation on emissions limits or planning controls" (Turner, 2015). With reference to a prior study on transition to renewable energy, van den Bergh and Bruinsma (2008) envisage that there are three main levels of developments for the transition into renewable energy. We intend to apply such conceptual aspiration into our proposed framework for Asia.

At the global level, the cross-border trend for ESG creates the movement towards climate change solutions through multilateral agreements as well as a range of principles and guidelines for sustainability. At the regional level, there are various key institutional stakeholders facilitating the rules and regulations for sustainable energy investments in the Asian region. At the local level, the investment and allocation of resources into sustainable energy infrastructure would take place at a jurisdiction. Local financial institutions and investors could also play an important role in supporting the development and growth of local enterprises venturing into sustainable energy infrastructure with complementary efforts. Participation of a variety of funding sources around the globe via international capital markets is critical. Complementary policy for lowering financing risk and operating expenses, payment incentives, increasing R&D for sustainable infrastructure and accounting for the risk of environmental degradation needs to be formulated in support of developing sustainable energy infrastructure. Our proposed holistic framework is illustrated in Figure 19.1.

Hong Kong as an international financial center in Asia demonstrates a holistic approach for financing sustainable energy infrastructure. As a financial hub that embraces international standards and approaches, it provides a platform for various financial institutions, enterprises and professionals of local and international origins to enable sustainable energy financing. The two examples of green energy investments and financing by CLP and MTR indicate Hong Kong's initial success with active endorsements by its government for green financing.

## Concluding remarks

Retiring and converting the legacy energy infrastructures into sustainable ones is a complex challenge that would take a number of years to complete given their lifespans and scales. It involves lengthy financing and investment evaluation processes as well as technical due diligence before substantial amounts of capital expenditures are allocated into such infrastructure development. It requires comprehensive feasibility studies to ensure proven and technologically sound solutions for deployment into the next generation of sustainable energy infrastructure, let alone to overcome the political and social economic hurdles. Considering ongoing climate change developments, there is, however, a time constraint for countries around the world, in particular in Asia, to respond with effective and proven solutions. Energy is a highly critical sector for its implications for climate change, but it is also crucial for reinforcing economic growth and development in the region.

There is an opportunity to reinforce sustainable growth and development through adopting the principles of responsible financing and investments in order to encourage the world to allocate resources into assets that will in turn reduce air pollution and greenhouse gas emissions that are causing the current trajectory of climate change. It takes cross-country efforts to develop a range of public finance policies as severally adopted by the regional governments. Investing in sustainable energy infrastructure and related clean tech requires support from a range of funders from the private sector capable of evaluating and managing risks associated with emerging renewable energy technologies, while integrating the proven ones into the future sustainable energy infrastructure.

We observe that to deal with such a complex challenge, there is a compelling need to consider jointly formulating complementary economic, finance, energy and environmental policies with an emphasis on a cross-disciplinary approach in dealing with the legacy of unsustainable infrastructure in the first place. Such policies also need to be developed and disseminated across the region in order to be effective for implementation. There is apparently no immediate remedy as it involves both regional and local political concerns as well as interconnected economic interests among the jurisdictions. For instance, detailed arrangements in complementarity between ADB and AIIB in financing sustainable energy for the region are yet to be concretely formulated. Harmonization of these multilateral development institutions would initiate a scalable sustainable development in Asia. The policies advocated by ADB and AIIB could be exemplary for individual Asian countries to develop their own specific ones taking into consideration their social and economic situations. However, one should not underestimate both the limitation of these institutions and the potentials of market-based financing through utilization of traditional and innovative financial instruments in the form of debt and equity.

We have proposed a holistic framework that can be utilized by local policy makers to evaluate the three levels of initiatives for incentivizing development of sustainable energy capacities through a regional public–private initiative for allocating resources into building infrastructure for sustainable energy. As the Asia Pacific is a diverse region in terms of social economic and cultural background, the efficacy of this framework needs to be tested in different jurisdictions in the region to help shape recommendations that will best reflect the local contexts. Characteristics of local financial institutions and investment communities in each country would need to be taken into account in adapting the proposed framework so as to allow multiple sources of financing for local sustainable infrastructure developments.

## Notes

1 For instance, Japan is reconsidering the weighting of nuclear power in its electricity power generating sector after the disaster that took place within the Fukushima nuclear power generating facilities in 2011.
2 Private equity is typically backed by pension and endowment funds as well as high-net-worth investors.
3 In 2015, representatives from 57 countries in Beijing signed the Bank's Articles of Agreement. These countries included Australia, Austria, Azerbaijan, Bangladesh, Brazil, Brunei Darussalam, Cambodia, China, Egypt, Finland, France, Georgia, Germany, Iceland, India, Indonesia, Iran, Israel, Italy, Jordan, Kazakhstan, Republic of Korea, Kyrgyz Republic, Lao PDR, Luxembourg, Maldives, Malta, Mongolia, Myanmar, Nepal, Netherlands, New Zealand, Norway, Oman, Pakistan, Portugal, Qatar, Russia, Saudi Arabia, Singapore, Spain, Sri Lanka, Sweden, Switzerland, Tajikistan, Turkey, the United Arab Emirates, the United Kingdom, Uzbekistan and Vietnam.

## References

ADB. (2016). [Online], available: www.adb.org/ [24 July 2016].
AIIB. (2016). [Online], available: www.aiib.org/ [24 July 2016].
Attig, N., Ghoul, S. E., Guedhami, O. and Suh, J. (2013). Corporate social responsibility and credit ratings. *Journal of Business Ethics*, 117(4), pp.679–694.
Baker & McKenzie (2014). Snapshots of key Asia-Pacific power markets. [Online], available: http://bakermckenzie.co.jp/pdf/Snapshots_of_Key_Asia-Pacific_Power_Markets_A5_spread_E.pdf [20 July 2016].
Bebbington, J., Larringaga, C. and Moneva, J. M. (2008). Corporate social reporting and reputation risk management. *Accounting, Auditing & Accountability Journal*, 21(3), pp.337–361.
Bhattacharya, A., Oppenheim, J. and Stern, N. (2015). Driving sustainable development through better infrastructure: key elements of a transformation program. *Global Economy & Development Working Papers*, Brookings Institute. July, pp.1–38.

Bhattacharya, C. B., Korschun, D. and Sen, S. (2008). Strengthening stakeholder–company relationships through mutually beneficial corporate social responsibility initiatives. *Journal of Business Ethics*, 85(2), pp.257–272.

Burer, M. J. and Wustenhagen, R. (2009). Which renewable energy policy is a venture capitalist's best friend? Empirical evidence from a survey of international cleantech investors. *Energy Policy*, 37(12), pp.4997–5006.

Chen, W. and Xu, R. (2010). Clean coal technology development in China. *Energy Policy*, 38(5), pp.2123–2130.

Donovan, C. W. (2015). *Renewable Energy Finance: Powering the Future*. London: Imperial College Press.

Fthenakis, V. M. and Kim, H. C. (2007). Greenhouse-gas emissions from solar electric and nuclear power: a life-cycle study. *Energy Policy*, 35(4), pp.2549–2557.

Hopwood, A., Unerman, J. and Fries, J. (2010). Introduction to the accounting for sustainability case studies, in Hopwood, A., Unerman, J. and Fries, J. (eds), *Accounting for Sustainability*. London/Washington, DC: Earthscan, pp.1–26.

IEA. (2014). *World Energy Investment Outlook Special Report*. Paris: IEA/OECD Publishing.

Ioannou, I. and Serafeim, G. (2015). The impact of corporate social responsibility on investment recommendations: analysts' perceptions and shifting institutional logics. *Strategic Management Journal*, 36, pp.1053–1081.

Neshiem, J. L. (2000). *High Tech Start Up*. New York: Free Press.

Perrini, F. and Tencati, A. (2006). Sustainability and stakeholder management: the need for new corporate performance evaluation and reporting systems. *Business Strategy and the Environment*, 15(5), pp.296–308.

Plumlee, M., Brown, D., Hayes, R. M. and Marshal, R. S. (2015). Voluntary environmental disclosure quality and firm value: further evidence. *Journal of Accounting and Public Policy*, 34, pp.336–361.

Potskowski, B. and Hunt, C. (2015). The growing role for private equity, in Donovan, C.W. (ed.), *Renewable Energy Finance: Powering the Future*. London: Imperial College Press, pp.225–244.

Railway Gazette. (2016). MTR Corp Issues First Green Bond. [Online], Available: www.railwaygazette.com/news/business/single-view/view/mtr-corp-issues-first-green-bond.html [13 November 2016].

Razavi, H. (1996). *Financing Energy Projects in Emerging Projects in Emerging Economies*. Tulsa, OK: PennWell.

Renneboog, L., Horst, J. T. and Zhang, C. (2008). Socially responsible investments: institutional aspects, performance and investor behavior, *Journal of Banking and Finance*, 32, pp.1723–1742.

Robins, N. (2008). The emergence of sustainable investing, in Krosinsky, C. and Robins, N. (ed.), *Sustainable Investing: the Art of Long-term Performance*. London/Sterling, VA: Earthscan, pp.3–18.

Schaltegger, S. and Wagner, M. (2006). Integrative management of sustainability performance, measurement and reporting. *International Journal of Accounting, Auditing and Performance Evaluation*, 3(1), pp.1–19.

Schroder, M. (2014). Financial effects of corporate social responsibility: a literature review. *Journal of Sustainable Finance and Investment*, 4(4), pp.337–350.

Sovacool, B. K. (2008). Valuing the greenhouse gas emissions from nuclear power: a critical survey. *Energy Policy*, 36(8), pp.2950–2963.

Tan, X. (2010). Clean technology R&D and innovation in emerging countries: experience from China. *Energy Policy*, 38(6), pp.2916–2926.

Turner, G. (2015). How much renewable energy will the global economy need?, in Donovan, C. W. (ed.), *Renewable Energy Finance: Powering the Future*. London: Imperial College Press, pp.47–76.

UNEP. (2016). Paris Agreement Signing on Earth Day Brings New Hope for People, Planet. [Online], available: www.unep.org/stories/ParisAgreement/Paris-Agreement-Signing-on-Earth-Day.asp [12 June 2016].

UNOPS. (2012). UNOPS Policy for Sustainable for Sustainable Infrastructure (1st Ed.), June.

UNPRI. (2016). United Nations Principles for Responsible Investments. [Online], available: www.unpri.org/ [25 April 2016].

van den Bergh, J. C. J. M. and Bruinsma, F. R. (2008). *Managing the Transition to Renewable Energy*. Cheltenham: Edward Elgar.

van Dijk-de Groota, M. and Nijhof, A. H. J. (2015). Socially responsible investment funds: a review of research priorities and strategic options. *Journal of Sustainable Finance and Investment*, 5(3), pp.178–204.

Viederman, S. (2008). Fiduciary duty, in Krosinsky, C. and Robins, N. (eds), *Sustainable Investing: the Art of Long-term Performance*. London/Sterling, VA: Earthscan, pp.189–200.

World Bank. (2016). [Online], available:ewww.worldbank.org/ [24 July 2016].

Wut, E. and Ng, A. (2015). CSR practice and sustainable business performance: evidence from the global financial centre of China. *Procedia: Social and Behavioral Sciences*, 195, pp.133–141.

Yue, L. (2012). Dynamics of clean coal-fired power generation development in China. *Energy Policy*, 51, pp.38–142.

# 20

# DEVELOPING ASIA'S RESPONSE TO CLIMATE CHANGE

## Reshaping energy policy to promote low carbon development

*Nandakumar Janardhanan and Bijon Kumer Mitra*

### Introduction

The Asian region will witness remarkable growth in energy demand in the years to come. It is expected that developing Asia, including India and China, will account for 44% of global GDP in the next three decades, which indicates the need for higher energy consumption (ADB, 2013b). It is estimated that coal consumption will increase by 81%, oil consumption will be doubled and natural gas consumption will be tripled by 2035 (ADB, 2013b) in this region. The surge in energy consumption will also lead to increasing energy-related emissions which are estimated to reach more than 22 billion tonnes by 2035 (ADB, 2013b). This period will also witness a significant portion of the energy demand from the top six primary energy consumers in Asia, namely China, India, Indonesia, Japan, Korea and Thailand. Although East Asian countries including China, Japan and Korea contribute the largest share of $CO_2$ emissions in the continent, the future projection shows a downward trend of contribution to the total $CO_2$ emissions due to mainstreaming of low carbon development in energy sectors in these countries. The contribution of South and Southeast Asian countries to $CO_2$ emissions is expected to increase from 12.9% in 2010 to 17.9% in 2035 and from 8% in 2010 to 11% in 2035, respectively (ADB, 2012). In developing Asian countries fossil fuels will remain the largest consumed energy sources, which will lead to greater contribution to greenhouse gas (GHG) emissions. Therefore, significant policy measures are needed among these countries to ensure that the region moves on a low carbon pathway. Balancing energy demand growth and environmental objectives is a critical policy challenge to all governments in developing Asia.

The energy policy of developing Asia varies significantly from one country to another as the region reflects diverse socio-political, economic and environmental situations as well as energy demand scenarios. While some of the countries depend heavily on fossil fuels (India and Indonesia), others have a hydropower-dominated energy mix (e.g. Bhutan, Nepal, Lao-PDR). Out of the total oil consumption of 1,501.4 Mt, 385.12 Mt was consumed by India and Indonesia together (BP, 2016). As domestic demand has been surging, the demand specifically for fossil fuels is likely to increase in the region. The cleaner energy mix in these countries also reflects varying degrees of development. India evinced remarkable progress in solar and wind energy resource development while many other countries are investing heavily in biomass and biofuel development. Malaysia is a leading player in the region in developing biofuels based on palm oil (Kuntom, 2014), while the Philippines has made significant inroads in developing geo-thermal resources (IEA, 2011(a)). Among other energy sources, nuclear power is playing a key role in Pakistan among the countries in focus here; however, it has made limited progress in other countries primarily due to the sensitivity related to the economic, political and security factors.

This chapter focuses on eight countries from South and Southeast Asia. For the convenience of the analysis, these countries are termed 'developing Asia'. Economies from East Asia such as China (though considered developing economies) are not included in the study conducted here.

## Energy and climate scenario in developing Asia

Developing Asia is expected to be one of the largest blocs of fossil fuel consumers in the world by the 2030s, with consumption of coal, oil and natural gas increasing significantly in the years to come if the current consumption pattern continues. According to estimates by British Petroleum (BP),

> commercially traded primary energy consumption in selected countries including India, Indonesia, Thailand, Malaysia, Pakistan, Vietnam, Philippines, and Bangladesh has increased from 833.1 million tons of oil equivalent (Mtoe) in 2006 to 1326.6 Mtoe by 2015 reflecting nearly 60% growth in ten years. However, the change in primary energy consumption during the same period in other regions is notably lower. North America witnessed a growth of −1.1%, South and Central America together witnessed 24.37% increase, Europe and Eurasia together witnessed a negative growth of −6.2%, Middle East reflected 49.4% and African region reflected 28.8% increase.
>
> *(BP, 2016)*[1]

Primary energy consumption trends in developing Asia show that India, Indonesia and Malaysia are the top three consumers that together account for more than 66% of total primary energy consumption in developing Asia (see Table 20.1).

The table also shows an interesting trend whereby the share of fossil fuels remains high in the energy mix of these countries, leading to growing emissions that in turn contribute to global warming. According to various Assessment Reports of the International Panel on Climate Change (IPCC), evidence of climate impacts is more frequent which points to the fact that climate change is real and happening much more strongly than previously observed by scientists. The increase in energy-related emissions from developing Asia, and the resulting larger impact on the region as well as the planet, is alarming.

Within developing Asia, India has been the largest GHG emitter for several years, which is a clear reflection of its fossil-dominated energy consumption pattern (see Figure 20.1). Though alternative energy sources are increasing in the domestic energy mix, the share of the same would not be able to replace conventional fuel sources.

The power sector is responsible for the largest chunk of global carbon emissions, primarily due to its dependence on coal and gas. According to the OECD, 'the electricity sector is still far from being low-carbon as it continues to be dominated by fossil fuels, with 41% power generated

*Table 20.1* Primary energy consumption and $CO_2$ emissions in developing Asia

| Countries | Primary energy consumption | $CO_2$ emissions |
|---|---|---|
| Bangladesh | 30.7 | 72.9 |
| India | 700.5 | 2,218.4 |
| Indonesia | 195.6 | 611.4 |
| Malaysia | 93.1 | 246.9 |
| Pakistan | 78.2 | 179.5 |
| Philippines | 37.7 | 106.5 |
| Thailand | 124.9 | 295.9 |
| Vietnam | 65.9 | 169.0 |
| Total Asia Pacific[a] | 5,498.5 | 16,066.7 |

*Data source*: BP Statistical Review of World Energy (2015).

*Note*: [a]The estimate includes data on Australia, Bangladesh, China, China Hong Kong SAR, India, Indonesia, Japan, Malaysia, New Zealand, Pakistan, Philippines, Singapore, South Korea, Taiwan, Thailand, Vietnam and other smaller economies in the region.

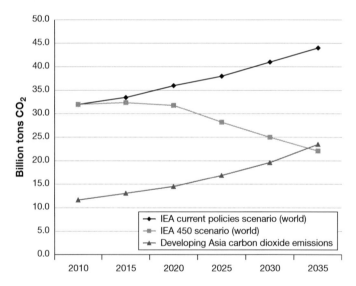

*Figure 20.1* Projected energy-related carbon dioxide emissions from developing Asia

*Data source*: www.adb.org/publications/asian-development-outlook-2013-asias-energy-challenge.

at the global level by coal and 22% by gas' (NEA, 2015). With growing demand for electricity, developing Asian countries are seen as major energy demand centres of the future. Primary energy demand is projected to increase at 3% per year in South Asia and 2.8% in Southeast Asia over the outlook period (2010–2035)—faster than the projected world average growth rate of 1.5% per year during the same period. With this growth, primary energy demand in South and Southeast Asia will reach 2,668 Mtoe by 2035, up from 1,300 Mtoe in 2010 (ADB, 2013a). The projected demand analysis also evinces that demand for oil, gas and coal will increase enormously. Eight countries (India, Indonesia, Malaysia, Pakistan, Philippines, Thailand and Vietnam) in developing

Asia together account for roughly 21% of the total primary consumption of the Asia Pacific region. This increased to 24% in ten years, indicating a growth in their respective share of primary energy consumption in the region (BP, 2016). This trend is likely to continue as the economic activities and corresponding energy demand are likely to increase among these economies. According to ADB,

> demand for coal in Asia and the Pacific will increase by 52.8% from 2010 to 2035, reaching 3,516.3 million tons of oil equivalent (Mtoe) by 2035; Oil demand in Asia and the Pacific is projected to increase by 1.9% yearly over the outlook period and reach 1,973.0 Mtoe by 2035, 59.3% higher than the 1,238.2 Mtoe in 2010; and demand for natural gas is projected to increase at 3.9% per year, reaching 1,463.2 Mtoe in 2035, 2.6 times the 2010 level of 566.7 Mtoe.
>
> *(ADB, 2013a)*

The developing Asian countries will also reflect this trend in the years to come.

In view of the increasing demand for electricity, investment is increasing in developing Asia. However, most of the investment is aimed at increasing power generation facilities that depend on either coal or gas. The International Energy Agency (IEA) analysis of the cumulative investment in energy supply and energy efficiency during the period 2014–2035 finds that Asia and Oceania together will account for approximately US$12,571 billion worth of investment. Of this, approximately US$7,871 billion will be invested in the electricity generation sector, mostly based on thermal facilities (IEA, 2014). The huge demand for fossil fuels for electricity generation highlights that the hazards of emissions will increase for developing Asia as well.

## Responses to climate change in developing Asia

A truly global and integrated energy technology revolution is essential to address the intertwined challenges of energy security and climate change while also meeting the growing energy needs of the developing world (IEA, 2011b). Hence a transition from conventional fuel dependency to a new, low carbon energy mix is important for Asian developing economies to address the challenge. Though moving away from a fossil fuel based economy may pose challenges to domestic economic targets, climate change mitigation policies can be the driving force in promoting low carbon energy development. Strong domestic policies on energy efficiency improvement and for promoting electricity generation from non-conventional sources are important to overcome the decades-old concerns about technology lock-ins. These measures also have the potential to address both supply-side and demand-side energy concerns in the region. Some of the major emerging economies such as India, Indonesia, Malaysia and Thailand declared a specific target of reduction of $CO_2$ emission by 2020 at the UNFCCC Copenhagen conference in 2009. Subsequently, relevant targets have been articulated in the national and sectoral policies of these countries. A key focus of the Indian government's mitigation target was to promote solar energy to replace fossil fuels. Indonesia also mainstreamed climate change issues in its national development planning and midterm development plan 2010–2014 and set a target to increase geothermal energy generation by 9GW by 2025.

The Paris Agreement was a turning point whereby parties adopted a historic agreement to curb growing emissions:

> So far, 117 Parties have ratified of 197 Parties to the Convention. Agreement entered into force on 4 November 2016, thirty days after the date on which at least 55 Parties to

the Convention accounting in total for at least an estimated 55% of the total global greenhouse gas emissions have deposited their instruments of ratification, acceptance, approval or accession with the Depositary.

*(UNFCCC, 2016b)*

Developing Asia's major economies are strengthening their commitments to reduction of global GHG emissions by increasing their respective domestic emission reduction targets for 2030. Some countries like Bangladesh outline the commitment to contribute to global mitigation efforts by setting up an unconditional target of a 5% $CO_2$ emission cut by 2030, although its share of total global emissions is very low. It will be achieved mainly from energy sectors (power, transport and industries). Countries like Indonesia and Bangladesh have demonstrated their willingness to concentrate on mitigation efforts, conditional on international support.

To achieve ambitious deep cuts in $CO_2$ emissions, a large number of countries have shown their interest in increasing their share of clean energy. An analysis of 156 INDCs (Stephan et al., 2016) found that 108 nations plan to increase renewables, 75 countries have set up quantitative targets for the renewable share in their energy mix and 8 countries plan to generate 100% power from renewable sources (see Table 20.2). The analysis also recognised that Asia's two giant emerging economies, China and India, committed to the largest increase in the share of renewable energy.

It is noticed that as part of the climate mitigation programmes, greater focus has been given to increasing alternative energy generation in many countries. As mentioned earlier, developing Asia has several policy initiatives that aim to enhance energy efficiency as well as renewable energy generation. The renewable energy promotion which is aimed at meeting short-term energy needs while strengthening long-term sustainable development is one of the major examples of legal initiatives. Other economies in developing Asia also have rolled out legal mechanisms that support various aspects of energy transition. Measures to address transportation sector challenges and to enhance industrial as well as residential energy efficiency are noticeable examples from other economies. However, one of the common issues which is witnessed across developing Asian countries is the low implementation rate of these legal tools. As an example one can note that the energy efficiency norms prescribed for residential buildings are often flouted. One may also notice that energy efficiency guidelines prescribed for residential buildings in many urban areas are too weak to make any ambitious energy efficiency improvement.

## Meeting climate objectives: the search for low carbon development

Despite all nations showing willingness to curtail global carbon emissions, assessment of current commitments and policies reveals that unconditional commitments that nations have made in the INDCs and existing policies will not be able to limit global warming below 2°C (CAT, 2016b; UNEP, 2015). Data shown in Table 20.3 indicates that ambitious plans are necessary to limit global warming below 2°C, beyond the existing policies and INDCs in the energy sector.

Actions in the energy sector as part of the countries' responses towards climate mitigation can be important catalysts in energy transition. Promotion and diffusion of low carbon energy technologies can also be part of this strategy, together with regional energy integration and cross-border energy trade through which these economies can share common resources and mutually beneficial electricity generation facilities and transmission infrastructure. With regard to external factors, international finance and technology will also play a key role in meeting energy demand.

As Asia played a vital role in setting up universal agreements on climate change, it is expected that it will remain at the forefront in taking proactive mitigation policies and actions. In INDCs,

*Table 20.2* Emission reduction plans in developing Asia

| Country | Copenhagen Commitment (2020 targets) | INDC (2030 targets) | Clean energy targets |
|---|---|---|---|
| India | Reduction of $CO_2$ emissions intensity of GDP by 20–25% below 2005 levels by 2020 | Reduction of $CO_2$ emissions intensity of GDP by 33–35% below 2005 levels by 2030 | INDC builds on its goal of installing 175 gigawatts (GW) of renewable power capacity by 2022 by setting a new target to increase its share of non-fossil-based power capacity from 30% today to about 40% by 2030 (with the help of international support) |
| Indonesia | Emission reduction of 26% below business-as-usual (BAU) by 2020 | Conditional 41% reduction below BAU by 2030 (with sufficient international support) | No specific target on clean energy/renewable energy/non-fossil fuel |
| Thailand | Thailand proposes action in the energy and transportation sectors to reduce emissions 7–20% below projections for 2020. (UNFCCC, 2015(a)) | By 2030, reduce energy intensity (energy use per unit of GDP) to 25% below 2005 levels, across all economic sectors (IIP, 2013) | Although no specific quantitative targets on renewable energy have been set in INDC, the PDP sets a target to achieve a 20% share of power generation from renewable sources in 2036 |
| Malaysia | Reduction of GHG emissions up to 40% by 2020, based on carbon intensity (UNCRD, 2009) | Malaysia intends to reduce its GHG emissions intensity of GDP by 45% by 2030 relative to the emissions intensity of GDP in 2005 (UNFCCC, 2015(b)). | Although no specific quantitative targets on renewable energy have been set in INDC, the National Renewable Energy Policy and Action Plan (2009) set a target to increase share of renewable energy in the total electricity mix to 11% of total electricity generated by 2030 |
| Pakistan | No specific targets on emission reduction | Reduce emissions by 10% by domestic actions (Dawn, 2015) | No specific quantitative targets on renewable energy |
| Vietnam | No specific target on emission reduction; showed support for emission reduction initiatives of UNFCCC based on CBDR principle. | Reduce GHG emissions by 8% compared with BAU and emission intensity per unit of GDP will be reduced by 20% compared with 2010 levels (Huy, 2010). | Increase the share of renewable energy in total primary energy supply by 3.4% in 2020 and 7% in 2050 (Nguyen et al., 2012). |

(*Continued*)

*Table 20.2* (continued)

| Country | Copenhagen Commitment (2020 targets) | INDC (2030 targets) | Clean energy targets |
|---|---|---|---|
| Philippines | No specific targets on emission reduction | The Philippines intends to reduce GHG (carbon dioxide equivalent) emissions by about 70% by 2030 relative to its business-as-usual estimate (UNFCCC, 2015(c)). | Increasing renewable energy installed capacity three-fold by 2030 from that of 2010 level (SEPO, 2014). |
| Bangladesh | No specific targets on emission reduction | An unconditional reduction of GHG emissions by 5% in the power sector, and a conditional (with international support) 15% reduction by 2030 | Improved energy efficiency in production and consumption of energy, renewable energy development, lower emissions from agricultural land (MoEF-Bangladesh, 2015) |

*Data sources*: Various sources.

*Table 20.3* Emission gaps under current policies and INDCs scenarios

| Scenario | Emission reduction compared with baseline (Gt. $CO_2e$) | Required additional emission reduction to stay below 2°C (Gt. $CO_2e$) |
|---|---|---|
| Baseline | – | 23 |
| Current policy | 5 | 18 |
| Unconditional INDCs | 9 | 14 |
| Conditional + unconditional INDCs | 11 | 12 |
| Below 2°C | 23 | 0 |

*Data source*: Adapted from UNEP (2015).

most of the countries recognised the energy sector (power, transport and industry) as a critical part of the mitigation efforts. Many countries in Asia have taken steps to cut down carbon emissions through both soft and hard interventions across all areas, from energy generation to end use. Table 20.4 shows a summary of existing interventions in selected developing Asian countries. Although in most of the cases development plans and policies are yet to be integrated with mitigation targets, interventions at both supply side and demand side have led to co-benefits in cutting down carbon emissions over the couple of decades. Efficiency improvements in power plants, reducing distribution loss and increasing the share of non-fossil sources in the energy mix will reduce $CO_2$ emissions.

The developing economies in the Asian region have various advantages and disadvantages with regard to their energy transitions. Regarding advantages, the first and most important one is that many of the smaller economies still belong to low energy consuming strata. This gives them enough space to change the energy consumption pattern to a clean energy mix rather than adopting the energy transition pathway followed by many in the developed world. As per

*Table 20.4* Selected low carbon policies in emerging economies in Asia

| Sectors | | India | Indonesia | Thailand |
|---|---|---|---|---|
| Energy supply | Advanced technologies for fossil to power generation | ✓ | | |
| | Minimise distribution loss through improvement of the transmission system | ✓ | ✓ | |
| | Retrofitting or retiring inefficient old power plants | ✓ | | |
| | Increasing share of renewable energy | ✓ | | |
| | Economic incentives | ✓ | ✓ | ✓ |
| Energy demand | Efficiency labels | ✓ | ✓ | ✓ |
| | Economic incentives | ✓ | ✓ | ✓ |
| Transport | Mass transit goals | ✓ | | |
| | Vehicle fuel efficiency goals | ✓ | | |
| | Promote biofuels | ✓ | ✓ | ✓ |

*Data source*: Adapted from ADB-ADBI (2012).

the upcoming demand in different sectors these countries can focus on developing locally available energy resources. Distributed energy systems appear to be more meaningful in these cases. Electricity generation facilities based on wind, solar and biomass resources have enormous potential in meeting the long-term energy demand in these economies. On the other hand, the economics also encounter numerous limitations and challenges to energy transition. The first and most critical challenge these countries face is a lack of adequate financial support. The use of advanced technology for building power generation facilities based on non-conventional sources requires huge investment as well as operational and governance capacity. The developing economies already challenged with financial and technology limitations will find it more difficult to make the transition smoothly because of these additional challenges. Here several domestic as well as external factors can play a critical role in energy transition.

Investment in the energy sector has been increasing remarkably in view of the challenges to overseas fuel dependency. While investment in energy efficiency is greater in OECD countries, in cumulative terms 63% of energy supply investment, US$25.2 trillion needs to be made in non-OECD countries by 2035 (IEA, 2014). The majority of investment in developing Asia will focus on domestic renewable energy infrastructure and supply facilities. The share of renewable energy will increase significantly in the energy mix of Asia's emerging economies, which can potentially contribute to a low carbon energy mix. However, additional efforts are necessary to achieve the globally adopted global warming limit. It is unlikely, some crucial low-carbon solutions such as carbon capture and storage (CCS), electric vehicles, advanced biofuels, sustainable urban planning to be developed under the INDCs, but essential to achieve 2°C target. Therefore, promoting technological development and innovation in the energy sector is key to low carbon development in the region.

## Regional energy integration

A combination of insufficient supply, uneven distribution, and the region's potential for cleaner and alternative sources of energy is encouraging development of new transportation corridors of energy, and greater investments in science and technology within the Asia-Pacific region. However there is a need for countries to consider new forms of energy cooperation and connectivity, which will help to balance the gaps in supply and demand across countries and to change the energy mix.

*(UNESCAP, 2016)*

Developing Asia has enormous potential for increasing alternative energy generation and supply infrastructure. However, limitations in the availability of technological support and financial backup have adversely affected energy sector development in many of the smaller countries in the region. One of the major factors that can possibly promote energy sector development and access to modern forms of energy among these countries is regional integration. It is often stated that a well-developed energy supply or transmission network within this region will not only help meet the growing energy demand but also help explore the resource potential of these economies.

More than the geographical difficulties, often security concerns limit efficient regional integration. Considering the surging demand for clean energy sources, especially renewable energy sources, regional connectivity of electricity transmission lines appears relevant, which would create a corridor to use unexploited hydropower potential for modern energy supply for all in the region with a relatively low $CO_2$ footprint. In Southeast Asia two regional energy cooperation initiatives supported by the Asian Development Bank and other donor agencies are the Mekong Power Grid and ASEAN Grid. Similarly, the South Asia Regional Energy Initiative (SARI) can be mentioned as another example of such an initiative in developing Asia. Many feel that to provide a long-term solution to the energy challenges cross-border energy trade will be of great importance. Experts have also highlighted the relevance of an Asian Energy Highway network which could combine different types of energy transmission networks, including pipelines and cross-border power grids (UNESCAP, 2016). This has been proposed in the context of enhancing energy security rather than clean energy access. However, considering the importance increasingly being attributed to clean energy access over conventional sources, it is likely that greater preference will be given to decarbonising any such regional energy transmission links.

## Southeast Asia

There is a growing need for regional power integration to address regional energy security and to address a low carbon development agenda, which is particularly important and beneficial for countries in Southeast Asia, particularly the Greater Mekong Subregion (GMS). This region has huge hydropower potential and development of hydropower under a regional cooperation agreement can be a 'win-win' for the concerned countries in distributing their pooled resources fruitfully towards low carbon regional development without hurting national development. Expanding the interconnection of GMS power systems alone can save US$14.3 billion by 2030, mostly coming from the substitution of fossil fuel generation with hydropower; it is also expected to result in slower growth of carbon emissions compared with a business-as-usual scenario (ADBI, 2013). Power trading in the GMS is probably the most advanced example in Asia. Rapid economic growth in the region, particularly in Thailand during 1980s and the early 1990s, and resolution of several regional armed conflicts led to exploitation of the abundant hydropower potential in China, Vietnam, Lao PDR and Myanmar to reduce the dependency of the region on expensive fossil fuels (ECA, 2010). Countries with abundant hydropower potential like Lao PDR and Myanmar have invested in export-oriented hydropower generation projects based on the power trading commitments of high economic growth countries like China, Thailand and Vietnam. Based on the commitment of 10,000 MW of power imports by Thailand (ECA, 2010), a number of hydropower plants such as Theun Hinboun and Houay Ho have already been commissioned in Lao PDR. Like Thailand, Vietnam also started importing hydropower from neighbouring countries to meet its demand for electricity, which was growing at double digit rates. As mentioned earlier, apart from energy security and economic benefit, exploitation of hydropower also brings an opportunity to reduce the $CO_2$ footprint of the energy sector. For example, a number of large-scale hydropower projects in the region have been approved as CDM projects (IGES, 2017a).

## South Asia

Like in Southeast Asia, cross-border power flows have already been taking place in South Asia, but with a relatively slower pace. Although only in a bilateral manner, a number of cross border power trading initiatives have emerged in the region, as shown in Table 20.5. These initiatives have demonstrated a willingness among member countries for regional integration for energy security.

While electricity generation in South Asia depends heavily on fossil fuels, upscaling of current bilateral efforts to a regional power integration would create the opportunity to exploit the abundant untapped hydropower potential in Nepal and Bhutan. Hydropower-based regional power cooperation can generate multiple benefits for member countries, including diversified energy mix, reduced investment cost for new power generation capacity and huge revenues for the host countries of hydropower projects. For example, expansion of low-cost hydropower generation in Nepal would allow India to replace fossil fuel based thermal power plants, particularly in the high power demanding Northern Grid. An energy model based study (Timilsina et al., 2015) argued that full operation of a cross-border power trade in South Asia would allow the replacement of 63 GW fossil fuel based thermal power capacity. As a result, regional power integration in South Asia would have a positive impact on mitigation efforts in the energy sector through substantial change from dependency on fossil fuel based power generation in Bangladesh, India and Pakistan to hydropower supply from Nepal and Bhutan. Replacement of 63 GW fossil fuel based capacity would contribute to an 8% $CO_2$ emission reduction relative to the baseline. As shown in Figure 20.2, Bangladesh would have the highest $CO_2$ emission reduction percentage ($-32\%$), followed by Pakistan ($-7\%$) and India ($-6.5\%$) by 2040.

## International finance and technology transfer

Addressing energy sector concerns requires multiple policy initiatives as well as efficient support from domestic as well as external sources. In this regard the international finance which forms a critical element in global climate mitigation has a major role in helping the developing world.

*Table 20.5* Cross-border power transmission capacity in South Asia

| Member countries | Transmission interconnection | Capacity (MW) |
|---|---|---|
| Bhutan → India | Grid reinforcement to evacuate power from Punatsangchhu I and II | 2,100 |
| India ↔ Nepal | Dhalkebar-Muzaffarpur 400kV line | 1,000 |
| Bangladesh ← India | 400kV HVDC back-to-back asynchronous link | 500 |

*Data source*: Compiled from various sources.

305

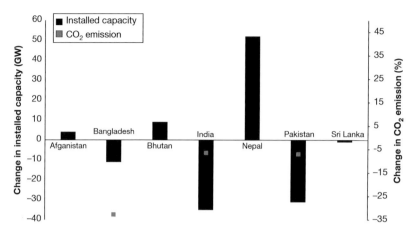

*Figure 20.2* Impact of regional power trading on energy-related $CO_2$ emissions in South Asia, 2015–2040

*Data source*: Prepared by authors based on Timilsina et al. (2015).

Climate finance is a key financial instrument in this regard: 'Climate finance refers to local, national or transnational financing, which may be drawn from public, private and alternative sources of financing and is critical to addressing climate change because large-scale investments are required to significantly reduce emissions, notably in sectors that emit large quantities of greenhouse gases' (UNFCCC, 2016(a)). Improving energy efficiency can reduce energy demand and GHG emissions, and deliver a range of other benefits such as improved air quality, enhanced economic competitiveness and, at the national scale, a higher degree of energy security (IEA, 2012). Many of the economies in the region have been investing heavily in improving energy efficiency.

Countries such as India have rolled out several mechanisms to promote energy efficiency in industrial, transportation and residential sectors. India's efforts to reduce the same to about 35% by 2013 from that of the 2005 levels bring along several policy measures directly targeting Energy Efficiency improvement in these two countries. However, for meeting these targets and addressing the climate impacts, domestic financial resources alone will not be sufficient. The need for international support is critical in meeting the mitigation as well as adaptation targets (see Table 20.6).

The climate mitigation actions in developing countries vary widely. Despite having a huge energy supply/demand gap there has been an interest in decommissioning outdated fossil based energy generation facilities. In India coal power plants with a total of 36,000 MW and more than 25 years (Jog, 2015) of service will be replaced mainly due to the old inefficient technology and partly due to huge water demand at these facilities.[2] In Southeast Asia too, similar efforts have been made by governments: 'According to IEA there is significant potential for deploying more efficient coal-fired power plants: while the average efficiency of Southeast Asia's coal-fired power plants increases by more than five percentage points over the projection period, by 2040, sub-critical technologies still comprise over 50% of total coal-fired installed capacity in the region' (IEA, 2015). Several efforts have been initiated by governments to ensure that banks offer preferential loan schemes with subsidised interest rates for clean energy and energy efficiency projects, such as the Bangladesh Bank which launched a refinance scheme for renewable energy and green industry. This refinancing scheme provides support for several categories of projects including a solar home system, biogas, a solar battery recycling plant, an LED manufacturing company, modernisation of the brick industry, etc.

*Table 20.6* Climate financing needs of select countries in developing Asia

| Country | Financing needs |
| --- | --- |
| Bangladesh | Bangladesh estimated that US$27 billion is needed for mitigation and US$42 billion for adaptation between 2015 and 2030. |
| India | US$206 billion is needed to implement adaptation programmes, US$7.7 billion for energy sector projects and US$834 billion for low-carbon development. |
| Indonesia | US$5.92 billion is required to reach the 41% target in Draft INDC (removed in final version). |
| Malaysia | The country continues to allocate financial resources for the implementation of climate change mitigation programmes through both public and private sector initiatives. |
| Nepal | The direct cost of current climate variability and extreme events is 1.5–2% of current GDP/year (US$270–360 million/year in 2013 prices). |
| Pakistan | The investment costs for adaptation interventions are being determined in consultation with the provinces and other stakeholders, and will also be conveyed in due course. |
| Philippines | Full implementation of INDC requires adequate, predicable and sustainable financing. |
| Thailand | The need for adaptation finance is expected to substantially increase, consequently creating an extra burden on an already scarce government fiscal budget. |
| Vietnam | The cost of adaptation is estimated to exceed 3–5% of Vietnam's GDP by 2030. Vietnam can cover only 30% of adaptation needs. |

*Data source*: USAID (2016).

Similarly, there is huge scope for international bilateral mechanisms in improving the energy scenario in developing Asia. Though the importance of the Clean Development Mechanism (CDM) has been fading in many countries, bilateral and multilateral joint mechanisms are gaining importance in Asia. International bilateral market mechanism schemes such as the Joint Crediting Mechanism (JCM), introduced by the government of Japan, have been well accepted in developing Asia. The main objective of this particular programme is to 'facilitate diffusion of leading low carbon technologies, products, systems, services and infrastructure as well as implementation of mitigation actions, and contributing to sustainable development of developing countries' (GEC, 2016). The JCM scheme provides up to 50% of initial investment costs for low carbon technology projects, aiming to contribute to GHG emission reductions through the diffusion of low carbon technologies in developing countries. As of July 2016 a total of 16 developing countries had signed the bilateral agreement of the JCM, among them 9 countries from South and Southeast Asia. According to recent data, a total of 95 projects have been approved for JCM schemes, which are expected to contribute nearly 100,000 tonnes of $CO_2$ reductions (IGES, 2017b). Among 95 JCM projects, 60% are energy efficiency related projects and 30% renewable energy projects.

## Are the climate actions of developing Asia adequate?

One of the most intriguing questions in the context of the climate mitigation actions of any country regards the adequacy of the policy measures. It is often mentioned by western scholars that the climate mitigation actions taken by many of the developing economies are not contributing to any substantive progress in global climate mitigation. However, there are divergent views and debate on this perception (see Table 20.7).

The climate debate and policy making towards mitigation in developing countries are largely based on general beliefs that the developed world has contributed significantly to anthropogenic emissions since the pre-industrial period. This perception has undeniably limited the extent to

*Table 20.7* Ranking of climate actions by different countries

| Country | Rating[a] |
|---|---|
| India | Medium |
| Indonesia | Medium |
| Philippines | Medium |
| Bhutan | Sufficient |

*Data source*: http://climateactiontracker.org/countries.html.

*Note*: [a]The rankings are defined as follows. Inadequate = if all governments put forward inadequate positions warming is likely to exceed 3–4°C; Medium = not consistent with limiting warming below 2°C as it would require many other countries to make a comparably greater effort and much deeper reductions; Sufficient = fully consistent with below 2°C limit; Role Model = more than consistent with below 2°C limit.

which developing economies commit to newer and innovative climate mitigation actions at domestic level. Lack of adequate financial resources and technological limitations, coupled with inadequacy in institutional capacity to manage climate mitigation actions efficiently, restrict the measurability and success of the policies of developing economies in this direction. Highlighting the fact that 'Common but Differentiated Responsibilities' (CBDR) need to be considered while measuring the climate mitigation actions and commitments of development economics, one may argue that whatever actions have been taken by developing nations' economies are in tune with the available optimum capacity and potential of the respective governments.

It is also important to note that the climate actions and commitment of many of the developed economies are far from being sufficient in arresting the global temperature rise. According to the assessment of INDCs, countries such as Australia, Japan, Russia, Singapore and South Korea (CAT, 2016a), which are technologically and economically stronger, have inadequate targets. This means that the climate mitigation targets proposed by these countries would not make any positive contribution to the global fight against climate change. While most of the developing countries of the Asian region have made reasonable commitments, the lack of substantial actions from major economies sends the wrong signal to the global community and highlights the disparity in interest among countries to address the biggest challenge to humanity and the planet.

## Inferences

The developing Asian countries together form one of the largest energy-consuming blocs in the world. These countries face significant challenges in terms of balancing energy demand growth and environmental health. With high fossil fuel consumption, emissions increase tremendously leading to adverse impacts on the environment. Reducing fossil fuel consumption in the immediate future is also a tough policy decision for many of these economies; any such a plan can seriously affect their domestic developmental objectives. This means that both the energy security and environmental health of these countries are equally challenged. However, the increasing dependency on fossil fuel, as well as financial and technological limitations to developing alternative energy sources among many of the developing economies in this region, poses serious challenges to achieve climate change targets.

Three aspects have been specifically highlighted in this chapter. First, the importance that needs to be attributed to the development of low carbon technologies and their efficient diffusion into the region. For development of clean energy sources, domestic policies have been playing a pivotal role. However, the scale of such energy development is still not sufficient to replace growing fossil fuel consumption. In this context technology corporation among overseas energy sector players will be of key importance to the region.

Second, the chapter highlights the need for greater energy integration within the developing Asian region. This is especially important because not all the countries in the region possess the same level of resource development potential. Networks of electricity transmission lines, joint energy development plans, possible cross-border pipelines, etc. are important. However, such a framework of efficient regional energy integration can develop only with greater cooperation among these countries at the political level.

Third, this chapter also notes the relevance of international finance and technology transfer. Climate finance as well as technology transfer mechanisms such as Joint Crediting Mechanism (proposed by Japan) will be of critical importance. Considering the fact that developing Asia will play a key role in the global energy market as a major consumer bloc, and for its potential to contribute heavily to global energy-related emissions, concerted efforts among policy makers from these countries are paramount. This chapter also points to the importance of stronger regional collaboration networks to address energy/climate concerns that countries are facing in the Global East.

While the chapter recognises that the climate mitigation actions of developing economies in Asia have not made any substantial progress, their proposed actions and commitments so far indicate enormous policy interest towards contributing to the climate mitigation efforts that are in progress worldwide. Specific actions from many of these countries rank much higher in terms of progressiveness than those of many leading economies (CAT, 2016a). As the climate debate has been increasingly focusing on developing country action rather than on developed country commitments in recent times, it appears to many observers that the climate mitigation responsibility has also been pushed from the Global North to the Global South. This current trend of pushing responsibilities to the developing and least developed countries may lead to adverse impacts on the global climate regime, unless the initiatives of the South are supported technologically and financially by the North.

## Notes

1 Detailed discussion on primary energy consumption is given in Chapter 2 of this volume.
2 The Indian government has proposed an additional capacity of 84,600 MW during its 13th plan (2017–2022) through super critical units which are expected to be less polluting power facilities.

## References

ADB-ADBI, 2012. *ADB-ADBI Study on Climate Change and Green Asia: Policies and Practices for Low-Carbon Green Growth in Asia—Highlights*, Manila: ADB & ADBI.
ADB 2013a. *Energy Outlook for Asia and Pacific*, Manila: Asian Development Bank.
ADB 2013b. *Asian Development Outlook*, Manila: Asian Development Outlook.
ADBI 2013. *Connecting South Asia "South Asia" and Southeast Asia*, Tokyo: ADBI.
BP 2016. *Statistical Review of World Energy*, London: British Petroleum.
CAT 2016a. *Rating Countries*. [Online] Available at: http://climateactiontracker.org/countries.html [Accessed 12 November 2016].
CAT 2016b. *Effect of Current Pledges and Policies on Global Temperature*. [Online] Available at: http://climateactiontracker.org/global.html [Accessed 23 July 2016].
ECA 2010. *The Potential of Regional Power Sector Integration: Greater Mekong Subregion (GMS) Case Study*, New York: ESMAP.

GEC 2016. *The Joint Crediting Mechanism*. [Online] Available at: http://gec.jp/jcm/about/index.html [Accessed 16 July 2016].

IEA 2011a. *Technology Roadmap: Geothermal Heat and Power*, Paris: International Energy Agency.

IEA 2011b. *Energy Transition for Industry: India and the Global Context*, Paris: International Energy Agency.

IEA 2012. *Plugging the Energy Efficiency Gap with Climate Finance*, Paris: International Energy Agency.

IEA 2014. *World Energy Investment Outlook*, Paris: International Energy Agency.

IEA 2015. *Southeast Asia Energy Outlook*, Paris: International Energy Agency.

IGES 2017a. *IGES CDM Project Database*, Hayama: IGES.

IGES 2017b. *IGES Joint Crediting Mechanism (JCM )*, Hayama: IGES.

Jog, S., 2015. *CEA Identifies Old Power Plants with 36,000 MW to be Replaced by Super Critical Units*. [Online] Available at: www.business-standard.com/article/economy-policy/cea-identifies-old-power-plants-with-36-000-mw-to-be-replaced-by-super-critical-units-115091300419_1.html [Accessed 11 November 2016].

Kuntom, A. 2014. *Malaysian Palm Oil Board*. [Online] Available at: www.mpoc.org.my/upload/IPOSC-2014-Malaysian-Sustainable-Palm-Oil-Current-Status-Dr-Ainie-Kuntom.pdf [Accessed 12 November 2016].

NEA 2015. *Nuclear Energy: Combating Climate Change*, Paris: OECD.

Nguyen, T. H., Gomi, K., Matsuoka, Y., Hasegawa, T., Fujino, J., Kainuma, M., Nguyen, T. T. D., Nguyen, T. L., Nguyen, L., Nguyen, V. T., Huynh, T. L. H., Tran, T., Nguyen, Q. K. & Tsujihara, H. 2012. *A Low Carbon Society Development towards 2030 in Vietnam*, Kyoto: Kyoto University. [Online] Available at: http://2050.nies.go.jp/report/file/lcs_asia/Vietnam.pdf [Accessed 18 August 2017].

Stephan, B., Schurig, S. & Leidreiter, A., 2016. *World Future Council*. [Online] Available at: www.world-futurecouncil.org/inc/uploads/2016/03/WFC_2016_What_Place_for_Renewables_in_the_INDCs.pdf [Accessed 1 December 2016].

Timilsina, G. R., Toman, M., Karacsonyi, J. & de Tena Diego, L. 2015. How Much Could South Asia Benefit from Regional Electricity Cooperation and Trade?. *Policy Research Working Paper*, No. 7341. World Bank, Washington, DC. [Online] Available at: https://openknowledge.worldbank.org/handle/10986/22224.

UNEP 2015. *The Emissions Gap Report 2015*, Nairobi: United Nations Environment Programme.

UNESCAP 2016. *Asia-Pacific Regional Connectivity and Integration*. [Online] Available at: www.unescap.org/speeches/asia-pacific-regional-connectivity-and-integration [Accessed 7 July 2016].

UNFCCC 2016a. *Climate Finance*. [Online] Available at: http://unfccc.int/focus/climate_finance/items/7001.php [Accessed 21 July 2016].

UNFCCC 2016b. *Paris Agreement*. [Online] Available at: http://unfccc.int/paris_agreement/items/9485.php [Accessed 2 December 2016].

USAID 2016. Analysis of Intended Nationally Determined Contributions (INDCs), United States Agency for International Development. [Online]. Available at: www.transparency-partnership.net/sites/default/files/analysis_of_intended_nationally_determined_contributions_indcs.pdf [Accessed 18 August 2017].

# PART 4

# Energy in a carbon-constrained world

# 21

# INTERACTIONS OF GLOBAL CLIMATE INSTITUTIONS WITH NATIONAL ENERGY POLICIES

## An analysis of the climate policy landscape in China, India, Indonesia, and Japan

*Takako Wakiyama, Ryoko Nakano, Eric Zusman, Xinling Feng, and Nandakumar Janardhanan*

### Introduction

Energy generated from fossil fuels has driven much of the world's economic development since the industrial revolution. In consequence, as countries industrialized and urbanized they often became more dependent on fossil fuels (Chow et al. 2003). In recent years, the adverse impacts of climate change have caused many countries to rethink this once popular developmental model (OECD 2007; Zhang et al. 2012). In many parts of the world, this reassessment has given rise to policy reforms that help conserve energy and harvest renewable resources. The region where these policy reforms have arguably made the most headway of late is Asia (ADB 2016). This progress is welcome because China sits first on the list of the world's greenhouse gas (GHG) emitters, while India, Japan and Indonesia are ranked third, fifth and fourteenth respectively (data from IEA 2015a). As such, China, India, Japan and Indonesia will be pivotal players in international efforts to stabilize the climate.

This chapter focuses on the climate and energy policies that these four countries adopted to help mitigate climate change. More concretely, it surveys the climate policy landscape in China, India, Indonesia and Japan, linking the effects of milestone policy reforms to key drivers of emissions identified thorough a decomposition analysis. Beyond contributing 40% of global $CO_2$ emissions, these countries are selected to argue a more general point. Namely, although all four countries are formulating policies due to an expanding interest in mitigating climate change, targeted sectors and regulatory approaches differ owing to varying development levels, resource endowments and policymaking institutions. The chapter therefore illustrates that, in an era when climate change is becoming a global imperative, resulting policies often reflect interactions between international agreements pushing for policy convergence and national institutional structures pulling domestic policy responses in distinct directions.

The structure of the chapter is as follows. The trend of $CO_2$ emissions in those four countries and policies to enhance mitigation actions are examined in the second section. The third section reviews energy and climate policies since mid-2000 when a growing interest in international climate negotiation contributed to stronger national energy and climate policies. The fourth section discusses the reduction potentials for energy use and $CO_2$ emissions. The fifth section concludes by reiterating and elaborating on key findings.

## Literature review: global institutional changes and national policy responses

Many countries in Asia have aimed to reduce $CO_2$ emissions and energy demand without slowing economic growth. Decoupling production from $CO_2$ emissions is critical to achieving this goal (IEA, 2015b). Decoupling can occur from reductions in energy per unit of production (energy intensity) and/or reductions in $CO_2$ emitted per unit of energy (carbon intensity). Reducing energy intensity involves changing energy consumption patterns and deploying energy efficient technologies on the demand side (Gillingham et al. 2009; Hammond and Norman 2012; Sun 1998). Carbon intensity, on the other hand, can be improved by decarbonizing energy supply, such as through shifts from $CO_2$ intensive energy sources to sources emitting less $CO_2$ or by changing the ratio of fossil fuel to renewable (Kawase et al. 2006; Takase and Suzuki 2011).

Figure 21.1 employs decomposition analysis to identify whether and to what extent the four case study countries witnessed changes in population growth and economic production as well as the carbon intensity and energy intensity drivers that could lead to decoupling (Fan and Lei 2016; Xu 2013). The analysis reveals that the global aggregate $CO_2$ emissions doubled from 1990 to 2013—a finding that is consistent with research showing that international emissions grew at 1.1% during 1990 to 1999 and 3.2% from 2000 to 2004 (Raupach et al. 2007). The analysis further demonstrates that from the 1990s to 2013, India's and Indonesia's emissions doubled; China's emissions tripled; and Japan's emissions grew by about 20% (from a notably higher starting point). The decomposition analysis also suggests that the increases in China, India and Indonesia were attributable to GDP growth, while population increases played a relatively greater role in India and Indonesia. Finally, it shows that, although carbon intensity rates across the four countries differed, all four countries registered notable improvements in energy intensity around 2005 and 2007 (Figure 21.2).

The decomposition analysis therefore raises questions that are central to climate and energy policy research. One such question is: what explains the similarities in the timing of the improvements in energy intensity across these countries? Answering this first question involves turning to literature on the internationalization of environmental policy. This literature argues that international institutional reforms can bring about the passage and diffusion of domestic policy reforms. Since the United Nations Conference on Human Environment (UNCHE) in Stockholm in 1972, a growing awareness of the interdependence of global ecosystems, freer flows of eco-friendly technologies, and the consequent formation and spread of sustainable development norms have helped drive domestic environmental policy reforms (Finnemore and Sikkink 1998).

The Kyoto Protocol of the United Nations Framework Convention on Climate Change (UNFCCC) helped move this process forward specifically for climate policy. Under the Kyoto Protocol many of the world's developed countries committed to time-bound emission reduction targets when it was ratified in 2005. It was arguably between 2005 and 2007, when climate negotiations began to consider a successor to the Kyoto Protocol (which was scheduled to expire in 2012), that climate negotiations would have a more pronounced impact on both developed and developing countries. This was because at the thirteenth session of the Conference of the Parties (COP 13) to the UNFCCC in 2007 negotiators agreed on a Bali Action Plan that marked the first time that all countries were willing to take nationally appropriate mitigation actions (NAMAs) in a measurable, reportable and verifiable manner (UNFCCC 2008). This decision—which was implemented with the pledge

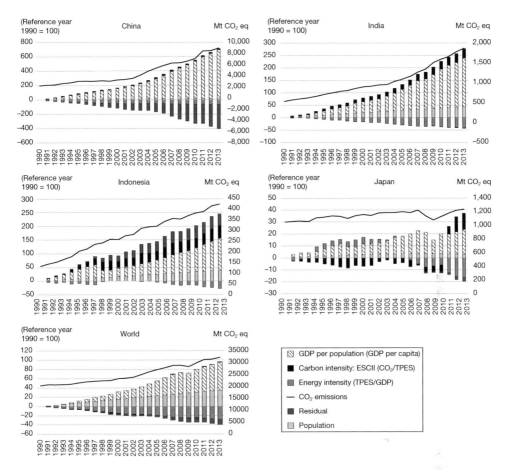

*Figure 21.1*    Trends in CO$_2$ emissions and decompositions of CO$_2$ emissions in four countries and the world

*Data source*: Developed by authors based on IEA (2015a).

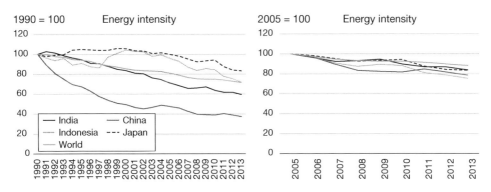

*Figure 21.2*    Trends in energy intensity in four surveyed countries and the world (1990 and 2010 baseline)

*Data source*: Developed by authors based on IEA (2015a). See the methodology in the Appendix.

and review of voluntary climate actions—was later codified in agreements such as the Copenhagen Accord (2015) that sought to strengthen existing targets towards 2020 by developed nations (Annex I Parties) as well as the implementation of NAMAs for developing nations (non-Annex I Parties) (UNFCCC 2010). These reforms would also foreshadow a similar pledge and review system for nationally determined contributions (NDCs) that currently sits at the core of the Paris Agreement.

In the wake of these agreements, all four case study countries would pledge emission reduction targets—albeit with varying baselines and ambition levels. Like many countries, all four governments also underwent significant institutional reforms, consisting, *inter alia*, of the formation of high-level climate policymaking bodies, the allocation of more capable staff and resources to climate change, and the creation of interagency coordination mechanisms (Hongyuan 2007; Indriyanto et al. 2011; Sengupta 2012). For some observers, the main incentives for these reforms were technological and financial in nature; the Kyoto Protocol's Clean Development Mechanism (CDM) had begun to bring investments and low carbon technologies to many countries following the ratification of the Kyoto Protocol in 2005 (Andonova 2008; Phillips and Newell 2013). The prospects of a Green Climate Fund—initially mentioned in the 2015 Copenhagen Accords—that would allocate up to 100 billion dollars in climate finance for developing countries annually by 2020, substantially increased the financial incentives for reforms. At the same time, a growing understanding that low carbon technologies could help modernize energy systems and deliver a range of other national and locally desirable benefits, such as green jobs, improved air quality and healthier communities, also helped make the above reforms politically feasible (Michaelowa and Michaelowa 2012).

Yet, as will be illustrated in the case studies, the sectors different countries targeted and the regulatory approaches that led to the greatest changes in energy and carbon intensity varied considerably across countries. This variation raises a second key question: why did policy responses differ? This divergence was partially a function of development levels. For instance, $CO_2$ emissions per capita are much higher in Japan (9.7 $tCO_2$) than China (6.6 $tCO_2$), Indonesia (1.7$tCO_2$) and India (1.49 $tCO_2$). By the same token, energy intensity is much lower in Japan with 0.095 megaton of oil equivalent (Mtoe) per GDP than in China (0.619 Mtoe), India (0.521 Mtoe) and Indonesia (0.472 Mtoe) (data from IEA 2015a). The variation is also related to different resource endowments: the limited availability of fossil fuels in Japan has led to a greater reliance on imported energy and nuclear power before 2011; the relative abundance of coal in China, India and Indonesia has made their economies more dependent on that natural resource.

The varied reasons for the responses across countries were related to existing economic structures and policymaking institutions (Economy and Schreurs 1997). The international change has influenced domestic policymaking in different ways in these four countries. For example, in Japan institutions that promoted industrial interests helped to shape the implementation of the Kyoto Protocol and subsequent emission reduction pledges; the result was an emphasis on voluntary programs, a focus on nuclear power, and a reliance on forest sinks and the Kyoto crediting mechanism (Schreurs and Tiberghien 2007). Meanwhile, the decision to put the climate portfolio under powerful line agencies in China and Indonesia helped ease the passage and implementation of ambitious energy policy reforms. The picture that thus emerges from this chapter is therefore one of countries adopting energy policies to mitigate climate change at a similar juncture with different institutions rendering varying policy outputs and outcomes. This next section reviews energy and climate policies that simultaneously began to change, but change differently, around 2005 to 2007.

## China

Over the past decade, China has introduced a series of institutional reforms that signalled the seriousness the leadership attached to combating climate change. These included the decision to appoint the National Development and Reform Commission (NDRC) as the lead agency on climate policy (a superagency that merged the lead planning ministry and the main energy agency in 1998) as well as the creation of a National Climate Change Commission led by the prime minister. These reforms facilitated the passage of several high-profile policy statements headlined by the National Climate Change Program (CNCCP) (National Development and Reform Commission 2007) and the White Paper on Climate Change (State Council 2008). As with the other cases covered in this chapter, it was not a coincidence that many of these reforms took shape around 2007—the period when agreement brought many countries to introduce their NAMAs. For example, in the lead up to the COP 15, China proclaimed its intentions to reduce its carbon intensity by 40–45% from a 2005 base year by 2020. But as is also the case with the other countries, existing economic and political institutions impacted these reforms.

The policy changes where this influence stood out the most involved China's Five-Year Plans. The Five-Year Plans once provided a blueprint for most commodities in China's state-run command-control economy prior to market-oriented reforms that began in 1978 (see Fei et al. 2009). But even with more than three decades of liberalizing economic reforms, the plans still offered a window into the political and economic priorities of the leadership. As such, the references to energy-saving goals in the 11th Five-Year Plan were a clear sign of leadership support of a proactive stance on climate change (see Table 21.1). Some of the targets in the 11th Five-Year Plan demonstrated this commitment by calling for improvements in carbon intensity such as a mandated increase in non-fossil fuel use for primary energy from 7.5% to 10% as well as greater reliance on renewables (hydropower, wind power, solar power, biomass and nuclear production capacity). China's 11th Five-Year Plan, which called for a reduction of 20% in energy use per unit of gross domestic product (GDP) between 2005 and 2010, was arguably the most important of these reforms as it affected other sectoral policies.

Because industry accounts for about 70% of China's total energy consumption, many of the sectoral actions adopted for the 20% energy efficiency target were aimed at inefficient industries. These included the energy conservation power generation dispatch program and the 1,000 Energy-Consuming Enterprises Program (Top-1,000 Program). The Top-1,000 Program allocated energy-saving targets to China's 1,000 highest energy-consuming enterprises in nine key sectors (iron and steel, petroleum and petrochemicals, chemicals, electric power generation, non-ferrous metals, coal mining, construction materials, textiles, and pulp and paper sectors). The enterprises were then called upon to establish an energy conservation organization, energy efficiency goals, energy utilization reporting systems, energy conservation plan, energy conservation incentives, and energy efficiency improvement options. Participating entities were further required to make quarterly energy consumption reports to the National Bureau of Statistics (NBS) and sign conservation agreements with local governments. The local government agreements, in turn, were used to evaluate the job performance of enterprise managers with commensurate rewards and sanctions affecting salaries and administrative ranks.

There were other evidences of the effect of existing institutions leaving a mark on China's climate policy response. For example, the majority of energy conservation measures focused on capturing efficiency gains from large energy-intensive sources. At the time, the primary instruments for achieving these reductions were command-control regulations and provincial and national government investment programs (Price et al. 2009). An important complement to the above described reforms was a system that evaluated enterprise managers on their performance. This system was part of larger architecture that evaluated provincial and city leaders on achieving the 20% energy efficiency goals. It further meant that, just as the Top-1,000 Program held managers accountable for enterprise targets, subnational leaders were accountable for local targets

*Table 21.1* References to energy efficiency, renewables and climate change in China's 11th and 12th Five-Year Plans

| 11th Five-Year Plan | 12th Five-Year Plan |
| --- | --- |
| Chapter 2<br>Develop industrial structure which promotes economic growth and simultaneously uses limited resources efficiently | Chapter 2<br>Promote sustainable development<br>(Keywords: carbon intensity; circular economy; low-carbon technology) |
| Chapter 12<br>Section 2: Promote optimal development of the energy sector<br>(Keywords: ultra-supercritical power plant; air cooling power plant; nuclear power plants; tear down inefficient small thermal power units)<br>Section 4: Develop policies to promote renewable energy<br>(Keywords: preferential tax; investment policies; mandatory market share) | Chapter 9<br>Section 1: Promote structural changes in key industries to facilitate energy efficiency and environmental protection |
| Chapter 22<br>Section 1: Energy conservation<br>(Keywords: reduce the number of old transport vehicles; remove inefficient coal-fired industrial boilers (kilns))<br>Section 6: Facilitate policy measures to promote conservation | Chapter 10<br>Section 1: Incubate new energy-saving technology, preserve the environment and use recycled resources<br>(Keywords: nuclear; wind; solar; biomass; heat; smart grid) |
| Chapter 24<br>Section 2: Facilitate strict environmental control in large and medium cities and their suburbs<br>(Keywords: limit new (expanded) coal-fired power plants without cogeneration ability; prohibit new (expanded) construction of steel, smelting and other high energy consuming enterprises; enhance the city's soot, dust, fine particulate matter and automobile exhaust control efforts) | Chapter 11<br>Section 1: Diversify the energy mix by increasing clean energy<br>(Keywords: efficient coal-fired plants; cogeneration; security ensured nuclear power plants; wind power; solar energy; biomass; geothermal) |
| | Chapter 21<br>Section 1: Control GHG emissions<br>Section 3: Extensive international cooperation |
| | Chapter 22<br>Section 1: Promote energy savings<br>(Keywords: renovate infrastructure to promote energy savings; promote energy savings in engineering) |

*Data source*: Authors, from 11th Five-Year Plan (available from www.gov.cn/english/special/115y_index. htm) and 12th Five-Year Plan (available from www.britishchamber.cn/content/chinas-twelfth-five-year-plan-2011-2015-full-english-version).

with rewards or punishments indexed to varying levels of performance (APERC 2009; Wang 2009). These incentives were paired with reforms that would arguably be less palatable without such incentives. For example, over the lifetime of the 11th Five-Year Plan there was a significant effort to close or consolidate smaller and typically inefficient power plants and industries. The result was the elimination of 76 million kW of power from small thermal plants, 12,000 tons of iron-smelting production capacity, 72 million tons of steel production capacity, and 370 million tons of cement production capacity.

Reports suggest the above reforms would leave China within 1 percentage point short of meeting the 11th Five-Year Plan goal—a modest shortfall given the ambition of the target (Price et al. 2011). It was this relative success that arguably led China to put forward a 16% target for reductions in energy consumption per unit GDP in the 12th Five-Year Plan. The new target was accompanied by reforms that aimed for reductions in the percentage of coal use as a proportion of primary energy and a greater reliance on hydropower and nuclear power as well as wind, solar and biomass (to collectively reach 200 million kW by 2020). It also found support from additional reforms modelled after the initial run of policy changes under the 11th Five-Year Plan. These included the now expanded Top-10,000 as well as efforts to expedite the elimination of "backward" or inefficient production capacity and introduce energy-saving technologies. The emphasis on deepening many of the 11th Five-Year Plan reforms during the 12th Five-Year Plan can be seen in Table 21.1, which shows the growing references across the plans to energy efficiency, renewables and mitigating climate change.

China appears poised to continue to forge links between domestic energy and global climate targets. In June 2015, China submitted its NDC to the UNFCCC. The NDC pledged to reduce carbon intensity by 60–65% from 2005 levels. With the NDC, China has also expanded the scope of its climate policy reforms, including the gradual spread of a domestic emissions trading scheme as well as a wide range of reforms at the city level that seek to capture reductions in energy demand.

## Indonesia

Indonesia has earned a global reputation as a first mover on climate change. This was evidenced by the decision of Indonesian President Yudhoyono to commit to a 26% emission reduction target from business as usual (BAU) by 2020 that could reach 41% with international technological and financial support. Much like China, its policy reforms were facilitated by changes to national institutional arrangements. Most notably, Indonesia established the National Council on Climate Change (DNPI) in 2008 and appointed the President to oversee its 17 ministries and line agencies, including the powerful Economic Planning Ministry (BAPPEDA). The institutional arrangement has changed after President Joko Widodo succeeded President Yudhoyono, but among the national movements to facilitate climate change measures the then DNPI helped institute, the National Action Plan for Reducing Greenhouse Gas Emissions (RAN-GRK) was particularly noteworthy in its coverage. The RAN-GRK defines GHG mitigation actions up to 2020 in five priority sectors: 1) forestry and peat land; 2) agriculture; 3) energy and transportation; 4) industry; and 5) waste.

Deforestation and peat account for most of the emissions reductions, followed by energy and transport. One of the reasons for the early movement in this policy area was the potential for climate finance through a mechanism called the Reduced Emissions from Deforestation and Forest Degradation (REDD). Emissions from deforestation and land use changes accounted for 47% of the annual GHG emissions in Indonesia in 2000 (Ministry of Environment Government of Indonesia 2010) and peat fires accounted for 13%. Thus international support for decreasing deforestation promised a sizable amount of funding. Indonesia reported a significant drop in the rate of loss of forest from 2000 to 2005 relative to the 1990s. And although the rate increased again from 2005 to 2010, it still stood at less than half of that during the period shortly after the peak of the large-scale transmigration program for landless citizens to the main island of Java from less populated islands in the 1980s and early 1990s (FAO 2010). Significant declines in the conversion of rain forests to farmland, a reduction in the number of forest fires and improved forest policies all contributed to these reductions.

While forestry and peat emissions are presently moving in the right direction, population and economic growth have increased energy consumption in the energy and transportation sectors. The transport sector emitted 68 $MtCO_2$-eq in 2005 or 23% of all energy-related emissions—with road transport consuming 91% of primary energy. Vehicle ownership is projected to more than double over the next 25 years due to population growth and related demands for two-wheelers and light-duty vehicles. Hence, Indonesia has resolved to use its NAMAs to facilitate emission reductions and earn climate finance for the implementation of bus rapid transport (BRT) and other mode-shifting transport options.

Another area drawing attention from the RAN-GRK is energy. As shown in Figure 21.3, the primary energy supply in Indonesia has increased by 162% from 2000 to 2013 (Ministry of Energy and Mineral Resources 2014). Oil, natural gas and coal continue to dominate energy supply (74% as of 2013) with hydropower and other renewable energy accounting for 26% of the total energy resources. Indonesia still depends heavily on fossil fuels despite the finite availability of these resources. Coal has been the most abundant and fastest-growing resource, but renewables (hydropower, geothermal, biomass and biofuel) have also picked up noticeably since the COP 15 Copenhagen pledges.

Energy conservation policies in Indonesia—first introduced in 1982—have further seen a notable increase with the past decade of developments in international climate change negotiations that have continued to facilitate voluntary actions from developing countries (Table 21.2). The Master Plan of National Efficiency Energy of 2005 (Energy Rencana Induk Konservasi Energi Nasional or RIKEN) served as a roadmap for central and regional government to participate in energy conservation activities. The primary targets were to reduce energy intensity by at least 1% annually through 2025. This was followed by two presidential decrees (Presidential Decree no. 10 of 2005 and no. 2 of 2008) that detailed plans for implementing energy efficiency measures. In 2009, the coverage of these plans was increased to large energy consumers (with energy consumption over 6,000 TOE per year). Actions ranged from requiring

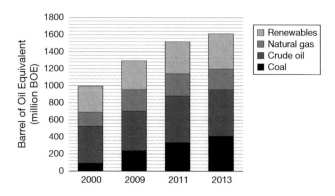

*Figure 21.3*   Indonesia's energy mix for primary energy supply

*Data source*: Adapted from *Handbook for Energy and Economic Statistics Indonesia*, 2014.

*Table 21.2* Indonesia's energy conservation and efficiency policies

| Year | Legislation | Measures | Target |
|------|-------------|----------|--------|
| 1982 | Presidential Instruction no. 9 | Government institutions become subject to monthly energy consumption reporting | |
| 2005 | National Masterplan for Energy Conservation (RIKEN) | • Identifies the energy–saving potentials and formulates a plan to achieve them with fiscal incentives, training and energy audits. Saving potentials are: Industry – 15–30% <br>• Commercial buildings – 25% <br>• Households – 10–30%. | Decrease energy intensity by 1% annually until 2025 |
| 2005 | Presidential Instruction no. 10 | Executive government officials are made accountable for energy efficient consumption (line ministers, governors, mayors and others) | |
| 2005 | MEMR Ministerial Regulation no. 31 | Energy efficient processes and policies are promulgated <br>A reporting mechanism for energy consumption is developed | |
| 2007 | Energy Law no. 30/2007 | National and regional governments introduce energy conservation programs <br>Government incentives for producing energy efficient equipment are created | |
| 2008 | Presidential Instruction no. 2 | A national energy and water saving team <br>National and regional governments developed and monitored energy and water conservation measures in their institutions (lighting, air conditioning, electrical appliances and official vehicles) | |
| 2009 | Government Regulation no. 70 | Energy consumers of 6,000 tons of oil equivalent and over are subject to energy management (i.e. energy conservation program, energy managers, energy audit) <br>Standards and labelling for energy efficient products are introduced to encourage manufacturers to increase the level of energy efficiency of products | |
| 2011 | Presidential Instruction no. 13 | Instructions to measure water and energy savings are shared with national and local leaders <br>Membership and functions for the national team for energy and water are put in place | Overall targets 20% saving—electricity10% saving— fuel10% saving—water |
| 2012 | MEMR Ministerial Regulation no. 12 | Regulations are phased in to improve fuel quality (gasoline/octane number 88, diesel fuel) in government and state-owned enterprise vehicles, freight and logistics vehicles, and cargo ships | |

*(Continued)*

*Table 21.2* (continued)

| Year | Legislation | Measures | Target |
|------|-------------|----------|--------|
| 2012 | MEMR Ministerial Regulation no. 13 | Electricity savings targets (20%) are introduced for national and local government offices (air-conditioning, lighting, etc.) and state-owned companies' offices | Overall target 20% saving—electricity |
| | | The targets are supported by programs that encourage energy savings (improved street lighting) and monitoring | |
| 2012 | MEMR Regulation no. 14 | Mandatory energy management energy consumers of over 6,000 tons of oil equivalent are introduced with a view to improving monitoring of energy management practices and strengthening compliance incentives | |

*Data source*: Adapted from The energy efficiency policies and measures database, IEA.

periodical energy audits to enforcing energy conservation programs that would enhance capacity and designating an energy manager to oversee these plans.

A final critical part of Indonesia's climate policy has been the decision to expand energy conservation to households through the development of energy efficiency standards and labelling for electric appliances. Labelling first started in 2011 for compact fluorescent lamps; plans for additional appliances followed (e.g. TVs, refrigerators, air conditioning, fans). The lack of incentives or penalties for such measures nonetheless remains a sizable challenge in implementing these reforms.

# India

Climate change debate began to gain significant attention in India's policy agenda following the Bali Action Plan which opened the door for greater engagement in climate negotiations regarding nationally appropriate actions by the developing world (Atteridge et al. 2012). With these shifts in the international policy landscape, India's government created the Prime Minister's Council on Climate Change in 2007 and charged the Ministry of Environment and Forestry (MOEF) with coordination of this new inter-ministerial body. The subsequent release of the "National Action Plan for Climate Change" is viewed as the key document giving policy direction on the climate change agenda of the country. The passage of the Copenhagen Accord two years after the Bali Action Plan led India to promulgate a series of national climate policy actions (ENVFOR 2010). Many of these actions were outlined in India's National Action Plan of Climate Change (NAPCC). This document emphasized aligning energy supplies and demands and referred to capping of GHG emissions at an unspecified future date (GOI 2008; MoEFCC 2014). The NAPCC further introduced the National Mission for Enhanced Energy Efficiency (NMEEE), which aimed to improve the market for energy efficient goods and services and earmarked around Rs. 74,000 crore (or USD 12 billion) toward these ends (Kolhe and Khot 2015). In addition, the Bureau of Energy Efficiency (BEE) and the Ministry of Power (MoP) introduced several regulatory schemes and economic incentives to promote national and state-level energy efficiency in key sectors. According to CII (2008), under the Energy Efficiency target for the 11th Five-Year Plan, 10,200 MW is expected to be reduced, including 4,000 MW from the domestic efficient lighting program, 3,000 MW from standards and labelling, 2,000 MW from agricultural and municipal demand management practices, and 500 MW from building codes as the policy instruments introduced for these purposes.

Also similar to the other case study countries, many of the global climate goals were outlined in national planning documents. India's 11th Five-Year Plan (2007–2012), for example, noted that clean energy (including clean coal and renewables) was essential to combat climate change and enhance long-term development prospects (GOI 2007b). Moreover, the plan offered a more detailed run-down of how to ensure accessibility and affordability of energy supply, while keeping the economy moving forward and curbing associated GHG emissions. In this connection, the plan stated that energy efficiency needed to be substantially improved across many sectors (GOI 2007b). To ensure it met these objectives, it suggested that of the increases in 78,700 MW over the 10th Five-Year Plan (GOI 2007b), 21,760 MW would come from renewable energy (GOI 2013a) (Table 21.3) (24,504 MW was actually achieved by the end of the 11th Plan).[1]

There were essentially two paths that India hoped to move down to achieve other key energy targets. The first was retiring inefficient coal-fired power plants and introducing supercritical and ultra-supercritical technologies. Supercritical and ultra-supercritical coal power plants have efficiencies of 45% and 46%, whereas a sub-critical plant's efficiency operate at 34% (GOI 2013b).[2] By the close of the 11th Five-Year Development Plan, eleven supercritical coal plants were operating and generating 7,400 MW (GOI 2013a) (Figure 21.4). During the 12th Plan, about 50% of coal-based capacity out of an additional 69,569 MW was expected to come from supercritical technologies. The other path aimed to reduce carbon intensity through renewables. Given significant needs for energy access for off-grid populations, decentralized renewable energy sources were particularly important for India (OECD 2012). As per current policies under the government, a significant increase in the renewable energy installations is planned with a target of 175,000 MW of installed capacity by 2022—six times that of the 2012 level (NITI 2015).

For all intents and purposes, India appears poised to deepen the targets and reforms that initially gained momentum with the Copenhagen pledge. For its post-2020 pledge (outlined in the NDC), for instance, India will aim for an expanded focus on tapping renewables and improving energy intensity by 30–33% from 2005 by 2030. Further, according to the national Energy Statistics, India's

*Table 21.3* Targets and achievements during the 11th National Development Plan

|  |  | 2005–2006 achievements | 11th target | 2011–2012 achievements | 12th target |
|---|---|---|---|---|---|
| GDP growth rate (%) |  | 7.6 | 9.0 | 8.0 | 9.0 |
| Access to electricity (%) | Urban households | 92.3 |  | 93.9 |  |
|  | Rural households | 54.9 |  | 67.3 |  |
| Energy demand | Energy intensity (MJ per rupee) | 0.466 |  | 0.427 (–8% from 2005) |  |
| Energy supply (installed capacity) | Hydro (MW) | 34,654 | 51,207 | 38,990 | 49,887 |
|  | Thermal (MW) | 86,015 | 144,659 | 131,603 | 203,943 |
|  | Nuclear (MW) | 3,900 | 7,280 | 4,780 | 10,080 |
|  | Renewables (MW) | 7,761 | 21,760 | 24,504 | 54,504 |
|  | Additional total capacity (MW) |  | 78,700 |  | 1,18,537 |

*Data source*: Authors, from data from GOI (2007a), GOI (2007b), GOI (2013a), GOI (2013b), GOI (2015a) and GOI (2015b).

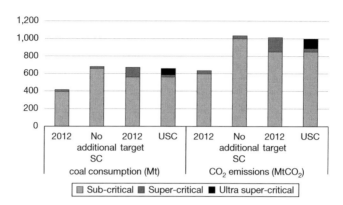

*Figure 21.4* Comparison of coal consumption and $CO_2$ emissions with different coal-based power plant technology

*Data source*: Developed by authors using data from GOI (2013b).

*Note*: USC: 50% of additional capacity of coal plan is built with ultra-supercritical technology during 12th Plan.

energy intensity was 0.466 MJ/rupee in 2005 (GOI 2015b). Thus it is expected to reduce to 0.312–0.303 MJ/rupee of energy intensity with energy efficiency measures.

## Japan

Unlike the other three developing country cases, the stimulus for Japanese climate and energy policy was the ratification of the Kyoto Protocol in 2005. In line with the Kyoto Protocol, the government of Japan set up several relevant institutions. First, it created institutions to support the GHG inventory led by the Ministry of the Environment, Japan (MOEJ) with support from other relevant ministries, agencies and organizations. Second, it established the Global Warming Prevention Headquarters inside the Cabinet to implement the Kyoto Protocol. It is worth highlighting that, although the headquarters consisted of all ministers including MOEJ and the Ministry of Economy, Trade and Industry (METI) (as deputy chief of the headquarters), these ministries frequently held different perspectives on the best direction for climate policy (Kawashima 2001; Sofer 2016). Several committees were then set up to discuss climate policy, with METI frequently leveraging its power and close ties with the industry to hold sway in policy debates when there were contrasting views (Hattori 2000; Pajon 2010).

A review of the Kyoto Protocol Target Achievement Plan was released two years after the Kyoto agreement's commitment period in 2012 and offers some insights into the compromises made to craft climate policy. It demonstrates that Japan achieved its 6% Kyoto target by bringing down emissions by 8.4% from 1990 levels (see Figure. 21.5). However, if only domestic reductions are counted, emissions actually increased by 19 $MtCO_2$. Forest and sinks contributed to reductions of 47 $MtCO_2$, and 74 $MtCO_2$ of the reduction came from emissions reduction credits that were purchased from the Kyoto flexibility mechanisms such as the CDM—the aforementioned project offset mechanism that allowed developed countries to purchase emission reduction credits from abroad.

In terms of key sectors, while the industry and transportation sectors fell more than estimated by 15 and 12 $MtCO_2$, emissions from commercial, household and energy conversion sectors exceeded targets with an additional 28, 38 and 17 $MtCO_2$ respectively. A sectoral review of the Kyoto Protocol Action Plan (MOE 2014) demonstrated that each sector set up a target by reducing either $CO_2$ emissions, energy consumption, carbon intensity or energy intensity. Reviews from 164 industry associations demonstrate the following: while 77% of manufacturing sectors could achieve

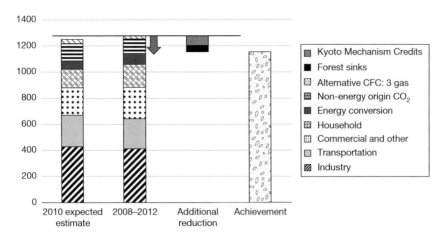

*Figure 21.5* Kyoto Protocol target and achievement in Japan

*Data source*: Authors, data from review report of Kyoto Protocol Action Plan (MOE 2014).

*Note*: Y-axis unit: Mt of $CO_2$ eq.

the target in commercial sectors, only 36% could do the same for industrial sectors, suggesting the majority of industries failed to achieve the target based on 2008–2012 averages (Figure 21.6).

A review of the Kyoto target demonstrates that Japan reduced more emissions than expected in the industrial and transportation sectors. This was largely attributable to sector-specific policies, especially those aiming for energy efficiency improvements during the Kyoto commitment period. For instance, a 2008 revision to the Energy Use Law (ECCJ 2009) strengthened energy efficiency policy by regulating reporting of energy use, initiating an energy manager system at factories, requesting a voluntary target for industry sectors to reduce more than 1% per year of energy intensity, and requiring the submission of a mid- to long-term energy efficiency plan. The energy efficiency policy was further revised to extend the coverage of factories and sectors. The main change to the policy was that the mid- to long-term benchmark index was set for high-energy consuming sectors (such as iron and steel, cement and power supply industry (coverage of 10 sectors, 6 industries as of 2012)). However, according to the review report of benchmark index,[3] the levels of achievement varied across companies and over time. The review further shows that only the soda chemical industry achieved average reductions that were at the benchmark index from 2010 to 2015; other industries failed to come up to that level.

Another significant revision to the energy policy in 2005 required that some shipping, freight transport and passenger transport companies develop energy-saving plans and report energy use (ECCJ 2011). These reforms may have contributed to reductions in energy intensity and $CO_2$ emissions in the industry and transportation sector. Yet an additional factor that had an important but opposite effect was limits on public funding. From 2007, outlays for petroleum production and distribution as well as domestic coal production were eliminated (Figure 21.7).[4] In fact, although there were efforts to step up energy conservation during the Kyoto commitment period, government budgets allocated towards these ends did not change markedly. The share of government budget for energy measures as a proportion of gross domestic product (GDP) stayed relatively constant at 0.5–0.6% from 2005 to 2014 (the slight trough in budgeted resources illustrated in Figure 21.7 were partially due to the 2008 financial crisis).

Though Japan implemented energy conservation measures to reduce $CO_2$ emissions and achieve the commitment of the Kyoto Protocol by setting voluntary targets, a central feature of

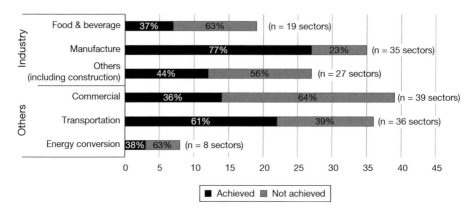

*Figure 21.6*  Achievement of Kyoto Protocol Action Plan by sectors

*Data source*: Made by authors from data from review report of Kyoto Protocol Action Plan (MOE 2014).

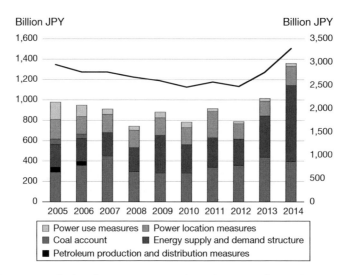

*Figure 21.7*  Government budget for energy measures (special account of energy)

*Data source*: Developed by authors from data retrieved from the Ministry of Finance, Budget and Financial Statements Database.

that plan was Kyoto credits. Further, only the voluntary targets for the manufacturing industry and transportation in the Kyoto Protocol Action Plan were achieved. Japan currently has set a post-2020 $CO_2$ emission target (in its NDC). This is backed up by the Plan for Global Warming Countermeasures that the Cabinet approved based on the Paris Agreement. That target aims to reduce 26% of GHG emissions from 2005 in 2030 (MOE 2016; UNFCCC 2015).

## Conclusion

The largest energy consuming countries in Asia all witnessed significant changes in $CO_2$ emissions (chiefly from improvements in energy intensity) from around 2006 and 2007. The chapter began with a decomposition analysis that breaks down emissions into four key drivers to shed light on the reasons why these countries experienced changes to energy intensity at a comparable juncture. The chapter then described some of the global institutional and national policy changes that shaped the trends in this analysis. The main finding was that international agreements can help provide a stimulus for policy changes in both developed and developing countries. However, the varying types of changes were influenced by different development levels, resource endowments and policymaking institutions. One of the more important policy implications from the chapter is that the international climate regime will need to continue to provide the flexibility to account for national differences while still offering the material incentives and normative framing that make such reforms feasible.

The recent changes to the global climate regime appear to be getting better at striking a balance that is internationally desirable and nationally appropriate. To illustrate, while many developing nations introduced NAMAs as part of national efforts from 2010 to 2020, both the number of countries and the ambition levels have increased markedly for the pledges for the NDC that go into 2030. At the same time, there has also been a proliferation of informal networks and partnerships that at the international level have helped to encourage context-appropriate variations of low carbon reforms. Networks like the International Council on Local Environmental Initiatives (ICLEI), for example, provide city-level policymakers with the tools and knowledge to identify context-appropriate solutions to climate change. Both formal and informal structures are making low carbon development more feasible.

At the same time, it is important to bear in mind a linkage that was not discussed in this chapter: that is, how domestic institutions influence international agreements. To a certain extent, the chapter illustrated that the institutional arrangements that eased national policy reforms lent greater legitimacy to the global climate regime. This, in turn, created a virtuous cycle where ambition levels for climate mitigation could be gradually ratcheted up. However, there are also signs that domestic political institutions could have precisely the opposite effect. The recent Brexit decision in the United Kingdom and election of a United States president who appears unwilling to support the Paris Agreement are signs that the world may be moving further apart on global goals such as climate due to institutions that allow people to participate in the political process. Identifying ways to appeal to constituencies who may appear to lose out from energy policy changes will be increasingly important to stop other countries from turning inward.

A final set of implications involves policies needed from the surveyed countries to strengthen engagement in the international climate regime. Most notably, the chapter shows that improving not only energy efficiency but also carbon intensity is crucial. The spread and scaling of renewables will hence be essential as countries contemplate future reductions. The chapter further underlines that the eventual transition to low carbon development and post-carbon development will be imperative to avoid dangerous climate changes. The chapter suggests that innovative and transformative policies are unlikely to come from existing national institutional architectures. It will hence be important to examine closely the interactions between international and national institutional changes in Asia.

# Appendix

## *Index decomposition analysis*

Index decomposition analysis of $CO_2$ emissions is used to identify the different drivers of global industry related emissions. In the decomposition analysis of Ang (2004, 2005), $CO_2$ emissions can be divided into production level, structural change, energy intensity, fuel mix and carbon intensity (Eq. 1)

$$C = \sum_{ij} C_{ij} = \sum_{ij} Y \frac{Y_i}{Y} \frac{E_i}{Y_i} \frac{E_{ij}}{E_i} \frac{C_{ij}}{E_{ij}} \qquad \text{(Eq. 1)}$$

Here $Y$ is total production level for all industry sectors, $C$ is total $CO_2$ emissions for all industry sectors, $C_{ij}$ is $CO_2$ emissions generated from $j$ fuel in $i$ industrial sector. $E_{ij}$ is energy consumptions in $j$ fuel in $i$ industrial sector, and the change of fuel mix is indicated as $E_{ij}/E_j$, and $CO_2$ emission factor is shown as $C_{ij}/E_{ij}$.

This chapter aims to provide an overview of key drivers. It focuses on four main drivers that affect $CO_2$ emissions: energy intensity, carbon intensity, production level and population; other factors being regarded as residuals using Kaya identity (IPCC 2000; Kawase et al. 2006) (Eq. 2).

$$F = P \times \frac{G}{P} \times \frac{E}{G} \times \frac{F}{E} = p \cdot g \cdot e \cdot c \qquad \text{(Eq. 2)}$$

where: $F$ is $CO_2$ emissions from human sources. $P$ is population and $G$ is GDP. $E$ is energy consumption. $p$ is population, $g$ is GDP per capita, $e$ is energy intensity and $c$ is carbon intensity.

The index set up the reference year in 1990 and describes the changes of four driving factors including population, carbon intensity, energy intensity, GDP per capita and residuals from 1990 level (Eq. 4). Residuals is other factors to drive changes of $CO_2$ emission except for population, carbon intensity, energy intensity and GDP per capita.

$$\frac{\Delta F}{F_{1990}} = \frac{\Delta p}{p_{1990}} + \frac{\Delta g}{g_{1990}} + \frac{\Delta e}{e_{1990}} + \frac{\Delta c}{c_{1990}} + residual \qquad \text{(Eq. 3)}$$

## Notes

1  However, it still only accounts for 12% of the total installed capacity.
2  The replacement of a sub-critical with a super or ultra-supercritical plant could lead to savings of 0.054 kg and 0.116 kg coal per calorific value. Similarly, if a sub-critical coal plant is replaced by a solar plant, 0.613 kg of coal can be saved for every kWh generated (GOI 2013a).
3  Data is available from METI: www.enecho.meti.go.jp/category/saving_and_new/benchmark/.
4  The government budget for domestic coal production (coal account) supported domestic production of coal until 2001. From 2001 to 2006, it was used for its loan redemption.

## References

ADB. (2016). Asian Development Outlook 2016: Asia's Potential Growth. Retrieved from www.adb.org/publications/asian-development-outlook-2016-asia-potential-growth.

Andonova, L. B. (2008). The climate regime and domestic politics: the case of Russia. *Cambridge Review of International Affairs*, 21(4), 483–504.

Ang, B. (2004). Decomposition analysis for policymaking in energy: which is the preferred method? *Energy Policy*, 32(9), 1131–1139. doi:10.1016/S0301-4215(03)00076-4.

Ang, B. W. (2005). The LMDI approach to decomposition analysis: a practical guide. *Energy Policy*, 33(7), 867–871. doi:10.1016/j.enpol.2003.10.010.

Asian Pacific Energy Research Center (APERC). (2009). *Understanding Energy in China–Geographies of Energy Efficiency*. September 2009, Japan. APERC#209-RE-01.3.

Atteridge, A., Shrivastava, M. K., Pahuja, N., & Upadhyay, H. (2012). Climate policy in India: what shapes international, national and state policy? *Ambio*, 41(SUPPL.1), 68–77. doi:10.1007/s13280-011-0242-5.

Chow, J., Kopp, R. J., & Portney, P. R. (2003). Energy resources and global development. *Science*, 302 (5650), 1528–1531.

ECCJ. (2009). Energy saving for office building. (Japanese). *Energy Conservation Center Japan (ECCJ)*. Retrieved from www.eccj.or.jp/office_bldg/img/office2.pdf.

Economy, E., & Schreurs, M. A. (1997). Domestic and international linkages in environmental policies. *Internationalization of Environmental Protection.* Cambridge University Press.

ECCJ. (2011). *Guidance of Energy Saving Promotion of Shippers.* Ministry of Economy, Trade and Industry. Retrieved from www.eccj.or.jp/law06/pamph_shipper-guide/ninushitebiki.pdf.

ENVFOR. (2010). *India: Taking on Climate Change.* Ministry of Environment & Forests, Government of India, 1–8.

Fan, F., & Lei, Y. (2016). Decomposition analysis of energy-related carbon emissions from the transportation sector in Beijing. *Transportation Research Part D: Transport and Environment,* 42, 135–145. doi:10.1016/j.trd.2015.11.001.

FAO. (2010). *Global Forest Resources Assessment 2010,* p. 19. Food and Agriculture Organization of the United Nations.

Fei, T., Wang, Y., Gu, A., Xu, R., Seligsohn, D., & McMahon, H. (2009). Mitigation actions in China: measurement, reporting and verification. WRI Working Paper. Retrieved from: http://pdf.wri.org/working_papers/china_mrv.pdf.

Finnemore, M., & Sikkink, K. (1998). *International Norm Dynamics and Political Change. International Organization, Vol. 52, No. 4, International Organization at Fifty: Exploration and Contestation in the Study of World Politics (Autumn, 1998).* The MIT Press, pp. 887–917.

Gillingham, K., Newell, R. G., & Palmer, K. (2009). *Energy Efficiency Economics and Policy. Resources for the Future, Discussion Papers.*

GOI. (2007a). *Eleventh Five Year Plan (2007–2012), Inclusive Growth,* Volume 1. Government of India (GOI), Planning Commission, I, 306. Retrieved from http://planningcommission.nic.in/plans/planrel/fiveyr/11th/11_v1/11th_vol1.pdf.

GOI. (2007b). *Eleventh Five Year Plan (2007–2012) Volume III: Agriculture, Rural Development, Industry, Services, And Physical Infrastructure.* Government of India, Planning Commission, III.

GOI. (2008). *National Action Plan on Climate Change.* Government of India, Prime Minister's Council on Climate Change. doi:10.1111/j.1746-1561.1993.tb06065.x.

GOI. (2013a). *Twelfth Five Year Plan (2012–2017), Economic Sectors,* Volume II. Government of India (GOI), Planning Commission, II, 1–438. Retrieved from http://planningcommission.gov.in/plans/planrel/12thplan/pdf/12fyp_vol2.pdf.

GOI. (2013b). *Twelfth Five Year Plan (2012–2017): Faster, More Inclusive and Sustainable Growth,* Volume I. Government of India (GOI), Planning Commission, 1, 1–392. Retrieved from http://planningcommission.gov.in/plans/planrel/12thplan/pdf/12fyp_vol1.pdf.

GOI. (2015a). *India's Intended Nationally Determined Contribution* [Online]. 2015. Retrieved from: www4.unfccc.int/submissions/INDC/Published Documents/India/1/INDIA INDC TO UNFCCC.pdf.

GOI. (2015b). *Energy Statistics 2015.* Central Statistics Office Ministry of Statistics and Programme Implementation Government of India. https://doi.org/Twentyfirst issue.

Hammond, G. P., & Norman, J. B. (2012). Decomposition analysis of energy-related carbon emissions from UK manufacturing. *Energy,* 41(1), 220–227. doi:10.1016/j.energy.2011.06.035.

Hongyuan, Y. (2007). International institutions and transformation of China's decision-making on climate change policy. *Chinese Journal of International Politics,* 1, 497–523. doi:10.1093/cjip/pom009.

Hattori, T. (2000). Integrating policies for combating climate change: role of the Japanese Joint Conference for the Kyoto Protocol. *Environmental Economics and Policy Studies,* 3(4), 425–445. doi:10.1007/BF03354049.

IEA (2015a). $CO_2$ emissions from fuel combustion highlights. OECD/IEA. Retrieved from https://www.iea.org/publications/freepublications/publication/CO2EmissionsFromFuelCombustion-Highlights2015.pdf.

IEA. (2015b). *Energy and Climate Change. World Energy Outlook Special Report.*

Indriyanto, A. R. S., Fauzi, D. A., & Firdaus, A. (2011). The sustainable development dimension of energy security, in *The Routledge Handbook of Energy Security,* edited by Benjamin K. Sovacool. London: Taylor and Francis Group, pp. 96–112.

IPCC. (2000). *Special Report on Emissions Scenarios. A Special Report of Working Group III of tiie Intergovernmental Panel on Climate Change.* Retrieved from www.ipcc.ch/pdf/special-reports/emissions_scenarios.pdf.

Kawase, R., Matsuoka, Y., & Fujino, J. (2006). Decomposition analysis of $CO_2$ emission in long-term climate stabilization scenarios. *Energy Policy,* 34(15), 2113–2122. doi:10.1016/j.enpol.2005.02.005.

Kawashima, Y. (2001). Japan and climate change. *Energy & Environment,* 12, 167–179.

Kolhe, M. R., & Khot, P. G. (2015). India's energy scenario: current and future. *International Journal of Management,* 6(7), 47–66.

Michaelowa, K., & Michaelowa, A. (2012). Negotiating climate change. *Climate Policy,* 12 (5), 527–533.

Ministry of Energy and Mineral Resources. (2014). *Handbook for Energy and Economic Statistics Indonesia.* Republic of Indonesia, pp. 20–21.

Ministry of Environment Government of Indonesia. (2010). Indonesia Second National Communication under the United Nations Framework Convention on Climate Change, p. II-7.

MOE. (2014). *Progress of the Kyoto Protocol Target Achievement Plan (Japanese).* Ministry of Environment, Global Warming Prevention Headquarters. Retrieved from www.env.go.jp/press/upload/24788.pdf.

MOE. (2016). *Plan for Global Warming Countermeasures (Japanese).* Ministry of Environment. Retrieved from www.env.go.jp/press/files/jp/102816.pdf.

MoEFCC. (2014). *India's Progress in Combating Climate Change.* Ministry of Environment, Forests and Climate Change, pp. 1–28.

National Development and Reform Commission. (2007). *China's National Climate Change Programme.* Retreived from www.ccchina.gov.cn/WebSite/CCChina/UpFile/File188.pdf.

NITI. (2015). *Report of the Expert Group on 175 GW RE by 2022, National Institute for Transforming India.* National Institution for Transforming India (NITI).

OECD. (2007). *What Policies for Globalising Cities? Rethinking the Urban Policy Agenda.* OECD International Conference.

OECD. (2012). *Linking Renewable Energy to Rural Development.* OECD Green Growth Studies. Executive Summary 19. doi:10.1787/9789264180444-en.

Pajon, C. (2010). *Japan's Ambivalent Diplomacy on Climate Change.* French Institute of International Relations (Ifri). Retrieved from www.ifri.org/downloads/japanambivalentdiplomacyonclimatechange_1.pdf.%5Cnhttp://www.ifri.org/downloads/japanambivalentdiplomacyonclimatechange_1.pdf.

Phillips, J., Das, K., & Newell, P. (2013). Governance and technology transfer in the clean development mechanism in India. *Global Environmental Change, 23*, 1594–1604.

Price, L., Wang, X., & Yun, J. (2009). The challenge of reducing energy consumption of the top-1000 largest industrial enterprises in China. *Energy Policy, 38*(11), 6485–6498. doi:10.1016/j.enpol.2009.02.036.

Price, L., Levine, M. D., Price, L., Zhou, N., Fridley, D., Aden, N., Lu, H., McNeil, M., Zheng, N., Qin, Y., & Yowargana, P. (2011). Assessment of China's energy-saving and emission-reduction accomplishments and opportunities during the 11th Five-Year Plan. *Energy Policy, 39*(4), 2165–2178. doi:10.1016/j.enpol.2011.02.006.

Raupach, M. R., Marland, G., Ciais, P., Le Quéré, C., Canadell, J. G., Klepper, G., & Field, C. B. (2007). Global and Regional Drivers of Accelerating $CO_2$ Emissions. *Proceedings of the National Academy of Sciences of the United States of America* (Vol. 104).

Schreurs, M. A., & Tiberghien, Y. (2007). Multi-level reinforcement: Explaining European Union leadership in climate change mitigation. *Global Environmental Politics, 7*(4), 19–45.

Sengupta, S. (2012). International climate negotiations and India's role. In *Handbook of Climate Change and India: Development, Politics and Governance,* edited by Navroz Dubash. Routledge.

Sofer, K. (2016). *Climate Politics in Japan.* Sasakawa Peace Foundation USA.

State Council. (2008). *China's Policies and Actions for Addressing Climate Change.* Retrieved from: www.gov.cn/english/2008-10/29/content_1134544.htm.

Sun, J. W. (1998). Changes in energy consumption and energy intensity: a complete decomposition model. *Energy Economics, 20*(1), 85–100. doi:10.1016/S0140-9883(97)00012-1.

Takase, K., & Suzuki, T. (2011). The Japanese energy sector: current situation, and future paths. *Energy Policy, 39*(11), 6731–6744. doi:10.1016/j.enpol.2010.01.036.

UNFCCC. (2008). *Report of the Conference of the Parties on its Thirteenth Session, held in Bali from 3 to 15 December 2007.* UNFCCC. 10.1016/j.biocon.2006.08.013.

UNFCCC. (2010). *Report of the Conference of the Parties on its Fifteenth Session, held in Copenhagen from 7 to 19 December 2009, Addendum, Part Two: Action taken by the Conference of the Parties at its Fifteenth Session.* UNFCCC. 10.1038/news.2009.1156.

UNFCCC. (2015). *Japan's Intended Nationally Determined Contribution.* Retrieved from www4.unfccc.int/submissions/INDC/PublishedDocuments/Japan/1/20150717_Japan's INDC.pdf.

Xu, X. (2013). *Index Decomposition Analysis of Energy Consumption and Carbon Emissions: Some Methodological Issues.* Singapore: National University of Singapore.

Wang, A. (2009). Environmental governance, transparency and climate change in China. Presentation given at 15th Conference of Parties to the United Nations Framework Convention on Climate Change, Copenhagen.

Zhang, J., Abbas, H., & Shishkin, P. (2012). Delivering environmentally sustainable economic growth: the case of China. *Asia Society Policy* (September). Retrieved from http://asiasociety.org/files/pdf/Delivering_Environmentally_Sustainable_Economic_Growth_Case_China.pdf.

# 22

# CLEAN ENERGY TRANSITION FOR FUELING ECONOMIC INTEGRATION IN ASEAN

*Venkatachalam Anbumozhi, Sanjayan Velautham,*
*Tsani Fauziah Rakhmah, and Beni Suryadi*

## Introduction

The Association of Southeast Asian Nations (ASEAN) was formed in 1967 by Indonesia, Malaysia, the Philippines, Singapore, and Thailand, with the aim of driving regional political and economic collaboration. The organization has expanded to ten countries, adding Brunei Darussalam, Cambodia, Lao PDR, Myanmar, and Vietnam. Over time, the ASEAN economy has taken the center stage and in 2015 the region reached a critical juncture with the launching of the ASEAN Economic Community (AEC). The AEC is premised on the free flow of goods, services, labor, and investment. What started merely as a straightforward push to reduce trade barriers has evolved into a vision for a dynamic and unified market that has implications for the future energy outlook of ASEAN.

The basic energy challenge facing ASEAN is three-fold. First, the region has to supply enough energy to meet demand and to help provide access to modern energy services for those currently without. Today, around 100 million people in the region have no access to electricity and still cook with polluting fuels such as kerosene, wood, charcoal, and dung. They need reliable, safe, and secure energy services to improve their living conditions. Second, it has to supply enough energy in an efficient and affordable way that balances the needs of industrial competitiveness and human development in relation to economic integration under the AEC framework. This also leads to the third energy challenge concerning the importance of sustainable energy that takes into account local and global environmental issues such as climate change.

The related set of energy goals will be challenging for the region and thus require innovative breakthrough and strong commitment from policy makers. Action is urgently required on multiple fronts. ASEAN must find ways to restrain its energy demand by making the most efficient use of its scarce resources. Investment in new infrastructures along with innovative financing mechanisms will be needed to achieve energy security. ASEAN also needs to prioritize renewable energy supplies and at the same time deploy the new clean energy technologies that can make conventional power more efficient and environmentally friendly. The global scope of climate change and cross-cutting nature of the energy sector provides much room for regional cooperation. ASEAN has enjoyed immense gains from intra-regional trade and cross-border energy cooperation schemes. This chapter is intended to provide better insights into the energy challenges, with the overarching objective of supporting the development of policies and implementation activities towards clean energy transition. It provides detailed information and analysis on energy supply and demand-side issues from the transition perspective and identifies opportunities to achieve greater sustainability at local, national, and regional levels, which could be applicable to other regions.

## Trends in ASEAN energy supply and demand

With combined economic growth of 6%, ASEAN is the third largest economy after China and India in developing Asia. Various studies have identified similar findings regarding the significant projection of energy demand for the region in the future. The Asian Century scenario estimates that ASEAN's share of world energy consumption is going to rise rapidly from barely one-tenth in 2000 to 20–25% by 2035 (ADB, 2011). The ERIA energy outlook predicts that the ASEAN energy supply will continuously increase and double from 2013 to 2040 (ERIA, 2015). In the business-as-usual (BAU) scenario, the total primary energy supply (TPES) of ASEAN is projected to increase steadily from 619 million tonnes of oil equivalent (Mtoe) in 2013 to 1,685 Mtoe in 2040, growing at an annual rate of 4.7%, mainly driven by integrated economic activities and a growing wealthy population. This projected growth is higher than the trends observed between 1990 and 2013, which averaged 4.2% per year. Carbon emissions during the period are estimated to grow at the rate of 4% per year.

In a recent study, the TPES in the Advancing Policy Scenario (APS) is projected to increase from the historic value of 619 Mtoe in 2013, to more than 998.2 Mtoe in 2025 and 1,468 Mtoe in 2035 (ACE, 2016). The difference between TPES in the APS and the BAU scenario (Figure 22.1) shows approximately the potential of energy saving that could be achieved by ASEAN through the implementation of its advance policies on energy efficiency in transport, residential, and industry sectors as well as in electricity power production and consumption. Although TPES under the APS in 2020 was only 3% lower than in the BAU scenario, energy efficiency policies are expected to contribute to a reduction of energy demand of 13% by the end of 2035.

As the ASEAN member states (AMS) have been actively pushing their targets on renewable energy, the growth of renewable energy under the APS will be higher than the average growth trend. Despite the growth of hydro and geothermal being only slightly better under the APS, other renewables (such as wind and solar) have the potential to reach the highest growth at 8.5% on average every year. This will contribute to the sharp increase of renewable energy in the TPES up to 18.5% by 2035, compared with only a 10.5% share under the BAU scenario by the same date.

However, without sufficient energy resources, the AEC would need to scale back its ambitions for further economic integration. The key issue here is whether ASEAN has sufficient energy resources on its own. Fossil fuels are still the main source of energy and it is projected that coal, oil, and gas will remain the dominant sources in the energy mix in 2035 (ACE, 2015). In this respect, there are some important oil producers in the region, namely Brunei Darussalam, Indonesia, Malaysia, and Vietnam. Those countries cumulatively produced 9.32 million tonnes of oil in 2015. Despite this, enormous growth in energy consumption has led to these countries becoming net oil importers. Only Brunei Darussalam has been predicted to remain as a net exporter in 2035 (IEA, 2014). On the other hand, ASEAN is still a net exporter of natural gas and coal. New prospects of these resources are located offshore in deep water or as unconventional gas, which is not easy to extract. Nevertheless, ASEAN will continue to be an important player in the global coal market in the coming decades, backed by Indonesia as one of the world's major producers and exporters.

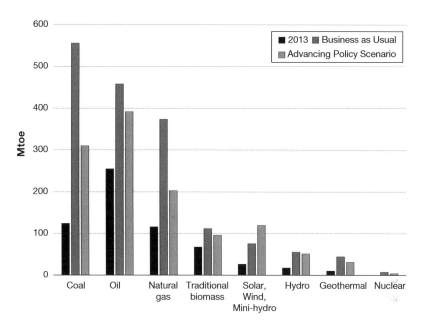

*Figure 22.1* Comparison of total primary energy supply in 2030 under Business as Usual (BAU) and the Advancing Policy Scenario (APS)

*Data source*: Prepared by authors with data sourced from ACE (2015).

Coal is cheaper and abundant compared with other sources of energy and hence going with coal is indeed the natural choice for the region to fulfill its sharp increase in energy demand to support economic development. At the same time, coal is the major source of air quality and greenhouse gas (GHG) emission concerns in this region, accounting for 18% of carbon dioxide ($CO_2$) emissions. To lessen the environmental impacts, ASEAN would need to utilize the latest and most efficient clean coal technologies. Although many advanced technologies that allow cleaner coal utilization are well developed, they have not been widely adopted due to the financing constraints and the limited technical knowledge in AMS. Apart from this concern, an additional challenge emerges from the increasing share of imported fossil fuel that obviously has alarming implications for $CO_2$ emissions and energy security. External costs related to air pollution from the combustion of fossil fuels will increase by 35%, from USD 167 billion annually in 2014 to USD 225 billion in 2025, which is also equal to 5% of the regional Gross Domestic Product (GDP) in 2025 (ACE & IRENA, 2016). Consequently, ASEAN will see rising costs for energy supply and for controlling pollution.

## Tapping cleaner and sustainable energy supplies

### *Building the momentum for renewable energy*

ASEAN's future energy supply will be bigger and the mix will be much cleaner than it is today. A greater use of renewable energy will play a key role in achieving cleaner and sustainable energy supplies. ASEAN has collectively achieved the target of 25% renewable share in the energy mix by 2013 (Figure 22.2). Following such a significant accomplishment, it is feasible that a more ambitious target of 30% renewable share that includes solar, wind, bio, and mini-hydro can be

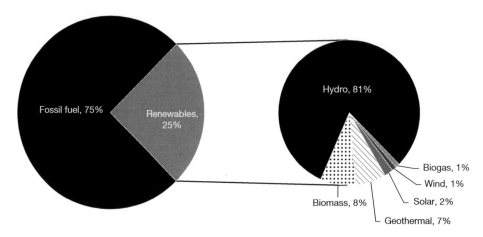

*Figure 22.2*   Installed capacity of renewable energy in ASEAN, 2013

*Data source*: Prepared by authors with data sourced from ACE (2015).

achieved. Moreover, hydropower is a well-established renewable energy source in ASEAN. The region has 350 gigawatt (GW) of hydropower capacity installed or under construction, with potential to quadruple the capacity to 1,204 GW by 2040. Hydro is one of the most affordable energy resources. Results from a levelized cost of electricity (LCOE) study showed that the average cost of a hydro project is 0.044 USD/kWh (kilowatt hour), with the lowest value reaching 0.029 USD/kWh (ACE, 2016). This is another reason for AMS to develop hydropower for electricity generation. Additionally, in 2016, utility-scale solar photovoltaics (PV) and onshore wind have seen record low prices of 0.03 USD/kWh. Such progress in renewable energy prices shows a large potential to be integrated into existing grids, while maintaining or improving power reliability and quality. This clean energy transition also comes at an important time for climate change action, with the Paris Agreement entering into force.

While various renewables have enjoyed significant cost reductions and are already economical in certain locations, including in remote areas, they have not entered into the mainstream. Being cost competitive should be understood as factoring in environmental and other social costs that are external to economic development. However, capturing the externalities with subsidies or the imposition of taxes and mandates could increase energy prices and cramps further economic activities if it is not planned properly. More innovative policy instruments, price structures, and inter-sectoral coordination are needed as a part of a comprehensive strategy capturing externalities. Also, the more expensive electricity costs derived from renewable sources could undermine affordability. Policy makers are struggling to find a tradeoff between non–market-based policy measures and strengthening old regulatory instruments that could bring a positive impact for renewable energy capacity (Table 22.1). Given the current trend in GHG emissions and the latest round of stalled global climate talks, the traditional ways of problem solving are no longer sufficient. Innovative actions that can accelerate the clean energy transition should be promoted to avoid the tragedy of the commons (Anbumozhi et al., 2016).

In terms of large hydropower projects, the cost is huge and addressing their environmental and social impacts remains a challenge. Nevertheless, hydropower projects can have strong positive outcomes if they are well planned and executed. For small-scale hydropower projects, finding the best business model to access finance is the major challenge because, if aggregated, small-scale hydropower can be more expensive than large hydropower with a longer payback period. In

*Table 22.1* Renewable energy targets and policy measures in selected ASEAN countries by 2030

| Country | Renewable energy targets | Programmes, measures, and incentives |
| --- | --- | --- |
| Indonesia | 23% share of renewable energy in the final energy mix | Feed-in tariff |
| Malaysia | 4,000 MW of installation capacity from renewable sources | Feed-in tariff, renewable energy standards |
| Philippines | 38.6% share of renewable energy in the primary energy supply | Feed-in tariff (planned), renewables portfolio standard, capital subsidies, tax incentives |
| Thailand | 20% share of renewable energy in power generation by 2036 | Feed-in tariff, feed-in premium, biodiesel blending mandate |
| Vietnam | 21% of renewable energy in power generation | Tax incentives |

*Data source*: Authors' compilation.

ASEAN, small-scale hydropower entities and solar power utilities are usually for the purpose of rural electrification, which makes them unattractive for institutional investors. International investment with guarantees from the government should be welcomed to overcome this challenge. Governments also need to create more investment-friendly policies, such as a long-term feed-in-tariff (FiT), tax holidays, and low interest loans by providing more incentives and financial supports for project developers. The government can encourage public–private partnership for this purpose. Another important challenge we may see in the development of hydropower is to increase the connectivity among the AMS. Grid integration across borders will boost hydro source deployment in the region. With better and higher grid integration, member states with abundant hydro potential can more easily export their electricity to other member states. This cross-border power trade is already in place on a bilateral basis among some AMS, such as Lao PDR and Thailand.

Further challenges for the uptake of renewable energy in countries like Cambodia, Lao PDR, Myanmar, and Vietnam include (i) involvement of governments in the promotion of renewable energy development, (ii) skilled labor to develop the renewable energy sources, (iii) intellectual property rights enforcement that has hampered technology transfer from developed countries, and (iv) affordability and accessibility of renewable sources as well as local sources (Anbumozhi and Tuan, 2016).

## Offsetting fossil fuel with alternative technologies

With growing energy demand, limited resource endowment, and also a need to minimize GHG emissions from energy production and utilization, it is vital to adopt all necessary measures to ensure that coal and other fuels are used with the highest efficiency possible. Coal demand in ASEAN is set to triple by 2040, making coal the largest fuel in the energy mix (IEA, 2015). This projected growth in coal provides much opportunity for ASEAN to deploy the most efficient coal technologies that are commercially available and to reduce the environmental footprint of coal-based electricity generation. Figure 22.3 shows the trend of coal production and consumption[1] over the period 1990 to 2014. As illustrated, increasing coal use gives ASEAN opportunities to deploy the most efficient coal technologies commercially available, reducing the environmental footprint of coal-based electricity generation. The increasing gap between coal production and consumption shows that ASEAN has a strong export potential.

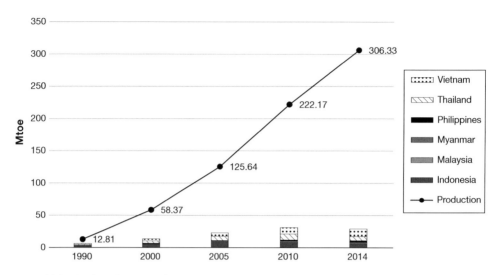

*Figure 22.3* Coal production and consumption in major ASEAN countries, 1990–2014

*Data source*: Prepared by authors with data sourced from IEA (2016).

There are several alternative technologies for improving the efficiency of coal utilization, for example, the supercritical and ultra-supercritical technologies that are commercially available. Even higher efficiencies will be possible when advanced ultra-supercritical becomes available. Power plants using low-grade coals, such as lignite which is abundant in ASEAN, are candidates for more efficient power generation. Integrated gasification combined cycle (IGCC) may also offer higher efficiency and emissions reduction if this technology is more widely deployed in the region. High efficiency low emission (HELE) coal-fired power plants are already being built in some AMS. Before 2020, Malaysia is expected to have 1,000 megawatt electrical (MWe) of supercritical and 4,080 MWe of ultra-supercritical capacity. The Philippines currently have 500 MWe of supercritical, Thailand has 600 MWe of supercritical and ultra-supercritical, and Vietnam has 19,136 MWe of supercritical and 1,200 MWe of ultra-supercritical. Such HELE technologies paired with carbon capture and storage (CCS) would be an important low–carbon emissions option for ASEAN countries that are building up coal and natural gas-fired power plant capacity. The cost of reducing carbon emissions would be dramatically higher without CCS technology, estimated as up to 70% higher internationally. The cost of deploying CCS in coal-fired power plants ranges from USD 60 to USD 90 per tonne of carbon abated, which is cost competitive with other forms of low carbon energy systems.

As shown in Table 22.2, in the spirit of cooperation following the COP21 Paris Climate Agreement, emissions reductions that are consistent with the international climate objectives must be advanced. For a developing region like ASEAN, climate mitigation efforts must be integrated with goals related to energy security, economic development, and poverty alleviation. Thus, ASEAN countries have included HELE coal-based power generation in their Intended Nationally Determined Contributions (INDCs). In the midst of pressing social development needs and resource constraints, it is useful to think of investments in clean technologies as a staged process where adjustment are made based on the level of development. Given the large heterogeneity of clean technology needs among the ASEAN countries, regional cooperation can help countries with a similar development stage to overcome their barriers to clean technology deployment (Anbumozhi et al., 2016).

*Table 22.2* Carbon mitigation targets set by regional economies in the Paris Climate Agreement

|  | *High income* | *Upper middle* | *Lower ncome* |
|---|---|---|---|
| Reduction below BAU | Korea: 37% | Thailand: unconditional reduction by 20% and conditional by 25% | Vietnam: unconditional reduction by 8% and conditional by 25%<br>Indonesia: unconditional reduction by 29% and conditional by 41% |
| Absolute reductions | Australia: 26–28% below 2005 level<br>Japan: 26% below 2013 level<br>New Zealand: 30% below 2005 level |  | The Philippines: 70% relative to its BAU scenario of 2000–2030 |
| Emission intensity | Singapore: 36% below 2005 level | China: 60–65% below 2005 level<br>Malaysia: 40% | India: 33–35% below 2005 |

*Data source*: ERIA (2016b).

There are several encouraging attempts at regional cooperation to foster clean technologies. For example, under the framework of the ASEAN Plan of Action for Energy Cooperation (APAEC) 2016–2025, the region has set collective actions to strengthen cooperative partnerships in the promotion and utilization of coal and clean coal technologies. In supporting this plan and to further scale up clean technology utilization, the implementation of clean technologies needs to be embedded in the long-term national energy programs that combine supply- and demand-side measures, cross-cutting functional and administrative boundaries, regional cooperation, and innovation processes.

## Trade potentials in natural gas

Natural gas is one of the region's most abundant energy resources, which poses the best possible low carbon alternative to coal and oil. Table 22.3 shows proven reserves and total production of natural gas in the region. Within the ASEAN countries, Indonesia has emerged as a key source of natural gas with total proved reserves amounting to 2.8 trillion cubic meters in 2015. The APAEC framework also focuses on the Trans-ASEAN Gas Pipeline (TAGP) project, which identified eight possible gas interconnection projects. Four of them would originate from gas fields operated by Indonesia. Apart from Indonesia, Myanmar and Vietnam are also emerging as key players in natural gas. However, in recent years, they have been losing their market share to other suppliers such as Australia, Qatar, and Russia. This has been attributed to the lack of coordinated strategy to improve gas infrastructure and attract new investments that would create sources of production. A high incidence of piracy in the Malacca Straits, the unregulated and insecure nature of maritime trade, and any disruption of global trade will have a divesting impact on liquefied natural gas (LNG) supply in the region, wherein local production is dwindling amidst the increasing demand for LNG.

According to ERIA (2016a), cumulative gas demand in Asia will be tripling from 208 billion cubic meters in 2010 to 672 billion cubic meters in 2030. In the ERIA projection, demand for

*Table 22.3* Natural gas potential in ASEAN, 2015

| Country | Total proved reserves (trillion $m^3$) | Share of world total (%) | Reserve to production ratio |
|---|---|---|---|
| Brunei Darussalam | 0.3 | 0.1 | 21.7 |
| Indonesia | 2.8 | 1.5 | 37.8 |
| Malaysia | 1.2 | 0.6 | 17.1 |
| Myanmar | 0.5 | 0.3 | 27.0 |
| Thailand | 0.2 | 0.1 | 5.5 |
| Vietnam | 0.6 | 0.3 | 57.9 |

*Data source*: BP (2016).

natural gas from China and India will increase remarkably. The developing markets of ASEAN are overshadowed by more mature markets such as Japan and South Korea as well as energy giants like China and India. Asian gas prices are not reflecting the regional supply demand fundamentals. ASEAN pays around USD 6 more for LNG compared with US and European buyers, and this has been the case for decades. This so-called Asian premium has emerged as a result of the limited access to cheaper gas pipelines in ASEAN and Japan, and lack of a secondary market to provide cost competition. Consequently, the inflated long-term contract prices remain unchallenged. However, the shale gas revolution in the USA and massive new projects in Australia are exacerbating the supply glut. Spot prices for LNG in ASEAN have slipped to an all-time low on this glut. Yet many buyers are locked into long-term contracts that are linked to oil prices.

The abundant reserves of natural gas in ASEAN (Table 22.3) would allow the region to decouple its growth in electricity demand and its related emissions through changing the fossil fuel mix by using natural gas. The realization of the Trans-ASEAN Power Grid (TAPG) project and investments to develop maritime Southeast Asia as a LNG hub can help to facilitate greater trade in natural gas. Investments in infrastructure for integration of gas transmission will make natural gas more accessible to countries in the region and also could alleviate concerns associated with energy security through energy supply diversification. Integrated gas transmission can help other countries, such as Myanmar, to join the regional gas trade more easily. In order to realize such trade potential in natural gas, ASEAN governments should first address the political, institutional, and legal barriers. Governments can start by harmonizing the legal and institutional frameworks on LNG markets and formulating suitable financing options for the integrated gas pipeline and LNG hub. This harmonization has been started with connections between Malaysia, Singapore, and Thailand, along with the feasibility project for grid connection between those countries and Lao PDR.

## Nuclear power plants progress

ASEAN countries are attempting to reduce their over-dependence on fossil fuels. Exploring alternative energy sources, including nuclear energy, is part of their strategies to ensure that energy supplies are secure, affordable, and environmentally sustainable. In total, about 17 reactors (including 6 research reactors) are planned. Countries exploring the nuclear energy option include Indonesia, Malaysia, the Philippines, Thailand, and Vietnam. In Vietnam, two Russian reactors of 2,000 MWe in total are planned to start operations by 2020, followed by another additional 2,000 MWe using Japanese technology. The government plans to further add 6,000 MWe by 2020 and therefore in total Vietnam would have 10,000 MWe by 2030. The government of Indonesia has changed its plan from building large units for the Java–Bali grid to building an initial small reactor near Jakarta, while in Thailand, 5,000 MWe of nuclear capacity is envisaged under the Thailand Power Development Plan 2010–2030, with 1,000 MW units between 2020 and 2028. The Philippines are also considering two 1,000 MWe Korean Standard Nuclear plant units using equipment from a North Korean project. At the same time, Malaysia plans to develop three or four reactors to supply 10–15% of its electricity by 2030.

Currently, some AMS have plans to build nuclear reactors and there are crucial reasons why the region has a vested interest in ensuring nuclear security, safety, and safeguards. First, any nuclear accident in the region would threaten public health through radioactive plume, which knows no borders. Second, the region's fast-growing economies can be easily jeopardized by a nuclear accident as the operations of key economic sectors, including the supply chain, would be disrupted. Third, the region's vital sea lanes, which radioactive materials will likely pass through, are not tightly guarded by maritime security forces. Hence, there is a challenge for AMS to be able to fully secure all the radioactive materials and waste from their future nuclear power plants and prevent terrorist threats. Having these circumstances, there is a need for AMS to start discussing possible regional mechanisms on nuclear safety. Lastly, the region is home to the world's major food producers and exporters. A nuclear accident would result in the radioactive contamination of farmlands and marine resources and in turn disrupt the food supply chain. It is therefore in the interests of the AMS to collectively institutionalize nuclear safety and security, including the safe and secure transport of radioactive materials.

Several advanced countries like Japan, South Korea, and Russia have bilateral cooperation with ASEAN countries through the Civilian Nuclear Energy Sub-Sector Network (CNE-SSN). CNE-SSN is an ASEAN network for nuclear matters under the directive of ASEAN Ministers on Energy Meeting through its APAEC. The CNE-SSN was tasked by ASEAN energy ministers to continue promoting and intensifying capacity-building efforts, in collaboration with the relevant partners, so that the region will be well informed and updated on the latest nuclear safety standards, developments, and technologies. In addition, ASEAN has embarked on various cooperation activities with dialogue partners to help the region build its nuclear energy capacity. Collectively, these efforts would play an important role in addressing issues related to high upfront capital costs, limited access to financing, and uneven public support in nuclear power plant development.

## Containing the burgeoning energy demands in ASEAN

### *Achievements and long-term plans in energy efficiency*

In spite of strong economic growth, the region's energy intensity – the amount of energy used to produce each dollar of GDP – is declining steadily over the years. During the past five years, ASEAN has implemented policies and programs directed toward increasing energy efficiency in energy intensive industry or transport sectors. These efforts have led to 10% of energy intensity reduction in 2015 against the 2005 level, which exceeded the target to reduce energy intensity by 8% in 2015 as stipulated in the previous plan. As now ASEAN stands at a 13% reduction, it is on track to achieve its new target under APAEC 2016–2025 for a 20% reduction in 2020 based on the 2005 level. Such reduction in energy intensity is largely determined by efforts being made in the individual member states, but also by collaborative partnership at regional level, such as the harmonization of standards and labeling for various household appliances.

Energy efficiency and conservation have been on the national agendas of AMS with a number of innovative plans and policies developed to facilitate improvement in energy efficiency. Targets to improve energy efficiency in Indonesia are part of the government's ambition to reduce GHG emissions by 26% by 2020, as committed in its NDC submission. To this end, the government has been working with local institutions and regional agencies through a number of programs, which include initiatives such as awareness raising, development of local capacity of Energy Management System (EnMS) and ISO 50001, as well as capacity building for energy managers in factories and companies. Energy efficiency programs are also being implemented with the support of both government and external agencies, targeting the industrial (e.g. textiles and garments, food and beverage, pulp and paper, and chemical industries), building, residential, and commercial (e.g. tourism) sectors. Support given to the industrial sector is primarily in the form of capacity development focusing on EnMS and ISO 50001. For the commercial sectors, the programs offer energy-auditing services, while the residential sector has benefited from the free distribution of compact fluorescent lamps.

Similarly, the Malaysian government has been actively fostering energy efficiency through a number of programs and initiatives. High-energy consumers (with energy consumption of 3 million kWh in six months) are required to demonstrate energy management in their operations, for example by appointing an energy manager. Malaysia also has an energy efficiency rating labeling programthat requires all manufacturers of selected commercial appliances and equipment to affix the energy efficiency labels on to their products. Programs implemented with the support of local and regional agencies focus on the industrial, building, transport, and commercial sectors. Types of intervention include the implementation of national energy management standards, the application of system optimization in manufacturing industries, energy efficiency improvements in buildings through retrofitting, and stimulation of sales of energy-efficient vehicles. Currently, the government is working on the National Energy Efficiency Plan to introduce mandatory requirements to achieve energy efficiency in all sectors. The draft plan is now going through a stakeholder consultation process.

The Philippines have yet to introduce a national policy to enforce energy efficiency activities. Nevertheless, there are multiple programs currently being implemented with support from donors and regional agencies, including improvement of energy efficiency through capacity building training for industries, replacement of CFC-based chillers with energy efficient non-CFC chillers, and promotion of high-energy efficient motors for the sugar industry. The Department of Energy is also actively promoting energy efficiency in the country through various programs, such as reduction of energy consumption in government buildings and operations by 10% annually.

In Thailand, the Department of Alternative Energy Development and Efficiency is the key agency driving energy efficiency. The recent introduction of its 20-year Energy Efficiency Development Plan 2011–2030 aims to achieve 20% reduction in final energy consumption by 2030 compared with the 2010 level. The government has designated about 2,800 buildings and 5,400 factories, which are required to reduce and report their energy consumption on a regular basis. The government has also implemented a revolving fund to strengthen the capacity of commercial banks to finance energy efficiency projects, developed the Energy Service Companies (ESCO) fund to enable smaller companies to access energy efficiency financing, and works with the Bureau of Investment to provide tax and duty exemptions for energy efficiency products. The private sector plays an important role in providing energy efficiency services under this mechanism.

Vietnam mandates strict requirements for all sectors to improve energy efficiency as regulated by the Energy Efficiency Conservation Law 2010. The Vietnam National Energy Efficiency Program sets out a comprehensive plan to implement measures for improving energy efficiency and conservation across the economy. Similar to other AMS, donors and regional agencies

*Table 22.4* Energy efficiency improvement measures in ASEAN

| Country | Reference document | Energy efficiency target |
|---------|-------------------|--------------------------|
| Brunei Darussalam | Energy White Paper 2014 | • 45% energy intensity reduction by 2030 (baseline 2005) |
| Cambodia | National Policy, Strategy and Action Plan on Energy Efficiency in Cambodia 2016 | • 20% reduction in energy demand by in 2035 (baseline BAU) |
| Indonesia | National Master Plan for Energy Conservation (RIKEN) 2014 | • 1% energy intensity reduction annually until 2025 <br> • Energy savings of 17% in final consumption by 2025 (industrial sector, 17%; transport sector, 20%; commercial sector and households, 15%) |
| Lao PDR | Energy Efficiency and Conservation Plan (under development) | • 10% energy saving by 2030 (baseline 2009) |
| Malaysia | Eleventh Malaysia Plan (2016–2020) | • 8% reduction of energy consumption in residential, commercial, and industry sectors by 2025 (baseline 2008) |
| Myanmar | Energy Efficiency and Conservation Policy, Strategy and Roadmap for Myanmar (under development) | • 16% energy efficiency improvement by 2030 |
| Philippines | National Energy Efficiency and Conservation Program (NEECP) | • 45% energy intensity by 2035 (baseline 2005) |
| Singapore | Sustainable Singapore Blueprint 2015 | • 35% energy intensity improvement by 2030 (baseline 2005) |
| Thailand | Energy Efficiency Development Plan 2015–2036 | • 30% energy intensity improvement by 2036 (baseline 2010) |
| Vietnam | Vietnam National Energy Efficiency Program 2005–2015 | • Initial energy savings of 3–5% (2006–2010) and a further 5–8% (2011–2015) |

*Data source*: Authors' compilation.

support a number of energy efficiency and conservation programs, including in the building sector and clean production.

As presented in Table 22.4, most AMS have adopted strategies or action plans to improve energy efficiency at all stages of the energy chain with specific defined objectives to achieve. Given that some AMS are in the process of adopting energy efficiency measures, the region has the potential to achieve significant energy savings in the foreseeable future.

With energy security and carbon emission reductions taking precedence in INDCs, ASEAN governments are seriously looking at energy efficiency. As the energy efficiency potential in ASEAN is largely untapped, efforts to increase energy efficiency can power the ASEAN economies to achieve sustainable economic growth, ensure universal energy access for all, and mitigate the onset of climate change (Anbumozhi et al., 2016). The demand-side management efforts continue across the transport, residential, and industrial sectors. Reduced energy demand would mean less public infrastructure expenditure on power generation plants in the long run, while the impact of more efficient power generation in existing plants will be seen in the medium term. Over the longer term, as efficiency in building, industry, and residential sectors becomes a more widespread priority, new regulations can be formulated to take efficiency standards to the next

*Table 22.5* Policy approaches for enhanced energy efficiency in selected ASEAN countries

| Country | Management | Standard/labeling | Financial support |
|---|---|---|---|
| Brunei Darussalam | Voluntary | Not available | Not available |
| Cambodia | Regulatory | In planning | Not available |
| Indonesia | Regulatory | Standard, labeling | Grant |
| Lao PDR | Voluntary | Standard (in planning) | Tax |
| Malaysia | Voluntary | Not available | Grant, tax, loan |
| Singapore | Regulatory, voluntary | Standard, labeling | Grant, tax |
| Vietnam | Regulatory | Labeling | Tax |

*Data source*: Authors' compilation.

level. Singapore, which already has a mandatory efficiency labeling for appliances, introduced an Energy Conservation Act in 2016 requiring energy-intensive companies to appoint an energy manager. Responsibilities include monitoring and reporting energy use, reporting GHG emissions, and submitting energy efficiency improvement plans to the government. Different policy approaches taken by ASEAN countries to improve industrial energy efficiency are described in Table 22.5. Regulatory and voluntary approaches are often twinned with financial incentives to promote effectiveness.

AMS are aiming at energy efficiency as a low hanging fruit. They begin to institute building codes and fuel efficiency standards in the transport sector, switch to light-emitting diode (LED) lights for street lighting and traffic lights, and label appliances to promote the purchase of efficient appliances. Countries can draw on existing multilateral funding to promote greater consumer awareness in energy efficiency. The Asian Development Bank (ADB), for instance, has consistently funded energy efficiency projects between 2005 and 2011. Its investments in projects with a demand-side energy efficiency component amount to USD 1.8 billion. ADB has also set up a new Energy Efficiency Technical Support Unit to provide technical policy and financial support in accelerating energy efficiency investments in developing member countries.

ASEAN has also embarked on various collaborative programs with other countries or regions. For example, ASEAN is partnering with Japan in the ASEAN–Japan Energy Efficiency Partnership (AJEEP) in an effort that focuses on energy efficiency labeling. The project objective is mainly to support AMS overcoming barriers to the implementation of measures on energy efficient home appliances by adopting the Japanese approach. This project also serves as a platform to discuss and exchange experiences on approaches to enhance energy efficient market transformation in the AMS and the region. The European Union has supported ASEAN through ASEAN Standards and Harmonization Initiatives for Energy Efficiency (SHINE), which aims to increase the market share of more efficient air-conditioners (ACs) through the harmonization of test methods and energy efficiency standards, adoption of common minimum energy performance standards, and changing consumer purchasing attitudes in favor of energy efficient ACs.

Although several ASEAN countries have existing policies and market incentives to promote energy efficiency, they need to ensure better enforcement and development of the requisite monitoring and evaluation systems. There are also numerous best practices in the region, which have addressed consumer and industrial energy demand while at the same time ensuring that legal mechanisms and new schemes are in place to support energy efficiency. Several innovations are also taking place in the region. In the 1990s, a fund was set up to support energy efficiency projects in Thailand, using proceeds from taxes on petroleum products. Simultaneously, demand-side management plans such as public awareness campaigns and energy efficiency standards for buildings and appliances were launched. Some AMS provided financing schemes, such as the Green Technology Financing Scheme, Green Technology Incentive, and the Energy Efficiency Revolving Fund Project (Malaysia), and the ESCO Fund Project (Thailand) to create a market and attract private investment. With these efforts, it is expected that the market players and private sector can also benefit from the buzz around the ESCO concept and reach their full energy efficiency potential.

Starting from the 1993 promotion of the Electricity Energy Efficiency Project, in 2014 the total amount of energy efficiency and renewable energy investment was Thai Baht 3,964.49 million, coming from 101 Energy Performance Contracts where the cumulative investment since 2009 reached THB 19,836.80 million by 466 EPCs.

### Behavioral changes to reap energy efficiency benefits

Driven by a huge energy saving potential, the region has implemented energy efficiency strategies for commercial and residential buildings. In the residential sector particularly, energy efficiency is becoming a great concern for member countries where the urbanization rate is high and the number of electrical home appliances is increasing rapidly. Without any action, residential energy use will increase tremendously. Market transformation towards increased energy efficiency could be a remedy to balance this situation. One way to progress the market transformation is to implement appropriate policies and measures for key players in the market, namely manufacturers and importers, retailers, and consumers. Effective functioning of the triangular relationship between these three key players is vital to foster market transformation in energy efficiency. One common barrier to market transformation is lack of awareness and knowledge of energy efficient products among consumers and retailers. Providing information on energy efficiency in home appliances is very useful for consumers' reference in choosing more efficient appliances.

### Fostering regional energy market integration

ASEAN has developed a master plan for integrating the energy markets by building on the progressive achievements of previous work on clean energy transition. With a blueprint on "Enhancing Energy Connectivity and Market Integration in ASEAN to Achieve Energy Security, Accessibility, Affordability and Sustainability for All," the APAEC outlines outcome-based strategies and action plans. Extended over a longer period of 10 years, the plan will be implemented in two phases. Phase I (2016–2020) will focus on the short- to medium-term strategies required to achieve energy security cooperation and move towards greater connectivity and integration. A mid-term review of Phase I will be conducted in 2018 to guide the region charting the roadmap for the next phase (Phase II: 2021–2025). Phase I will continue to focus on the seven program areas as in the previous APAEC. These include the ASEAN Power Grid (APG), Trans-ASEAN Gas Pipeline (TAGP), coal and clean coal technology, energy efficiency and conservation, renewable energy, regional energy policy and planning, and civilian nuclear energy.

### Increasing interconnectivity through infrastructure development

The main objective of regional power infrastructure development is to reap the full benefits of reduced investment cost, improved energy supply stability, and reduced emissions. There are two main priorities for energy infrastructure projects in ASEAN, namely the TAGP and APG. By the end of 2016, there were 13 bilateral gas pipeline interconnection projects in operation with a total

*Table 22.6* Regasification terminals operating in ASEAN

| Regasification terminals | Size (metric tonnes per annum) |
|---|---|
| Thailand (Ma Ta Phut) | 5.0 |
| Indonesia (West Java) | 3.0 |
| Singapore (Singapore LNG) | 6.0 |
| Malaysia (Sg Udang Melaka) | 3.8 |
| Indonesia Arun Regas | 3.0 |
| Indonesia Lumpung FSRU | 1.7 |

*Data source*: ASCOPE (2016).

length of 3,673 km connecting six countries – Indonesia, Malaysia, Myanmar, Singapore, Thailand, and Vietnam. As the gaps between demand and supply in ASEAN will become larger since the indigenous reserves are depleting, TAGP now also includes LNG as an option to secure energy supply. Many countries have initiated infrastructure construction to flow gas from outside ASEAN. Currently, LNG terminals are already in operation in Indonesia and Thailand. The construction of LNG receiving terminals has been progressing well in Indonesia, Malaysia, Singapore, and Thailand. In April 2016, there are six regional gas terminal operations with the total capacity of 22.5 metric tonnes per annum (Table 22.6).

In terms of electricity infrastructure, ASEAN has identified three APG priority projects for completion and three additional APG projects that will commence construction shortly. Through these projects, power exchange and purchase are expected to triple from 3,489 MW in 2014 to 10,800 MW in 2020 and further increase to 16,000 MW post-2020.

## Maximizing the benefits and reducing the costs of integrated energy markets

Energy market integration within the ASEAN region will allow the optimum utilization of energy resources and further improve energy security. The region can supply energy and at the same time provide investment opportunities for its dialogue partners, such as Japan and South Korea. Brunei Darussalam, Indonesia, Malaysia, and Vietnam have considerable potential to produce oil and gas. Countries such as Myanmar, Lao PDR, and Cambodia have large untapped hydropower potential which provides many opportunities for investment and technology transfer from developed countries. There is also pressing demand to use energy sources that reduce the carbon footprint and hence energy trade across the borders through hydropower generation is considered as an appropriate choice. Table 22.7 shows the current level of energy trading within ASEAN.

The existing power trade in ASEAN is mostly established on a bilateral basis. Given that progress has been largely focused on bilateral interconnections, the new strategy for ASEAN is to embark on multilateral interconnections. The regional power trade in the Greater Mekong Subregion is estimated to save 19% of the total energy costs or about USD 200 billion. The savings resulting from the interconnection of regional power systems alone are estimated at USD 14.3 billion, mainly from the substitution of fossil fuel generation with hydropower. Further, if power interchange takes place between countries with different times in peak demands, then the investment needed could be reduced to maintain the reserve margin. Such regional grid interconnections will generate economic benefits for the entire region. However, the size of investment for linking different markets is mostly considerable. Prioritization of

*Table 22.7* Current regional power trade in ASEAN (in units of MW, data for 2016)

| Country | Imports | Exports | Total trade | Net Imports |
| --- | --- | --- | --- | --- |
| Cambodia | 1,546 | – | 1,546 | 1,546 |
| Lao PDR | 1,265 | 6,944 | 8,209 | −5,679 |
| Myanmar | – | 1,720 | 1,720 | 1,720 |
| Thailand | 6,938 | 1,427 | 8,365 | 5,511 |
| Vietnam | 1,720 | 1,318 | 6,917 | 4,281 |

*Data source*: HAPUA (2016).

construction, taking into account the benefits and feasibility of each route, would help to avoid situations where construction of all planned routes commences at the same time and creates implementation deficits.

## Barriers to energy market integration

Numerous potential barriers confront energy market integration, including technical, political, and environmental barriers. First, technical barriers are prevalent, ranging from grid synchronization and grid codes to electric power and natural gas pipeline technology. Second, power transfer among ASEAN countries is mainly hampered by reluctance to give up sovereignty for energy security (Kimura et al., 2013). The fundamental reason is that they are in fact seeking maximum protection for their interests amidst the market opening and restructuring processes. Third, negotiations for a trading agreement are commonly affected by unequal starting positions and differing energy security concerns. Political barriers and lack of political trust in coordination among member countries can also hinder such negotiations. Fourth, regulatory barriers and distorted energy pricing from existing subsidy regimes in many ASEAN countries discourage trading of energy on commercial terms, as the entities selling energy at a subsidised rate will have to pay for the energy at cost, with negative financial consequences. IEA (2014) suggests that to a certain extent (on the basis of LCOE) many renewable technologies are cost competitive compared with conventional sources even without subsidies for their generation. Finally, hydropower generation and the construction of multi-purpose projects are considered to have significant environmental repercussions. The construction of multipurpose projects, which include large reservoirs, means a disruption of riverine fauna and displacement of human settlements and agriculture. An integrated approach to energy market integration is needed, which would include building institutions to improve coordination and address multiple barriers.

## Conclusions

An alignment of economic integration, demands for improved living standards, and pressing climate change issues have set in motion ongoing energy transformation in ASEAN. The foregoing analysis makes clear that several actions on energy security, environmental, and equity fronts are being taken at the national and regional levels. Promoting clean energy means providing a secured energy supply to support economic integration, improve trade balances, as well as create local value and jobs. With the current technologies and new policies, the transition to a sustainable energy future by 2030 is economically feasible. ASEAN must aspire to create a greater AEC to realize maximum efficiency in power generation and utilization. The first step is to get the economic and energy ministers together and agree on the way forward to achieve regional integration. Then they should set up a ministerial task force under the East Asia Summit Energy Cooperation Task Force framework to promote the political will to share more openly information on national power sectors that aim to move towards better harmonized policies.

For this remarkable transformation to take place, the following actions are needed in the coming years.

- Strengthen policy commitment to stimulate markets for clean energy. Enabling market-oriented policies and regulatory frameworks to create stable and predictable investments will help overcome barriers and ensure predictable revenue streams for projects. Setting clean energy sector targets, such as clean coal, renewable energy, and energy efficiency, and formulating dedicated policies to implement them, for example, provides strong market signals that reflect the government's commitment to energy sector development and economic integration. Depending on the national context, complementary measures such as FiT, feed-in premiums, removal of pervasive subsidies, and appropriate energy pricing can level the playing field for clean energy.
- Mobilize private sector investments for clean energy transition. Public funding will remain an important catalyst and will need to increase, but a major share of new investment will have to come from the private sector. To mobilize private investment, the strategy pursued must focus on risk mitigation instruments and structured finance to develop sustainable cross-border energy connectivity projects. To scale up investments in international projects, traditional public finance channels need to combine with new private sector investments.
- Harness the cross-cutting impact of sustainable energy development. Access to reliable and cost effective clean technology can have multiplier development impact in both advancement and access contexts. In particular, renewable energy and energy efficiency solutions can expand the electricity access, increase productivity, create jobs, and bolster poverty alleviation in least developed countries in ASEAN. The wider development of clean energy must be taken into account when implementation strategies for Sustainable Development Goals and Paris Agreement on climate change are formulated.
- Build institutional and human capacity to support clean fossil fuel development. From economic policy and regulations on international project development, a wide array of skills needs to be built up in the ministries, financing institutions, and agencies for promoting clean coal technologies. Coordination is also vital between the different stakeholders in order to ensure, for instance, that physical infrastructure and complementary soft measures and standards, such as grid codes, keep pace with accelerating new energy infrastructure development.
- Enhance regional cooperation approaches. Regional approaches to connect the grids and common initiatives to improve energy efficiency and standards can bring competiveness, attract more investments, boost financial capacity, stimulate cross-border energy trade, and enable common progress in accelerating the deployment of clean technology sources region-wide. To meet the national renewable energy targets, ASEAN could benefit from coordinated action that APAEC offers. Governments should tap into opportunities for multi-level stakeholder engagement and international cooperation on APG and TAGP.

# Note

1 Total final coal consumption excluding coal used in power generation and other transformations.

# References

ACE (ASEAN Centre for Energy) (2016), *Renewable Energy Policies*, ASEAN Centre for Energy, Jakarta.

ACE (ASEAN Centre for Energy) (2015), *The 4th ASEAN Energy Outlook 2013–2035*, ASEAN Centre for Energy, Jakarta.

ACE (ASEAN Centre for Energy) and IRENA (International Renewable Energy Agency) (2016), *Renewable Energy Outlook for ASEAN: a REmap Analysis*, ASEAN Centre for Energy, Jakarta.

ADB (Asian Development Bank) (2011), *Asia 2050: Realizing the Asian Century*, Asian Development Bank, Manila.

Anbumozhi, V and Tuang, T (2016), *Integrative Strategies and Policies for Promotion of Appropriate Renewable Energy Technologies in Lower Mekong Basin Region*, ERIA Research Report, ERIA, Jakarta.

Anbumozhi, V, Kalirajan, K, Kimura, F, and Yao, X (2016), *Investing in Low-Carbon Energy Systems: Implications for Regional Economic Cooperation*, Springer, Singapore.

ASCOPE (ASEAN Council for Petroleum) (2016), ASCOPE Secretary in Charge report to the 34th ASEAN Ministers on Energy Meeting on the Progress of Trans-ASEAN Gas Pipeline Project 2016, Nay Pyi Taw, Myanmar, 23 September 2016.

BP (2016), *Statistical Review of World Energy June 2016*, British Petroleum, London.

ERIA (Economic Research Institute for ASEAN and East Asia) (2015), *Energy Outlook and Energy Saving Potential in East Asia*, Economic Research Institute for ASEAN and East Asia, Jakarta.

ERIA (Economic Research Institute for ASEAN and East Asia) (2016a), *The Energy Outlook and Energy Saving Potential in East Asia 2016*, Economic Research Institute for ASEAN and East Asia, Jakarta.

ERIA (Economic Research Institute for ASEAN and East Asia) (2016b), *INDCs and Globalization of Technologies: What's Next for Energy and Economic Communities in EAS Region?* ERIA Working Group Meeting on Globalization of Low Carbon Technologies and INDC, Bangkok.

HAPUA (Heads of ASEAN Power Utilities/Authorities) (2016), HAPUA Secretary-in-Charge Report to the 34th ASEAN Ministers on Energy Meeting on the Progress of ASEAN Interconnection Project 2016, Nay Pyi Taw, Myanmar, 23 September 2016.

IEA (International Energy Agency) (2014), *South East Asia Energy Outlook – World Energy Outlook Special Report*, International Energy Agency, Paris.

IEA (International Energy Agency) (2015), *Energy and Climate Change*, International Energy Agency, Paris.

IEA (International Energy Agency) (2016), *Statistics: Coal Information*, International Energy Agency, Paris.

Kimura, F, Phoumin, H, and Jacob, B (2013), *Energy Market Integration in East Asia: Renewable Energy and Its Deployment into the Power System*, Economic Research Institute for ASEAN and East Asia, Jakarta.

# 23

# IMPORTANCE OF REGIONAL CLIMATE POLICY INSTRUMENTS TOWARDS THE DECARBONISATION OF ELECTRICITY SYSTEM IN THE GREAT MEKONG SUB-REGION

*Akihisa Kuriyama and Kentaro Tamura*

## Introduction

The decarbonisation of electricity systems is a common agenda under the climate change policies in developed countries as well as middle-income countries. The Greater Mekong Sub-region (GMS) countries, consisting of Cambodia, the Lao PDR, Myanmar, Thailand, and Vietnam, have been working on this issue as well. Though there are differences in the amount of electricity demand and renewable energy potential among the GMS countries, those potentials need to be considered in an integrated manner because the economies in this region are being integrated by enhancing regional initiatives such as the ASEAN Economic Community (AEC) and the extension of the electricity grid beyond the countries' boundaries.

Also, the least developed countries in the GMS, i.e. the Lao PDR, Cambodia, and Myanmar, enjoyed high GDP growth rates in 2014, i.e. 7.5%, 7.1%, and 8.5%, respectively (World Bank, 2016). Since their economic growth is greater than that of other GMS countries, they are catching up with other relatively developed GMS countries, which will result in increasing fossil fuel use. Therefore, it is important to develop a strategic climate policy for GMS countries as soon as possible.

In fact, all of the GMS countries have implemented climate change policies. Under the Cancun Agreements of 2011, developing countries are requested to develop mitigation actions up to 2020 with so-called nationally appropriate mitigation actions (NAMAs). Furthermore, the Paris Agreement of 2015 sought for all the countries to implement their post-2020 mitigation contributions and submit them every five years after 2015. As a consequence, all the GMS countries have already submitted their intended nationally determined contributions

(INDCs) in response to this initiative. One common feature of their INDCs is to reduce the electricity produced by fossil fuel power plants even though there are some construction plans for coal- and gas-fired power plants in the GMS countries owing to the increase in electricity demand. Therefore, effective climate strategies and incentives to promote greenhouse gas (GHG) reduction in the electricity sector, in particular, are a key element in the decarbonisation of the society.

Therefore, this chapter summarises the key factors for achieving decarbonisation in the electricity systems. The first section estimates the renewable energy potential and electricity demand in 2035 on the basis of existing model scenarios for the GMS countries. By comparing those two potentials and highlighting the current initiatives to extend transmission lines across the GMS countries, it discusses the feasibility of decarbonising the electricity system.

The next section shows the possible institutional barriers presented by existing country-based climate policies. To highlight this, it reviews the experience of the Clean Development Mechanism (CDM), in which all the GMS countries participated actively. It also provides case studies demonstrating how the identified country-based grid emission factors created institutional barriers to the installation of a hydropower plant in the Lao PDR. The final section summarises the findings and leads to the conclusions of this chapter.

## Prospective electricity market and renewable energy potential in the Greater Mekong Sub-region

### *The approach to identifying electricity demand forecasts and renewable energy potentials*

This section reviews the renewable energy potential and electricity demand forecast for 2035 in the GMS countries as well as the current status of the grid structure. In particular, it compares the estimated electricity demand in 2035 and technical renewable energy potential in the GMS countries. Through this comparison, it highlights the feasibility of decarbonising the electricity system through the use of renewable energy sources and increased energy efficiency, which could fill the gap between electricity demand and renewable energy potential.

### *Electricity demand*

Since electricity demand depends on assumptions about GDP growth, population, and energy efficiency improvement, it required three studies to review the electricity demand forecast. Each study has a business-as-usual (BaU) scenario and an alternative (Alt) scenario.

Energy Supply Security Planning for ASEAN (ESSPA) applied the methodology for final energy demand forecasting using econometrics. While the estimation of primary energy consumption used an engineering-based model, the energy development programmes of each member state served as the major inputs used in the models (IEEJ & ACE, 2011). The study by Kimura (2013) used the World Energy Outlook Model developed by the Institute of Energy Economics, Japan's IEEJ model and the LEAP model. It focussed on analysing the additional energy savings that might be achieved under the mitigation goals and action plans of each GMS country. The analysis also includes the scenario beyond the current policy level using the assumptions confirmed by their respective working group members. The Asian Development Bank (ADB, 2013) publishes "Energy Outlook for Asia and the Pacific", which also applies IEEJ and LEAP models. It reflects the diversity of regional economic development and population growth. The Alt scenario under the ADB study considers the potential for energy savings and

$CO_2$ emissions reduction: "With the deployment of advanced technologies, electricity demand of Asia and the Pacific will increase at an annual rate of 2.5% from 2010 to 2035—a slower rate compared with the growth rate of primary energy demand in the BAU case at 2.1% per year" (ADB, 2013).

Table 23.1 summarises the macroeconomic assumptions for the three models. Each study provides two scenarios for electricity demand forecasts in 2035. The growth of the economy and electricity demand would be affected by the economic status in other countries such as China, but the analysis of macroeconomic indicators is beyond the scope of this study.

## Renewable electricity potential

The estimation of renewable electricity from solar, wind, and biomass power is based on "Renewable Energy Developments and Potential in the Greater Mekong Subregion" published by the Asian Development Bank (ADB, 2015b). In this report, the technical potential for solar energy is based on the degree of solar irradiation, the efficiency of conversion technologies, the suitable land area, and other factors. The authors calculated the wind energy potential based on average wind speeds over specific land areas, in metres/second (m/s), and the wind turbine generator (WTG) installation capacity (or wind power density), in megawatts per square kilometre (MW/km$^2$). To estimate technical potential for wind power from these figures, they (1) exclude the wind energy potential in protected forest areas, mountainous and remote areas, and urban areas, and (2) take account of the current capacity of the grid electricity systems in each country because the capacity of the transmission line is so critical to maintaining stability (ADB, 2015b). In fact, the transmission capacity of the connected grid electricity system imposes a significant restriction to limit the wind potential, especially in the least developed countries. The ADB's model excluded the electricity potential of biomass power plants based on the findings where "the technical potential of agricultural residues is much less than the theoretical potential, partly because of the difficulty of residue collection" (ADB, 2015b). For electricity potential from hydropower plants, the model used the estimations by each government. Even though some GMS countries such as Vietnam plan to start the use of nuclear power plants, the potential of nuclear power is excluded from the technical potential of electricity supply due to the uncertainty of implementing the technology in these countries.

*Table 23.1* Macroeconomic assumptions of the three models for forecasting electricity demand

|  | *IEEJ & ACE (2011)* | *Kimura (2013)* | *ADB (2013)* |
|---|---|---|---|
| Model | LEAP | LEAP | LEAP |
| Annual GDP growth for ASEAN countries (%) | 5.2 | 4.9 | 4.1 |
| Population in GMS countries (million) | 277.9 | 257.2 | 273.3 |
| Assumption of Alt scenario | Consider the potential for energy saving | Reflect each country's goal | Use the energy saving goals set by the government |
| Energy demand forecast for all GMS countries (TWh) | BaU: 920 Alt: 752 | BaU: 939 Alt: 802 | BaU: 770 Alt: 672 |

*Data source*: Authors' compilation.

## Result of total electricity demand and renewable energy potential in GMS countries

The estimated electricity demand in all the GMS countries is from 757 TWh to 1,008 TWh under the BaU scenario. On the other hand, it ranges from 651 TWh to 757 TWh under the Alt scenarios. The technical renewable energy potential in the Mekong region was estimated to be 115 TWh for solar power, 10 TWh for wind, and 478 TWh for hydropower plants. The technical potential of wind power could be larger because the wind power potential is calculated based on the capacity of the current grid electricity system. From those figures, if we take account of the full technical renewable energy potential, the gap between the electricity demand forecast and the renewable energy potential is calculated to be 154–405 TWh under the BaU scenario. Furthermore, the Alt scenario shrinks the gap between electricity demand forecast and renewable energy potential to 48–233 TWh. Even though renewable energy potential is slightly less than the lowest electricity demand estimation under the Alt scenario, there is still the possibility that renewable energy could satisfy all the electricity demand at the 2035 level if we consider the upper range of energy efficiency potential.

First, in all the three scenarios, incremental energy efficiency potentials from the BaU scenario to the Alt scenario in Vietnam are estimated to be around 10%. However, according to ADB (2015a), the electricity saving in Vietnam can be enhanced to around 20% with proper policy implementation. The enhanced energy efficient potential is not only the case for Vietnam but also for other GMS countries. This finding means that a country's effort on energy efficiency is necessary to decarbonize its electricity system.

Second, the enhancement of transmission capacity would increase the technical wind power potential. As Figure 23.1 shows, there is huge theoretical wind power potential in the GMS countries, but it is not counted as the technical potential due to the limitation of transmission capacity according to ADB (2015b).

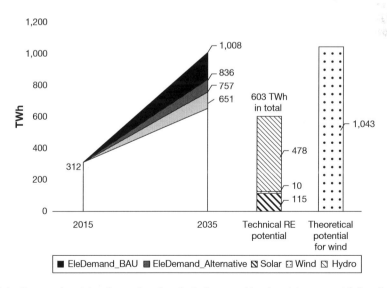

*Figure 23.1* Future electricity demand and technical renewable electricity potential for all the GMS countries

*Data source*: Authors, based on IEEJ & ACE (2011), ADB (2013), ADB (2015a) and ADB (2015b).

## Result of electricity demand and technical renewable energy potential by country

As shown in Figure 23.1, the renewable energy potentials in the GMS countries could be enough to meet the electricity demand in 2035 if the countries fully utilised their wind energy potentials. However, the location of renewable energy potentials is separated from the place where a significant amount of electricity is needed. Figure 23.2 shows the technical renewable energy potential and electricity demand by each GMS country.

For Cambodia, electricity demand is predicted to be 8–16 TWh under the BaU scenario and 7–14 TWh under the Alt scenario that reflects the potential of energy savings in 2035. As to renewable energy, potentials of solar power, wind power, and hydropower are 12 TWh, 0.2 TWh, and 61 TWh, respectively. Cambodia has substantial solar resources that could meet a significant portion of that country's electricity demand in 2035. Wind energy, on the other hand, is limited by low wind speeds and the weakness of the grid and load system. The potential for hydropower plant is so great that Cambodia could export electricity to neighbouring countries.

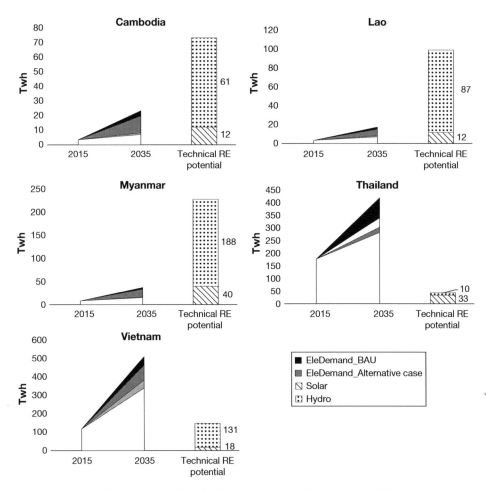

*Figure 23.2* Future electricity demand and technical renewable electricity potential

*Data source*: Authors, based on IEEJ & ACE (2011), ADB (2013), ADB (2015a) and ADB (2015b).

Regarding the Lao PDR, electricity demand in 2035 is predicted to be 8–17 TWh under the BaU scenario and 7–15 TWh under the Alt scenario in 2035. For renewable energy, the potential is predicted to be 12 TWh for solar power and 87 TWh for hydropower. Wind power potential is limited in the Lao PDR due to the limited capacity of the electric grid. There is, however, a considerable potential for small-scale solar and wind power electricity. Although wind power capacity is not huge, the theoretical potential for wind power in the Lao PDR is substantially higher than in the other GMS countries, which could lead to further investment in wind power, possibly beyond the capacity of the current grid electricity system. For example, a wind power project developer in Thailand, Impact Electrons Siam Co. Ltd (IES), published a press release to mark its Memorandum of Understanding (MOU) with Vestas Wind Systems to install 600 MW of wind turbines in the Dak Cheung and Sanxay districts of Sekong and Attapeu provinces. Also, the potential of hydropower is so enormous that electricity could be exported to neighbouring countries.

As to Myanmar, the electricity demand in 2035 is predicted to be 16–37 TWh under the BaU scenario and 15–34 TWh under the Alt scenario. Renewable energy potential in Myanmar is 40 TWh for solar power, 1 TWh for wind, and 188 TWh for hydropower plants. Myanmar has huge areas of high solar irradiation levels, but project developers cannot use some of the areas for solar power due to the mountainous terrain and protected areas. The potential for wind power is limited since average wind speeds in most areas are too low even if a developer applies the latest wind turbine technologies. Myanmar has the largest hydropower potential in the GMS countries, but the capacity of the grid electricity system is so small that further development of the system is required to transmit electricity to neighbouring countries.

Figure 23.2 shows that technical renewable potential, especially hydropower plants, in the Lao PDR, Cambodia, and Myanmar is much higher than the predicted electricity demand. At the same time, those three countries have less electricity saving potential. On the contrary, Thailand and Vietnam have a growing electricity demand, and it is expected to increase rapidly. The models predicted electricity demand in Thailand to be 340–419 TWh under the BaU scenario and 281–302 TWh for the Alt scenario. Renewable energy potential in Thailand is predicted to be 33 TWh for solar power, 5 TWh for wind power, and 9.8 TWh for hydropower plants. Thailand has plenty of solar power potential, and the government also has an ambitious target of nearly 2,000 MW of solar PV installations by 2021. On the other hand, weak wind speeds result in Thailand's modest wind power potential. There is less hydropower potential in Thailand because it has few mountainous areas. For Vietnam, electricity demand is forecasted to be 385–512 TWh under the BaU scenario and 341–465 TWh under the Alt scenario. Renewable energy potential in Vietnam is predicted to be 18 TWh for solar power, 7 TWh for wind power, and 131 TWh for hydropower. Obviously, there is a large gap between electricity demand and renewable energy supply in Thailand and Vietnam. Solar power potential in Vietnam is located in the southern half of the country owing to the relatively high solar irradiation levels. A large part of its wind power potentials exists in the southern coastal areas and offshore.

From those figures, it can be observed that Cambodia, the Lao PDR, and Myanmar have a huge potential to export renewable energy electricity to Thailand and Vietnam where the amount of renewable energy is not enough to meet their large electricity demands.

### *Expanding the grid system in GMS countries to fill the electricity demand and renewable energy supply gap*

The grid electricity systems in the GMS countries except for Myanmar are interconnected with one another as shown in Figure 23.3. One-fifth of the electricity supply for Lao's grid system consists of imported electricity from Thailand. Also, as discussed below, the operation of the grid

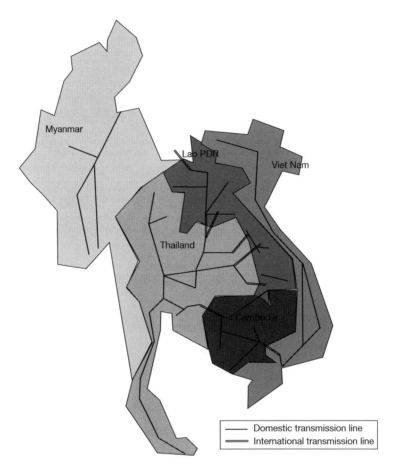

*Figure 23.3*   Grid map of the Mekong region

*Data source*: Authors.

electricity system in the Lao PDR highly depends on the grid electricity system in Thailand. For Cambodia, Vietnam supplies one-quarter of the electricity and Thailand supplies one-tenth. Also, there is an initiative of an ASEAN Power Grid (APG) that interconnects the electricity grids in all of the ASEAN countries, enabling cross-border electricity exchange, and could potentially increase the export capacity of electricity by renewable energies from the Lao PDR, Cambodia, and Myanmar to other countries. As has been discussed above, the Lao PDR and Cambodia have an enormous potential for renewable energy; the integration of the grid electricity systems would enable those countries to export more electricity to neighbouring countries.

## Towards an effective institutional framework for mitigation actions to promote renewable and energy efficiency potential

As shown above, there are enough renewable energy potentials to meet the electricity demand increase in the GMS countries. It is important to implement an effective institutional framework to promote mitigation actions that enable them to mobilise renewable and energy efficiency potentials in these countries. Mitigation actions could be mobilised by not only project-based mechanisms such as baseline and credit mechanisms but also national climate and energy policies under national emission reduction targets such as INDCs under the Paris Agreement. In any case, MRV (measurement, reporting and verification) for mitigation actions at the project level, sector level, or national level is becoming critical for enhancing the transparency of those mitigation actions including renewable energy projects. Therefore, this section highlights the lessons learned from the operation of MRV under the Clean Development Mechanism (CDM) to provide insights into developing an efficient institutional framework that harnesses mitigation actions, because CDM is the first international mitigation mechanism for developing countries under the United Nations Framework Convention on Climate Change (UNFCCC) as a project-based baseline and credit mechanism under the Kyoto Protocol. In fact, CDM has suffered from identifying the baseline of mitigation project activity by renewable energies because CDM project developers had to determine the amount of emission reduction within host countries as a principle. In this case, it creates a complicated process for the calculation of grid emission factors in countries where the grid electricity systems are interconnected with those of other countries. Therefore this section highlights this issue because a similar concern arises as long as the mitigation initiatives are conducted in a single country rather than in multiple countries.

## Calculation and accounting for emission reduction by mitigation actions

While mitigation policies or schemes provide an adequate incentive to implement renewable energies, this section highlights two technical issues which should be considered under a framework to mobilise mitigation actions: identification and accounting method of emission reduction by policy instruments.

First, Table 23.2 summarises the importance of identifying emission reductions. At the first stage of implementing mitigation policy instruments, a government often selects a baseline and crediting schemes such as CDM and the joint crediting mechanisms which Japan and the partner countries are jointly implementing. For those mechanisms, an emission factor is an essential number that quantifies the emission reduction by project activities of energy efficiency and renewable energy under this scheme.

A cap and trade scheme is another major instrument used to put a price on carbon by setting an emission cap for target entities. The basic design of the scheme covers direct emissions from target entities, but some cap and trade schemes cover indirect emissions such as the Tokyo Cap and Trade Programme. During the first commitment period of the Tokyo Cap and Trade Programme (FY 2010–2014), the average emission factor of 2005–2007 was applied (TMG, 2014). In this case, an emission factor is also needed to identify the achievement of emission reduction by through the scheme.

Mitigation efforts by all the countries are requested to be reported to the UNFCCC secretariat as biannual update reports. Therefore, the current climate regime requests each country to quantify its emission reduction amount by energy efficiency and renewable projects to communicate its efforts. While the mitigation outcome is reported by emission reductions which are calculated with direct emission data, countries that claim emission reduction efforts by mitigation action not by emission reduction targets, such as a goal of installing renewable energy or improving energy intensity per capita, may be recommended to quantify their emission reduction impacts to enhance transparency of their mitigation efforts.

Second, the accounting rule is also important to secure the environmental integrity of international mitigation policy instruments. In fact, accounting issues among developing countries have not been fully addressed under the CDM because no developing countries had mitigation commitments during the first commitment period of the Kyoto Protocol. However,

*Table 23.2* The role of grid emission factors for identifying impacts of mitigation policies and projects

| Mitigation policy instruments | Importance of identifying the grid emission factor |
| --- | --- |
| Baseline and crediting scheme | Calculation of emission reduction by energy efficiency project and renewable energies project applies an emission factor of grid electricity system. |
| Cap and trade scheme | Some schemes cover indirect emissions and use the emission factor for calculating $CO_2$ emissions. |
| National mitigation actions/target such as NAMA and NDC | Some NDCs describe renewable energy implementation or energy efficiency targets as mitigation actions. It is preferable to identify emission intensity to quantify policy impacts. |

*Data source*: Authors.

when the developing countries have emission reduction targets such as NAMAs and INDCs, accounting issues need to be addressed. IGES (2016b) points out four categories of accounting issues: double registration, double issuance, double usage, and double claiming. All of the issues are relevant to the identification of any mitigation mechanisms.

## Lessons learned for identifying and accounting for mitigation effects: a case in the Lao PDR

One feature of the electric power sector in the GMS countries is grid electricity connection across national borders. This connected grid system provides both challenges and opportunities for international collaborative actions for promoting electricity savings and renewable energy under the international climate change policy regime, as the experience of the CDM shows. Though the price of CDM credits has declined sharply due to the recent economic recession, the CDM has provided major incentives to promote energy savings in this region. A total of 303 CDM projects had been registered in the GMS countries up to the end of 2015 (IGES, 2016a).

At the initial stage of the CDM, a grid electricity system was defined along the national borders, when emission factors were determined. This approach, however, hindered the development of CDM projects, in particular in the Lao PDR and Cambodia. First, since those two countries import a significant amount of electricity, they cannot identify the real emission factor of the electricity system by using only domestic electricity generation data. In this case, the two countries had to use an overly conservative emission factor for imported electricity, which resulted in lower Certified Emission Reduction issuance and did not provide an incentive to develop projects. Second, a project that reduces the importing of electricity has to be approved by both countries, that in which the project is located and the one from which the electricity is imported, to communicate how much the CDM project will make an impact on the electricity system of the neighbouring countries and contribute to emission reduction in both countries. Since this process was difficult for CDM project developers, it was only after taking a new approach, having the option of an international grid system, that CDM projects increased in the Lao PDR. The following goes into more detail on how identification of the baseline was improved through the case of the Lao PDR.

### *Identification of emission factors of the electricity system*

The Lao PDR has plenty of hydropower sources owing to the abundance of water in the Mekong River Basin, which covers the entire area of the Lao PDR. As electricity demand in the Lao PDR and neighbouring countries such as Thailand and Vietnam has increased in proportion to economic development, especially for Thailand, the Lao PDR exported 200–700 TWh of electricity between 2009 and 2012, depending on the availability of water in rivers (EDL, 2009; 2010; 2011; 2012). Thus, the Lao PDR has attracted developers' attention as the "Battery of Mekong". However, the Lao PDR needs to import electricity from neighbouring countries such as Thailand despite having plenty of hydropower potential. This is because the capacity of hydropower plants with water reservoirs is not sufficient during the dry season.

Figure 23.4 shows the structure of the grid system in the Lao PDR. According to the grid map of Electricité du Laos (EDL), the Lao national grid electricity system is divided into four quadrants: the Northern Grid, Central Grid 1, Central Grid 2, and the Southern Grid. The Northern Grid covers five provinces and includes a population of 1.6 million people. The Northern Grid covers Sayaburi and Luang Prabang provinces. Central Grid 1 covers four provinces and has a population of 2 million people. This area includes the capital city, Vientiane, and it has the greatest electricity consumption among the four quadrants. The Central Grid 2 covers three provinces and has a population of 1.6 million people. Both Asian Highway 15, which passes Thakhek, and Asian Highway 16, which passes through Avannakhet, are located in this area. As a result, this area has clusters of industrial factories, which consume a fair amount of electricity. The Southern Grid covers four provinces and has a population of 1.3 million people.

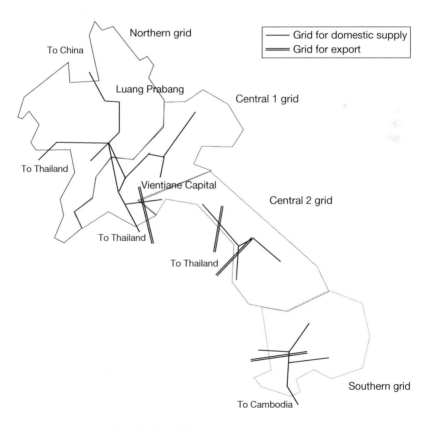

*Figure 23.4*   Four separate grids in the Lao PDR

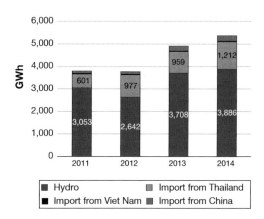

*Figure 23.5* Electricity mix in the Lao PDR

*Data source*: Authors, based on the data provided by EDL.

Figure 23.5 shows the electricity mix of the Lao PDR from 2011 to 2014. As noted above, all the domestic power plants in the Lao PDR are hydro, but the electricity supplies were not stable. In 2011, the amount of electricity generated by hydropower plants was 3,053 MWh, but this decreased to 2,642 MWh owing to insufficient water resources. To supply the difference between the electricity demand and electricity supply capacity, imported electricity from Thailand increased even after the new hydropower plant started operating during 2012 and 2013. In 2014, electricity imported from Thailand increased to 1,212 GWh.

To calculate emission reduction by electricity saving or renewable energies, which displaces the electricity supplied by a fossil fuel fired power plant, a project participant had to identify the emission intensity (i.e. $tCO_2$/MWh) of the grid electricity system to which the proposed project connects. To calculate this number, the CDM Executive Board (CDM EB) provided guidelines for a "tool to calculate the emission factor for electricity" at the 35th CDM Executive Board meeting in 2007. This guideline requireed the defined boundary of the grid electricity system to be a national boundary. However, in practice, this request made it difficult for project developers to promote any CDM project activities in the Lao PDR since all the domestic power sources are hydropower. For example, even though the Lao PDR imports electricity from Thailand and Vietnam where many fossil fuel power plants supply electricity, the guidelines also denote "the emission factor is 0 tonnes $CO_2$ per MWh for imports from connected grid electricity systems located in another host country(s)" (UNFCCC, 2007). In the end, the grid emission factor is calculated to be 0 $tCO_2$/MWh, which results in no emission reduction by any energy efficiency and renewable energy project.

However, setting the grid boundary by the national boundary does not reflect on the actual operation of the grid electricity system for the countries where electricity import and export are frequently transacted through the international transmission line. Figure 23.6 shows the electricity supply from domestic hydropower resources and the electricity imported to each local grid in the Lao PDR.

Northern Grid does not have a large hydropower plant; it largely depends on imported electricity from China to meet electricity demands. For Central Grid 1, electricity supply has continuously increased. Even though the EDL power generation marked the largest amount of electricity supply, imported electricity has a major role in supplementing it, especially when demand exceeds supply during the dry season. Electricity supply from the Independent Power

Producer (IPP) had been kept around 40 GWh in 2011 and 2012 owing to the long-term contract on electricity supply to Central Grid1 between the EDL and IPP owners.

For Central Grid 2, the amount of electricity supply for this region has also increased since 2010. Because this area harnesses some large IPP power plants, the main power source is electricity from IPP power generators. It should also be noted that, from the beginning of 2012, the imported electricity supplements the electricity supply when the electricity demand exceeds the supplying capacity of the IPP power plant for Central Grid 2. Even though it is not captured by Figure 23.6, there is also electricity export to Thailand by IPP plants through their original transmission line.

In Southern Grid, there are several large hydropower plants owned by the EDL, but those hydropower plants do not have a water reservoir; therefore, during the rainy season, the power stations can produce a lot of electricity and export it to Thailand and Cambodia. On the other hand, during the dry season, Central Grid 2 needs to import electricity from Thailand to meet the demand.

Owing to efforts to improve the guideline tool, some mitigation projects including energy saving projects in the Lao PDR come to the CDM project pipeline. Figure 23.7 shows the pathways to developing CDM projects in Mekong countries, reflecting the resulting share of the registered project of each year in the total registered project at 2015. The CDM project proponents extensively developed CDM projects in Cambodia, the Lao PDR, and Thailand around 2011. In contrast, around 60% of CDM projects in the Lao PDR were registered during 2013–2015. For this reason, until 2012, there had been a low incentive to develop CDM projects due to the low grid emission factors. However, after the government of the Lao PDR published the official grid emission factor at the beginning of 2014, the number of CDM projects in these countries significantly increased. It demonstrates good planning that the integration of the grid electricity system in this region will, by its very nature, contribute to promoting mitigation projects, including energy efficiency in the electricity sector and the implementation of renewable projects.

### *Reporting mitigation projects to countries exporting electricity*

In addition to identifying emission factors, project approval by both the host country and the country exporting electricity was an additional burden on project developers. According to the clarification of CDM EB in its 28th meeting (Paragraph 14, meeting report), the transnational electricity systems are eligible under ACM0002 methodology (i.e. the CDM methodology for grid connected renewables), and the designated national authority (DNA) of countries in the region which the electric system spans shall be considered as host parties and shall provide a letter of approval stating that the project activity assists them in achieving sustainable development. Even though nothing is specified, there are at least two reasons for this. First, because the displacement of imported electricity could affect the operation of power plants (the stability of the grid electricity system in the exporting country, for example), the project developers in importing countries are encouraged to communicate with companies exporting electricity. Second, the displacement of imported electricity could also reduce GHG emissions from power plants in the exporting countries, owing to reduced electricity generation. When exporting countries implement domestic mitigation policies for the power sector, those policies would also help reduce $CO_2$ emissions by improving heat efficiency, lowering the amount of generation, etc. Therefore, when policy makers in exporting countries evaluate mitigation policies, they should consider the accounting issues involved in mitigation actions. So far, accounting issues have been less important because there has been no emission reduction

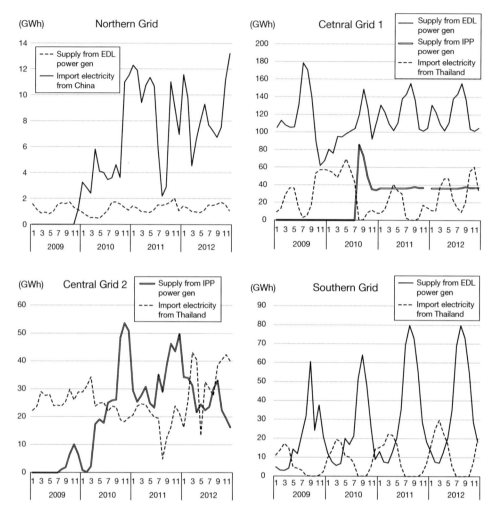

*Figure 23.6* Electricity generation and import of each grid system in the Lao PDR

*Data source*: Authors, based on the data provided by EDL.

commitment for non–Annex I countries during the first commitment period of the Kyoto Protocol, and the CDM has been the only international mitigation mechanism. However, it will be more important during the second commitment period of the Kyoto Protocol and even after 2020, when all countries including the GMS countries need to take mitigation actions and implement several international mitigation schemes.

## Examples of regional mitigation policy being consistent with integrated grid electricity systems in other regions

While grid electricity systems in the GMS countries are becoming integrated, it is reasonable to implement mitigation measures to reduce GHG emissions by achieving energy efficiency or increasing renewable energy production at the regional level. For example, the EU emission trading system (EU-ETS) covers all the grid electricity systems that are connected to one another. If some

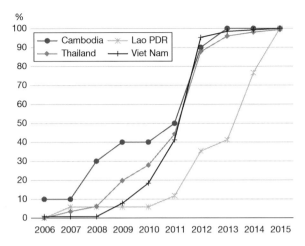

*Figure 23.7*    CDM project development pathways in GMS countries

*Data source*: Authors, based on IGES (2016a).

countries were not covered by the emission trading system, and fossil fuel power stations in these countries supplied electricity to the transnational electricity grids, power companies would have incentives to construct fossil fuel power plants in these countries. This disincentive to reduce GHG emissions is called "carbon leakage". For example, RGGI is the first mandatory emission trading system in the United States among the states of Connecticut, Delaware, Maine, Maryland, Massachusetts, New Hampshire, New York, Rhode Island, and Vermont to reduce $CO_2$ emissions from power plants. Banks et al. (2015) report the carbon leakage by transmitting electricity from the non-RGGI states, while RGGI achieved significant $CO_2$ emissions reductions.

On the other hand, if an ETS covers all the fossil fuel power plants that supply electricity to the transnational grid, zero emission power sources including renewable energy can be invested without carbon leakage. In the case of the US Clean Power Plan (CCP), which aims to reduce $CO_2$ emissions from power plants, each state develops its rules based on EPA guidelines. However, the CCP provides options for joint proposals by multi-states because grid electricity systems in some states are interconnected, and it is more efficient to develop an implementation plan by the states where power plants connect to the same grid electricity system (US EPA, 2015). Such cases imply that an integrated grid system is essential for robust market-based mitigation measures including a baseline and crediting system and a cap and trade system.

Therefore, once the regional emission reduction measures are also taken in the GMS countries, it is not necessary to address the identification and accounting issues as discussed above for the Lao PDR case because the area of emission sources corresponds to the area of emissions targets. The benefit of this integrated grid system approach is not only to promote mitigation projects but also to harness both domestic policies and the emerging international mitigation policies, including green finance and changing individual behaviour, by putting a price on carbon pollution (OECD, 2013).

## Conclusion

The GMS countries have the potential to decarbonise their electricity systems by promoting energy efficiencies and renewable energies. While the mitigation policy instrument would have a major role to play in maximising the renewable energy and energy efficiency potentials towards decarbonisation, the identification and accounting method for emission reduction in a country-specific manner would impose significant barriers in the GMS countries based on the experience of the CDM.

To tackle those barriers, the proposed integrated electricity system, known as the ASEAN Power Grid, would have a positive impact in two respects. First, it physically bridges the linkage between the large renewable energy potential area and the demand area. Second, it presents the opportunity to bring in regional mitigation initiatives that would eliminate the complexity of identifying and accounting for mitigation impacts.

# References

Asian Development Bank (ADB). (2013). *Energy Outlook for Asia and the Pacific 2013*. Manila.

ADB. (2015a). *Energy Efficiency Developments and Potential Energy Savings in the Greater Mekong Subregion*. Manila.

ADB. (2015b). *Renewable Energy Developments and Potential in the Greater Mekong Subregion*. Manila.

Banks, J. P., Boersma, T., & Ebinger, C. K. (2015). Does decarbonization mean de-coalification? Discussing carbon reduction policies. *Coal in the 21st Century, Brookings Institution*. Retrieved from www.brookings.edu/articles/does-decarbonization-mean-de-coalification-discussing-carbon-reduction-policies/#cancel.

Electricite Du Laos (EDL). (2009). *Electricity Statistics 2009*. Vientiane.

EDL. (2010). *Electricity Statistics 2010*. Vientiane.

EDL. (2011). *Electricity Statistics 2011*. Vientiane.

EDL. (2012). *Electricity Statistics 2012*. Vientiane.

IEEJ & ACE. (2011). *The 3rd ASEAN Energy Outlook*. ASEAN Centre for Energy, Jakarta.

IGES. (2016a). IGES CDM Project Database. Retrieved August 15, 2016, from http://pub.iges.or.jp/modules/envirolib/view.php?docid=968.

IGES. (2016b). IGES Submission to Views on Guidance on Cooperative Approaches Referred to in Article 6, Paragraph 2, of the Paris Agreement. Retrieved November 8, 2016, from https://unfccc.int/files/parties_observers/submissions_from_observers/application/pdf/689.pdf.

Kimura, S. (2013). *Analysis on Energy Saving Potential in East Asia*. Economic Research Institute for ASEAN and East Asia (ERIA), Jakarta.

OECD. (2013). *Climate and Carbon: Aligning Prices and Policies*. OECD Publishing, Paris. doi: 10.1787/5k3z11hjg6r7-en.

Tokyo Metropolitan Government (TMG). (2014). *Daikibo Jigyousyo Heno Onshitsukouka Gasu Haisyutsuryou Sakugen Gimu to Haisyuturyou Torishiki Seido* (GHG Emission Cap for Large Emission Entity and Introduction to Emission Tradiny System) (in Japanese). Tokyo.

United Nations Framework Convention on Climate Change (UNFCCC). (2007). *Tool to Calculate the Emission Factor for an Electricity System (Version 01.00)*. Bonn.

United States Environmental Protection Agency (US EPA). (2015). *Overview of the Clean Power Plan*. Washington, DC.

World Bank. (2016). *World Bank Open Data*. Washington, DC. Retrieved from http://data.worldbank.org/.

# 24

# COSTS AND BENEFITS OF BIOFUELS IN ASIA

*Shabbir H. Gheewala, Noah Kittner, and Xunpeng Shi*

## General introduction to biofuels – and in Asia

Agro-based liquid transportation fuels, or biofuels, have been promoted across Asia through public policies and technological innovations. This chapter takes a critical view of biofuel production, deployment, and policy – navigating through the environmental, economic, and social trade-offs involved in biofuel decision-making and providing a review of the experiences in terms of costs and benefits across Asia. Systems-scale analyses generally indicate the complex and nuanced nature of biofuel development. For instance, biofuels may be able to contribute to national energy security planning and regional energy security and trade (Gheewala et al. 2013a). Additionally, they often can reduce greenhouse gas (GHG) emissions relative to petroleum-based fossil fuels and there is a relatively large potential resource of feedstock biomass in Asia. However, their costs sometimes outweigh purported benefits, including management of land, freshwater resources, and impacts on food production. The costs often burden marginalized sections of society more so than others.

Empirically, it has been shown that biofuel development within Asian countries is motivated by several factors, mainly concerns for domestic energy security and socio-economic development, but not climate change mitigation (Kumar et al. 2013; Shi 2016; Zhou & Thomson 2009). In the future this may change; domestic energy security and socio-economic development from biofuels have not fulfilled expectations and yet climate change mitigation is becoming increasingly imperative.

Across Asia, a majority of the biofuel development occurred through the promotion of subsidies and public policy initiatives (Phalan 2009). The relatively large potential availability of feedstock biomass comes from sugarcane, cassava, palm oil, rapeseed oil, and other vegetable oils (Gheewala et al. 2013b; Kudoh et al. 2015; Phalan 2009). Their economic benefits have been questioned though, and impacts on food prices and food security have raised serious questions on the merits of subsidising biofuels. (Doshi et al. 2016; Silalertruksa & Gheewala 2012). Purported employment benefits, for instance, have not been realized to the extent biofuels have been promoted and developed by different national and local governments (Mukherjee & Sovacool 2014; Silalertruksa et al. 2012b; Zhou & Thomson 2009).

In this chapter, we outline the issues surrounding sustainability of biofuels and the socio-economic and environmental costs, identify pitfalls, and evaluate the conditions under which benefits are possible to improve biofuel agriculture and industry. We combine these through the lens of both theory and practice, particularly drawing on experiences from Asia.

## Sustainability issues

Biofuels remain critical to evaluating sustainable energy production across Asia. Recently, climate change has surfaced as another motivating factor, alongside energy security, for the promotion of biofuels in the public and private sectors. The development of new technologies poses challenges and the rise in demand for services requires careful attention to realize the potential benefits of bioenergy. The main premise driving biofuel promotion is that carbon dioxide emissions from the combustion of biofuels in transportation are balanced by atmospheric carbon dioxide sequestered during the growth of biomass, which in theory should offer substantial climate change mitigation benefits (Danielsen et al. 2009; Fargione et al. 2008). This can be achieved in practice and positive examples do exist throughout Asia (Nguyen et al. 2007; 2010; Woolf et al. 2010). Second-generation biofuels that derive from agricultural wastes can even become carbon-negative. As Asian countries prepare their nationally determined contributions (NDCs) following the Paris Agreement, improving the accounting of greenhouse gas (GHG) emission balances from biofuels is becoming more critical. This is especially due to the wide variety of feedstocks and emerging agricultural waste technologies that could become carbon-negative, which could play an integral role in limiting global warming. Understanding the extent to which greenhouse gas reductions are possible through the biofuel life-cycle is therefore an area for further improvement and research, and the tools and frameworks already exist for rapid and detailed evaluation. With the signing of the Paris Agreement, a sustainable energy system in Asia will require the evaluation of advanced and existing biofuel technologies in standardized frameworks to improve transparency and hold countries accountable.

Increasingly, we observe emerging cases of projects that are improving on and learning from previous experiences in terms of better greenhouse gas emission balances, reduced petroleum and water inputs, and less land competition with food. Academic, governmental, and industry critiques point to the previous failures of biofuels to deliver the perceived benefits in terms of GHG reductions, land use change and management, enhancing energy security, and reducing dependence on fossil fuel consumption. Despite lingering issues, biofuels will not disappear from the fuel mix anytime soon, as many countries throughout Asia still depend on critical biofuels for household, commercial, and industrial energy consumption. Instead, many Asian countries have targets and plan for bioethanol and biodiesel in transportation fuels since biofuels are the most readily available alternatives to fossil fuels in the transportation sector. Biofuels have a lower initial transition cost for users of internal combustion engine (ICE) vehicles than electric cars, especially for vehicles that use liquefied petroleum gas (LPG) or compressed natural gas (CNG) in Asia. Therefore, it becomes prudent to understand the risks associated with new generation biofuels and identify productive uses that can contribute toward achieving sustainability goals – including climate change mitigation, designing better land use policy, and improving the livelihoods of those depending on biofuels for agricultural production and daily life.

Some first-generation biofuels required significant fossil fuel inputs for cultivation and had negative energy balances (Farrell et al. 2006; Gasparatos et al. 2013; Searchinger et al. 2008; Tilman et al. 2009). Second-generation biofuels have improved as many use agricultural waste products as a feedstock or more efficient crops. However, further considerations beyond direct fossil fuel inputs and energy balances have questioned the sustainability of certain biofuel feedstocks. Greenhouse gas emissions from the conversion of tropical forests or peatlands for biofuel production confound potential benefits from direct fossil fuel substitution. This occurs because, although biofuels come from biological and organic materials, one should evaluate their environmental impact holistically across the entire life-cycle. If biofuels are grown on lands that were previously tropical forests or peatlands that represent large carbon sinks, then greenhouse gas emissions from biofuel production could be high as the biofuel growth may not fully replace the previous carbon stock from the tropical forest (Fargione et al. 2008; Gasparatos et al. 2013). At the same time, soil carbon stocks can change when lands are converted for biofuel production. Marginal lands that have little soil fertility and low carbon stocks can improve their carbon balances by becoming sites for future biofuel production. Because there are many steps to consider along a typical biofuel feedstock chain, we introduce a few assessment tools that can evaluate life-cycle greenhouse gas emissions, soil quality, non-GHG pollutants, water footprints, and energy balances, to name a few of the relevant environmental sustainability indicators. Increased dependence on fossil fuel inputs in the feedstock of biofuels can increase the greenhouse gas footprint and reduce energy security.

Additionally, under changing climate regimes, competition for agricultural land and water resources is increasing. Recent volatility in food prices raises the issue of whether land use is best allocated for biofuel or food cultivation. Scarce agricultural land competes directly with market forces whether to produce food or fuel. For instance, the experiment in India with *Jatropha curcas*, a drought-tolerant, low water crop with previous commercial viability, illustrates the significant land requirement on a unit-energy basis. Alternative crops like sweet sorghum may provide larger benefits to the poor and smaller landowners (Rajagopal 2008). Scarcity of available agricultural land for biofuels could simultaneously drive innovation, but marginalize the poor who use marginal lands to find fuel and fodder (Rajagopal et al. 2007). Many biofuel crops including sugarcane, cassava, palm oil, and other vegetable oils require significant freshwater inputs in addition to energy. Emerging techniques to evaluate the water footprint of biofuels are adding to the quantitative toolset available to policy-makers and practitioners to weigh these key trade-offs. Changing environmental conditions threaten water availability and supply in the region. Many biofuel technologies use agricultural lands and water resources as an input to development. However, the Mekong River, a river that sustains nearly 326 million people, suffered the worst drought in more than a hundred years in 2015, and declining water resources and increasing sedimentation pose threats to maintaining biofuel production (ADB 2016; Larson 2016). The need for water as an input to agricultural and biofuel processes is clear. Localized impacts from biofuel production on water resources could be the "Achilles heel" of biofuel promotion as increased development could place severe stress on water availability (Fingerman et al. 2010; Gheewala et al. 2011). Furthermore, fewer biofuel value chains utilize local resources, adding to the environmental burden in the production process. However, advances in waste biomass technologies including bagasse, stalks, stovers, and municipal and industrial wastes could improve overall sustainability issues, reduce freshwater from agriculture inputs, and stall the encroachment on tropical forest or agricultural lands. In tropical Asia especially, first- and second-generation biofuel production from palm oil seriously burdened ecosystems with the clearing of forests. This destroyed habitats and reduced biodiversity especially in Malaysia and Indonesia. Future research that evaluates the effects of climate change on biofuel production and competition with agricultural land remains paramount to assessing the sustainability of biofuel production in Asia.

## Energy security

The concept of promoting domestic energy production is quite attractive from an energy security standpoint. If countries can increase domestic productivity, especially by utilizing agricultural waste residues for either biofuel production or electricity generation, there can be substantial benefits. Especially in Southeast Asia, where domestic crude oil is limited, biofuels theoretically can provide security benefits by increasing domestic production – as long as water, land, and climate resources are not compromised in the process. One prominent co-benefit of sugarcane biofuel production to highlight opportunities for on-site electricity generation is bagasse (which comes from sugarcane). It can either be used in a dedicated plant co-located with a sugar refinery to provide power or co-fired in existing coal plants as a lower-carbon electricity transition option as countries begin to identify niche opportunities to reduce imports of coal or other fossil fuels and utilize more bioenergy within the country.

The energy security benefits of biofuel production and generation have been perceived, but not always realized. For instance, ASEAN countries that have led Asia in biofuel promotion as a way to reduce dependence on oil have reduced their overall diversification of energy resources and level of energy security from 2005 onwards (Tongsopit et al. 2016). Despite stated goals through the ASEAN Plan of Action for Energy Cooperation (APAEC) to promote renewable energy and regional energy policy and planning that would diversify energy in electricity and transportation sectors, fossil fuels remain prevalent due to increased production capacity, demand, and government direction. This historical increase in dependence on fossil fuels has facilitated fossil fuel driven inputs along the supply chain of new biofuel technologies. This has placed a renewed emphasis on the "renewability" metric, which measures the ratio of total net energy output to total fossil fuel energy input (Gheewala 2013). Though there is a diversity of biofuel feedstock production pathways, it remains unclear the extent to which overall renewability is increasing and fossil fuel energy inputs are decreasing. This questions the overall energy security benefits of biofuels, especially when fossil fuel energy imports are increasing in ASEAN. Though biofuels have the potential to significantly reduce dependence on imported fossil fuels, in practice the substitution has not been direct and oftentimes biofuel feedstock production can require fossil fuel inputs.

Integrated resource planning for energy, water, and food systems across the region will benefit individual countries' energy security through better outcomes from a regional cooperation standpoint. For transportation biofuels, the use of sugarcane for bioethanol becomes somewhat attractive due to the relatively lower water input requirement compared with other biofuel options. Advances in algae fuels for biodiesel could be good areas for research innovation and deployment as they have small land footprints on arable lands (Doshi et al. 2016). At the combustion end of the life-cycle, it is important to test the different fuel blends against current gasoline or ethanol blends for aldehyde formation, particulate matter emissions, black carbon, and volatile organic compounds that could also influence tropospheric smog formation (Venkataraman et al. 2005). In the future, evaluating security beyond simply energy– but to include water, land, climate change, and other environmental effects – will be a useful framework to understand biofuel costs and benefits.

## Trade and green certificates

Promoting open regional markets in biofuels could be a key strategy to promote sustainable development of biofuels by allowing the lower-cost and more efficient producers to produce more (Gheewala et al. 2013a). Total free trade, however, could lead to a race to the bottom when the least sustainable producers could expand their market share by employing cheap, but not sustainable practices.

Various green certificates, including carbon footprint labeling, which have been increasingly used as a policy to guarantee the sustainable development of biofuels, will change the cost–benefit analysis of biofuel. For example, the Roundtable on Sustainable Palm Oil (RSPO) developed a set of environmental and social criteria which companies must comply with to produce Certified Sustainable Palm Oil (CSPO).[1] From the consumer side, WWF uses the Palm Oil Buyer's Scorecard to assess the performance of companies that buy palm oil – and assesses customers on their use of CSPO. Since 2009, WWF has regularly scored company performance on core actions needed to ensure responsible action.[2] For example, in November 2015, the Singapore Environment Council (SEC) explained in the *Straits Times* that it would look to certify palm oil products from February 2016 onward under the Singapore Green Labeling Scheme (SGLS), yet no companies as of October 2016 have taken up the SEC's Green Label for Palm Oil.[3] Also, there are increasing carbon footprint labeling activities both within and beyond the Asian regions (Shi 2013). These green certificates are seeking to internalize the cost of biofuel development and thus increase the costs of biofuels, while allowing different fuels to compete on their ability to provide low-carbon energy. The life-cycle methodology, however, if used when accounting for green certificates, will have spillover effects beyond national boundaries (i.e. leakage) and may adversely impact the narrowing development gap (Shi 2013). Such spillover effects, however, could be managed in the case of biofuels such that their supply chain is limited and the system boundary of the calculation remains limited to local inputs. For example, imported fertilizers, regardless of their source of materials, should be assigned a standard GHG emission quantity.

Such green certificates could be abused as a barrier to free trade and may jeopardize inclusive development. For example, the EU Renewable Energy Directive[4] disqualifies palm oil as a source of biodiesel unless proven to be renewable (Commission of the European Communities 2009). The default value of GHG saving potential assigned for palm oil-based biodiesel is much lower than the GHG saving potential found in other studies. This has resulted in many controversies between the EU and palm oil producers, namely Indonesia and Malaysia (Choo et al. 2011; Malik et al. 2009). Broadly, biofuels from developing countries in Asia may face discrimination due to developed countries' domination of methodology development, a lack of quality carbon inventory data, and the size disadvantage of small firms (Shi 2013). To prevent adverse and unfair consequences, various stakeholders, in particular those from the developing countries and small firms, should remain closely engaged when developing and implementing the methodologies and regimes.

## The cost of government support

Various government support policies have been implemented in Asian countries to promote the penetration of biofuels. These policies could be sorted into two categories: mandatory targets and fiscal incentives. Mandatory targets often require fuel retailers to blend biofuels at a certain level, while fiscal incentives encourage users to consume biofuels. These policy interventions are often justified by market failure theory (Rajagopal and Zilberman 2007). Government support for many biofuel promotion programs has remained costly (Bell et al. 2011). The costs of some of these policies could be accountable, such as subsidies, but the costs of some remain unaccountable, such as costs of incurred market distortions in international trading schemes (Gheewala et al. 2013a).

In the face of the Paris Agreement and the critical climate change challenges facing Asia and the world, the role of biofuels in greenhouse gas mitigation and climate change adaptation becomes increasingly important. Improved management of policy programs is important to understand the extent to which biofuels can provide real emission reductions, and the assessment tools presented here follow different methodologies to compare biofuels for sustainability purposes.

## Assessment tools

A variety of assessment tools exist to evaluate the costs and benefits of biofuels. In this section we present a few of the options available and comment on their applicability in an Asian context. The tools used include energy return on investment (EROI), life-cycle assessment (LCA), sustainability assessment, and other social, economic, and environmental frameworks. These tools can provide systems-level comparisons of biofuels to allow better-informed decisions to be made using a consistent and fair methodology. Although the different tools available can answer a variety of research questions and evaluate different outcomes, some may be limited in their scope and application. For instance, energy return on investment does not provide a direct link to environmental costs or benefits (Kittner et al. 2016). Life-cycle greenhouse gas emissions do not necessarily give the full picture of the particular biofuel's economic cost or water footprint. LCA methodology is also sensitive to boundary definition and scenario settings (Kittner et al. 2013; Prapaspongsa et al. 2017). Differences in scenario settings caused the contradictory conclusions regarding the GHG saving potential of palm oil based biodiesel (Shi 2013; Silalertruksa et al. 2012a). Advanced tools incorporating computable general equilibrium (CGE) models are useful to assess how economies may react to changing policies with regard to biofuel promotion and also can be used in conjunction with life-cycle assessments and sustainability analyses. These can be combined with life-cycle assessments or social sustainability analyses to understand larger systems effects – for instance, the effect of sugarcane cultivation on global commodities markets or promotion of biodiesel on land tenure and labor rights (Phalan 2009). One reason this has emerged as a significant research issue for biofuels, particularly, is the concern that biofuel cultivation may compete for scarce land, which could drive up global food prices or induce further deforestation or land use change to meet other agricultural needs.

Productivity, economic benefits, profitability, and job creation are often used as key indicators for assessing economic and social benefits for biofuels. Input costs are major indicators of cost while opportunity costs are often employed to compare various options.

Net energy balance and renewability remain important concepts in the evaluation of biofuel sustainability (Gheewala 2013). These concepts refer to the notion of the amount of energy output divided by the amount of energy input to create biofuels over the life-cycle of biofuel production, usually from the "cradle-to-gate" or "well-to-wheel" for transportation fuels.

### GBEP

The Global Bioenergy Partnership (GBEP) developed one assessment tool that draws from public, private, and civil society stakeholders. They have established a suite of indicators, shown in Table 24.1, that are useful to evaluate the costs and benefits of biofuels and separated them into three main groups – environmental, social, and economic.

The advantages of the GBEP sustainability indicators are that the approach taken is informed through civil society, government, and industry partnerships and that it is a more comprehensive viewpoint compared with previous frameworks that focused solely on the environment, economy, or society. One disadvantage is that there are too many indicators without hierarchy which makes it difficult for policy-makers to make a decision. Further improvements could allow for interactions between the different areas and set priorities relative to the different indicator goals.

### FAO-BEFS

The Food and Agriculture Organization of the United Nations (UN-FAO) organized an evidence-based policy framework called the Bioenergy and Food Security Analytical Framework

*Table 24.1* Global Bioenergy Partnership (GBEP) sustainability indicators for bioenergy

| *Environmental* | *Social* | *Economic* |
| --- | --- | --- |
| Life-cycle GHG emissions | Allocation and tenure of land for new bioenergy production | Productivity |
| Soil quality | Price and supply of a national food basket | Net energy balance |
| Harvest levels of wood resources | Change in income | Gross value added |
| Emissions of non-GHG air pollutants, including air toxins | Jobs in the bioenergy sector | Change in consumption of fossil fuels and traditional use of biomass |
| Water use and efficiency | Change in unpaid time spent by women and children collecting biomass | Training and requalification of the workforce |
| Water quality | Bioenergy used to expand access to modern energy services | Energy diversity |
| Biological diversity in the landscape | Change in mortality and burden of disease attributable to indoor smoke | Infrastructure and logistics for distribution of bioenergy |
| Land use and land-use change related to bioenergy feedstock production | Incidence of occupational injury, illness, and fatalities | Capacity and flexibility of use of bioenergy |

*Data source*: GBEP (2011).

(BEFS-AF) to assess the sustainability of bioenergy options. This flexible approach contains an initial "rapid appraisal" and a secondary in-depth "detailed BEFS." The idea is that a rapid appraisal can provide a preliminary indication of sustainable bioenergy potential for specific countries or regions with potential risks and benefits, economic viability, and social indicators. Second, countries looking for high spatial resolution can perform a detailed analysis. This covers a diagnostic analysis including trends in domestic agricultural markets and food security, and a natural resource assessment identifying areas suitable for bioenergy production, supply and demand of fuelwood using the Woodfuel Integrated Supply/Demand Overview Mapping (WISDOM) model, and water footprint using the Water Evaluation and Planning System (WEAP). In addition, the framework incorporates a techno-economic bioenergy production cost model and life-cycle GHG analysis. Finally, the FAO framework investigates economy-wide impacts on household incomes, national economic growth, and labor. From a social perspective, household-level survey data can assess food security and vulnerability to price changes that may occur from new bioenergy feedstock production.

The advantage of using the FAO-BEFS framework is its versatility in conducting a rapid appraisal and then moving toward a detailed assessment if necessary. Also, the FAO-BEFS framework incorporates more implications of bioenergy production on price volatility and external markets than other frameworks. This framework has been applied at the ASEAN regional level and at the country level in Thailand and the Philippines (UN FAO BEFS 2016). The FAO-BEFS tool provides a critical framework that operates at multiple levels – providing flexibility to the analysis and allowing for more in-depth analysis that emphasizes economic and social indicators.

## ERIA

In addition to the FAO–BEFS framework, the Economic Research Institute for ASEAN and East Asia (ERIA) has developed a framework to evaluate the sustainability of different biofuel options. Through a series of region-specific assessments specifically to evaluate sustainability of bioenergy, ERIA has evaluated GBEP indicators within the Asian context, determining the level of relevance for different indicators in Asia (Kudoh et al. 2015). ERIA takes a more practical than academic approach to evaluating sustainability, by prioritizing theoretically sound and implementable methodologies that are simpler by design and rapid to assess. This is a pragmatic approach and tool available to sustainability practitioners. The approach tries to classify indicators so that some indicators are weighted as more important than others. The ERIA methodology also focuses on reducing reliance on fossil fuel resources as a main component of biofuel sustainability. This can be achieved through the biofuel value chain or deployment measures that directly reduce fossil fuel consumption.

## Innovation

Innovations in advanced biofuel technologies pose new opportunities and challenges in Asia. Developing a next generation biofuel industry in Asia will require investments along the value chain and targeted policies that promote sustainable production. Government subsidies in the past have not yielded their purported benefits. From a climate perspective, growing concerns over land use and agriculture-led deforestation of carbon-intensive forest stocks pose significant barriers to innovation. With ever limited land and water resources available for biofuel feedstock production, innovation is necessary, together with government expenditure on research to ensure future biofuel sustainability and certification programs can enable Asia to lead the world in finding new solutions, especially for resource-constrained environments.

As far as patenting extends as a proxy for innovation activities, Chinese, Japanese, and Korean firms dominate global patent activity for new and advanced biofuel technologies, including the Chinese Academy of Sciences (Albers et al. 2016). Second-generation co-product or waste stream based biofuels have been slow to meet the challenge of scaling up or to meet the expectations held after the surge of innovative activities with first-generation biofuels developed in the late 2000s due to high energy prices and other policy incentives. In fact, innovation activity for first-generation bioethanol, biodiesel, and biogas continues to dwarf that of advanced biofuels, especially with the rise of China's R&D programs focused on these technologies (Albers et al. 2016). Therefore, R&D programs and economic conditions in Asia will likely drive global activity for further innovation in biofuels, in terms of not only sustainability, but technological change.

At the same time, careful use of indicators remains important to evaluate technologies. Geographical location of the palm oil production plays a large role in determining its sustainability. For instance, most oil palm in Thailand is not planted on converted forest land. However, in Indonesia and Malaysia, the case may differ. Experience promoting *Jatropha curcas* in India and Malaysia has led to inefficient outcomes. Therefore, technologies and geographies should interact in concert with one another. Appropriate biofuel feedstocks should be promoted based on soil suitability, water availability, land, and energy density (UN FAO BEFS 2016).

Palm oil, controversial as it may be due to biofuel-induced land use change and soil carbon stocks, remains much less land intensive per unit energy as a crop than feedstocks such as rapeseed. These costs and benefits should be included in new sustainability frameworks that incorporate carbon and land as prices. Palm oil in Malaysia and Indonesia has received international criticism as plantations can replace tropical forests which represent large carbon sinks. Furthermore, the large potential revenues from logging and oil palm production often outweigh the cost of forest conservation (Fisher et al. 2011). However, palm oil may have some sustainable applications, if properly managed on appropriate lands. The high energy density reduces the amount of land needed, and many oil palm plantations in Thailand could enable biodiesel production within the country that is friendlier to the environment and conservation efforts (Silalertruksa & Gheewala 2012). Further analyses incorporating energy density, land efficiency, greenhouse gas emissions, and downstream applications for alternative biofuel feedstocks could improve sustainability of biofuels and work in congruence with tropical forest conservation efforts.

Advances in biofuel technologies have improved sustainability over time based on previous life-cycle and systems analyses. Based on indicators including greenhouse gas emission reductions, EROI, and improved utilization of waste products as feedstocks, technological advances are improving viability and innovation in biofuel value chains.

From a greenhouse gas perspective, the Intergovernmental Panel on Climate Change (IPCC) highlights the important role of biofuels for climate change mitigation and the heightened role of biomass for electricity generation (IPCC 2015). Biomass systems have the potential, when combined with carbon capture and storage (CCS) applications, to create negative emissions, a phenomenon that has not yet fully reached commercialization or testing, but adds an intriguing dimension to the technical possibilities of climate change mitigation (Anderson & Peters, 2016; Sanchez & Kammen 2016; Sanchez et al. 2015).

Electricity, heat, and chemicals all represent alternative pathways for biofuel feedstocks that can simultaneously contribute to decarbonization pathways in addition to using biofuel feedstocks as traditional agro-based liquid transportation fuels. Innovations in alternative applications of biofuels could further improve sustainability and security, and create new job opportunities in emerging industries in a more "bio-based" economy.

It is possible that future biofuel production could shift from predominantly transportation-based applications to electrification projects. Advances in integrated gasification combined cycle (IGCC) turbines and CCS technologies are highlighted by the IPCC in future greenhouse gas mitigation scenarios targeting 1.5 or 2 °C of warming.

## Food–energy–water nexus

The food–energy–water nexus provides a critical framework looking forward to the future development of biofuels in Asia. Biofuels uniquely impact food systems through competition for land and as an agricultural commodity (Gheewala et al. 2011). Third-generation biofuels may begin to address the issue of competition for agricultural land, but still require large non-arable land areas. They also require substantial water resources to develop and grow, yet in turn can also alter the geomorphological features of a watershed. At the same time, the demand for energy is increasing, and alternatives to traditional fossil fuels address security issues for food, energy, and water by increasing domestic production. Under a changing climate, the food–energy–water nexus provides a unique framework to evaluate biofuels for environmental sustainability, economic viability, and social protections. Prevalent biofuel feedstocks for biodiesel like palm oil grown in Southeast Asia compete with palm oil markets in the global food market. Using tools available at the nexus to evaluate policy decisions will allow for more sustainable palm oil production, for instance as in the case of Thailand where often oil palm lands are not competing with tropical forest and have a lesser effect on the global food market than oil palm in Indonesia or Malaysia.

From a food and land perspective, biofuels traditionally have competed with food for use of agricultural lands. Recent price volatility in the food sector warrants further economic research into the effects of biofuel promotion on food prices (Mukherjee & Sovacool 2014; Serra & Zilberman 2013). At the same time, certain biofuels can be considered co-products of agricultural food production. This could provide certain co-benefits, including waste reduction, land efficiency, and higher value agricultural products for economic development.

Next generation biofuels show promise. Bacteria and fungi can use lignocellulose for advanced energy return on investments. Microalgae may also reduce land requirements for biodiesel production. Alternative applications of biofuel production including the utilization of waste residues for electricity generation could provide new uses of feedstocks. It is possible that rapid adoption of electric vehicles in the near term could reduce demand for biofuels, but increase demand for electricity generation that biomass waste could provide – along with new pathways for advanced heat and chemicals.

Increasing demands for food and water place biofuels at a pivotal juncture. Can we find technologies that use less water, have greater land use efficiencies, and compete less with agricultural food production?

Despite the decreased availability of water for agricultural purposes, there are some lessons that have been learned about the water intensity of different biofuel crops. Cassava cultivation, in particular, provides more net benefit per unit of water compared with other feedstock crops (Kaenchan & Gheewala 2017). At the same time, systems-scale analysis will help identify appropriate pathways and feedstocks for the future of bioenergy including transportation fuels and electricity. Water requirements will likely pose serious challenges, along with competition for food and land (Gheewala et al. 2014). Therefore, food–energy–water nexus based approaches are becoming increasingly valuable to evaluate trade-offs and opportunities to improve the sustainability of biofuels.

On the issue of energy, food, and water security, future research may lead to clearer distinctions between the pursuit of different end uses for biofuels, especially when comparing bioethanol and biodiesel. Globally, most bioethanol is produced with domestic crops grown in a country, which is true for Asia. However, Southeast Asia disproportionately provides crops for a significant portion of biodiesel, and more than a third of global biodiesel consumption is supplied by globally traded feedstocks (Rulli et al. 2016). This provides an opportunity for further research and also opens the possibility to pursue second-generation and advanced generation biofuels that alleviate such concerns over competition for agricultural land and water resources.

Other articles have called for the inclusion of justice in the food–energy–water nexus framing of trade-offs and identifying where and who benefits from such policies. By identifying winners and losers in the biofuel sector through a justice lens, perhaps more sustainable development and a greener economy can be achieved (Middleton et al. 2015). As the competition between food and fuel especially resonates with more vulnerable sections of society, incorporating environmental justice into the food–energy–water nexus framework would identify opportunities to alleviate inequities.

A fossil fuel dependent energy sector necessitates a transition to renewable energy in the age of climate change. Biofuels can contribute to this transition.

## What have been the experiences of sustainability of biofuels in Asia?

The sustainability of biofuels in Asia shows a lot of promise for the future, especially as experiences in theory and practice are shared with the rest of the world. In Asia particularly, a number of cases stand out where perceived benefits of biofuel promotion failed to deliver. These policies created environmental, economic, and societal costs. Learning from such past experiences can help inform future policy-makers and stakeholders seeking to create a more sustainable biofuel industry in Southeast Asia. The case of palm oil is an example. A promising biofuel, it has been the subject of poor land management policies that led to severe tropical deforestation and carbon loss, leading to a negative image. However, as a biofuel, its high energy density means that it could provide potential greenhouse gas benefits with proper land management and when grown in a non-tropical forest or peatland setting.

We outline the challenges and issues along with a set of strategies to further improve biofuels from a sustainability perspective using theory and practice in Asia.

Employment generation still represents a key challenge and opportunity for improvement in the production of biofuels across Asia. Promises of increased jobs and employment opportunities have failed to deliver in practice. Opportunities for enhancing livelihoods through the use of waste feedstocks such as bagasse for electricity production could create new jobs (Wei et al. 2010). Eco-industrial-farming parks that can refine sugar on-site while using heat and power from sugar wastes in a more sustainable fashion may offer new solutions. Although biofuels may not generate the purported employment benefits to the extent policy-makers had previously touted, moving forward a focus on improving livelihoods, generating incomes for farmers from co-production of biofuel feedstocks, and offering extra revenue generation could provide indirect economic benefits even if it does not address labor shortages and generation.

## Looking forward

Future energy projections suggest that non-OECD countries will play a critical role in future emission reduction targets, paving a clear route for Asia to pay more attention to biofuels, their perceived benefits, and potential pitfalls in terms of greenhouse gas emissions. Therefore, sustainability assessments specifically targeted in Asia become prominent tools for decision-makers to help guide future policies and targets for biofuel production and applications.

Looking forward we should emphasize land use change as a key determinant of sustainability including issues of greenhouse gas emissions, water resources, and agricultural vulnerability. Job creation remains a critical feature. However, employment benefits have yet to be realized to the extent policies have promised local jobs and enhanced livelihoods. New policies must address the social and economic landscape surrounding biofuels development if biofuels are to provide benefits to society as idealized.

In the future, biofuel production may become more mechanized due to labor shortages, harming employment opportunities. Previous research has observed this phenomenon specifically in the case of sugarcane farming. Increased mechanization could reduce the costs of biofuels from an economic perspective, at the consequence of increased environmental impacts and reduced employment (Pongpat et al. 2017). However, another lingering issue remains the lack of human capital available as labor. Increasing yields, appropriately managing fertilizer and agrochemicals, and improved zoning for agricultural crops could remedy some of the negative societal trade-offs of biofuel production and allow for sugarcane-based biofuels to improve their sustainability, given their lower water and energy footprints (Prasara-A & Gheewala 2016; Silalertruksa et al. 2015). Looking forward, we may need to evaluate opportunities to improve employment benefits from sugarcane programs and investigate other feedstocks that could provide jobs and transition to a low-carbon economy.

Appropriate trade regimes with suitable policy instruments are necessary to make sure the increased trading of biofuels will be supplied using sustainable methods. Such institutional arrangements, however, should not be used as barriers to trade. Close engagement with various stakeholders in developing these institutions would be useful to secure sustainability and inclusive growth.

The large potential availability of biomass feedstock across Asia highlights the importance of evaluating the costs and benefits of biofuels and identifying pathways to improve production and deployment. This chapter has discussed some of the assessment tools in practice and the theory behind how biofuel promotion can simultaneously improve and contribute to energy security goals, greenhouse gas mitigation in the face of the Paris Agreement, and employment generation.

## Notes

1 www.rspo.org/about.
2 Http://wwf.panda.org/what_we_do/footprint/agriculture/palm_oil/solutions/responsible_purchasing/palm_oil_buyers_scorecards/palm_oil_buyers_scorecard_2016/.
3 Www.theonlinecitizen.com/2016/09/28/zero-palm-oil-green-label-certifications/.
4 The European Renewable Energy Directive allocates a GHG saving of 19% by default and 36% maximum. It may discriminate palm oil in two ways: first, set an arbitrary standard at 35%; second, allocate a low default value for palm oil (19%).

## References

ADB, 2016. Greater Mekong Subregion: Overview. www.adb.org/countries/gms/overview.
Albers, S.C., Berklund, A.M. and Graff, G.D., 2016. The rise and fall of innovation in biofuels. *Nature, 201*, p.6.
Anderson, K. and Peters, G., 2016. The trouble with negative emissions. *Science, 354*(6309), pp.182–183.
Bell, D.R., Silalertruksa, T., Gheewala, S.H. and Kamens, R., 2011. The net cost of biofuels in Thailand: an economic analysis. *Energy Policy, 39*(2), pp.834–843.
Choo, Y.M., Muhamad, H., Hashim, Z., Subramaniam, V., Puah, C.W. and Tan, Y., 2011. Determination of GHG contributions by subsystems in the oil palm supply chain using the LCA approach. *International Journal of Life Cycle Assessment, 16*(7), pp.669–681.
Commission of the European Communities, 2009. Directive 2009/28/EC of the European Parliament and of the Council of 23 April 2009 on the promotion of the use of energy from renewable sources and amending and subsequently repealing Directives 2001/77/EC and 2003/30/EC, *Official Journal of the European Union L, 140*, pp.16–62.
Danielsen, F., Beukema, H., Burgess, N.D., Parish, F., Bruehl, C.A., Donald, P.F., Murdiyarso, D., Phalan, B.E.N., Reijnders, L., Struebig, M. and Fitzherbert, E.B., 2009. Biofuel plantations on forested lands: double jeopardy for biodiversity and climate. *Conservation Biology, 23*(2), pp.348–358.
Doshi, A., Pascoe, S., Coglan, L. and Rainey, T.J., 2016. Economic and policy issues in the production of algae-based biofuels: a review. *Renewable and Sustainable Energy Reviews, 64*, pp.329–337. doi:10.1016/j.rser.2016.06.027.
Fargione, J., Hill, J., Tilman, D., Polasky, S. and Hawthorne, P., 2008. Land clearing and the biofuel carbon debt. *Science, 319*(5867), pp.1235–1238.
Farrell, A.E., Plevin, R.J., Turner, B.T., Jones, A.D., O'Hare, M. and Kammen, D.M., 2006. Ethanol can contribute to energy and environmental goals. *Science, 311*(5760), pp.506–508.
Fingerman, K.R., Torn, M.S., O'Hare, M.H. and Kammen, D.M., 2010. Accounting for the water impacts of ethanol production. *Environmental Research Letters, 5*(1), 014020.
Fisher, B., Edwards, D.P., Giam, X. and Wilcove, D.S., 2011. The high costs of conserving Southeast Asia's lowland rainforests. *Frontiers in Ecology and the Environment, 9*(6), pp.329–334.
Gasparatos, A., Stromberg, P. and Takeuchi, K., 2013. Sustainability impacts of first-generation biofuels. *Animal Frontiers, 3*(2), pp.12–26.
GBEP, 2011. The Global Bioenergy Partnership Sustainability Indicators for Bioenergy. The Global Bioenergy Partnership (GBEP). Food and Agricultural Organization of the United Nations (FAO) – Climate, Energy and Tenure Division, Viale delle Terme di Caracalla, Rome, Italy.
Gheewala, S.H., 2013. Environmental sustainability assessment of ethanol from cassava and sugarcane molasses in a life cycle perspective, In: Singh, A., Olsen, S.L. and Pant, D. (eds), *Life Cycle Assessment of Renewable Energy Sources*, Springer, London.
Gheewala, S.H., Berndes, G. and Jewitt, G., 2011. The bioenergy and water nexus. *Biofuels, Bioproducts and Biorefining, 5*(4), pp.353–360.

Gheewala, S.H., Damen, B. and Shi, X., 2013a. Biofuels: economic, environmental and social benefits and costs for developing countries in Asia. *Wiley Interdisciplinary Reviews: Climate Change, 4*(6), pp.497–511.

Gheewala, S.H., Silalertruksa, T., Nilsalab, P., Mungkung, R., Perret, S.R. and Chaiyawannakarn, N., 2013b. Implications of the biofuels policy mandate in Thailand on water: the case of bioethanol. *Bioresource Technology, 150*, pp.457–465.

Gheewala, S.H., Silalertruksa, T., Nilsalab, P., Mungkung, R., Perret, S.R. and Chaiyawannakarn, N., 2014. Water footprint and impact of water consumption for food, feed, fuel crops production in Thailand. *Water, 6*, pp.1698–1718.

IPCC, 2015. *Fifth Assessment Report (AR5)*, Cambridge University Press, Cambridge.

Kaenchan, P. and Gheewala, S.H., 2017. Cost–benefit of water resource use in biofuel feedstock production. *Journal of Cleaner Production, 142*(3), pp.1192–1199.

Kittner, N., Gheewala, S.H. and Kamens, R.M., 2013. An environmental life cycle comparison of single-crystalline and amorphous-silicon thin-film photovoltaic systems in Thailand. *Energy for Sustainable Development, 17*(6), pp.605–614.

Kittner, N., Gheewala, S.H. and Kammen, D.M., 2016. Energy return on investment (EROI) of mini-hydro and solar PV systems designed for a mini-grid. *Renewable Energy, 99*, pp.410–419.

Kudoh, Y., Sagisaka, M., Chen, S.S., Elauria, J.C., Gheewala, S.H., Hasanudin, U., Romero, J., Sharma, V.K. and Shi, X., 2015. A region-specific methodology for assessing the sustainability of biomass utilisation in East Asia. *Sustainability, 7*, pp.16237–16259.

Kumar, S., Shrestha, P. and Salam, P.A., 2013. A review of biofuel policies in the major biofuel producing countries of ASEAN: production, targets, policy drivers and impacts. *Renewable and Sustainable Energy Reviews, 26*, pp.822–836.

Larson, C., 2016. Mekong megadrought erodes food security. *Science.* doi:10.1126/science.aaf9880.

Malik, U.S., Ahmed, M., Sombilla, M.A. and Cueno, S.L., 2009. Biofuels production for smallholder producers in the Greater Mekong Sub-region. *Applied Energy, 86*, pp.S58–S68.

Middleton, C., Allouche, J., Gyawali, D. and Allen, S., 2015. The rise and implications of the water–energy–food nexus in Southeast Asia through an environmental justice lens. *Water Alternatives, 8*(1), pp.627–654.

Mukherjee, I. and Sovacool, B.K., 2014. Palm oil-based biofuels and sustainability in Southeast Asia: a review of Indonesia, Malaysia, and Thailand. *Renewable and Sustainable Energy Reviews, 37*, pp.1–12.

Nguyen, T.L.T., Gheewala, S.H. and Garivait, S., 2007. Energy balance and GHG-abatement cost of cassava utilization for fuel ethanol in Thailand. *Energy Policy, 35*(9), pp.4585–4596.

Nguyen, T.L.T., Gheewala, S.H. and Sagisaka, M., 2010. Greenhouse gas savings potential of sugar cane bio-energy systems. *Journal of Cleaner Production, 18*(5), pp.412–418.

Phalan, B., 2009. The social and environmental impacts of biofuels in Asia: an overview. *Applied Energy, 86*, pp.S21–S29.

Pongpat, P., Silalertruksa, T. and Gheewala, S.H., 2017. An assessment of harvesting practices of sugarcane in the central region of Thailand. *Journal of Cleaner Production, 142*(3), pp.1138–1147.

Prapaspongsa, T., Musikavong, C. and Gheewala, S.H., 2017. Life cycle assessment of palm biodiesel production in Thailand: impacts from modelling choices, co-product utilisation, improvement technologies, and land use change. *Journal of Cleaner Production, 153*, pp.435–447. doi:10.1016/j.jclepro.2017.03.130.

Prasara, A.-J. and Gheewala, S.H., 2016. Sustainability of sugarcane cultivation: case study of selected sites in north-eastern Thailand. *Journal of Cleaner Production, 134*, pp.613–622.

Rajagopal, D., 2008. Implications of India's biofuel policies for food, water and the poor. *Water Policy, 10*(S1), pp.95–106.

Rajagopal, D. and Zilberman, D., 2007. *Review of Environmental, Economic and Policy Aspects of Biofuels.* Policy Research Working Paper. The World Bank, Washington, DC.

Rajagopal, D., Sexton, S.E., Roland-Holst, D. and Zilberman, D., 2007. Challenge of biofuel: filling the tank without emptying the stomach? *Environmental Research Letters, 2*(4), 044004.

Rulli, M.C., Bellomi, D., Cazzoli, A., De Carolis, G. and D'Odorico, P., 2016. The water–land–food nexus of first-generation biofuels. *Scientific Reports, 6*, 22521.

Sanchez, D.L. and Kammen, D.M., 2016. A commercialization strategy for carbon-negative energy. *Nature Energy, 1*, p.15002.

Sanchez, D.L., Nelson, J.H., Johnston, J., Mileva, A. and Kammen, D.M., 2015. Biomass enables the transition to a carbon-negative power system across western North America. *Nature Climate Change, 5*(3), pp.230–234.

Searchinger, T., Heimlich, R., Houghton, R.A., Dong, F., Elobeid, A., Fabiosa, J., Tokgoz, S., Hayes, D. and Yu, T.H., 2008. Use of US croplands for biofuels increases greenhouse gases through emissions from land-use change. *Science*, *319*(5867), pp.1238–1240.

Serra, T. and Zilberman, D., 2013. Biofuel-related price transmission literature: a review. *Energy Economics*, *37*, pp.141–151.

Shi, X., 2013. Spillover effects of carbon footprint labelling on less developed countries: the example of the East Asia summit region. *Development Policy Review*, *31*(3), pp.239–254.

Shi, X., 2016. The future of ASEAN energy mix: a SWOT analysis. *Renewable and Sustainable Energy Reviews*, *53*, pp.672–680.

Silalertruksa, T. and Gheewala, S.H., 2012. Food, fuel, and climate change. *Journal of Industrial Ecology*, *16*(4), pp.541–551.

Silalertruksa, T., Bonnet, S. and Gheewala, S.H., 2012a. Life cycle costing and externalities of palm oil biodiesel in Thailand. *Journal of Cleaner Production*, *28*, pp.225–232.

Silalertruksa, T., Gheewala, S.H., Hünecke, K. and Fritsche, U.R., 2012b. Biofuels and employment effects: implications for socio-economic development in Thailand. *Biomass and Bioenergy*, *46*, pp.409–418.

Silalertruksa, T., Gheewala, S.H. and Pongpat, P., 2015. Sustainability assessment of sugarcane biorefinery and molasses ethanol production in Thailand using eco-efficiency indicator. *Applied Energy*, *160*, pp.603–609.

Tilman, D., Socolow, R., Foley, J.A., Hill, J., Larson, E., Lynd, L., Pacala, S., Reilly, J., Searchinger, T., Somerville, C. and Williams, R., 2009. Beneficial biofuels: the food, energy, and environment tri-lemma. *Science*, *325*(5938), pp.270–271.

Tongsopit, S., Kittner, N., Chang, Y., Aksornkij, A. and Wangjiraniran, W., 2016. Energy security in ASEAN: a quantitative approach for sustainable energy policy. *Energy Policy*, *90*, pp.60–72.

UN FAO, 2016. The Bioenergy and Food Security (BEFS) Approach. http://www.fao.org/energy/bioenergy/bioenergy-and-food-security/en/

Venkataraman, C., Habib, G., Eiguren-Fernandez, A., Miguel, A.H. and Friedlander, S.K., 2005. Residential biofuels in South Asia: carbonaceous aerosol emissions and climate impacts. *Science*, *307*(5714), pp.1454–1456.

Wei, M., Patadia, S. and Kammen, D.M., 2010. Putting renewables and energy efficiency to work: how many jobs can the clean energy industry generate in the US? *Energy Policy*, *38*(2), pp.919–931.

Woolf, D., Amonette, J.E., Street-Perrott, F.A., Lehmann, J. and Joseph, S., 2010. Sustainable biochar to mitigate global climate change. *Nature Communications*, *1*, p.56.

Zhou, A. and Thomson, E., 2009. The development of biofuels in Asia. *Applied Energy*, *86*, pp.S11–S20.

# 25

# FINANCING ENERGY ACCESS IN ASIA

*Binu Parthan*

## Background

Asia is home to a large number of people (543 million) without access to electricity as well as 1.9 billion (International Energy Agency, 2015a) who depend on solid fuel for thermal energy needs. The levels of electricity access also vary across Asia, from 100% electricity access in Central and East Asia to low-levels of electricity access (39%) in South-East Asia (International Energy Agency (IEA) and the World Bank, 2015). Similarly access to non-solid fuels also varies across Asia, with high levels of access of 99% and 95% in East and West Asia respectively to low levels of 36% in South Asia (International Energy Agency (IEA) and the World Bank 2015). The 2030 Agenda for Sustainable Development and the 17 Sustainable Development Goals (SDGs) in 2015, and in particular the SDG 7 on affordable and clean energy (UN 2015), present an opportunity for governments and stakeholders in Asian countries and development agencies to address this major challenge by 2030. However, significant levels of resources will be needed to address the energy access challenges by that date.

## The financing challenge

The estimates for the financial resources required for energy access vary significantly, from $12 billion to $279 billion per year, primarily owing to the varying assumptions about the levels of energy access, costs and technologies in the models used. The Sustainable Energy for All (SE4All), estimates that resources to the tune of $49.4 billion will be required on an annual basis to meet universal energy access, of which $40.3 billion will be incremental investments (SE4All, 2015). The focus of these investments in Asia will need to be in South Asia where energy access challenges are particularly severe. While a large share of unelectrified parts of Asian countries can be electrified through expansion of existing electrical networks, a number of alternative approaches involving mini-grids, off-grid systems, etc. will also be needed. To provide non-solid fuel access to more than 1.9 billion people in Asia, investments in transport and supply infrastructure for liquefied petroleum gas (LPG), kerosene, biogas and liquid biofuels as well as electric cooking will be needed. Efficient biomass stoves will also help to reduce the use of solid fuels during this transition.

Assuming the costs of energy access to be uniform globally, Asia will require total investments of $394 billion by 2030 to provide universal electricity and non-solid fuel based thermal energy access, of which the major share of about \$343 billion will be required for providing electricity access. The reminder of about \$51 billion will be required to provide non-solid fuel access for thermal energy. However, considering the current baseline investment levels, investments in non-solid fuel access will need to increase by a factor of 44 compared with a factor of 5 for electricity access (SE4All, 2015) to ensure universal access in Asia by 2030. Government resources augmented by borrowings, carbon finance and development assistance may not be able to meet such high levels of investment and there is a need for private finance to play a larger share in the investments. For private finance to play a role in financing energy access in Asian countries, there needs to be strong financing institutions and an ecosystem that can develop and implement energy access projects. There is also a need for national finance architecture consisting of national development banks, commercial banks and capital markets as well as structures to de-risk the required level of private investments. Policy and regulatory frameworks in the financial and energy sector that are progressive and encourage private sector participation and investment are also important. In countries where financial resources from local public institutions and the private sector are inadequate there may be a need to attract external development assistance or external borrowings. This could pose a challenge for a number of countries in Asia that are not considered investment grade and is particularly unfavourable to Least Developing Countries (LDCs) and Small Island Developing States (SIDS). Generally many Asian LDCs and SIDS tend to have less developed project development and implementation ecosystems, and limited national finance architecture and institutions, and energy and financial policy and regulatory frameworks. Therefore a particular challenge in many Asian LDCs and SIDS would be development of robust pipeline of projects and to syndicate local and external financial resources to support universal energy access.

## Institutional challenges

Investments in expansion of existing electrical grid networks in Asia will likely involve governments and energy utilities playing a central role. While providing electricity access to meet basic energy requirements of lighting may only require relatively low levels of resources, providing electricity access for lighting, entertainment, communication and other appliances would require much higher levels. A critical factor in the financing of grid-based electrification in Asia would be the credit worthiness of the electric utilities, many of which in Asia are vertically integrated and financially and operationally inefficient. Many utilities operate like government departments and often are loss-making or have a low return on their investments owing to low levels of collection of bills and high levels of losses. The high levels of losses for utilities are due to a combination of technical losses resulting from poor quality equipment and unchecked pilferage and theft of electricity at the distribution network level. Another reason for energy utilities in Asia making operational losses is that the electricity tariffs do not reflect the cost of supply and investments. In order to ensure that the electricity tariffs are cost reflective, electric utilities are financially sustainable and the interests of all other stakeholders are balanced, there also needs to be an independent electricity regulator. Building adequate capacity of government and regulatory agencies in Asia is important to ensure that policy and regulatory frameworks secure the financial viability of electricity utilities, thereby enhancing the credit worthiness of the utilities and associated stakeholders. Therefore there is a need for improvements in governance, regulation and management of the electricity sector in Asian developing countries to increase the level of financing available to achieve universal energy access by 2030.

Also important is the role of private entities in the electricity sector. In Asia, the role of the private sector in the electricity sector has been generally limited to electricity generation by Independent Power Producers (IPPs) or the manufacture and supply of equipment. The role of the private sector has been rather limited in the transmission and distribution verticals in the electricity sector. Where the private sector has a role in distribution, there may be an effort to make distribution infrastructure investments in areas with a high concentration of consumers, where the return on investments is likely to be better. This is likely to be a challenge for electrifying peri-urban and rural locations in Asia where the concentration of households and commercial or industrial consumers is likely to be less and electricity demand levels rather low, leading to low return on investment prospects. Therefore there is a role for government to provide incentive frameworks such as performance-based payments to encourage private sector electricity distribution companies to invest in low-voltage electricity distribution infrastructure and provide energy access to customer segments for which there is no clear business case.

In the off-grid energy sector, electrification options in Asia have included household-level electrification systems such as Solar Home Systems (SHSs) and gasoline-based generator systems as well as mini-grid systems powered by diesel, hydropower, biomass, solar, etc. Household-level systems may be purchased outright by the unelectrified households, while some systems are offered through finance, particularly for SHS. A range of ownership models exist for mini-grids such as energy utility, government agencies, community, private sector, NGOs, etc. Increasingly a number of Pico Solar Products (PSP), which are relatively inexpensive to purchase, are portable and can provide lighting and mobile phone charging, are also gaining popularity at the household level. Some PSPs and SHSs have also been sold using mobile money payment systems and sometimes on a Pay-As-You-Go (PAYG) mode, whereby the households pay regularly over a period of time to acquire the systems. While the contribution of the mini-grid-based systems and household electrification systems has generally been secondary to grid-based electrification, this area is seeing considerable interest and higher levels of financing of late. A number of innovative models of finance and product/service delivery arrangements are being implemented in Asian countries by the private sector, supported by the banking and finance sector and development agencies.

The provision of universal electricity access in Asia is likely to be a challenge in terms of resource mobilisation compared to larger economies and middle-income countries where the business models for providing centralised, decentralised and off-grid are relatively well developed. The deployment of the resources required for non-solid fuel access is likely to be a challenge in these middle-income countries in terms of financing and business models to deliver sustainable thermal energy services on the ground. The development of a pipeline and project implementation as well as mobilisation of local and external resources for both electricity and non-solid fuel access is likely to be particularly difficult in many Asian LDCs and SIDS. Availability of finance at an unprecedented scale, the business and implementation structures to deploy the financing to provide electricity and non-solid fuel based thermal energy access, and a conducive policy and regulatory environment will all be required to achieve universal energy access in Asia Pacific by 2030 and to achieve the universal energy access target of SDG 7.

## Existing financing mechanisms

Due to the high initial investments and low returns associated with energy access investments, Asian governments either have directly financed energy access efforts through public finance or have used regulatory instruments and incentive frameworks to encourage energy access investments by energy utilities. International development agencies such as the World Bank, Asian Development Bank (ADB), United Nations (UN) agencies, non-governmental organis- ations (NGOs), philanthropic and impact financing have all contributed in varying degrees to increasing energy access in Asia. However, the size of the problem and the scale of resources required of about $394 billion by 2030 will require more financial resources than are available from the combination of Asian governments, borrowings, carbon finance and development finance. These traditional sources of finance would need to be augmented by private sources of finance to provide universal electricity access and thermal energy from non-solid fuel sources.

The size and sophistication of the local financial sector and the domestic credit market are at a higher level in Asia than in other regions with an energy access deficit. For instance, Asia has a large local currency denominated bond market, which reached US$8.6 trillion in the second quarter of 2015 (ADB, 2015). Several energy utilities and renewable energy companies are also using local currency bonds to raise financial resources and in 2014 the total bond issuance by renewable energy companies was $18.3 billion. This local currency bond volume is comparable to the $20.2 billion incremental investments required annually in Asia for universal energy access. Issuance of 'green bonds' is also gaining momentum in Asia, and several Chinese and Indian financial institutions and banks[1] were actively issuing green bonds in 2015. The green bonds are a segment of the larger bond market and many more bonds are issued in Asia that could be classified as 'green' but are not actually so labelled. The early green bond issuances in Asia seem to be targeting renewable energy rather than energy access, although the market is at a nascent stage. The existence of a strong local financial sector and local currency bond markets, augmented by international development and carbon/climate finance, has the potential to provide the required scale of resources for sustainable energy for all by 2030 in Asia. However, achieving this unprecedented level of resource mobilisation for energy access is expected to be a major challenge.

Another important aspect to consider is the current government expenditure on energy subsidies, which primarily support consumption and production of fossil fuels for energy and transport. It is estimated that in 2014 the total annual global expenditure on fossil fuel subsidies was $510 million, of which Asia's share was $120 billion (International Energy Agency, 2015b). These fossil fuel subsidies are also not considered an effective public policy as only 8% of them reach the poorest 20% of the population in these countries (International Energy Agency, 2011). Considering these high levels of inefficient subsidies on fossil fuels in the context of resources for electricity and non-solid fuel access, a relatively small share of 20% of the current subsidies in Asia or more than $24 billion is needed annually in Asia for achieving universal energy access. Therefore reallocation of a small share of public finance within the energy sector in Asia should be able to provide the required resources for electricity and non-solid fuel access. However, at the country level the expenditure on fossil fuel energy subsidies may not always correspond to investment requirements for energy access, as some Asian countries spend more on fossil fuel subsidies than others. Asian countries that spend considerable resources on fossil fuel subsidies include China, India, Japan, Iran, Saudi Arabia, South Korea, Indonesia, Malaysia and Thailand (International Monetary Fund, 2015).

Assuming that the resources may indeed be available through public finance or from the markets, efforts need to be made at the country level to deploy these resources to energy access programmes and projects in electricity and thermal energy. Here countries' circumstances at the national and sub-national levels vary and not all countries in Asia will be able to attract the required level of resources. Many Asian countries are considered investment grade and many others are not attractive for investments. Levels of financial market development and business sophistication[2] vary among Asian countries, as well as the ease of doing business[3] and levels of transparency,[4] which are important factors in countries being able to attract investments. Many countries that have relatively low levels of energy access will need to establish enabling investment environments to attract financing for expanding energy access. In addition to the enabling environment, energy access programmes and projects will need to be conceived and developed and enterprises will need to be strengthened, re-oriented or even established to develop electrification and non-solid fuel based thermal energy access solutions. There is a clear role for innovation in financing structures and business models in this effort.

A number of examples of financing mechanisms for energy access exist in Asia. Some key examples of rural electrification and non-solid fuel access in the region are offered in the following sections.

## Financing household electrification in Bangladesh

The Rural Electrification and Renewable Energy Development Project in Bangladesh has been implemented from 2002 with World Bank support. A national development bank – Infrastructure Development Company Ltd (IDCOL) – received large-scale credit lines from the World Bank to support SHS-based solar electrification in Bangladesh. IDCOL empanelled Partner Organisations (POs) which channelled financing as well as delivering the SHS to end-users. The financing arrangements involved soft loans to be provided to POs with an interest subsidy allowing for lending below market rates. A capital subsidy component was also included in the programme to buy down the cost of the SHS to make it more affordable compared with the traditional kerosene-based lighting alternatives. The SHS programme in Bangladesh benefited significantly from the micro-finance ecosystem existing in the country spearheaded by Grameen Bank. In the Bangladesh programme, POs such as Micro Finance Institutions (MFIs) and NGOs offered both technical and financial services, with the SHS suppliers limiting their role to equipment supplies. Today the Bangladesh SHS programme is probably the largest such programme globally, with 3.83 million SHSs installed (Ruhul-Quddus, 2015), with an annual average of 780,000 installations and a target is to reach a total of 6 million SHSs by 2018 (Ruhul-Quddus, 2015). Subsequently, the Bangladesh SHS programme also received additional support from other governments such as those of Germany and the United Kingdom.

The Bangladesh SHS programme has benefited from the learning and refinement of the service and finance model from earlier Indian and Sri Lankan programmes which were implemented by the World Bank in the 1990s, and it was able to leverage the strong micro-finance ecosystem that exists in Bangladesh to accelerate the development of SHS markets. Through concessional funding from the World Bank and other donors, IDCOL was able to provide loans to the MFIs at 6–9%/year interest rates against the prevailing rate of 16%, and offer a loan tenure of 5–7 years against the prevailing maximum of 3 years (Parthan, 2015a). The current model for SHS financing in Bangladesh is shown in Figure 25.1.

The example of Bangladesh shows how an LDC in Asia can use international development assistance to leverage existing national strengths and local finance and implement the world's largest SHS programme. The Bangladesh SHS programme design also incorporated features and best practices from previous World Bank programmes in South Asia while customising the programme to the local environment and institutional structures.

## Subsidy reforms and financing transition to LPG in Indonesia

Indonesia in 2004 decided to reform its fossil fuel subsidies and promote the use of liquefied petroleum gas (LPG) as a cooking fuel. Although 48 of 52 million households in Indonesia were

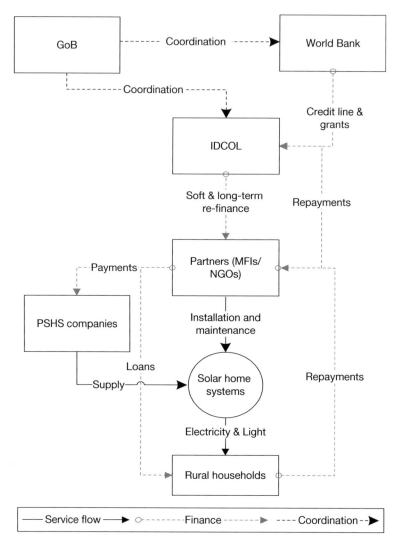

*Figure 25.1* Financing model for household solar electrification in Bangladesh

*Data source*: Author.

using kerosene for cooking, the subsidised kerosene was being diverted for blending with other fuels in industrial and commercial enterprises and was also being smuggled outside the country (PT Pertamina and World LPG Association (WLPGA), 2012). The high international prices and the increased levels of demand for subsidised kerosene resulted in fossil fuel subsidies accounting for 18% of Indonesian public expenditure in 2005, with 57% of the subsidies being for kerosene (PT Pertamina and World LPG Association (WLPGA), 2012). The government considered LPG as an alternative and 0.39 kg of LPG was considered to be equivalent to 1 litre of kerosene (PT Pertamina and World LPG Association (WLPGA), 2012) based on local cooking practices and as efficiency of an average LPG stove was 10% higher than that of a kerosene stove (International Energy Agency, 2015b). There were also advantages to using LPG over kerosene owing to reduced indoor air pollution and less time used for cooking. However, the initial cost of switching to LPG was higher due to the higher cost of LPG stoves and the LPG cylinders compared with the cost of kerosene-based cooking.

The programme was launched in 2007 and households were provided with an initial package consisting of a 3 kg cylinder filled with LPG, an LPG gas stove with one burner and accessories[5] free of cost subsidised by the government. Pertamina invested in the supply infrastructure of LPG terminals, depots, filling stations, etc. and made available adequate quantities of LPG through the supply channel. The private enterprises which were previously retailers and distributers of kerosene were converted to LPG retailers and distributers. The ministries of industry and enterprises were involved in creating the supply infrastructure for stoves, regulators and cylinders and for distribution of the initial packages. The ministry of women's empowerment launched a major sensitisation and social mobilisation campaign in the targeted areas. Once the distribution network was in place and the start-up package was distributed, the vast majority of users switched permanently from kerosene to the use of LPG for cooking. The geographical coverage of the programme was gradually increased in four phases (PT Pertamina, 2015). The finance flow for the LPG conversion programme is shown in Figure 25.2.

The financing for the establishment of the LPG supply infrastructure came from the Indonesian government and the government also financed the initial package, thereby offsetting the cost of transition. The distributers who were previously distributing kerosene were incentivised to convert to LPG distribution through financial support from government. Once the households converted to LPG, the LPG refills were then financed by households without the need for additional government support.

The results have been impressive. Currently the programme is probably the world's largest clean energy transition programme, and has converted 57 million households and micro-enterprises (PT Pertamina, 2015) from kerosene to LPG. The consumption of LPG increased six-fold from about 1 million kt in 2007 to 6 million kt (PT Pertamina, 2015). The consumption of kerosene also decreased from 8 million kl to less than 1 million kl. The Indonesian government has also made considerable savings in subsidies which in 2010 amounted to $6.9 billion, a small part of which was used to finance the conversion costs. The investment in the supply infrastructure has also resulted in an additional 38,000 jobs created in an LPG infrastructure consisting of refrigerated terminals, depots and filling stations (PT Pertamina and World LPG Association (WLPGA), 2012). More employment was also created in manufacturing facilities that produced LPG cylinders, LPG stoves and regulators. Indonesia also transitioned from a net importer of kerosene to a net exporter. The LPG conversion programme is an example of long-term consistent policy and programme implementation by a government backed by investments, leading to benefits to the users and to the economy.

### Financing electrification in China, South Korea and India

China is one of the few developing countries in Asia which has achieved 100% or universal electricity access. This has been achieved through a government-driven initiative which has lasted for more than five decades since the early 1960s. Energy consumption levels in China, especially in rural areas, were traditionally very low and in the 1950s electricity was often not available in rural areas, where consumption in 1957 was only 0.66% of national electricity consumption (Bhattacharya and Ohiare, 2012). From the late 1950s till the early 1970s, China promoted large amounts of hydropower-based electrification in rural areas, aimed at providing electricity to agriculture and mining applications as well as rural communities. These initiatives resulted in rural electricity access increasing to 66% in 1978 (Bhattacharya and Ohiare, 2012). The

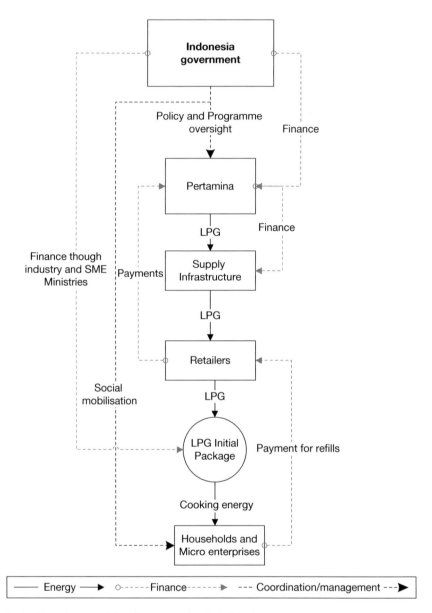

*Figure 25.2* Financing model for LPG conversion in Indonesia

*Data source*: Author.

economic reforms by the Chinese government subsequently resulted in increased levels of economic activities in the country which also increased the demand for electricity. This period saw a continued emphasis on hydropower and bioenergy as well as on agricultural development. The financing of electrification was also changed from a top-down approach from the federal government financed electricity infrastructure to a partnership between local governments and the federal government. This resulted in an acceleration of the electrification efforts and by late 1980s 78% of the rural population had electricity access. One aspect of this phase of electrification in China was the emphasis on Township and Village Enterprises (TVE) along with household electrification placing a clear emphasis on productive uses of energy.

However, due to inefficient and obsolete infrastructure, rural areas of China continued to be plagued by high levels of transmission and distribution losses and higher electricity tariffs. Therefore from the late 1990s major reforms were carried out by China in the electricity sector backed up by high levels of investments in electricity networks, especially in rural areas. China also launched, in 1996, the brightness programme using primarily solar and wind energy to provide electricity to unelectrified parts of China. The brightness programme was followed by the township electrification programme in 2002, aimed at providing electricity to 1,013 unelectrified townships in the western provinces using hydro, solar and wind electricity (Bhattacharya and Ohiare, 2012). The township electrification programme succeeded in providing electricity access to 840,000 people (Bhattacharya and Ohiare, 2012) by 2005. The township electrification programme was followed in 2005 by the village electrification programme which was aimed at electrifying the remaining 10,000 villages through renewable energy. As a result of a consistent policy direction by the Chinese government, targeted programmes and significant levels of resource allocation, China achieved universal electricity access in 2015. On 25 December 25 2015, the National Energy Administration (NEA) announced that the remaining 39,800 people in Qinghai province had been provided with energy access (Xinhua, 2015). The electrification efforts also benefited from the decentralised development structure in China, with the provinces having their own development frameworks and resources, as well as the industrialisation and growth that China witnessed as a result of the market reforms over the past three decades. In terms of the number of people with access to electricity, Chinese rural electrification has also been the largest in the world.

Similarly the experience of South Korea, which increased its rural electrification levels from 12% in 1965 to 98% in 1979 (van Gevelt, 2015), is another successful example of an Asian country achieving universal electricity access. The South Korean electrification efforts were part of the *Saemaul Undong* or the New Village movement which was a bottom-up rural development initiative that was enabled by the South Korean government through the cooperation, leadership and contribution of the villagers. The rural electrification efforts were implemented by Korea Electric Power Corporation (KEPCO) and financed by three sources: international development credit, internal resources from KEPCO, and end-user financing of households and electricity consumers by the ministry of commerce (MoC). About 81% of the financing was through long-term loans from the Asian Development Bank (ADB), World Bank, Japanese Overseas Economic Cooperation Fund (OECF), Korea Development Bank (KDB), etc. guaranteed by the government (Yim et al., 2012). These loans carried low interest rates and long grace periods and repayment periods. More than 11% of the financing was contributed by KEPCO from internal accruals and more than 8% was provided by MoC to households for financing internal wiring as a 35-year loan to be repaid along with electricity bills. The financing model for the Korean rural electrification project is shown in Figure 25.3. The Five-Year Electrification plan of 1965–1969 and the subsequent Long-Term Rural Electrification Project Scheme during 1970–1979 succeeded in raising the household electrification level to 98% (Yim et al., 2012). The island and remote regional electrification project using submarine cables and very tall towers succeeded in increasing the electrification rate to 99.3% in 1982 and 99.9% in 1988 (Yim et al., 2012).

Similarly India is currently implementing a large electrification initiative titled Deen Dayal Upadhaya Gram Jyoti Yojana (DDUGJY) which aims at electrifying all remaining 18,452 unelectrified villages in India by May 2018 (Government of India, 2016). The programme, which was launched in 2015, has a public finance outlay of $11 billion and government provides

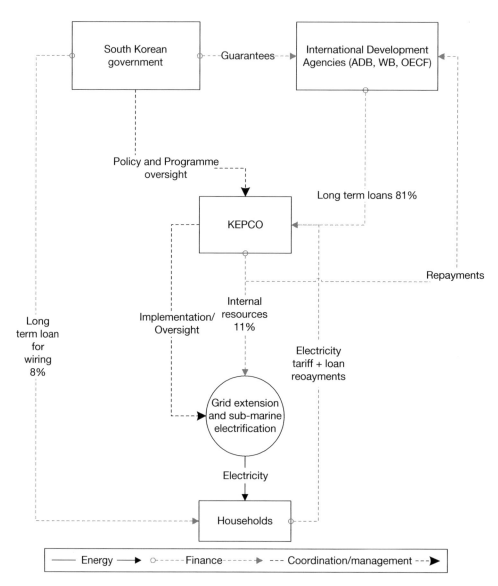

*Figure 25.3* Financing model for the South Korean rural electrification project

*Data source*: Author.

subsidies of either 75% or 90%[6] for electrification of villages through electrical network extension or through off-grid energy systems. The subsidies are routed through the Rural Electrification Corporation (REC) and the implementing entities are state electricity boards (SEBs), distribution companies (DISCOMs) including private DISCOMs, cooperatives, etc. and the implementing entities are expected to contribute 10% or 5% from their own sources with the rest being financed by REC or banks and financial institutions (Ministry of Power, 2014). The financing model of DDUGJY is shown in Figure 25.4. It is planned that 15,142 villages will be electrified through electrical grid extensions and 2,019 through off-grid systems. In early 2016 a total of 4,744 villages have been electrified under DDUGJY.

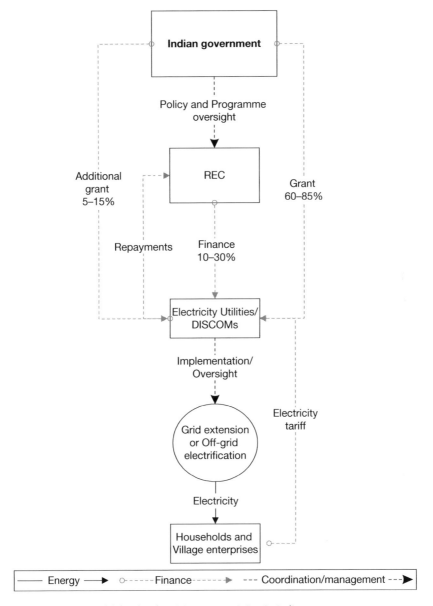

*Figure 25.4* Financing model for the electricity access mission in India

*Data source*: Author.

The financing for both the Chinese and Indian rural electrification programmes has been from government finances. Both governments have used large-scale grant finance and loans from the federal budgets, supplemented by contributions from provinces/states. The financing has been in the form of grants as well as soft loans. The loan finance was provided by development banks in China and REC in India, but was often a small share of the investment costs. Once the initial investments in the electricity infrastructure were made, consumers paid for electricity access charges and tariffs. In South Korea, a major share of funding for KEPCO came from international long-term low cost assistance which was backed by government guarantees. The remaining resources were provided by KEPCO and the South Korean government.

DDUGJY also follows the Chinese and South Korean approach for universal electricity access in a mission mode with specific time-bound targets and follow-up, driven by government finance and with a decentralised implementation approach. The DDUGJY progress can be tracked regularly through a web-based programme dashboard and a mobile phone app.[7] DDUGJY is therefore taking advantage of technological advances to increase the rate and transparency of government-led electricity access efforts.

## Public–private partnerships for financing

The role of public–private partnership (PPP) models has been steadily increasing in the infrastructure sector in Asia particularly in water and sanitation, transport, communications and energy. In the energy sector a number of countries have used PPP models such as Build-Own-Operate-Transfer (BOOT), Build-Own-Operate (BOO) and Build-Operate-Transfer (BOT). However, energy sector investments in Asia have generally been skewed towards power generation infrastructure in general rather than investments in transmission and distribution networks, LPG supply infrastructure or off-grid energy access solutions such as mini-grids or home energy systems. Many such PPPs have had an indirect impact on energy access as the additional electricity available can be used to extend electrical networks to unelectrified areas. An example of a PPP in electricity generation in Asia is the $1.2 billion Nam Theun 2 hydropower project in Lao PDR with shareholding from Electricité de France (EdF), Italian–Thai Development and the Lao PDR government (UN-ESCAP, 2015).

There have also been PPP models tried in the off-grid electricity access sector in Asia but their role has generally been limited. One example of such a PPP in off-grid energy access has been in the Philippines under the Qualified Third Party (QTP) model which is used for rural electrification in remote Barangays. The QTP model works on a reverse bidding mode for identified locations and the private sector entity which offers the lowest electricity tariff is provided with the electricity concession. The difference between the electricity tariff offered by the QTP and the regulated tariff is provided as viability gap funding (VGF) from a subsidy fund. There have been PPPs that have emerged as QTPs and one such QTP is in Barangay Rio Tuba in Palawan, Philippines which is a PPP involving PowerSource Philippines Inc. (PPI), the Department of Energy (DoE) and Palawan Electric Cooperative (PALECO) serving 1,132 households with diesel powered mini-grids (Kritika and Palit, 2011). There have also been PPPs in the Pacific such as the Pacific Islands Applied Geoscience Commission (SOPAC)'s project in the Solomon Islands where funding from the Renewable Energy and Energy Efficiency Partnership (REEEP) was routed to a private entrepreneur to provide solar lighting systems to unelectrified households. In the absence of a micro-finance ecosystem such as in Bangladesh and without access to banking services, the farming households paid for the electricity using their crops such as *dalo* (or taro, a root vegetable) and bananas (Parthan, 2009). This PPP arrangement tapped into the existing barter systems where villagers used to exchange crops for fish and allowed the public investment through SOPAC to be operated and maintained by the private operator.

In Indonesia another interesting model of a PPP has been supported by the United Nations Economic and Social Commission for Asia Pacific (UN-ESCAP). A cooperative named Mekar Sari, consisting of 450 members, was established in 2003 in the village of Cinta Mekar to implement a hydropower-based mini-grid. The financial investments in the hydro-mechanical and electro-mechanical equipment and civil structures were made by UN-ESCAP and a private company, Hidropiranti, on a 50:50 financing split. The members of the cooperative also contributed labour and local materials for the construction of civil structures and the power plant. An Indonesian NGO, Inisiatif Bisnis dan Ekonomi Kerakyatan (IBEKA), coordinated and managed the initiative. Once the Indonesian electricity utility Perusahaan Listrik Negara (PLN) extended the electricity network to Cinta Mekar, the hydropower plant was converted to run as an independent power producer (IPP) selling electricity to PLN. The members of the cooperative who were electricity consumers were taken over and serviced by PLN.

The model has now evolved into a pro-poor public–private partnership (5P) where the revenues from selling electricity to PLN are shared between Mekar Sari cooperative and Hidropiranti. Of the revenues, 20% is set aside for financing operation, maintenance and replacements and 40% is paid to Hidropiranti as a return on its investment. The 40% share that is received by the Mekar Sari cooperative is spent on 'pro-poor' initiatives (Parthan, 2015b). The funds have so far been used to support electricity access for the poorest households in the community and to provide land to households which did not have landholdings. Mekar Sari cooperative also provides scholarships to 360 school-going children and also pays allowances to old people in the community and to women during pregnancy and childbirth. Mekar Sari also plans to finance construction of toilets and drinking water supply points. The financing model of the Cinta Mekar hydro project is shown in Figure 25.5.

While there are community-based electrification initiatives using the cooperative model in Nepal, Bangladesh and India in Asia, there are not many examples like Cinta Mekar of a cooperative and the private sector collaborating for rural electrification investments with revenue sharing and the cooperative making significant pro-poor investments. UN–ESCAP is considering spreading this model in other countries in Asia and also in other parts of the world.

The financing for the initial investments of the Cinta Mekar 5P project was from UN–ESCAP and Hidropiranti on a 50:50 share, with Mekar Sari cooperative members contributing in-kind labour and local materials. The operation and maintenance is financed by a 20% share of the profits for electricity sales to PLN. For electricity generation projects such as the Nam Theun 2 projects in Lao PDR, the major shares of the investments have come from the private sector partners, with the government providing a relatively small financial contribution. In the PPP energy projects, the private sector partners are able to bring efficiency in construction and operation, ensuring better performance of assets, and the public sector partnership ensures limited risks arising out of policy and regulation. Often PPPs are able to provide better energy service and the government also benefits from the taxes and royalties as well as dividends from the PPP.

### Private finance for energy access

Husk Power Systems (HPS) is a private for-profit company which operates micro-grids that work using rice husk based gasifiers targeting unelectrified rural areas with availability of biomass residues. HPS, which was established in 2007, has operations in India, Nepal, Uganda and Tanzania and operates 90 mini-grids that service 35 villages (Schnitzer et al., 2014). The company normally provides 14–16 hours of electricity a day; however, it has in late 2015 moved to a hybrid energy technology model that utilises biomass and solar energy to provide 24 hours of electricity a day.

HPS initially used financing models such as Build-Own-Operate-Maintain (BOOM) and Build-Own-Maintain (BOM) where the ownership of the micro-grids was with HPS (Schnitzer et al., 2014). HPS has since started using Build-Maintain (BM) models where the ownership and

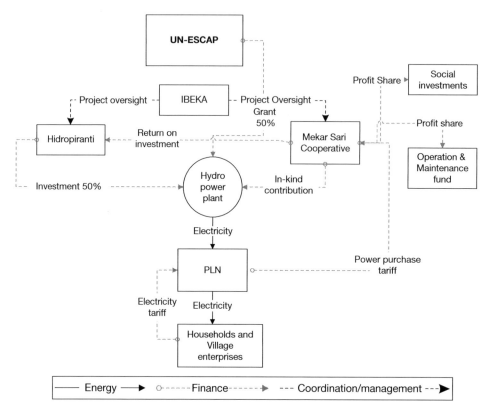

*Figure 25.5*   Financing model for 5P community electrification in Indonesia

*Data source*: Author.

operation of the micro-grid lies with a local entrepreneur. The consumers have to pre-pay for electricity and are metered through a pre-pay meter. To ensure steady availability of rice husks for the operation of the power plants, HPS have installed rice mills next to their power plants and offer free-of-cost dehusking of paddy in return for the rice husks. The customers pay $2.20–2.30 per month for a basic electricity connection with two lamps and a phone-charging point, but as electricity use increases with more appliances, the consumers end up paying more. The financing model for a BM model of HPS is shown in Figure 25.6.

During the early stages of establishment HPS was able to finance its operations with venture capital from impact investors[8] such as Acumen Fund, Bamboo Capital, the Shell Foundation, etc., which was supplemented with finance from government subsidies, prize monies, etc. During this phase HPS deployed BOOM and BOM models where finance for investment had to be generated by HPS. However, HPS has since moved on to the BM model which is essentially a franchise model where the capital investments are from a local entrepreneur. During 2015, HPS started implementing a hybrid model integrating solar and has also brought in a strategic investor, First Solar, which has picked up an equity stake in HPS and is providing preferential pricing for solar photovoltaic (PV) modules. HPS is a case of a private sector initiative using venture capital investments and incentives to establish and refine a for-profit business model and then scaling up and rolling out using a franchising model and private finance.

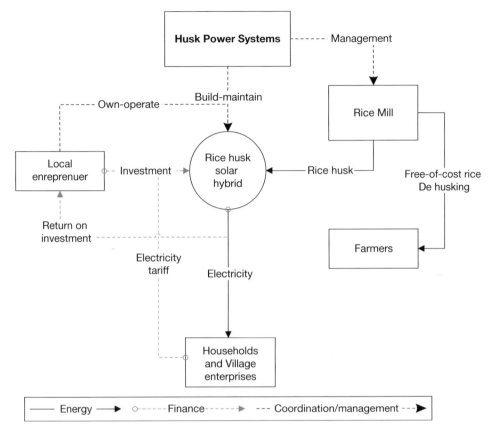

*Figure 25.6*  Financing model for Husk Power Systems

*Data source*: Author.

## Sources of finance

### *Public sector*

From the Asian experience showcased above, it is evident that the public sector has made major investments and impacts in energy access. The sources that have been used in Asia to mobilise public finance for energy access are illustrated in Figure 25.7 and explained below.

### *Public funding*

The major initiatives in energy access in Asia have been supported through public finance from governments. As illustrated in the Chinese, South Korean, Indian and Indonesian cases previously, the government finance that was targeted and at scale has been responsible for large-scale electrification and transition to LPG respectively in these countries. These and other governments in Asia have traditionally used government budgets to support rural electrification, but the rate of progress with energy access has been slow. In many Asian countries rural and unelectrified locations offer very low return on investments due to the low levels of expected energy demand and limited economic opportunities. These opportunities offer returns that are

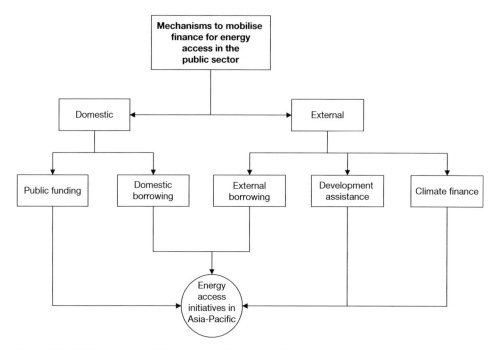

*Figure 25.7*  Existing sources of finance for public sector led energy access

*Data source*: Author.

below the investment return expectations of the private sector and the investment time horizons are also longer than the acceptable periods for private finance. Therefore public finance through government budgets is often the only financing option available for energy access. However, the challenge has often been the scale of the public finance available to address the energy access challenge. The experiences in China, South Korea and Indonesia show that energy access can be achieved on a mission mode supported by public finance at scale. The same approach is being implemented by India under DDUGJY. Public finance at scale has been a major factor in the energy access gains that some Asian countries have made in the past.

## Domestic borrowing

Domestic borrowing has played an important role in the energy sector in Asia. Governments have issued infrastructure bonds to finance electrification. Institutions such as REC in India and the development banks in China such as Agricultural Development Bank have issued public bonds to finance rural electrification. These bonds often have a long tenure and lower returns than other private investment opportunities but offered assured returns with lower risks. Some of the bonds, such as the REC bonds for electrification in India, were 'tax-free', where the retail investors could avail of reductions in their taxable income for the amount that was invested in the rural electrification bonds. Such tax incentives have also resulted in high levels of investor interest such as REC's 20-year $100 million bond issue in 2015 with a return of about 7% being oversubscribed 400% (The Financial Express, 2015). However, the use of domestic borrowing to finance energy access has been limited to Asian countries which have established bond markets and associated market regulatory frameworks to protect investments. Where such markets exist, long-term, low-return bonds linked to tax incentives appear to have significant investor interest and are being used by some Asian governments to raise finance for energy access.

## External borrowing

Several Asian countries also borrow from multilateral financial institutions such as the World Bank or Asian Development Bank (ADB) as well as from bilateral development agencies such as the Kreditanstalt für Wiederaufbau (KfW) or Japan Bank for International Cooperation (JBIC). External borrowing for energy access supplements public funding and domestic borrowing, and in some Asian countries is also able to offset the lack of local finance. In LDCs and Small Island Developing States (SIDS) in particular, where availability of local finance is limited and has low credit-worthiness, special funds from the World Bank and ADB such as the International Development Association (IDA) and the Asian Development Fund (ADF) respectively provide long-term, low interest finance. Bilateral development agencies also provide financial resources to Asian countries but there are often conditional, with the source of equipment and services limited to the country providing external finance.

The World Bank has been a major financer of energy access programmes and projects in Asia and some of its key initiatives in Asia include the Bangladesh IDCOL SHS project which was supported using low-cost, long-tenure IDA resources; the 100,000 solar *ger*[9] programme in Mongolia supporting household solar electrification of nomadic cattle herds; and development of the inter-regional grid network of Power Grid Corporation of India (PGCI). ADB has also been a major financier of rural energy in Asia though its projects and programmes, including electrification of rural villages in Bhutan and ofcoastal communities in Sindh in Pakistan using ADF resources. Similarly KfW has also supported energy access projects and programmes in Asia including the Bangladesh IDCOL project. Other bilateral agencies such as Agence Française de Développement (AFD) have also supported energy access projects in Asia. External borrowing from development agencies offering concessional and large-scale financing has been one of the key financing sources for energy access in Asia.

## Development finance

Official development assistance (ODA) routed through multilateral development agencies, bilateral development agencies, non-government organisations (NGOs), etc. has also supported energy access in Asia. Generally ODA has been in the form of grants which are available through these agencies to support energy access. The key multilateral agencies active in energy access in Asia have been the United Nations Development Programme (UNDP), European Union (EU), United Nations International Development Organisation (UNIDO), United Nations Environment Programme (UNEP), Food and Agricultural Organisation (FAO), UN-ESCAP, etc. A number of bilateral development agencies such as the Department for International Development (DfID), UK, Norwegian Agency for Development Cooperation (NORAD), Gesellschaft für Internationale Zusammenarbeit (GIZ), Germany, etc. have all supported energy access efforts in Asia. Some international NGOs that are involved in energy access initiatives in Asia include Practical Action UK, SNV Netherlands and Winrock International.

Development assistance has generally been in the form of grants and technical assistance to support energy access. Some of the development assistance support focuses on the policy and capacity building aspects to improve the environment for promoting energy access and build human and institutional capacity. Other development finance agencies and programmes have supported projects and programmes where electrical energy or thermal energy access was provided to communities and institutions. However, due to the limited availability of grant resources, development assistance supported energy access programmes have generally focused on piloting and demonstrating specific energy access technologies, models and approaches. However, many development assistance initiatives have played an important role in financing and supporting the progression of the energy access agenda in Asia.

## Climate finance

Climate finance has also played a significant role in supporting energy access initiatives using renewable energy in Asia. The major role has been played by the Clean Development Mechanism (CDM) which has enabled many Asian countries to get finance for greenhouse gas (GHG) emission reductions which were certified and issued under the CDM framework as Certified Emission Reductions (CERs). The revenue from the sale of CERs improved the financial viability of a number of energy access projects in Asia. The segment of CDM which bundled small projects together, named the Programme of Activities (PoA), was instrumental in several small energy access initiatives such as SHS, Cookstoves, Biogas digesters, etc. obtaining financing from the sale of CERs. However, with the end of the first commitment period of the Kyoto Protocol in 2012, the demand for CERs has reduced considerably, depressing prices and reducing the impact on energy access projects in Asia. A voluntary carbon finance market exists to offset emission reductions at institutional and individual levels through Verified Emission Reductions (VERs); however, the combined demand for CERs and VERs is currently at a low level which limits the ability of the carbon market to influence energy access efforts.

Other significant finance available under the United Nations Framework Convention on Climate Change (UNFCCC) has been through the oldest of its financing mechanisms, the Global Environmental Facility (GEF). GEF has financed the incremental costs[10] of a number of energy access projects that supported renewable energy based energy access in Asia. GEF support to Asian countries was routed through accredited implementing agencies such as UNDP, World Bank, UNEP, UNIDO, ADB, etc. The finance from GEF enabled projects which had potential to offer GHG mitigation to use renewable energy technologies for energy access. The GEF support enabled electricity access projects using off-grid approaches and thermal energy projects using biogas or efficient cooking technologies to be implemented.

Another climate finance mechanism available for renewable energy based energy access initiatives in Asia is the climate investment funds (CIFs) which finance low-carbon development and climate change adaptation initiatives in developing countries including in Asia through the multilateral development banks (MDBs) such as the World Bank and ADB. Of particular relevance is the Scaling-up Renewable Energy Programme (SREP), which currently supports renewable energy investments in countries such as the Maldives and Nepal in Asia with plans to support other countries.

## Private sector

From the Asian experience in energy access as articulated in the cases above, it is evident that the role of the private sector has in general been limited to manufacture and supply of systems and equipment for energy services. However, a number of initiatives have encouraged a role for the private sector beyond manufacture and supply of energy access hardware and software. Governments, financiers and development agencies have employed a number of mechanisms to attract private finance in energy access in Asia, some of which are illustrated in Figure 25.8 and explained below.

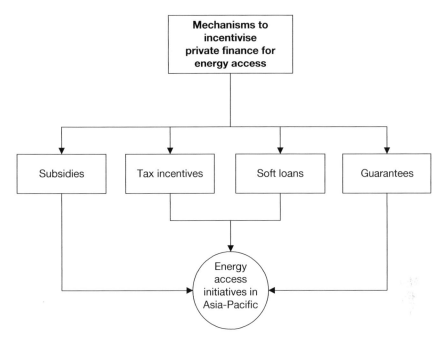

*Figure 25.8*  Existing mechanisms to incentivise private finance for energy access

*Data source*: Author.

## Subsidies

Subsidies are an important financial mechanism used by governments and energy regulators to ensure that energy access initiatives are able to leverage private finance. Different types of subsidies have been used in Asia to leverage private finance from enterprises and individuals for energy access investments. The most common type of subsidy that has been used in Asia has been the capital subsidy which has been provided as a share of the capital cost of investments. The 60–85% capital subsidy to private DISCOMs under the DDUGJY in India is such an example. Capital subsidies have also been part of the Bangladesh IDCOL programme as a grant to buy down the initial costs. These subsidies are aimed at capital cost reduction to encourage private investors to invest in energy access infrastructure that is otherwise not financially viable considering the returns on investment. Such subsidies are also relevant to induce rural and poor households to switch to modern energy options such as LPG from traditional biomass-based cooking as when the Indonesian government provided a 100% subsidy for the initial change from kerosene to LPG.

Another form of subsidy is the output or performance based subsidies which are available based on the energy access services delivered rather than the investment costs. The QTP model of electrification of Barangays in the Philippines is an example of such performance subsidies where the subsidy is administered based on the electricity supplied. In output or performance based subsidies the investments need to be made by the investor and energy services delivered before the subsidies are made available. There is an incentive to increase and improve the performance levels and utilise the energy access assets more efficiently in the case of output or performance based subsidies.

Another form of subsidy is the direct tariff subsidy where consumers are provided with the subsidy amount for energy consumption. In this model energy consumers from poor households are provided with coupons which can be used for purchase of energy or given direct cash transfers when energy services are utilised. India has such a direct benefit transfer (DBT) scheme effective from January 2015, where LPG consumers are paid the subsidy for LPG directly to their bank accounts by the government (Tripathi et al., 2015). To facilitate this transformation, the sale of subsidised LPG was discontinued and consumers pay the market price; consumer households with low incomes are eligible for the direct payment of LPG subsidy. This direct cash transfer mechanism is expected to save about 15% of the amount spent or $1.1 billion annually by the Indian government on LPG subsidy (Tripathi et al., 2015).

## Tax incentives

The mechanism of tax incentives has also been used by governments to mobilise private investments in energy access in Asia. One common tax exemption provided in Asian countries is the exemption from import and customs duties for energy access equipment. This mechanism allows for a reduction in the capital costs of systems and equipment for energy access that are imported into the country. Nepal has exempted generators for Micro Hydro Power (MHP) from import duties (Williams et al., 2015) and this opportunity to import high quality generators to pair them with locally manufactured quality hydro-mechanical equipment has helped Nepal to use MHPs for off-grid rural electrification of remote villages.

There can also be exemptions of sales tax or Value Added Tax (VAT) for components or complete energy systems that are used for energy access, which also reduces the capital expenditure of such systems. Such exemptions are also prevalent in Asian countries such as China and India and help incentivise private sector investment in energy access. Tax holidays for a specified period are used to encourage start-ups in energy access and to encourage private investment in new enterprises. India has a five-year tax holiday for renewable energy enterprises that is intended to encourage private investments. Another tax incentive is the use of accelerated depreciation, which allows private enterprises that invest in renewable energy equipment to avail of 60–100% depreciation for the capital equipment in the first year. This helps reduce the income tax payable for private companies that are making investments in clean energy based energy access systems. India also has an accelerated depreciation for renewable energy equipment that can be utilised by private companies investing in renewable energy based energy access systems.

## Soft loans

Soft loans are loans that have loan parameters like interest rates, loan tenures and moratoria[11] that are more favourable than the terms of loans available to the private sector for similar investments. Generally for soft loans, interest rates are lower than the market rates which are available for specific types of investments. The availability of soft loans for energy access investments helps in improving the viability of such investments. The loans can also be made attractive by increasing the loan tenure so it is significantly longer than the loan periods available for commercial activities. The moratorium on the loan can also be increased to make loans for energy access investments attractive for the private sector. Soft loans have been used by development financial institutions in Asia to support energy access investments by the private sector. The Bangladesh IDCOL SHS programme offers interest rates in the range of 6–9% against the prevailing interest rate of 16% and also offers a loan tenure of 5–7 years against the prevailing term of 3 years.

## *Guarantees*

Guarantees have been used to help private investors obtain debt financing from commercial banks and financial institutions. Such mechanisms are relevant in countries where potential commercial finance is available but the financing of energy access investments is considered to be risky by prospective financiers. Such loan guarantees are offered by development agencies or through the establishment of a guarantee mechanism whereby part of any loan defaults will be compensated by the guarantor. Loan guarantees have been used in Asia during the early stages of market development to create track-record and build confidence with banking and finance institutions. International Finance Corporation (IFC), the private sector window of the World Bank, offered a loan guarantee to electrify parts of eastern India through SHS as part of the Photovoltaic Market Transformation Initiative (PVMTI) (IEA PVPS, 2003). Similarly partial risk guarantees can also be offered by a development agency if the revenues are dependent on the government or a public entity. In the event that the government or the public agency is unable to fulfil the payment obligations, the guarantor will perform the payment obligations. Such guarantees have been used by MDBs to de-risk private sector investments in energy access. Such partial credit guarantees allow the private investors to have access to commercial finance sources and often to obtain financing at more favourable terms as the risk is mitigated.

## *Barriers and lessons*

Analysis of the existing experience in the Asian region in financing energy access and an examination of the financial mechanisms used and the cases as presented in the previous sections lead to the following inferences on the barriers to finance and the lessons from existing experience.

## *Barriers*

There are some barriers and bottlenecks that constrain an accelerated level of energy access financing in the Asia region, the key among which are:

- A number of Asian countries have a low credit rating which limits these governments' ability to raise resources that are low cost and long term, which affects available resources for energy access investments. This is a particular challenge for Asian LDCs.
- Absence of a local finance ecosystem of financial institutions and banks, micro-finance institutions, rural and agricultural development banks as well as local debt and equity markets in some Asian countries constrains the ability of both the public and private sector to raise local finance for infrastructure investments including energy access. This is a particular challenge for SIDS and LDCs in Asia.
- In countries that have been making major energy access gains at scale, namely China, South Korea, India, Indonesia, Sri Lanka, etc., the governments have provided either high levels of subsidies or full finance to create the infrastructure required to provide energy access. Not all governments in Asia have the resources and capability to make large-scale investments.
- There is a need in Asia for a 464%[12] increase from the current annual levels of investments in energy access to achieve the universal energy access targets under SDG 7. Such a scale-up of resources would require major changes in the current energy access finance architecture.
- The scale of financing needs for universal energy access in Asia means that it will not be met through external borrowings or external grant financing. However, the levels of financing needs will vary between Asia countries.

- The emphasis of the energy access financing has primarily been on electrification and not enough investments are being made in non-solid fuel access for thermal energy use. This has meant that the progress with non-solid fuel energy access has lagged behind electrification.
- The role of the private sector and the leverage of private finance for energy access investments in Asia has been limited. While there is early experience with non-grant financing mechanisms, the scale of leverage is considerably lower than what is needed to achieve universal energy access by 2030.
- The level of investments in providing clean thermal energy of non-solid fuel access to cooking and space heating remains the biggest energy access financing challenge as the level of need is set to increase by a factor of 44.
- A major bottleneck for financing non-solid fuel thermal energy access is that the current business models and financing are mostly focused on supporting cooking and space heating devices at a smaller scale than the large-scale service or fuel sales based model.
- The role of the private sector and PPPs has been leveraged in electricity generation as IPPs but not significantly in electrification and non-solid fuel access. There are a number of barriers and bottlenecks at the policy and institutional levels which need to be addressed to increase the role and leverage finances from the private sector.
- Capabilities for developing and structuring energy access projects that can attract direct private finance or PPPs are likely to be a bottleneck in many Asian countries. Such capabilities will be important in reaching the financial resource mobilisation requirements to achieve the SDG 7 goal on energy access.
- While a number of innovations have been attempted to engage private sector investments in energy access, these haven't been scaled up or replicated to significant levels at the country or regional levels.

### Lessons

There are a number of lessons that can be learnt from the existing experience in financing energy access in the Asia, the key among which are:

- Domestic financial sources such as public finance and direct government subsidies are the major source of finance for energy access investments in Asia. This is followed by external borrowing assistance from MDBs, development assistance and climate finance.
- The role of the private sector needs to significantly increase if Asian countries are to achieve universal energy access by 2030. However, the low returns from such investments and the absence of a progressive policy and incentive framework have so far constrained the role played by the private sector.
- While subsidies and grants from public finance have been used to promote electricity access in a number of countries, these financial instruments are limited by government budgets and are not sustainable in the long term.
- External financial sources like borrowings, development aid and climate finance have played an important role in increasing energy access in Asia. While their relative contribution to energy access infrastructure creation has been limited, these finance flows have played an important role in demonstrating technologies, business and financial models and providing policy support.
- Due to a wide range of government and public sector capabilities, local financial institutions, private sector capabilities and local finance ecosystems, the type of resource mobilisation models required in Asian countries is likely to vary.

- Due to the lack of a business case for investments in electricity and modern non-solid fuels to rural or decentralised populations, there is a role for mechanisms such as grants and subsidies from the governments or development finance to support energy access in Asia.
- Private sector participation and finance in energy access can be increased through de-risking energy access investments and improving financial returns and capital recovery of the investments. Risk transfer instruments such as guarantee funds as well as mechanisms such as incentives, soft loans, etc. to improve the investment prospects have been used and hold promise.
- The PPP structure has been used effectively in infrastructure financing in water and sanitation, transport, communications and energy generation, but has had a limited role in energy access. Under the right conditions, PPPs could play an increased and highly significant role in enhancing energy access in Asia.
- While the scale of the financial resources required is considerably smaller, non-solid fuel thermal energy access is likely to be the major challenge for financial resource mobilisation owing to limited capabilities, under-developed business models and financing structures.
- There appears to be a need to look beyond existing structures, business models and financing mechanisms for energy access to scale-up finance for achieving universal energy access by 2030 in Asia. There may be alternative and innovative financing mechanisms in Asia and beyond that may offer possibilities.

## Alternative innovative financing

A number of alternative and innovative approaches are being tried out to finance and implement energy access initiatives in Asia and beyond, some of which may offer potential for application to scale-up energy access finance in Asia. Some of these mechanisms are examined below.

### *Crowdfunding*

Crowdfunding is the process of financing a project or an initiative using contributions from a large number of people. Crowdfunding whereby financing was raised through a benefit event or through mailshots has been in existence for a considerable period of time. However, in the past decade, and in particular in the past few years, internet platforms have been used for crowdfunding, expanding reach, increasing transparency and reducing transaction costs. Crowdfunding involves either equity or loan investment in a specific project, the details of which and the financing required are published on the website. Depending on the finance sought, large numbers of investors commit to provide equity or loans through the internet platform. If the financing target is met, the funds are made available to the promoter of the initiative through the internet platform to enable the project to be implemented. Upon successful completion of the project/initiative, the promoter pays returns to the equity investors or pays back the loans with interest. Some platforms use field partners to carry out due diligence on each project/initiative and to monitor and route the finance to the promoter and facilitate recovery of loans or returns on equity. Crowdfunding platforms are also used for making donations. It is estimated that global crowdfunding volume in 2015 was $16 billion, of which the share of renewable energy was a meagre $170 million, supporting more than 300 projects (Versteeg, 2015).

Initial crowdfunding efforts have focused on artistic creations like music and movies as well as technology ventures. Increasingly crowdfunding platforms are also supporting clean energy and energy access issues. Many of the platforms are based in Europe and the United States like Trillion Fund, Abundance, Village Power, Solar Schools, etc., which mainly crowdfund energy projects

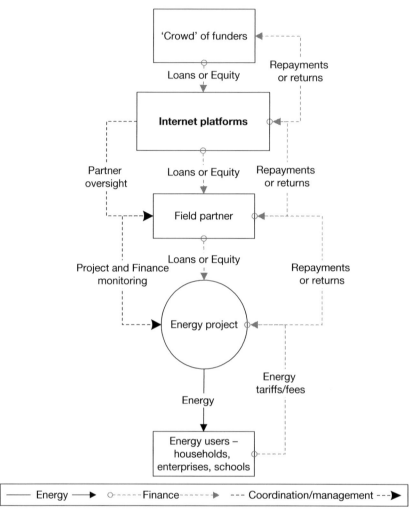

*Figure 25.9* Crowdfunding energy access projects

*Data source*: Author.

in the countries where they are located. However, there are sites like SunFunder which focuses on off-grid solar covering renewable energy and has raised $7 million investment in 58 loans supporting 368,000 people (SunFunder, 2016). The $700 million Kiva Microloans which supports micro-loans in developing countries also offers green loans for solar lighting, home systems and clean Cookstoves. Kiva is also active in a number of Asian countries through its field partners. Figure 25.9 shows how crowdfunding works in the energy access context. Crowdfunding platforms like Kiva have high rates of loan recoveries as well. The financial regulatory frameworks for crowdfunding are emerging.

While crowdfunding platforms are growing as an alternative means of finance for projects and initiatives which may not be supported by the commercial banking sector, the share of energy access projects so far has been quite small. Crowdfunding of energy access projects and initiatives in Asia could be a significant prospect in the future. Small-scale financiers in Asia and elsewhere could use such platforms to support energy access projects in their countries or others. While crowdfunding holds potential, its role will be determined by private sector entrepreneurs and public sector agencies. If integrated into the policy and institutional frameworks for energy access, crowdfunding could provide significant supplementary resources to support universal energy access in Asia.

## New carbon finance

As mentioned above, carbon finance in the form of monetisation of emission reductions from energy access projects as CERs and VERs did support a number of energy access projects in Asia up until 2012. However, lack of demand during the second commitment period of the Kyoto Protocol has depressed CER prices and has also affected the demand for VERs. Under the current low demand and low-prices scenario, very few projects are being supported by carbon finance. This situation is also true for energy access projects in Asia where additional investments in development of documentation for CDM or voluntary carbon projects and validation, registration, verification and issuance of CERs or VERs are not justified due to low process for CERs and VERs. The outlook for the carbon market is also unclear, although the Paris Agreement has established a new market-based mechanism which will come into effect after 2020. This new mechanism may create a demand for internationally transferred mitigation outcomes (ITMOs). The impact of the proposed new mechanism will depend on the demand and the rules that will be developed. In the short term the voluntary carbon market for VERs may provide some limited finance for energy access projects in Asia.

## Green bonds

Green bonds are debt instruments that are used by entities to raise financial resources from the markets, the proceeds of which are exclusively used for 'green' projects. Bonds have traditionally been used by both public and private entities for financing infrastructure projects such as energy projects, including bonds that were used for energy access like the REC bonds mentioned above. 'Green bonds' constitute a new global effort to establish a labelling system, involving a number of organisations such as Climate Bonds Standard and Ceres Green Bond Principles, as well as national regulators such as China Central Bank's green bond guidelines. It is estimated that only 11% of the bonds that can be classified as 'green' are actually labelled as such and the major locations where bonds are so labelled include Europe and North America (Climate Bonds Initiative, 2015). The MDBs and bilateral development banks active in Asia such as the World Bank, ADB and KfW have made green bond issues. Private companies, commercial banks and municipalities have also issued green bonds. The 'green' projects being supported by green bond proceeds totalling $65.9 billion include energy, buildings and industry, transport, water and waste management (Climate Bonds Initiative, 2015).

Green bonds represent labelling of a small segment of the global bond market and even the local currency denominated bond market in Asia was $8.6 trillion in 2015 (ADB, 2015). There are no significant financial advantages offered to the issuers by labelling the bond green, although a green label may indeed attract environmentally conscious investors. In the Asian context, green bonds may represent only the labelling of existing bonds, but if green bonds induce investors to consider investing in these green debt instruments rather than in options of equity, property, precious metals, etc., this would represent an important source of energy access finance. Existing Asian green bond issues have targeted renewable energy investments which are consistent with the global trend and it is also possible that proceeds of green bond

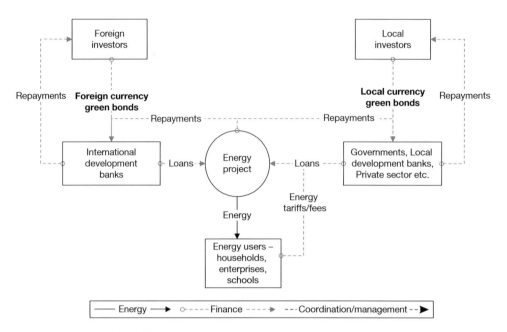

*Figure 25.10*  Green bonds for energy access projects

*Data source*: Author.

issues by the World Bank, ADB, KfW, etc are supporting energy access projects in Asia. It is indeed possible to label local currency and foreign currency denominated bonds that target renewable energy based energy access projects as green bonds. If such green bonds result in additional foreign or local currency resources and increase the scale of resources mobilised, they could augment the financial resources needed for universal energy access in Asia. It is likely that green bonds are more relevant in countries that are credit worthy and have a strong financial ecosystem, whereas the challenge for energy access financing is likely to be in countries that are not credit worthy or do not have a strong local finance ecosystem, such as some LDCs and SIDS. However, it is possible that MDBs and bilateral development agencies wanting to support renewable energy based energy access projects in Asian LDCs and SIDS may use green bonds for resource mobilisation from the European and North American markets. This approach is also applicable to bilateral development agencies in Asia from countries such as Australia, Japan, South Korea, etc. which may use domestic or internationally issued green bonds to support energy access investments in other Asian countries including LDCs and SIDS. Figure 25.10 shows the possible use of green bonds to support universal energy access in Asia.

## Expatriate remittances

Asian countries have a large number of nationals who work outside their countries and make remittances back to their home countries to support their families. The size of global remittances to developing countries has been increasing steadily, overtaking the development assistance flows in 1997, and was estimated at $435 billion in 2015 compared with about $160 billion of total development assistance. Remittances are also considered to be more stable than private capital flows (World Bank, 2015a). A number of countries in Asia receive very high levels of remittances as a share of GDP, such as Tajikistan (42%), Kyrgyz Republic (30%), Nepal (29.9%), Bangladesh (8.6%), Pakistan (6.9%), as well as the Philippines, Vietnam, Indonesia, Cambodia, Timor-Leste, etc. (World Bank, 2015a). The scale, predictability and increasing nature of the remittances offer an opportunity to use these finance flows to supplement public sector and private sector resources to finance energy access investments. An example of such an initiative is the '3 x 1 programme' in Mexico where remittances were used to support infrastructure development. Under the programme, infrastructure such as roads, water supply and sanitation in Mexican provinces was financed by a combination of government and expat remittances. For every peso contributed by remittances, the government contributed 3 pesos – 1 peso each from the federal, state and municipal budgets. The programme, which started in 2003, supported more than 2,500 projects up to 2011.

It is estimated that South Asia, East Asia and the Pacific received close to $250 billion in remittances in 2015 (World Bank, 2015a). Even if only a small share of these remittances can be used to support energy access it could make a significant resource at scale. It is also likely that a significant share of the remittances is being used for subsistence including payment of electricity and fuels. In many Asian countries a significant share of migrants are from locations without energy access and could be encouraged to contribute to providing electrification and non-solid fuel supplies in their home towns and villages. The $50 million UNDP project in Afghanistan, Afghanistan Sustainable Energy for Rural Development (ASERD), which began implementation in early 2016, aims to capture these remittances to support rural energy access in that country. Annual remittances to Afghanistan total about 3.1% of its GDP (World Bank, 2015b), which in 2012 was estimated to be $445 million (Parthan, 2015c). The ASERD

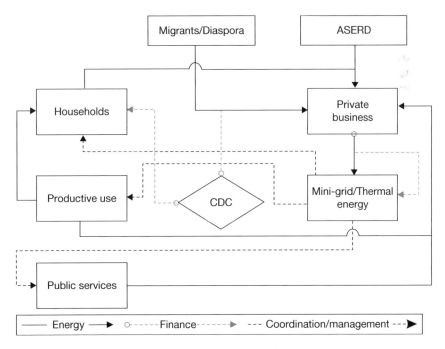

*Figure 25.11*  Expatriate remittance financed rural energy pilot in Afghanistan

*Data source*: Author.

project, which will be implemented during 2016–2019, will pilot a rural energy model where public funding through the project and remittances will be used to support a rural energy enterprise in one of the migrant corridor villages in Afghanistan. The details of the pilot are shown in Figure 25.11.

Attracting migrant finances for energy access can be achieved through programmes like the ASERD or through government policies that offer taxation and other incentives. If implemented strategically diaspora finance could be a significant factor helping Asia achieve universal energy access by 2030.

## New climate finance

In December 2015, 197 countries attending the Paris Climate Change Talks agreed on the Paris Agreement which aims to limit the increase in the global average temperature to well below $2^{\circ}$C above pre-industrial levels. The Paris Accord also agreed on mobilising $100 billion per year to support climate change mitigation and adaptation in developing countries by 2020 and to agree to a bigger resource mobilisation goal by 2025. The Green Climate Fund (GCF) is the new and the

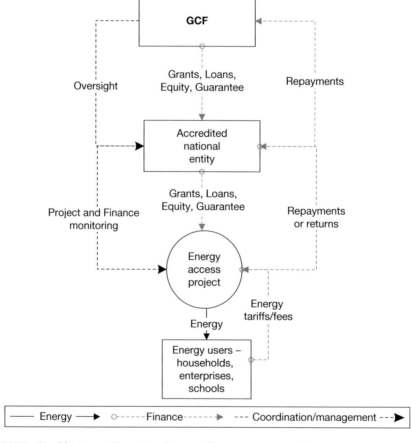

*Figure 25.12*    Possible Green Climate Fund support for energy access projects

*Data source*: Author.

largest financing mechanism of the UNFCCC and the Paris Accord and has already mobilised more than $10.3 billion. The GCF also has energy as one of the eight strategic priority areas and will be financing energy access projects under this priority area. The GCF can be accessed by organisations that are based in developing countries through a 'direct access' route and the GCF also has a Private Sector Facility (PSF) which will make climate finance investments that directly involve the private sector. The GCF will also use a range of financial instruments – grants, debt, equity and guarantees – to support climate change projects in developing countries. It also prioritises investments in LDCs and SIDS.

The scale of resources, direct access by national institutions, range of financial instruments and prioritisation of energy as well as LDCs/SIDS by the GCF is advantageous from an Asian energy access perspective. Electricity access and non-solid fuel access initiatives in Asia using renewable energy and low-carbon energy sources like LPG have the potential to benefit from GCF at scale. The emergence of the GCF could be of importance to LDCs and SIDS as there will not be a cap on resources that are allocated to smaller countries. However, accessing large-scale resources from the GCF for energy access will require mobilisation of strong national institutions including the private sector and the ability to structure and develop a pipeline of projects that can take advantage of loans, equity and guarantees. The emergence of the GCF provides an opportunity to access international climate finance at scale to achieve universal energy access at scale. Figure 25.12 shows the possibilities offered by GCF finance in increasing energy access in Asia countries.

## Recommendations

Based on an examination of the barriers, bottlenecks and lessons from existing financing mechanisms, and inferences from new and alternative financing mechanisms, the following recommendations can be made:

- In most Asian countries public finance alone would be insufficient to achieve universal energy access by 2030 and there is a need to leverage private finance using the appropriate structures including PPPs.
- In addition to government and private sector resources, domestic or international borrowings provide a mechanism to supplement resources. Labelling domestic or international debt as 'green bonds' in mature financial markets could be an interesting resource mobilisation option for development agencies raising resources to invest in LDCs and SIDS.
- Remittances from the Asian diaspora offer a hitherto untapped, steady and significant finance stream that is very relevant to Asian infrastructure and energy access investments. Government policies and development programmes need to leverage this financing stream for energy access investments.
- More effort is needed on business models and financing structures for a transition to non-solid fuels for thermal energy. The financing challenge is likely to be significant in this area due to limited government policies and absence of models and track-record in this segment.
- Climate finance from the GCF offers a new and significant source for energy access in Asia especially for LDCs and SIDS. However, accreditation by strong national organisations from Asia and development of strong project pipelines is needed to take advantage of this opportunity. While carbon finance from international emissions trading may not have a significant impact in the short and medium term, crowdfunding could turn out to be an interesting source of finance.

## Conclusions

The following conclusions can be drawn from an assessment of efforts to finance universal energy access in Asia, existing financing mechanisms, alternatives and innovative mechanisms:

- Asia has a large number of people without energy access, with significant variations across countries. Asia will need to significantly increase the current level of finance for energy access to achieve universal energy access by 2030. While this is possible in many countries, it could present a challenge to Asian LDCs and SIDS.
- There are a number of examples of models in Asia such as the Chinese, South Korean and Indian electrification programmes and the Indonesian LPG transition programme where governments have deployed large-scale public finance and guarantees to support energy access resources on a mission mode. In these and other countries, governments have been the major financing source for energy access, a role which is expected to continue in future as well.
- The Bangladesh solar household electrification programme, the 5P model in Indonesia and the HPS BM model illustrate examples of how private sector engagement and resources have been used to provide energy access in Asia. However, the role of the private sector and the contribution of private finance need to increase significantly to accelerate the rate of energy access achievements in Asia.
- Asian governments and public sector institutions have utilised public finance, domestic and international borrowings, development finance and climate finance to support energy access investments thus far. Among these financing mechanisms, pubic finance has played the major role to date.
- Asian countries have used mechanisms such as subsidies, tax incentives, soft loans and guarantees to ensure private sector participation in energy access projects and initiatives. However, a number of barriers and bottlenecks still continue to limit private sector participation in providing energy access services in Asia.
- Among alternatives to existing finance mechanisms, carbon finance may not have a major role in the short term whereas crowdfunding could turn out to be an interesting source of finance. While green labelled bonds in local and international markets are an option, the immediate potential could be for green bonds to be used by development agencies as a resource mobilisation tool to support energy access projects in Asian countries.
- Remittances from the Asian diaspora for energy access investments represent a predictable, stable and significant financial stream that requires serious consideration. There are examples from Mexico and Afghanistan that demonstrate how remittances can be utilised for infrastructure development and energy access.
- The new climate finance architecture within the framework of the Paris Accord, with the GCF as one of the key financing mechanisms, is likely to provide finance at scale for energy access that can be directly accessed through national organisations and using a range of financial instruments such as grants, loans, equity and guarantees. Asian countries will need to build national institutional capacities and a quality energy access pipeline to access this climate finance window.
- Asian governments also need to put in place progressive policy and regulatory frameworks that engage and leverage private sector business capabilities and financial resources. Such frameworks could also be made consistent with SDG 7 using the SE4All framework and to leverage remittances, climate finance, green bonds and crowdfunding if relevant.

Efforts to achieve universal energy access in Asia offer a unique opportunity to combine development frameworks, national policy and finance and leverage private financial resources, through a strategic mix of existing financing mechanisms and alternative and innovative mechanisms. To seize this opportunity, Asian governments will need to lead these efforts in partnership with the private sector and supported by the development assistance ecosystem.

## Notes

1  Xinjiang Goldwind Science & Technology, Agricultural Bank of China, CLP India, Yes Bank India, Export-Import Bank of India, Indian Renewable Energy Development Agency, etc.
2  As measured by the Global Competitiveness Index (GCI) of the World Economic Forum (WEF).
3  As measured by the Ease of Doing Business Index by the World Bank Group.
4  As measured by the Corruption Perceptions Index by Transparency International.
5  A rubber tube and a gas flow regulator connecting the LPG cylinder and the stove and controlling the gas flow.
6  The higher level of subsidy is available to states in the north and north-eastern parts of India which are in difficult terrain and remotely located.
7  The dashboard is available at http://garv.gov.in/dashboard and the GARV App can be downloaded from Google Play, iTunes and Windows app stores.
8  Investments which are made with an objective of generating social and/or environmental impacts alongside financial returns, often below market returns.
9  Traditional tents used by nomadic herders in Mongolia.
10  The additional costs of pursuing a low-carbon energy access technology or approach compared with the baseline business-as-usual approach.
11  The initial grace periods during which the loan repayments need not be made.
12  Calculated from SE4All (2015), Advisory Board's Finance Committee Report on Scaling up Finance for Sustainable Energy Investments.

## References

Asian Development Bank (ADB), 2015, *Asia Bond Monitor*, September 2015.
Bhattacharya, SC and Ohiare, S, (2012), The Chinese Electricity Access Model for Rural Electrification, *Energy Policy*, 49 (October), pp. 676–687.
Climate Bonds Initiative, (2015), *Bonds and Climate Change: The State of the Market in 2015*, London.
Government of India, (2016), Press Release, http://pib.nic.in/newsite/PrintRelease.aspx?relid=134587.
IEA PVPS, (2003), 16 Case Studies on the Deployment of Photovoltaic Technologies in Developing Countries.
International Energy Agency, (2011), *World Energy Outlook 2011*, Paris.
International Energy Agency, (2015a), *World Energy Outlook 2015*, Paris.
International Energy Agency, (2015b), *World Energy Outlook Special Report: Energy and Climate Change*, Paris.
International Energy Agency (IEA) and the World Bank, (2015), *Sustainable Energy for All 2015—Progress Toward Sustainable Energy* (June), World Bank, Washington, DC. doi: 10.1596/978-1-4648-0690-2 License: Creative Commons Attribution CC BY 3.0 IGO.
International Monetary Fund, (2015), *Country Level Data on Energy Subsidies*, Washington, DC.
Kritika, PR and Palit, D, (2011), *Review of Alternative Participatory Business Models for Off-Grid Electricity Services*, Working Paper 09, OASYS SOUTH ASIA Research Project, The Energy and Resources Institute, New Delhi.
Ministry of Power, India, (2014), Deen Dayal Upadhaya Gram Jyoti Yojana: Office Memorandum, Government of India, New Delhi.
Parthan, B, (2009), Solomon Islands Solar: A New Microfinance Concept Takes Root, *Renewable Energy World*, 12(2).
Parthan, B, (2015a), *Evaluation of the Solar Home System Service Models and Financing Modalities*, African Development Bank, Abidjan.
Parthan, B, (2015b), The 'Real 5P Model' in Cinta Mekar, https://sustainableenergyassociates.wordpress.com/2015/03/18/the-real-5p-model-in-cinta-mekar/.

Parthan, B, (2015c), *Afghanistan Sustainable Energy for Rural Development*, UNDP Project Document.

PT Pertamina, (2015), *Kerosene to LPG Conversion Programme*, Jakarta.

PT Pertamina and World LPG Association (WLPGA), (2012), *Kerosene to LP Gas Conversion Programme in Indonesia, A Case Study of Domestic Energy*, Paris.

Ruhul-Quddus, M, (2015), *After Sales Service of Solar Home System (SHS) Program under IDCOL in Bangladesh*, IDCOL, Dhaka.

Schnitzer, D, et al, (2014), *Micro-grids for Rural Electrification: A Critical Review of Best Practices Based on Seven Case Studies*, UN Foundation.

SE4All, (2015), *Advisory Board's Finance Committee Report on Scaling-up Finance for Sustainable Energy Investments*, Sustainable Energy for All, Vienna.

SunFunder, (2016), www.sunfunder.com, accessed February 2016.

The Financial Express, (2015), www.financialexpress.com/article/markets/indian-markets/despite-low-yields-public-issue-of-rec-tax-free-bonds-evokes-good-response/157637/.

Tripathi, A, Sagar, AD and Smith, KR, (2015), Promoting Clean and Affordable Cooking. *Economic & Political Weekly*, 48, pp. 81–84.

United Nations (UN), (2015), http://www.un.org/sustainabledevelopment/energy/, accessed November 2015.

UN-ESCAP, (2015), *Infrastructure Financing, Public–Private Partnerships, and Development in the Asia-Pacific Region*, United Nations Economic Social Commission for the Asia and the Pacific, Bangkok.

van Gevelt, T, (2015), Rural Electrification and Development in South Korea. *Energy for Sustainable Development*, 23, pp. 179–187.

Versteeg, K, (2015), *Tracking Renewable Energy Crowdfunding, Solarplaza*, www.solarplaza.com/channels/finance/11417/tracking-renewable-energy-crowdfunding/.

Williams, NJ, et al., (2015), *Renewable and Sustainable Energy Reviews*, 52, pp. 1268–1281.

World Bank, (2015a), *Beyond Connections: Energy Access Redefined*, International Bank for Reconstruction and Development/World Bank, Washington, DC.

World Bank, (2015b), *Migration and Development Brief 25*, International Bank for Reconstruction and Development/World Bank, Washington, DC.

Xinhua, (2015), China realizes universal power access, http://news.xinhuanet.com/english/2015-12/25/c_134949385.htm.

Yim, HB, Park, MJ and Lee, SW, (2012), *2011 Modularization of Korea's Development Experience: Rural Electrification Project for Expansion of Power Supply*, Ministry of Knowledge Economy and Korea Electric Power Corporation, Ministry of Strategy and Finance, Seoul, Republic of Korea.

# 26

# SOCIO-TECHNICAL INNOVATION SYSTEMS

## A new way forward for pro-poor energy access policy and practice

*David Ockwell and Rob Byrne*

### Introduction

International policy commitments such as the UN's Sustainable Energy for All (SE4All) by 2030 and the Sustainable Development Goals (SDGs) imply an unprecedented transformation in sustainable energy access in Asia and beyond. But what do interventions in policy and practice that are able to facilitate such transformations look like? How can they intervene in ways that change the long-term, systemic conditions that provide the context within which poor people gain and sustain access to clean and reliable energy? SE4ALL is framed around three key policy priorities, or the "energy trilemma" (Goldthau, 2012) – energy security, climate change and energy poverty – but whose vision of security does SE4ALL and its attendant processes and institutions represent? Whose climate change? Whose energy poverty? Is SE4ALL necessarily pro-poor? With trillions of dollars in climate finance coming online following the Paris climate agreement, now is a germane time to ask such questions. Moreover, how do we avoid the mistakes of past international climate policy, which has not served the interests of lower income countries or even poor people in rapidly developing countries?

Building on previous empirical work in India (Ockwell and Mallett, 2012; Ockwell et al., 2008), China (Watson et al., 2015) and Malaysia (Hansen and Ockwell, 2014), as well as more recent work in Kenya (Ockwell and Byrne, 2016b; Rolffs et al., 2015) and policy articulations thereof (Ockwell and Byrne, 2016a; 2016b), we describe a conceptual framework that can inform such pro-poor, systemic interventions in policy and practice. The approach builds on insights from innovation studies and uses this to extend socio-technical transitions theory in ways that enable policy to better attend to the social practices that energy access might facilitate; the co-evolutionary nature of technological change and social practices; and the challenge for new sustainable energy technologies to compete with existing – often fossil-dominated – regimes of social practice and technology use. We also translate this conceptual framework into two interlinked and complementary proposals that can inform both inter-national and domestic policy and practice in developing Asia and elsewhere.

## The scholarly deficit and recent "socio-cultural turn" in sustainable energy access research

Theorising on sustainable energy access is failing the kinds of transformative policy ambitions noted above. Certainly, as work elsewhere testifies, a huge amount of analysis has been done on two dimensions of the problem, those relating to technological hardware and finance, with attendant disciplinary biases towards engineering and economics (Watson et al., 2012). So, as Gollwitzer (2016) summarises, we now know a lot about financial issues such as relative costs and benefits of different technological options for rural electrification (e.g. Levin and Thomas, 2014; Nerini et al., 2015; 2016; Anderson et al., 2012; Lee et al., 2014); willingness and ability of potential electricity consumers to pay (e.g. Abdullah and Jeanty, 2011; Bacon et al., 2010; Alfaro and Miller, 2014); and the nature and appropriateness of different mechanisms for financing investments in rural electrification infrastructure (e.g. Levin and Thomas, 2014; Bhattacharya, 2013a; UNDP, 2011; Monroy and Hernández, 2005). From a technology hardware perspective, we also know much about technology selection methods (e.g. Barry et al., 2011; Brown and Sovacool, 2007; Mainali and Silveira, 2015); applicability of particular technologies to specific contexts (e.g. Azimoh et al., 2016; Kishore et al., 2013; Narula et al., 2012; Kaundinya et al., 2009); spatial mapping to determine technological applicability and least-cost scenarios (e.g. Deichmann et al., 2011; Ohiare, 2015; Szabó et al., 2011; Zeyringer et al., 2015); how to improve the efficiency of existing technology applications (e.g. Sebitosi et al., 2006); issues around technology adoption, such as the idea of the energy ladder (e.g. Hosier and Dowd, 1987; Campbell et al., 2003; Murphy, 2001; Hiemstra-van der Horst and Hovorka, 2008); and ideas around different scenarios involved in energy transitions (e.g. Elias and Victor, 2005: not to be confused with socio-technical transitions).

Importantly, as other volumes demonstrate, several scholars with expertise in the technology and finance dimensions of sustainable energy access have done us great service by taking essential steps towards translating relevant insights into practical lessons for policy and practice (see, e.g., Bhattacharyya, 2012; 2013b; Bhattacharyya and Palit, 2014; Bazilian et al., 2012; Terrado et al., 2008). Nevertheless, as Watson et al. (2012) demonstrate in their systematic review of the available literature on barriers to increased use of modern energy services in Sub-Saharan Africa, we know little about the socio-cultural dimensions of sustainable energy access and almost nothing about its political/political-economy dimensions. It is this lacuna that we have recently referred to as "the scholarly deficit" in research on sustainable energy access (see chapter 1 of Ockwell and Byrne, 2016b). Whilst political/political-economy aspects of sustainable energy access remain woefully underrepresented in the literature – save for a handful of contributions that could be of relevance[1] (see, e.g., Shen and Power, 2016; Tyfield et al., 2015; Ahlborg et al., 2015; Newell and Phillips, 2016; Baker et al., 2014) – recent years have witnessed what might be described as a "socio-cultural turn" (Ockwell and Byrne, 2016b) in the sustainable energy access literature. This has included a number of contributions that operationalise theoretical insights from the socio-technical transitions literature (Ahlborg and Sjöstedt, 2015; Rolffs et al., 2015; Sovacool et al., 2011; Ulsrud et al., 2011; 2015; Hansen and Nygaard, 2014; Nygaard and Hansen, 2016), social anthropology (Winther, 2008; Campbell et al., 2016; Campbell and Sallis, 2013), and common pool resource management (Gollwitzer, 2016; Gollwitzer et al., 2016). Although less obviously socio-cultural in focus, a number of contributions that seek to operationalise insights from innovation studies might also be linked to this socio-cultural turn, not least due to its emphasis on indigenous capabilities, albeit more at a firm than community level (Hansen and Ockwell, 2014; Watson et al., 2015; Hansen et al., 2015; Ockwell and Mallett, 2012; Bell, 2009). In our own recent work we have tried to bridge these two fields of socio-technical transitions and innovation studies to develop a hybrid conceptual framework that, we argue, stands a better chance of informing policy options that will deliver against the self-determined needs of poor and marginalised women and men in the Global South (see Ockwell and Byrne, 2016a; 2016b). In the remainder of this chapter we focus on articulating how a hybrid of these two fields can better inform effective policy in Asia and the rest of the developing world, before setting out what such policy interventions might look like in practice.

## Understanding innovation, technological change and development

The field of innovation studies has evolved over several decades, focussing on developing more sophisticated understandings of the role of technology and technological innovation in relation to economic growth and industrialisation. Within this field, a strong area of focus has been on the role of policy in driving innovation and technological change, giving rise to a more systemic perspective on how such change occurs and can be nurtured. Of particular relevance to sustainable energy access is the broad base of empirical and theoretical work that has studied structural change in the world economy, with significant research effort focussed on understanding industrialisation and technological change in developing countries. Importantly, a large part of the empirical evidence base upon which this literature has drawn has focussed on technological change and industrialisation in developing Asia, and the Asian Tiger economies in particular (e.g. D'Costa, 1994; 1998; Gallagher, 2006; 2014; Freeman, 1987; Lall, 1992; Kim, 1998; Ockwell et al., 2008; Watson et al., 2015; Hansen and Ockwell, 2014; Bell, 1997; Bell and Pavitt, 1993). Whilst only a more recent part of this literature focusses on low carbon energy technologies (e.g. Watson et al., 2015; Gallagher, 2014; Ockwell et al., 2008; Hansen and Ockwell, 2014), there nevertheless exists a rich base of empirical and theoretical work from which we can learn when trying to understand how policy might intervene to encourage the uptake of new technologies in developing-country contexts in Asia and beyond.

When thinking about access to sustainable energy technologies in developing Asia and else-where, there are arguably two key areas of insight that can be taken from the innovation studies literature. The first concerns the central role of knowledge flows and the accumulation of technological capabilities in underpinning broader technological change within given sectors in an economy (e.g. Ockwell and Mallett, 2012; Hansen and Ockwell, 2014). This also relates to the often incremental nature of innovation and technological change that has been observed to underpin such processes (e.g. Bell, 2012). The second key insight is the role of "National Systems of Innovation" (NSIs) in creating the fertile context within which technological change is more likely to occur (e.g. Freeman, 1997). We elaborate on each of these insights and their relevance to sustainable energy access below.

### *Knowledge, technological capabilities and incremental innovation*

One of the key observations from the innovation studies literature is that long-term, more widespread technological change tends to occur over time as a result of the accumulation of technological capabilities. These capabilities develop along a continuum from productive to innovative, the latter being fundamentally underpinned by flows of knowledge. These knowledge flows are of far more importance to defining the extent to which the accumulation of technological capabilities leads to either broader economic development or broader processes of technology uptake elsewhere in an industry or sector (for an elaboration of this continuum, see

Hansen and Ockwell, 2014). Sometimes, this knowledge is codified (e.g. trade secrets and intellectual property/patents). The more important type of knowledge, however, tends to be tacit knowledge (i.e. knowledge gained by doing), often resulting from inter-firm interactions, particularly with international technology suppliers, implying a need for a shift in the emphasis of policy interventions away from technology hardware towards knowledge flows and technological capability building.

This raises the question as to how knowledge flows and technological capability building might be nurtured via policy interventions. There is no easy answer to this question. The issue lies with the fact that, with development goals like sustainable energy access, there is a public good incentive to try to deliberately facilitate such capability building under conditions of urgency (Ockwell and Mallett, 2012). These conditions are largely absent when such knowledge flows and capability building occur under market conditions. Nevertheless, it is clear that some countries, such as China, have pursued deliberate policy-driven approaches to achieving rapid technological capability building around specific technologies, including low carbon energy technologies (Watson et al., 2015). It is therefore possible to learn from these experiences to inform interventions elsewhere. Some of the clues as to how to achieve this lie in the area of innovation system building and the extension of these ideas via insights from the socio-technical transitions literature (see below).

Another key insight of relevance is the evidence from the field of innovation studies that radical innovation or technological change is unusual – the exception rather than the rule. Technological change is far common as a result of longer processes of incremental innovation over time (c.f. Bell, 2012). Moreover, "innovation" does not imply "invention". It is also innovative for a firm to make an incremental change to an existing technology or process, be the first firm in a sector to apply a new technology or process, or to use a new technology or process for the first time (Bell, 2009; OECD and Eurostat, 2005). This applies whatever the firms or actors concerned: e.g. whether farms or small holdings, or households. For example, a farmer might, for the first time, experiment with the use of a well-established rainwater harvesting technique, but in doing so the farmer is being innovative. In this way, technology *adoption* is as much innovation as early stage research and development (R&D). Indeed, as Arnold and Bell (2001) report, the contribution of publicly funded R&D to processes of technological change and innovation – even in OECD countries – is, in their words, "vanishingly small".

Several empirical studies from developing Asia have been used to support this idea of the centrality of incremental innovation. For example, incremental innovations have been shown to have characterised the efficiency improvements observed in the Korean steel industry, which resulted in it eventually occupying the international technological frontier (D'Costa, 1998; Gallagher, 2006). This has been contrasted with progress in the steel industry in India which was for some time characterised by fragmentation and the coexistence of a diverse range of different technologies that slowed down technological progress (D'Costa, 1994). Such observations enabled authors such as D'Costa to make recommendations on the nature of industrial policy with greater potential to support and create new knowledge in order to rejuvenate "older industries, such as steel, in socially acceptable ways and organically [create] new knowledge for national development and social welfare" (D'Costa, 1998: 271). D'Costa is, however, careful in attending to potential critiques around dependency theory, e.g.:

> Global competitiveness of South Korea does not refute the technological dependence of developing countries. Rather it suggests the kind of socio-institutional and economic policy contexts that are often necessary to foster competitive industries.
>
> *(D'Costa, 1994: 44)*

Elsewhere, adaptive innovation has also been shown to be of critical importance, for example in the adaptations to the internal combustion engine that facilitated the international technological leadership Brazil has exhibited in relation to transport-related biofuels (Lehtonen, 2011).

## National systems of innovation

The second insight of relevance to this chapter emerging from the innovation studies literature relates to work around the idea of "National Systems of Innovation" (NSIs). This area of the literature demonstrates how innovation and technological change takes place within a systemic environment, referred to in different parts of the literature as the "innovation system" or "NSI". This concept has its roots in the emergence of an initial body of work in the 1980s and 1990s (e.g. Freeman, 1987; 1997; Lundvall, 1988; Patel and Pavitt, 1995; Nelson and Winter, 1982), which mounted an empirical and theoretical challenge to a neoclassical economics account of economic growth. The latter was challenged on the basis of its treatment of innovation and technological change as exogenous to economic models (for a historical overview of the NSI literature, particularly in relation to development policy, see Watkins et al., 2015). Instead, the NSI perspective offers a broader, evolutionary perspective on technological change and development, focussing on the network of actors (e.g. firms, universities, research institutes, government departments, NGOs) within which innovation and technological change occurs, and the strength and nature of the relationships between them. It has been applied across various industrialised and developing-country contexts over time. One example of its application in Asia is the work of Kim (1993), who uses the NSI perspective to explain the success of the Korean semi-conductor industry.

From the NSI perspective, then, we are offered a systemic approach through which policy might seek to engage in strengthening the networks of relationships that provide the context within which technological change and innovation occurs. It is essentially these relationships on which a particular elaboration of the NSI literature has focussed. This is the "Technological Innovation System" (TIS) literature (e.g. Bergek et al., 2008a; 2008b; 2015), which is concerned with providing policy prescriptions for managing innovation systems to improve the chances of the development and uptake of specific technologies. Despite providing a tantalising approach for policy makers and policy analysts, however, the TIS literature has been subject to critique. Kern (2015), for example, discusses how the TIS approach can be overly instrumental, technocratic and deterministic in its understanding of policy processes and its focus on the various rational "functions" of an innovation system that policy makers might engage with and seek to nurture, as well as its failure to properly deal with the political realities within which technology adoption occurs. Such instrumental approaches might not be particularly helpful in understanding technology adoption and technological change in the real world. But the idea of focussing on NSIs and adopting a systemic policy perspective provides significant analytical and practical purchase in thinking about what kind of policy interventions would facilitate the nurturing and strengthening of such innovation systems around sustainable energy technologies. Certainly, it goes a long way beyond the usual two-dimensional focus on technology and finance that has dominated the literature and policy to date.

## Insights from socio-technical transitions

Before outlining specific policy recommendations, it is important to emphasise another area of literature that has potential to enhance both our understanding of how new technologies end up in widespread use and how interventions through policy and practice might promote such processes. A weakness of the innovation studies literature relates to its firm-centric focus.

This might be useful at some levels, but the sustainable energy access issue demands consideration of how technologies are adopted by poor women and men in developing countries. The conceptual frameworks that underpin the innovation systems literature are based on (albeit impressively in-depth) empirical studies that focus on firms and industries. Technology users tend only to be considered in terms of providing knowledge feedback mechanisms into firms' innovation processes via their consumption patterns in relation to marketed goods. This tells us little about the social contexts within which these goods are consumed, or the social practices they are intended to facilitate. When considering how access to new technologies (e.g. solar home systems or clean cookstoves) might increase "energy access", it is necessary to consider the social practices that consumption of energy via these technologies facilitates (e.g. cooking, or reading and working at night) and the existing technologies (e.g. wood stoves or kerosene lamps) that currently support these social practices. Politics aside, without consideration of the social practices that new energy technologies are intended to facilitate, there is a risk that interventions will be overly managerial and ultimately fail to effect any kind of long-term change (Shove, 2010). As Stirling (2015, p.1) asserts, innovation isn't just about technological invention, "it involves change of many kinds: cultural, organisational and behavioural as well as technological".

Here, the socio-technical transitions literature offers us a more sophisticated perspective on processes of (socio-)technical change. This rapidly growing literature, with its roots in historical empirical studies of widespread change in industrialised countries, understands such change to be socio-technical in nature, as opposed to just technical (e.g. see Berkhout et al., 2004; Geels, 2002; Geels and Schot, 2007; Raven, 2005; Smith et al., 2010). Importantly in the context of this book, recent work in this field has included work on energy in developing Asia as one of the early moves in the direction of exploring the application of this approach in the context of developing countries (e.g. Ulsrud et al., 2011; Berkhout et al., 2010). At its core is an understanding of innovation as a process that co-evolves with the social practices that technologies facilitate. Think of the internal combustion engine and how cities, road infrastructure, cultural norms around driving, rules and regulations about using roads, all co-evolved around this break-through innovation. This also helps to explain path dependency. These socio-technical "regimes" of technologies, social practices and institutions serve to support incumbent technologies, making it hard for new, often more sustainable, technologies to compete. Socio-technical regimes can be understood as rules (knowledge-base, belief systems, mission, strategic orientation) shared by multiple actors, from firms, to technology users, to policy makers and beyond (Hoogma, 2000). In this sense, the focus of analysis is no longer the technology itself, which is simply understood to be an artefact of socio-technical co-evolutionary processes. Instead, the social processes that technologies facilitate become an important focus of analysis.

New technologies therefore have to compete with and somehow influence existing socio-technical "regimes". This has led to efforts to introduce new technologies being theorised as socio-technical "niches" (e.g. solar mini-grids in India). Beyond the niche, and sitting above the level of the regime, socio-technical scholars using what is known as the "Multi-Level Perspective" (MLP) (Geels, 2002) also conceptualise a "landscape" level, which they consider to include exogenous factors such as war, economic growth or recession, climate change and more broad scale changes in social and cultural preferences. Through this conceptualisation of these multiple levels – niche, regime and landscape – the socio-technical transitions literature also offers a theory of change. Change might occur through successful nurturing of niches of sustainable energy technologies to the extent that they compete with fossil-based socio-technical regimes; through landscape-level forces such as widespread changes in social and political demands for low carbon energy; or through the destabilisation of current socio-technical regimes.

In the context of energy access, a socio-technical perspective therefore provides us with a new focus. Our attention is refocused on what constitutes the regimes of energy provision for poor people in developing countries and how any attempt to introduce sustainable energy technologies to these contexts needs to compete with these existing socio-technical regimes. New technologies can then be understood as part of niches that need to influence existing regimes. A socio-technical perspective also allows us to understand technologies as co-evolving with the social contexts within which they are used, recognising that new technologies will be widely adopted not simply because they successfully harness technical principles, but also if their form and function are "aligned" or "fit" with dominant social practices, or offer opportunities to realise new practices (or "stretch" existing socio-technical practices) that are attractive in particular social-cultural settings (Hoogma, 2000; Raven, 2007).

Similar to the TIS literature's emergence as a "managerial" version of the innovation systems approach, a body of work has emerged within the socio-technical transitions literature that seeks to understand how to nurture new technology niches in ways that might successfully influence existing regimes. This is known as the Strategic Niche Management (SNM, or niche) literature (Kemp et al., 1998; Smith, 2007). This literature, based on extensive empirical analysis, has developed a number of core principles through which niches might be strategically managed in order to influence existing socio-technical regimes, including:

1  Creating "*protective spaces*" for experimentation and development of new technologies within a supportive environment (Smith et al., 2014).
2  *Experimenting* with new technologies in different contexts to generate lessons about socio-technological configurations, their social and technological potential and broader social implications of adoption (Smith and Raven, 2012; Smith et al., 2014).
3  Creating broad *actor-networks* for building robust support for new socio-technical practices, for facilitating knowledge-exchange, for enabling interactions between stakeholders and for providing access to resources (Schot and Geels, 2008).
4  Creating robust, collectively shared *expectations and visions* about how niche socio-technical configurations might inform different potential future(s) (Raven, 2005).
5  Building *institutions* (including laws, regulations and policies as well as social practices, norms and conventions) around new socio-technical configurations.

## Policy options for "socio-technical innovation system" building

So far, we have focussed on identifying the gaps that exist in the literature on energy access (the "scholarly deficit") and laying out how two areas of literature (innovation studies and socio-technical transitions) might contribute towards a more systemic, socio-culturally cognisant perspective on how access to sustainable energy technologies might be facilitated for the 1.1 billion women and men who lack access to electricity, or the 2.9 billion who lack access to non-solid fuels (SE4All, 2015). These literatures have cast analytic attention on the following key dimensions:

1  *The systemic environment within which technologies emerge and become established in different national contexts.* Despite the limitations of the innovation studies literature due to its firm-centric focus, it is difficult to imagine a thriving market for sustainable energy technologies amongst new consumer groups without a well-functioning NSI emerging to support this new market: innovation systems are often weak or absent in many low income countries, or around sustainable (as opposed to conventional) energy technologies in most countries.

It should be emphasised that "innovation" in this case does not refer to "invention" or things that are new to the world. The adoption of globally, or even nationally, well-established technologies (e.g. solar PV) in new contexts (from countries down to communities, households and businesses) is as innovative as, say, a technology-leading firm developing a new variant of a technology via the investment of millions of dollars in R&D.

2   *The social practices of poor women and men in consuming, producing, managing and paying for energy services.* As the example of mobile payments for solar PV has demonstrated, better understanding of the social practices of poor women and men in consuming and paying for energy services in East Africa has enabled a potentially transformative new approach to facilitating access to electricity via solar PV (see Rolffs et al., 2015). Starting from an understanding of socio-culturally specific social practices around energy-service consumption may therefore provide a powerful starting point in understanding the potential for existing or new socio-technical configurations around energy production and use that hold more transformative potential than approaches that privilege a focus on either (or both of) the technology hardware or finance dimensions of the sustainable energy access challenge. The socio-technical transitions literature's emphasis on the co-evolutionary nature of technological innovation and social change also provides a more nuanced and sophisticated focus for understanding how new socio-technical configurations of energy-service consumption might evolve and change in future. In turn, this raises potential for better considering who might gain or lose – and how and why – from different evolving pathways of sustainable energy access (Leach et al., 2010; Newell and Phillips, 2016; Ockwell and Byrne, 2016b).

3   *The regimes of socio-technical practice that reinforce the stability of existing energy technologies and form powerful barriers (political, economic and social) to the uptake of new technologies.* Even at the level of energy-service consumption by poor and marginalised women and men, these existing socio-technical energy regimes can be intertwined with difficult-to-disrupt cultural practices and, perhaps even more so, existing political economic power structures (e.g. the interests of kerosene or diesel supply networks or charcoal producers).

In an attempt to fuse and extend these insights via a hybrid theoretical framework with potential to inform future policy and practice in this field, we have developed the idea of using interventions in policy and practice to build "socio-technical innovation systems" (Ockwell and Byrne, 2016b). Whilst there is doubtless considerable work to be done in developing this perspective, not least in terms of attending to political and political-economy considerations and the governance concerns they raise, it offers a potential way forward for designing policy interventions at both the national and international level. We therefore end this chapter by briefly elaborating on two linked policy proposals that might work to facilitate socio-technical innovation system building around pro-poor sustainable energy technologies in practice (for more detail see Ockwell and Byrne, 2016a; 2016b).

## Proposal 1: Climate Relevant Innovation-system Builders (CRIBs) and/or Sustainable Energy Access Relevant Innovation-system Builders (SEA-RIBs)

This involves creation of specific institutions (preferably based within existing organisations) within different countries that are focussed either on building socio-technical innovation systems around climate technologies more broadly (CRIBs) or around sustainable energy technologies specifically (SEA-RIBs) (see Ockwell and Byrne, 2016a; 2016b). CRIBs or SEA-RIBs would play a strategic facilitating role within countries, acting as the focal/convening point for a national network of actors across the spectrum of those involved in relevant or potential socio-technical innovation systems (from users, through supply chains, to NGOs and policy makers) and championing the development of socio-technical innovation systems around different technologies. Their core remit would be to link together national actors around a strategic, long-term, nationally defined vision (cognisant of national policy goals and local realities). They would develop detailed knowledge of national capabilities, key areas where opportunities exist for rapid development and growth, and identify areas where international expertise and knowledge-sharing is required. CRIBs or SEA-RIBs would provide strategic oversight, advising on how to target sustainable energy access (or broader climate technology) programmes and projects in a coordinated way that responds to identified priority areas for both rapid growth and long-term capability building. As implementers of Proposal 2 below, such institutions would also work to ensure that all projects and programmes are used strategically as ways to build and strengthen socio-technical innovation systems. This approach could be pursued as a flagship initiative under the UN's SE4All programme. It could also be integrated with an extension of the United Nations Framework Convention on Climate Change (UNFCCC) architecture (specifically through the Climate Technology Centre and Network, CTCN) and/or link with various centre-based initiatives under the Global Environment Facility (GEF) (for more detail on the nature of such integration, see Ockwell and Byrne, 2016a). It is also perfectly feasible for it to be pursued bi- or unilaterally via donor or individual government-driven activities, or via the activities of NGOs. Such centres would, however, be likely to have greater impact if networked across different countries and regions (e.g. via the UNFCCC's CTCN).

### Proposal 2: Using projects and programmes to build socio-technical innovation systems

This proposal focusses on designing project- and programme-based investments in sustainable energy access to maximise impacts on building socio-technical innovation systems. So, for example, with every investment in a sustainable energy or broader climate technology project, policy or programme, careful consideration should be given to the wider impacts it could be designed to have on the existing socio-technical innovation system in any given context. Could it be used to create new, or strengthen existing, actor-networks? Might it be used to create protected spaces for experimenting with and learning from new socio-technical configurations? Might it bring diverse actors together, serving to convene them around new expectations and visions of what these new socio-technical configurations might deliver? Could it create momentum towards the creation of new, or revision of existing, legal or social institutions? Might it focus on building technological capabilities amongst indigenous technology importers, suppliers, vendors, technicians and consumers? All such considerations could play a role in ensuring that new projects and programmes add up to more than the sum of their parts, having a greater long-term impact on building capabilities and creating or strengthening socio-technical innovation systems around sustainable energy or broader climate technologies. The result would be far greater potential for investments to lead to more widespread socio-technological change, with attendant benefits both for poor women and men who gain access to new technologies and for the broader economies of the developing-country contexts where such socio-technical innovation system building occurs. This second proposal could be pursued by any actors engaged in work on sustainable energy access or climate technology transfer, uptake and development. Again, however, it is likely to have far greater impact if actors pursuing such approaches coordinate via new or existing networks, such as those that would be championed under Proposal 1 above.

Proposal 1 could be adopted in a way that integrates with Proposal 2, working to ensure that as many project-based interventions as possible are pursued in ways that maximise opportunities for socio-technical innovation system building. Proposal 2, on the other hand, could be pursued separately from Proposal 1 and could represent a strategic commitment by actors involved in efforts to improve sustainable energy access to integrating considerations concerning the impact of their activities on socio-technical innovation system building. Clearly, however, the transformative impact of activities under Proposal 2 is likely to be higher if pursued as part of nationally and internationally coordinated efforts via the kinds of institutions Proposal 1 advocates.

## Conclusion

We have attempted to chart some of the key contributions to date in the field of sustainable energy access. As we have seen, this field is presently dominated by a range of important insights that focus on technology/engineering and finance/economics. In other words, a multi-dimensional problem (i.e. sustainable energy access) is mainly analysed via a two-dimensional perspective, with subsequent implications in terms of how policy has tended to approach the problem to date (and, we would argue, explaining to a large extent why policy to date has failed to achieve transformative impacts). Whilst a literature is emerging that focusses on socio-cultural aspects of the problem, almost nothing has been done to address the political/political-economy challenges.

Our argument has been that it is imperative for both scholarship and policy/practice in this field to adopt a more multi-dimensional, systemic approach to this problem. By introducing insights from the literatures on innovation studies and socio-technical transitions, many of which have been developed via decades of empirical work in developing Asia, we have highlighted a number of imperatives that policy interventions need to attend to if they are to live up to the transformative ambitions of international policy commitments like SE4All, the SDGs or the Paris climate agreement. These imperatives include:

1 A need to understand sustained processes of technological change as being catalysed by the development and strengthening of innovation systems around the specific technologies of interest (in our case, sustainable energy technologies).
2 A need to understand the social practices of poor women and men in consuming and paying for energy services (the emphasis being on services, not technologies – so light and cooking rather than, say, solar PV or biogas stoves). It is such a focus that has underpinned the early success of mobile-enabled, pay-as-you-go finance models for solar PV in countries like Kenya (Rolffs et al., 2015). So, instead of focussing on technological innovation, begin by understanding social practices and their implicit needs and then think how new socio-technical configurations might meet these needs.
3 A need to appreciate the barriers that existing socio-technical regimes of energy production and use imply for introducing sustainable energy technologies. This means learning from the socio-technical transitions literature on how to nurture technology niches so that they can influence/change existing socio-technical energy regimes.

All of these imperatives will vary with context, nationally, regionally and locally. The two policy proposals articulated above are designed in such a way as to respond directly to these imperatives and to do so in ways that are context-specific and sensitive, both of which are essential if interventions through policy and practice are to have transformative potential. As we acknowledge, there is still work to be done in analysing the political and political-economic dimensions of the policy proposals we set out above: for example, many governance issues arise from the kinds of institutions we propose, prompting questions about how they might be addressed in the interests of poor and marginalised women and men. Nevertheless, we hope these proposals provide the beginnings of an architecture through which more transformative, pro-poor interventions might emerge, with potential to deliver against the laudable ambitions of the emerging international policy focus on achieving sustainable energy for all.

## Note

1 Note, however, that only Ahlborg's work engages directly with sustainable energy access rather than sustainable energy and climate change more broadly.

## References

Abdullah S and Jeanty PW. (2011) Willingness to pay for renewable energy: Evidence from a contingent valuation survey in Kenya. *Renewable and Sustainable Energy Reviews* 15: 2974–2983.

Ahlborg H and Sjöstedt M. (2015) Small-scale hydropower in Africa: Socio-technical designs for renewable energy in Tanzanian villages. *Energy Research & Social Science* 5: 20–33

Ahlborg H, Boräng F, Jagers SC, et al. (2015) Provision of electricity to African households: The importance of democracy and institutional quality. *Energy Policy* 87: 125–135.

Alfaro J and Miller S. (2014) Satisfying the rural residential demand in Liberia with decentralized renewable energy schemes. *Renewable and Sustainable Energy Reviews* 30: 903–911.

Anderson R, Ræstad MJ and Sainju P. (2012) Cost–benefit analysis of rural electrification. *Norplan Policy Brief Norplan.* pp.1–6 (www.noexperiencenecessarybook.com/o80LO/policy-brief-cost-benefit-analysis-of-rural-electrification-norplan.html) accessed 20 August 2017.

Arnold E and Bell M. (2001) Some new ideas about research for development. *Partnerships at the Leading Edge: A Danish Vision for Knowledge, Research and Development.* Copenhagen: Commission on Development-Related Research Funded by Danida, 279–319.

Azimoh CL, Klintenberg P, Wallin F, et al. (2016) Electricity for development: Mini-grid solution for rural electrification in South Africa. *Energy Conversion and Management* 110: 268–277.

Bacon R, Bhattacharya S and Kojima M. (2010) *Expenditure of Low-Income Households on Energy: Evidence from Africa and Asia* (Extractive Industries for Development Series Working Paper #16). Washington, DC: World Bank.

Baker L, Newell P and Phillips J. (2014) The political economy of energy transitions: The case of South Africa. *New Political Economy* 19: 791–818.

Barry M-L, Steyn H and Brent A. (2011) Selection of renewable energy technologies for Africa: Eight case studies in Rwanda, Tanzania and Malawi. *Renewable Energy* 36: 2845–2852.

Bazilian M, Nussbaumer P, Eibs-Singer C, et al. (2012) Improving access to modern energy services: Insights from case studies. *The Electricity Journal* 25: 93–114.

Bell M. (1997) Technology Transfer to Transition Countries: Are There Lessons from the Experience of the Post-war Industrializing Countries? In: Dyker DA (ed.) *The Technology of Transition: Science and Technology Policies for Transition Countries.* Budapest: Central European University Press.

Bell M. (2009) *Innovation Capabilities and Directions of Development*, STEPS Working Paper 33. Brighton: STEPS Centre.

Bell M. (2012) International Technology Transfer, Innovation Capabilities and Sustainable Directions of Development. In: Ockwell DG and Mallett A (eds) *Low-Carbon Technology Transfer: From Rhetoric to Reality.* Abingdon: Routledge, 20–47.

Bell M and Pavitt K. (1993) Technological accumulation and industrial growth: Contrasts between developed and developing countries. *Industrial and Corporate Change* 2: 157–210.

Bergek A, Jacobsson S, Carlsson B, et al. (2008a) Analyzing the functional dynamics of technological innovation systems: A scheme of analysis. *Research Policy* 37: 407–429.

Bergek A, Jacobsson S and Sanden BA. (2008b) "Legitimation" and "development of positive externalities": Two key processes in the formation phase of technological innovation systems. *Technology Analysis & Strategic Management* 20: 575–592.

Bergek A, Hekkert M, Jacobsson S, et al. (2015) Technological innovation systems in contexts: Conceptualizing contextual structures and interaction dynamics. *Environmental Innovation and Societal Transitions* 16: 51–64.

Berkhout F, Smith A and Stirling A. (2004) Socio-Technological Regimes and Transition Contexts. In: Elzen B, Geels F and Green K (eds) *System Innovation and the Transitions to Sustainability: Theory, Evidence and Policy.* Cheltenham: Edward Elgar, 48–75.

Berkhout F, Verbong G, Wieczorek AJ, et al. (2010) Sustainability experiments in Asia: Innovations shaping alternative development pathways? *Environmental Science and Policy* 13: 261–271.

Bhattacharyya S. (2013a) Financing Electrification and Off-Grid Electricity Access Systems. In: Bhattacharyya S (ed.) *Rural Electrification through Decentralised Off-Grid Systems in Developing Countries.* London: Springer.

Bhattacharyya S. (2013b) *Rural Electrification through Decentralised Off-Grid Systems in Developing Countries.* London: Springer.

Bhattacharyya SC. (2012) Energy access programmes and sustainable development: A critical review and analysis. *Energy for Sustainable Development* 16: 260–271.

Bhattacharyya S and Palit D. (2014) *Mini-Grids for Rural Electrification of Developing Countries: Analysis and Case Studies from South Asia.* London: Springer.

Brown MA and Sovacool BK. (2007) Developing an "energy sustainability index" to evaluate energy policy. *Interdisciplinary Science Reviews* 32: 335–349.

Campbell B and Sallis P. (2013) Low-carbon yak cheese: Transition to biogas in a Himalayan socio-technical niche. *Interface Focus* 3: 1–11 (www.ncbi.nlm.nih.gov/pmc/articles/PMC3638286/pdf/rsfs20120052.pdf) accessed on 20 August 2017.

Campbell BM, Vermeulen SJ, Mangono JJ, et al. (2003) The energy transition in action: Urban domestic fuel choices in a changing Zimbabwe. *Energy Policy* 31: 553–562.

Campbell B, Cloke J and Brown E. (2016) Communities of energy. *Economic Anthropology* 3(1): 133–144.

D'Costa AP. (1994) State, steel and strength: Structural competitiveness and development in South Korea. *Journal of Development Studies* 31: 44–81.

D'Costa AP. (1998) Coping with technology divergence policies and strategies for India's industrial development. *Technological Forecasting and Social Change* 58: 271–283.

Deichmann U, Meisner C, Murray S, et al. (2011) The economics of renewable energy expansion in rural Sub-Saharan Africa. *Energy Policy* 39: 215–227.

Elias R and Victor D. (2005) *Energy Transitions in Developing Countries: A Review of Concepts and Literature,* Working Paper #40. Stanford University - FSI Stanford, Program on Energy and Sustainable Development (PESD).

Freeman C. (1987) *Technology and Economic Performance: Lessons from Japan.* London: Pinter.

Freeman C. (1997) The national system of innovation in historical perspective. *Cambridge Journal of Economics* 19: 5–24.

Gallagher KS. (2006) Limits to leapfrogging in energy technologies? Evidence from the Chinese automobile industry. *Energy Policy* 34: 383–394.

Gallagher KS. (2014) *The Globalization of Clean Energy Technology: Lessons from China.* Cambridge, MA: MIT Press.

Geels F. (2002) Technological transitions as evolutionary reconfiguration processes: A multi-level perspective and a case-study. *Research Policy* 31: 1257–1274.

Geels FW and Schot J. (2007) Typology of sociotechnical transition pathways. *Research Policy* 36: 399–417.

Goldthau A. (2012) From the state to the market and back: Policy implications of changing energy paradigms. *Global Policy* 3: 198–210.

Gollwitzer L. (2016) All together now: Institutional innovation for pro-poor electricity access in Sub-Saharan Africa. Doctoral thesis. Brighton: University of Sussex.

Gollwitzer L, Ockwell D and Ely A. (2016) *Institutional Innovation in the Management of Pro-Poor Energy Access in East Africa,* SPRU Working Paper Series. Brighton: SPRU, University of Sussex.

Hansen UE and Nygaard I. (2014) Sustainable energy transitions in emerging economies: The formation of a palm oil biomass waste-to-energy niche in Malaysia 1990–2011. *Energy Policy* 66: 666–676.

Hansen UE and Ockwell D. (2014) Learning and technological capability building in emerging economies: The case of the biomass power equipment industry in Malaysia. *Technovation* 34: 617–630.

Hansen UE, Pedersen MB and Nygaard I. (2015) Review of solar PV policies, interventions and diffusion in East Africa. *Renewable and Sustainable Energy Reviews* 46: 236–248.

Hiemstra-van der Horst G and Hovorka AJ. (2008) Reassessing the "energy ladder": Household energy use in Maun, Botswana. *Energy Policy* 36: 3333–3344.

Hoogma R. (2000) *Exploiting Technological Niches.* Enschede: University of Twente.

Hosier RH and Dowd J. (1987) Household fuel choice in Zimbabwe. *Resources and Energy* 9: 347–361.

Kaundinya DP, Balachandra P and Ravindranath NH. (2009) Grid-connected versus stand-alone energy systems for decentralized power: A review of literature. *Renewable and Sustainable Energy Reviews* 13: 2041–2050.

Kemp R, Schot J and Hoogma R. (1998) Regime shifts to sustainability through processes of niche formation: The approach of strategic niche management. *Technology Analysis and Strategic Management* 10: 175–196.

Kern F. (2015) Engaging with the politics, agency and structures in the technological innovation systems approach. *Environmental Innovation and Societal Transitions* 16: 67–69.

Kim L. (1993) National Systems of Industrial Innovation: Dynamics of Capability Building in Korea. In: Nelson R (ed.) *National Innovation Systems: A Comparative Analysis*. Oxford: Oxford University Press.

Kim L. (1998) Crisis construction and organizational learning: Capability building in catching-up at Hyundai Motor. *Organization Science* 9: 506–521.

Kishore VVN, Jagu D and Nand Gopal E. (2013) Technology Choices for Off-Grid Electrification. In: Bhattacharyya S (ed.) *Rural Electrification through Decentralised Off-Grid Systems in Developing Countries*. London: Springer.

Lall S. (1992) Technological capabilities and industrialization. *World Development* 20: 165–186.

Leach M, Scoones I and Stirling A. (2010) *Dynamic Sustainabilities: Technology, Environment, Social Justice*, Abingdon: Routledge.

Lee M, Soto D and Modi V. (2014) Cost versus reliability sizing strategy for isolated photovoltaic micro-grids in the developing world. *Renewable Energy* 69: 16–24.

Lehtonen M. (2011) Social sustainability of the Brazilian bioethanol: Power relations in a centre–periphery perspective. *Biomass and Bioenergy* 35: 2425–2434.

Levin T and Thomas VM. (2014) Utility-maximizing financial contracts for distributed rural electrification. *Energy* 69: 613–621.

Lundvall B-A. (1988) Innovation as an Inter-Active Process: From User-Producer Interaction to the National System of Innovation. In: Dosi G (ed.) *Technical Change and Economic Theory*. London: Pinter.

Mainali B and Silveira S. (2015) Using a sustainability index to assess energy technologies for rural electrification. *Renewable and Sustainable Energy Reviews* 41: 1351–1365.

Monroy CR and Hernández ASS. (2005) Main issues concerning the financing and sustainability of electrification projects in rural areas: International survey results. *Energy for Sustainable Development* 9: 17–25.

Murphy JT. (2001) Making the energy transition in rural East Africa: Is leapfrogging an alternative? *Technological Forecasting and Social Change* 68: 173–193.

Narula K, Nagai Y and Pachauri S. (2012) The role of decentralized distributed generation in achieving universal rural electrification in South Asia by 2030. *Energy Policy* 47: 345–357.

Nelson R and Winter SG. (1982) *An Evolutionary Theory of Economic Change*. Cambridge, MA: Harvard University Press.

Nerini FF, Dargaville R, Howells M, et al. (2015) Estimating the cost of energy access: The case of the village of Suro Craic in Timor Leste. *Energy* 79: 385–397.

Nerini FF, Broad O, Mentis D, et al. (2016) A cost comparison of technology approaches for improving access to electricity services. *Energy* 95: 255–265.

Newell P and Phillips J. (2016) Neoliberal energy transitions in the South: Kenyan experiences. *Geoforum* 74: 39–48.

Nygaard I and Hansen UE. (2016) Niche development and upgrading in the PV value chain: The case of local assembly of PV panels in Senegal. Paper presented at the Annual Conference of the EU-SPRI Forum 2016, Lund, Sweden, 7th-10th June 2016.

Ockwell D and Byrne R. (2016a) Improving technology transfer through national systems of innovation: Climate relevant innovation-system builders (CRIBs). *Climate Policy* 16: 836–854.

Ockwell D and Byrne R. (2016b) *Sustainable Energy for All: Innovation, Technology and Pro-Poor Green Transformations*. Abingdon: Routledge.

Ockwell DG and Mallett A. (2012) *Low Carbon Technology Transfer: From Rhetoric to Reality*. Abingdon: Routledge.

Ockwell DG, Watson J, MacKerron G, et al. (2008) Key policy considerations for facilitating low carbon technology transfer to developing countries. *Energy Policy* 36: 4104–4115.

OECD and Eurostat. (2005) *Guidelines for Collecting and Interpreting Innovation Data*, 3rd Edition, Paris: OECD Publishing.

Ohiare S. (2015) Expanding electricity access to all in Nigeria: A spatial planning and cost analysis. *Energy, Sustainability and Society* 5: 1–18.

Patel P and Pavitt K. (1995) Patterns of Technological Activity: Their Measurement and Interpretation. In: Stoneman P (ed.) *Handbook of the Economics of Innovation and Technological Change*. Oxford and Malden: Blackwell.

Raven R. (2005) *Strategic Niche Management for Biomass: A Comparative Study on the Experimental Introduction of Bioenergy Technologies in the Netherlands and Denmark*. Eindhoven: Technische Universiteit Eindhoven.

Raven R. (2007) Niche accumulation and hybridisation strategies in transition processes towards a sustainable energy system: An assessment of differences and pitfalls. *Energy Policy* 35: 2390–2400.

Rolffs P, Ockwell D and Byrne R. (2015) Beyond technology and finance: Pay-as-you-go sustainable energy access and theories of social change. *Environment and Planning A* 47: 2609–2627.

Schot J and Geels FW. (2008) Strategic niche management and sustainable innovation journeys: Theory, findings, research agenda, and policy. *Technology Analysis & Strategic Management* 20: 537–554.

SE4All. (2015) *Progress toward Sustainable Energy, Global Tracking Framework 2015 Summary Report*. Washington, DC: Sustainable Energy for All.

Sebitosi AB, Pillay P and Khan MA. (2006) An analysis of off grid electrical systems in rural Sub-Saharan Africa. *Energy Conversion and Management* 47: 1113–1123.

Shen W and Power M. (2016) Africa and the export of China's clean energy revolution. *Third World Quarterly* 38: 1–20.

Shove E. (2010) Beyond the ABC: Climate change policy and theories of social change. *Environment and Planning A* 42: 1273–1285.

Smith A. (2007) Translating sustainabilities between green niches and socio-technical regimes. *Technology Analysis & Strategic Management* 19: 427–450.

Smith A and Raven R. (2012) What is protective space? Reconsidering niches in transitions to sustainability. *Research Policy* 41: 1025–1036.

Smith A, Kern F, Raven R, et al. (2014) Spaces for sustainable innovation: Solar photovoltaic electricity in the UK. *Technological Forecasting and Social Change* 81: 115–130.

Smith A, Voß J-P and Grin J. (2010) Innovation studies and sustainability transitions: The allure of the multi-level perspective and its challenges. *Research Policy* 39: 435–448.

Sovacool BK, D'Agostino AL and Bambawale MJ. (2011) The socio-technical barriers to Solar Home Systems (SHS) in Papua New Guinea: "Choosing pigs, prostitutes, and poker chips over panels". *Energy Policy* 39: 1532–1542.

Stirling A. (2015) *Towards Innovation Democracy? Participation, Responsibility and Precaution in the Politics of Science and Technology*, STEPS Working Paper 78. Brighton: STEPS Centre.

Szabó S, Bódis K, Huld T, et al. (2011) Energy solutions in rural Africa: Mapping electrification costs of distributed solar and diesel generation versus grid extension. *Environmental Research Letters* 6: 034002.

Terrado E, Cabraal A and Mukherjee I. (2008) *Designing Sustainable Off-Grid Rural Electrification Projects: Principles and Practices (Operational Guidance for World Bank Group Staff)*. Washington, DC: World Bank.

Tyfield D, Ely A and Geall S. (2015) Low carbon innovation in China: From overlooked opportunities and challenges to transitions in power relations and practices. *Sustainable Development* 23: 206–216.

Ulsrud K, Winther T, Palit D, et al. (2011) The Solar Transitions research on solar mini-grids in India: Learning from local cases of innovative socio-technical systems. *Energy for Sustainable Development* 15: 293–303.

Ulsrud K, Winther T, Palit D, et al. (2015) Village-level solar power in Africa: Accelerating access to electricity services through a socio-technical design in Kenya. *Energy Research & Social Science* 5: 34–44.

UNDP. (2011) *Towards an "Energy Plus" Approach for the Poor : A Review of Good Practices and Lessons Learned from Asia and the Pacific*. Bangkok: UNDP-APRC.

Watkins A, Papaioannou T, Mugwagwa J, et al. (2015) National innovation systems and the intermediary role of industry associations in building institutional capacities for innovation in developing countries: A critical review of the literature. *Research Policy* 44: 1407–1418.

Watson J, Byrne R, Morgan Jones M, et al. (2012) What are the major barriers to increased use of modern energy services among the world's poorest people and are interventions to overcome these effective? [online] www.environmentalevidence.org/wp-content/uploads/2014/07/CEE11-004.pdf: Collaboration for Environmental Evidence.

Watson J, Byrne R, Ockwell D, et al. (2015) Lessons from China: Building technological capabilities for low carbon technology transfer and development. *Climatic Change* 131: 387–399.

Winther T. (2008) *The Impact of Electricity: Development, Desires and Dilemmas*, Oxford: Berghahn Books.

Zeyringer M, Pachauri S, Schmid E, et al. (2015) Analyzing grid extension and stand-alone photovoltaic systems for the cost-effective electrification of Kenya. *Energy for Sustainable Development* 25: 75–86.

# 27

# CONCLUDING REMARKS

*Subhes C. Bhattacharyya*

## Summary of main conclusions

This Handbook, in four parts spread over 26 chapters, has considered various aspects of the Asian energy sector. The present century has been variously termed as the Asian Century and over the past 15 years, the region has made significant improvements economically, socially and of course energy-wise. Between 1990 and 2015, Asian economies grew faster than the world economy, thereby increasing the Asian share in the global economic output. As a result, most countries transited to the middle income category in purchasing power parity terms (2011 base) by 2015 and millions of people have come out of poverty. Asian economic development has transformed the economies through rapid industrialisation, unprecedented growth in urbanisation and the emergence of the middle income class. Simultaneously, the population has grown due to a relatively youthful population structure but now population growth is slowing down and the proportion of the older population is growing.

The economic and social transformation of the region has influenced its energy use tremendously. The region has emerged as a global leader in energy use, demanding more than 40% of global primary energy consumption. East Asia has dominated the energy demand in Asia in the past, accounting for more than 70% of the regional energy needs. However, going forward, South Asia and South East Asia will make significant contributions to the region's energy use as these economies grow, urbanise and take millions out of poverty. The region is highly dependent on its local coal resources, particularly for electricity generation and industrial energy needs. Coal use has dramatically increased over the past 15 years, bringing air pollution and other environmental concerns at the local level and contributing to climate change worries.

The region has emerged as a major player in the international oil market and regional natural gas market due to growing demand for these fuels in the transport sector and the power sector, respectively. Most of the incremental global demand for oil came from Asia in the past decade and the effect of growing demand was felt on international trade due to the limited local availability of oil. Major oil exporters are now competing to improve their Asian market share. Relatedly, the Asian national oil companies have emerged as important players, particularly in upstream asset acquisitions. However, the region lacks its own crude reference price and the region is likely to continue with its presence reliance on Price Reporting Agencies in the absence of any plausible alternative options. In the case of natural gas, Liquefied Natural Gas has emerged as the preferred supply option due to long distances separating the sources from the demand centres. The demand for gas is expected to grow in the future, which in turn will require significant investment in developing the required infrastructure.

Industry is the main user of energy in Asia which is unlike many other regions where residential and commercial sectors tend to dominate. Rapid industrial growth in China was responsible for a high share of industry in China's final energy use but in other countries and sub-regions, industry played an important role in energy demand. Although countries have taken steps to improve energy efficiency in the sector, there is significant variation in industrial energy intensity and the overall regional gross domestic product (GDP) elasticity of industrial demand remains high. There is the potential for experience sharing and learning from one another, particularly because Japan is the leader in energy efficiency and offers a role model for others in the region. Since 2012, China has been slowly adjusting to its new normal economic condition where the economy is expected to grow at a slower rate. Chinese energy demand as a result is likely to grow slowly compared to the speed of growth in the past 15 years or so. This also offers a window of opportunity to reduce dependency on coal and to promote renewable energies.

The residential sector is the second major user of energy in Asia, after the industrial sector. Although modern energy use is growing and the preference for electricity is clearly visible, the region still depends on traditional energies to a great extent. The growing trend of urbanisation and the emergence of the middle income class is changing the residential energy use in Asia and as economies become better off, the demand for energy intensive appliances (such as air-conditioners, washing machines, dishwashers, etc.) will rise. This is expected to result in a higher growth of residential energy demand compared to the historical rates. However, there are significant opportunities for demand management through efficient energy pricing and energy efficiency programmes.

The transport sector is the third major user of energy and the demand has grown several fold between 1980 and 2013. Population growth in the region and income growth over time have contributed significantly to the growing transport energy demand. The level of motorisation is improving in the region but developing Asia lags behind developed Asia in this respect. With growing income, the demand for personal transport is expected to increase in the future, which will aggravate environmental impacts and other externalities.

Despite significant progress in expanding energy services to the population, Asia remains one of the two most important regions where energy access remains a challenge. While electricity access has improved in many countries, the same cannot be said about clean cooking energy access. Significant investment will be required to achieve the target of universal sustainable energy provision by 2030 and Asian countries will have to leverage private capital through a strategic mix of existing and innovative funding mechanisms. Moreover, the process of universal energy access has to go beyond its technology focused solution to a people-centric approach where the needs and practices of the users are understood to develop a solution.

Although the past economic growth has been supported by fossil fuels, the environmental impacts have become clearly visible in Asia with many cities and urban areas exceeding the safe air quality limits set by the World Health Organisation. The region has taken initiatives to harness renewable forms of energies such as hydropower, solar energy, wind and biofuels. Asian countries had installed more than 83 GW of solar photovoltaics (PV) and some concentrated solar power capacity by 2015. Cost competitiveness of solar PV, favourable policies in the region and innovative financing options have supported this growth. Similarly, Asia is a major player in wind power development, and China, India, Japan and South Korea are major manufacturers of wind turbines and their components. Asian governments have adopted supporting policies to promote wind power and China is the world leader in terms of installed capacity. The region has significant hydro-resource potential as well but only a small part of it has been developed so far. The environmental impacts of large hydropower development and the geopolitical constraints of connecting resource areas with demand centres have affected hydro resource development in the region. The region has significant biofuel potential but managing the life-cycle impacts remains important. As the countries try to balance economic growth and energy sector development, renewable energy resources and technologies will play an important role in reducing the carbon intensity of the sector.

Facing the challenges of ensuring energy security while minimising environmental degradation and climate change threats, the energy sector in Asia has been evolving in terms of governance and policy regimes. The growing energy needs of the region will require large investments in developing infrastructure and the region needs comprehensive policies to support sustainable infrastructure development. This in turn requires adapted governance arrangements that create an enabling environment for investment in smart infrastructure, moving away from the legacy technologies. Countries in the region have undertaken varying degrees of energy sector reform. For example, in the electricity sector, the traditional vertically integrated structure has been replaced with single buyer models or some competitive elements in the structure and in many cases, independent regulation has been introduced. However, most countries have been unsuccessful in implementing the reform processes fully. A common set of threats to reforms and energy security include uneconomic pricing policies being followed in many countries of the region and fragmented markets with limited integration and coordination. As a result, the individual country efforts towards decarbonisation and energy security remain fragmented and weak. The need for greater coordination and regional integration comes out as a main theme of this book. The successful example of Greater Mekong Sub-region where a number of beneficiary countries working together to develop low-carbon energy resources for mutual benefits stresses the importance of a regional approach. The ASEAN Power Grid initiative or South Asia energy integration could help achieve wider regional and national benefits.

Conversely, the region has significant diversity in terms of economic development, population growth, urbanisation, resource endowments, social structures and human factors. Consequently, one solution will not fit all cases and the unsuccessful sector reform initiatives confirm that the standard design approach does not work. The local conditions, political support and long-term policy objectives will always influence the governance mechanisms and countries will have to search for the best option through a trial and error process. The Chinese reform highlights this approach and through a gradual process of experimentation and careful untangling, the country is moving towards a more competitive environment while pushing aggressively for the mitigation of the environmental effects of energy use.

## Final reflections

As Asian countries progress with their economic development in the 21$^{st}$ Century, the socio-economic changes that shaped its development, namely population growth and demographic transition, rapid economic growth, unprecedented urbanisation and the rise of the middle income class, will continue to manifest their influences. However, the pace and level of change will not be the same or similar in all countries. It is anticipated that the remaining low income countries of the region will try to move to the middle income category while the others will inch forward to emerge as high income countries. These transformations will have immense implications for the energy sector of the region.

While the past economic growth was driven by industrialisation and export-led demand, the next phase of development will also focus on the needs of a growing urban population and the burgeoning middle income class. It is also likely that the manufacturing base spreads across the low and medium income countries of the region to take advantage of the cheap workforce and perhaps weaker regulations and enforcements. Consequently, the composition of the GDP will change for many countries: high value addition from industry will be taken over by service-based activities in existing export-dominant countries while the industry share will rise in countries where industrialisation picks up. With more than 60% of the population living in urban areas by 2050, many of whom are relatively affluent, the domestic market will be lucrative for product manufacturers and industries. This section of the population is likely to imitate the lifestyles of their international counterparts and this has the potential of dramatically increasing the energy demand in the residential sector and the transport sector. The demand for commercial fuels will increase and the preference for electricity and flexible fuels will amplify. In contrast, the rural population will expect better access to energy services, where location of residence will not hinder their development potential. This section of the population along with the rising share of the older population in the region can have a somewhat dampening effect on demand, as they are likely to continue with less energy intensive habits.

Clearly, security of energy supply ensuring affordable and reliable provision of energy will remain a major policy issue and preference for cheap local resources such as coal cannot be ruled out. However, the dangers of degrading the local environmental, the expectation of a better quality of life commensurate with new-found affluence by a larger section of the population and the threats of changing climate will militate against heavy reliance on a dirty fuel. Thus the decarbonisation efforts of the energy sector will have to be pursued with greater conviction and political will. This will in turn demand a suitably adapted governance mechanism that better coordinates the development of the energy sector, greater integration with neighbours and eventually other sub-regions to exploit the benefits of economies of scale and coordinated sector development, and learning from one another. Surely, the energy sector of the region holds a fascinating future. As the countries of the region adjust and adapt to unfolding challenges in an attempt to define an appropriate sector development strategy, newer opportunities for further research will be presented.

# INDEX

Printed and bound by CPI Group (UK) Ltd, Croydon, CR0 4YY

30/10/2024

01781414-0001